QUANTITATIVE MICROANALYSIS WITH HIGH SPATIAL RESOLUTION

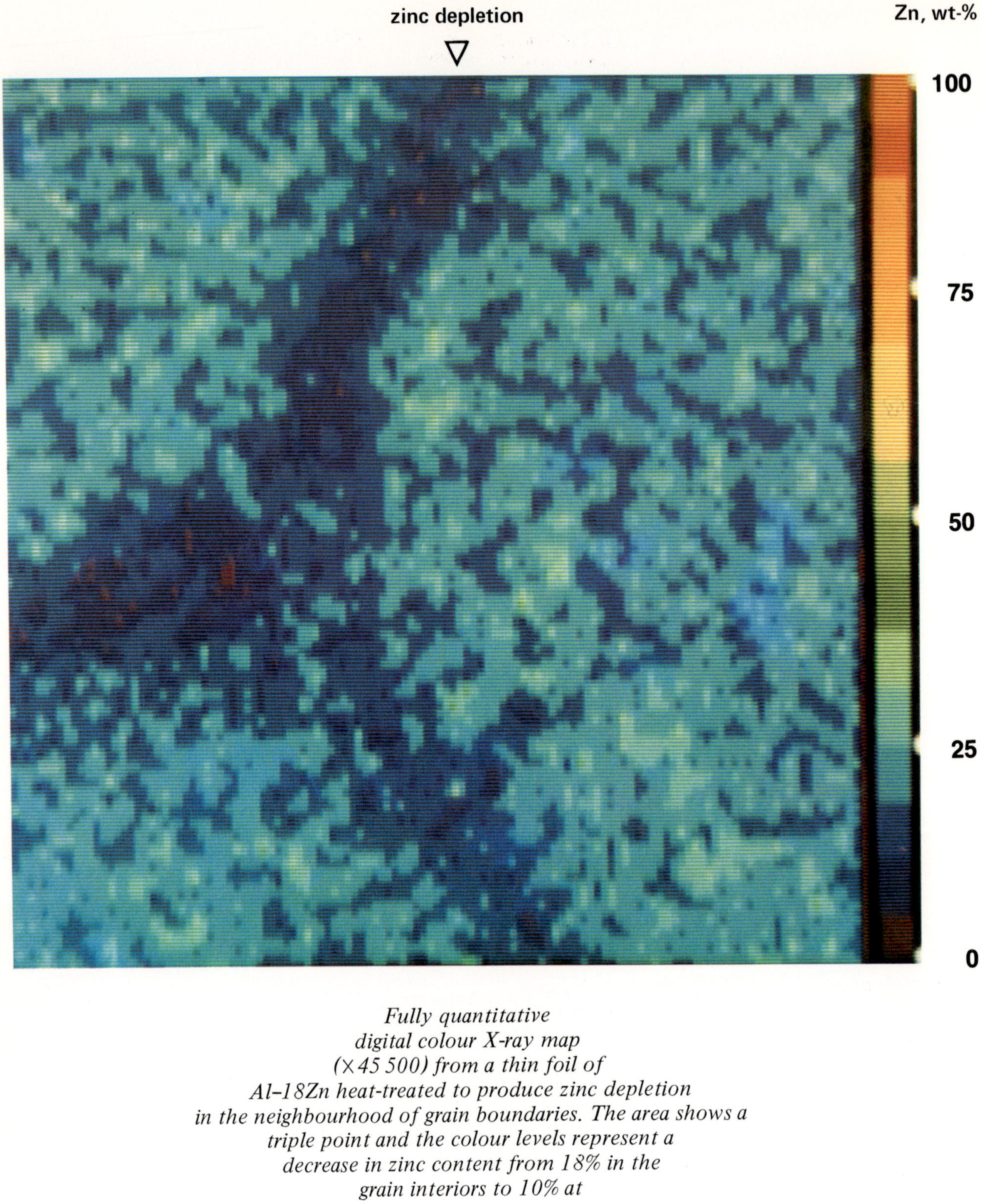

Fully quantitative digital colour X-ray map ($\times$45 500) from a thin foil of Al–18Zn heat-treated to produce zinc depletion in the neighbourhood of grain boundaries. The area shows a triple point and the colour levels represent a decrease in zinc content from 18% in the grain interiors to 10% at the grain boundaries.

See HUNNEYBALL, JACOBS, and LAW in this volume on pages 195–201

QUANTITATIVE MICROANALYSIS WITH HIGH SPATIAL RESOLUTION

Proceedings
of the conference,
jointly sponsored by
the Metals Science Committee of
The Metals Society and the EMAG Group
of The Institute of Physics, which was
held in the Renold Building of the
University of Manchester Institute
of Science and Technology
on 25-27 March 1981

The Metals Society
London

Book 277
published Autumn 1981 by
The Metals Society
1 Carlton House Terrace
London SW1Y 5DB

ISBN 0 904357 38 4

Consultant editors Drs G W Lorimer, M H Jacobs, P Doig
Production D J Hatfull

Composition by Unicus Graphics Ltd, Horsham, West Sussex
Printed and made in England by J W Arrowsmith Ltd, Bristol

CONTENTS

***DISCUSSIONS* immediately follow papers**

DISCUSSIONS **immediately follow papers**

***DISCUSSIONS* immediately follow papers**

PREFACE

These Proceedings contain the papers and discussion presented at a meeting on 'Quantitative Microanalysis with High Spatial Resolution' held in Manchester from 25 to 27 March 1981 jointly sponsored by the Metal Science Committee of The Metals Society and the EMAG Group of The Institute of Physics.

The techniques and application of high resolution microanalysis have developed rapidly during the last five years and it was considered timely to hold a meeting to discuss two areas of particular interest: X-ray microanalysis and electron energy loss spectrometry of thin specimens. In addition to papers concerned with recent theoretical and instrumental advances in these techniques, examples of their application to the investigation of materials were described. Papers were also presented which illustrated the use of other high resolution microanalytical techniques – in particular convergent beam diffraction and field ion atom probe spectrometry – to complement or supplement X-ray microanalysis and energy loss spectrometry.

The Organizing Committee acknowledges the financial support of the European Research Office of the United States Army and Philips Analytical Department of Pye Unicam and would like to express their thanks to Mr F. Hasan for his assistance with the Discussion and Mr D. J. Hatfull of The Metals Society for organizing the rapid publication of this volume.

G. W. LORIMER
M. H. JACOBS
P. DOIG

Opening address

by P Duncumb

Dr P. Duncumb, FRS, is
Director and General Manager of
Tube Investments Research Laboratories,
Hinxton Hall, Saffron Walden, Essex.

One of the things which has fascinated me most about the electron probe field is the way it seems to be continually renewing itself. Time after time it appears to be settling into respectable middle age, and then a spurt of new growth occurs. The same can be said of electron microscopy, and indeed one of the main leaps forward in both areas occurred when they came together for the purpose of 'Quantitative microanalysis with high spatial resolution' – the subject of this meeting.

The general reason for this self-renewal is not hard to find. It is essentially that there are at least four different ways of looking at the subject:

Theory – ranging from basic principles to the actual interpretation of results
Instrument – the means of testing the theory experimentally and getting those results in practice
Technique – the way of using the theory and the instrument to best advantage
Application – the combination of theory, instrument and technique to acquire new knowledge in quite another discipline, such as metallurgy or biology.

These four approaches are essentially interactive, each stimulating developments in the others, so that the subject as a whole is continually undergoing change and evolution. This point is well illustrated in history, as I shall try to show in this brief survey. What follows is by no means intended as a complete record, but rather as an aperitif for the newcomer.

Most of the papers in this conference concern the impact of an electron beam on a thin film specimen, thin enough for most of the electrons to pass through with little deviation from their original direction. It has long been appreciated in the history of electron microscopy that in these circumstances it should be possible to analyse an area smaller in lateral extent than the film thickness, without the spreading of the beam which occurs in a bulk sample. This analysis could be carried out either by using the characteristic X-ray emission or the energy loss spectrum of electrons passing through the film. It was not until the 1950s, however, with the advent of the electron probe microanalyser that microanalysis by X-rays could be demonstrated in practice. So the arrival of an *instrument* allowed a test of the *theory*.

During the middle 1950s there was a strong interest in electron probe instruments in Cambridge, following the work of Cosslett and Nixon in the Cavendish and Oatley and McMullan in the Engineering Laboratory. Castaing had demonstrated the static probe microanalyser in Paris, and I had the interesting task under Cosslett of developing the first scanning probe instrument. We looked initially at thick samples, achieving a resolution of 1 μm for microanalysis, but turned our attention in 1958 to the study of carbon film extraction replicas to see how far we could improve this performance. We demonstrated the analysis by X-ray emission of single particles down to

0.3 μm, confirming the particle size by transmission microscopy. Unfortunately the analysis and the microscopy had to be carried out on two different instruments, which was quite impracticable as a technique for routine analysis. So the shortcomings of the *technique* established the need for a new *instrument* – one which combined the capability of a microanalyser with that of an electron microscope.

In 1959 I moved to Tube Investments Research Laboratories to build such an instrument, which we called EMMA: a combined electron microscope and microanalyser. Initial interests lay with extraction replica samples obtained in the study of carbide precipitation in stainless steel, and we were able to improve the resolution for analysis down to 0.2 μm, and to offer simultaneous transmission microscopy and electron diffraction. In 1962 we were approached by Howie and Whelan from the Cavendish to confirm a theory of electron interaction with single-crystal thin films; their theory predicted an enhancement of X-ray emission at the extinction contours visible in transmission. We verified this *theory* with our *instrument*, the only one at the time capable of providing the necessary *techniques*.

Meanwhile at IRSID in Paris the static probe microanalyser had been in use for a number of years for metallurgical applications, and from 1958 to 1961 Philibert, Henry and Plateau had studied the effect of heat treatment on the composition of $M_{23}C_6$ carbides in stainless steel. They measured the ratio of iron to chromium successfully in terms of X-ray emission, but increasingly found the need to convert these measurements in terms of concentration ratios. A *theory* was needed, stimulated by the requirements of the *application*. Indeed we at TI had felt the same need, but because of preoccupation in both laboratories with the more basic theory for bulk specimens no progress was made for several years. It is interesting to record that when it came, it occurred in both places simultaneously, as described by Tixier and Philibert in 1968. This early work established the foundation of the so-called 'ratio method' of quantitative analysis. Subsequently, the pressure of applications demanded a more practical way of using the theory, and in 1972 Cliff and Lorimer presented their '*k* factor' approach, a development of the ratio method which is in common use today.

One final illustration of the interaction between different parts of the subject is to be found elsewhere in this volume in the paper by Hunneyball, Jacobs and Law. The quantitative imaging of metal specimens thinned from the bulk is confused by thickness variation; at first sight an increase in thickness is indistinguishable from an increase in concentration. By processing the image point by point, however, it can be normalized to a constant intensity of background from the continuum, revealing compositional variations, for example at grain boundaries, which were previously invisible. We thus have an example of the *application* stimulating developments in the *instrument*, the *theory*, and indeed in the *technique* of using them. Many other developments can no doubt be analysed in this way.

I have said nothing of electron energy loss spectrometry, applications of scanning transmission electron microscopy, the atom probe, or other techniques referred to in the conference. With the excellent range of commercial instruments that now exists there is a growing body of experience in using them, as evidenced by the variety of applications presented. The conference is thus continuing this rich mixture of theory, instrumentation, technique and applications which has been apparent in the past and I have no doubt will continue in the future.

Part 1
Energy-dispersive X-ray analysis: theory

X-ray microanalysis of thin specimens

J I Goldstein and D B Williams

Both recent developments and areas of active research in the techniques of thin film X-ray microanalysis are reviewed. Among the recent developments which are discussed are the ratio method, causes and cures of extraneous X-rays, the measurement of X-ray peak and background intensities and improvements in instrumentation. A number of techniques are still being developed. Among these active areas of research are the collection of data for (1) k_{AB} *factors, particularly for the light elements, (2) detectability limits, and (3) beam spreading/spatial resolution. In addition better techniques for the measurement of the path length for X-ray absorption and for handling composition measurements at interfaces need to be developed. These topics are discussed in detail since they are currently the limiting factors for future progress in thin film X-ray microanalysis.*

The authors are in the Department of Metallurgy and Materials Engineering at Lehigh University, Bethlehem, Pa, USA.

During the last four years there has been an expanding amount of work utilizing the technique of X-ray microanalysis of thin films in the analytical electron microscope (AEM). The major reasons for the interest in this technique are the small volume of material which can be analysed and the relative simplicity by which quantitative analyses can be obtained. There are, however, a number of problems which make X-ray microanalysis in the AEM a challenge. Among these are the corrections for X-ray absorption, the determination of beam spreading and therefore spatial resolution, and measurement of the composition at phase interfaces and grain boundaries. The purpose of this paper is twofold: (1) to review advances in instrumentation and in compositional analysis and (2) to discuss areas of active research in techniques for thin film X-ray microanalysis. Although most of the ideas presented in this paper are applicable to biological problems, the material is specifically oriented to the physical scientist.

RECENT DEVELOPMENTS IN TECHNIQUES

Ratio method

Cliff and Lorimer[1] have outlined a simple technique for quantitative analysis of thin foils in an AEM. For thin foils in which X-ray absorption and fluorescence can be neglected, the following expression may be used:

$$\frac{C_A}{C_B} = k_{AB}\frac{I_A}{I_B} \qquad (1)$$

where I_A and I_B are the measured characteristic X-ray intensities, and C_A and C_B are the weight fractions of any two elements A and B in the thin film.

To obtain the scaling factor k_{AB} one can use either measured or calculated values. Cliff and Lorimer[1] and Lorimer *et al.*[2] have obtained calibration curves of k_{ASi} using various silicate standards with known compositions. Figure 1 shows such a measured calibration curve for k_{ASi} obtained at 100 kV in an EMMA-4 instrument. Calibration curves should differ from one instrument to another primarily due to variations in the characteristics of the EDS. Until recently very few k_{AB} factors have been measured.[3,4]

Various authors have attempted to calculate k_{AB} directly.[5–7] Figure 1 also shows an example of a k_{ASi} calibration curve calculated by Goldstein *et al.*[6] The agreement is reasonably good except for the elements Na, Mg and Al. Problems with the measurement of k_{NaSi}, k_{MgSi} and k_{AlSi} will be discussed below under 'Areas of active research/ratio method/(b)'. Calculated k_{AB} values vary from one investigator to another due to differing choices of the ionization cross-section and the fluorescence yield for a given inner shell.[6–8] The largest uncertainties occur from the choice of the ionization cross-section[6,9] when L/K or M/K combinations are considered. Further progress in the calculation of k_{AB} factors awaits the measurement of ionization cross-sections for the 80–200 kV range.

Extraneous X-rays

Investigators have been aware for some time that the measured EDS spectrum may contain X-rays generated from other than the sample region excited by the focused electron beam. These additional X-rays are called 'spurious' or 'extraneous' radiation and this problem has received much attention.[7,10–12] It appears that the spurious X-ray signal is caused by two major effects: (1) stray radiation in the illumination system and (2) specimen interaction with the primary electron beam. The resultant extraneous signal is specific to the AEM because, in contrast to the electron microprobe (EMP), the system uses high kV electrons and the specimen is electron-transparent.

The elimination of extraneous X-rays due to stray radiation in the AEM illumination system is a problem which is largely instrument-dependent. With the latest AEM instruments, various modifications[13–16] are now being incorporated which reduce extraneous X-rays from electron and X-ray sources to acceptable levels. Various authors[7,10,17]

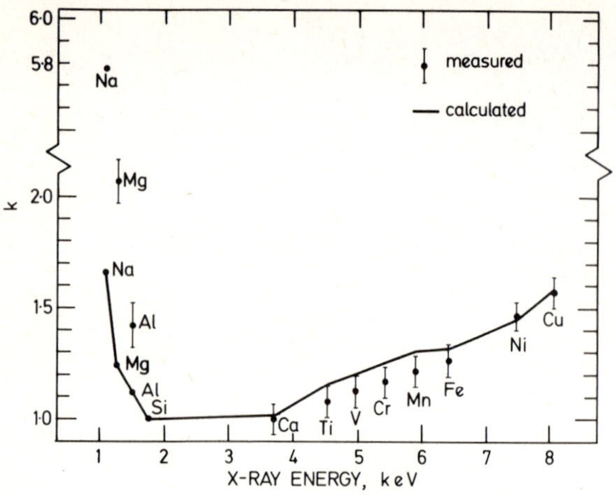

1 **Calculated and experimentally measured k values at 100 kV as function of energy at k_α lines (error limits for measured Na k value were not reported); solid curve represents calculated k values. From Mehta *et al.*[26]**

have suggested subtracting the spurious spectrum from the measured spectrum. In many cases the spurious spectrum may not be localized and may not be representative of the region of analysis. Such a situation argues against a simple subtraction of the spurious spectrum.

The sources of X-rays that can reach the EDS due to specimen interaction with the primary electron beam are summarized for a thin disc specimen in Fig. 2. X-ray fluorescence is probably the most significant problem. As shown in Fig. 2, the possibility of self-fluorescence by the specimen-generated continuum exists, particularly when the specimen is tilted from the horizontal. If a specimen support grid is used, then this radiation will fluoresce the grid. The presence of specimen-generated continuum probably accounts for the observation of Cu peaks in otherwise 'clean' AEM systems, when a Cu support grid is used. Clearly if the grid, some distance from the point of interest, is fluoresced, then thick portions of the specimen, also some distance from the analysed region, must be fluoresced. There is currently no way to distinguish this spurious contribution to the EDS spectrum from the desired X-rays generated at the point of interest.

Extraneous X-rays produced by scattered electrons (Fig. 2) are still somewhat instrument-dependent, but may be minimized by suitable light element inserts which either reduce the backscatter coefficient or produce X-rays which are not detectable in the EDS. The effect of self-fluorescence can be minimized by using thin film specimens rather than discs and by using high take-off angle detectors (standard on some STEMs or AEMs and being developed on others) and the specimen as close to the horizontal as possible.[12,18] With such ideal microanalytical configurations, it is possible to reduce spurious contributions to the X-ray spectrum to levels at which they do not limit the accuracy of quantification.

Measurement of EDS peak and background intensity

Once the sources of spurious or extraneous X-rays are minimized, the measured EDS spectrum will contain the information generated from the desired region of interest. EDS systems have been used successfully on SEM and EMP instruments for some years. Problems of spectrometer resolution and artifacts such as peak broadening, Si escape peaks, absorption edges, etc. are similar to those described for the SEM and EMP and are well documented.[19,20]

Of particular interest in the high voltage AEM is the increased possibility of high energy backscattered electrons which may penetrate the Be detector window, exciting SiK_α radiation in the detector as well as X-ray continuum from the Be window, Si detector, collimator, etc. An electron trap or possibly a mylar film placed in front of the detector[21] may be the best method to remove the undesired radiation. High energy X-rays generated in the specimen may limit the acceptable count rate allowed for the EDS detector from the ≤10 keV X-rays. However, this is rarely a problem since in most AEM instruments the analyst faces the problem of obtaining a sufficient count rate from the thin specimen.

Once the X-ray spectrum is obtained, the values of I_A, I_B, etc. to be used in the ratio method (equation (1)) for obtaining the local chemistry are measured. Methods for background modelling,[22] background filtering,[23] and handling peak overlaps, etc. are well documented[20] in the SEM-EMP literature and can be applied directly to X-ray microanalysis of thin specimens. These methods produce characteristic peak intensities with the continuum background intensity removed. When the characteristic intensities approach those of the background, detectability problems are encountered as discussed in a following section ('Data handling'). Such methods do not account for crystallographic orientation effects which may affect the production of X-rays. In general the simplest way of minimizing such orientation effects is to use a high convergence beam and to avoid strong Bragg diffraction orientations.[24]

Improvements in instrumentation

In order to obtain statistically reliable values of I_A and I_B, sufficient X-ray counts must be accumulated for each X-ray peak. Until recently the effects of contamination of

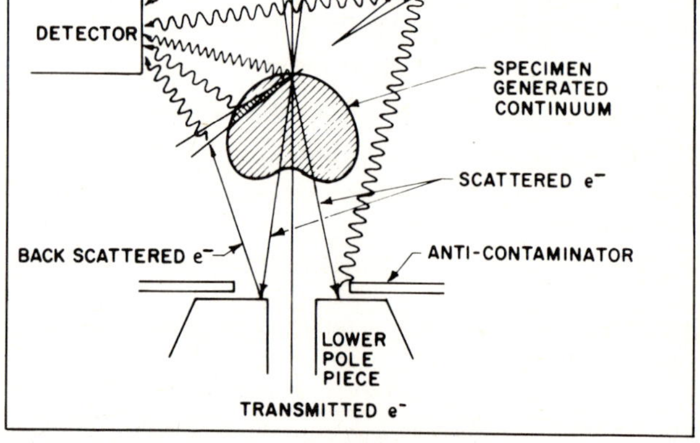

2 **Sources of spectral artifacts due to electron beam–thin disc specimen interaction in AEM[12]**

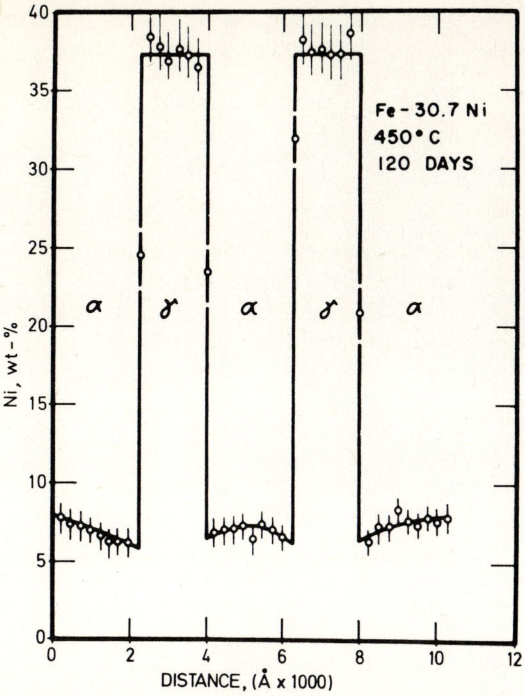

3a Concentration profile across two-phase region of Fe–30.7%Ni aged 120 days at 450°C, showing high spatial resolution[28]

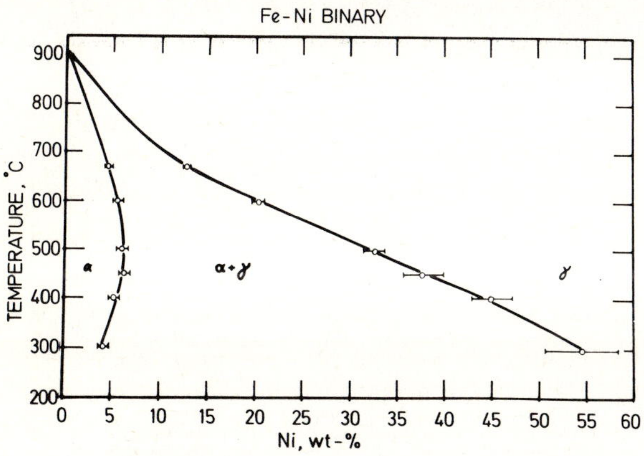

3b Low temperature Fe–Ni phase diagram[28] **determined from concentration profiles measured in STEM (similar to Fig. 3a)**

the analysis point limited not only the analysis time but also the amount of electron current in the focused beam. The various mechanisms of contamination have been summarized recently by Hren.[25] The effects of contamination on quantitative analysis as discussed by Bentley *et al.*[10] include: (1) generation of continuum background from the contaminant and (2) absorption of low energy X-ray lines, particularly Na, Mg and Al[26] within the contaminant.

To minimize contamination, two distinct approaches are taken. The first is to minimize the contributions from the instrument. The manufacturers have recognized this problem and have developed and used oil-free vacuum systems, specific cryo-pumping around the specimen area, metallic O-rings, etc. The second approach is to minimize specimen-borne material. Besides specimen surface cleaning before insertion in the instrument, specimen heating, cooling and immobilization of surface contaminants by electron flooding have been suggested.[10] With contamination effects minimized, the analyst can accumulate an X-ray spectrum for analysis times of several minutes to obtain good X-ray statistics.

In modern AEM instruments using a tungsten filament, sufficient X-ray counts can be obtained from a 5 nm diameter focused electron beam. A field emission gun will provide two to three orders of magnitude more current and therefore a similar increase in X-ray count rate at the same probe size.[27] With such small beam sizes, specimen drift must be <1 nm in the typical counting time of ~ 100 s or the spatial resolution of the instrument must be sacrificed. Most modern instruments have sufficiently stable stages but precautions must be taken to ensure that specimen drift is minimized before quantitative analyses are obtained.

Applications

Although the technique of X-ray microanalysis of thin specimens is still developing, certain well-defined problems in materials are clearly amenable to solution by this technique. Among these problems are partitioning between phases, grain boundary segregation and phase identification. This section briefly highlights some of the areas where successful analyses have been made. No attempt has been made to develop a comprehensive bibliography.

(a) Alloy partitioning

Figure 3*a* shows a composition profile across a two-phase α/γ region of an Fe–Ni specimen.[28] The high spatial resolution is clearly shown, and the points are fitted to a line calculated from known diffusion data for Ni in Fe. From the data in Fig. 3*a*, assuming local equilibrium at the α/γ interface, the solvus lines on the Fe–Ni phase diagram may be determined at low temperatures, as shown in Fig. 3*b*.[28] Other alloy partitioning studies include an investigation of the ferrite transformation in an Fe–2%Cu–5%Ni alloy,[29] measurement of composition variations in ferroelectric and non-ferroelectric regions of $BaTiO_3$,[30] the determination of composition variations in precipitate-free zones in

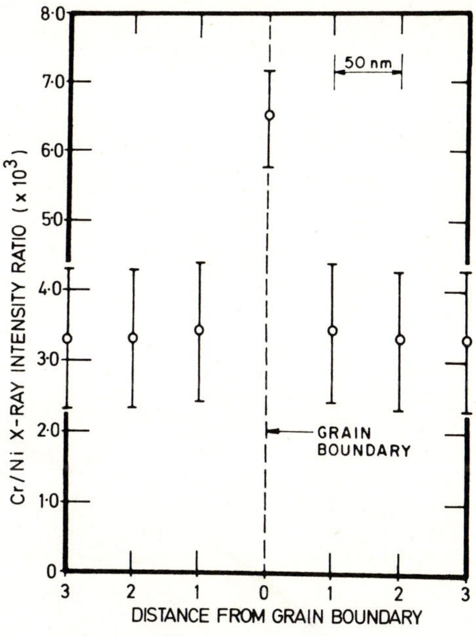

4 Concentration profile across grain boundary in hot pressed NiO, showing segregation of Cr that occurs during densification[33]

Field of view: 1.6 × 1.2 μm

5a **Grain boundary precipitate in carbon steel weldment[37]**

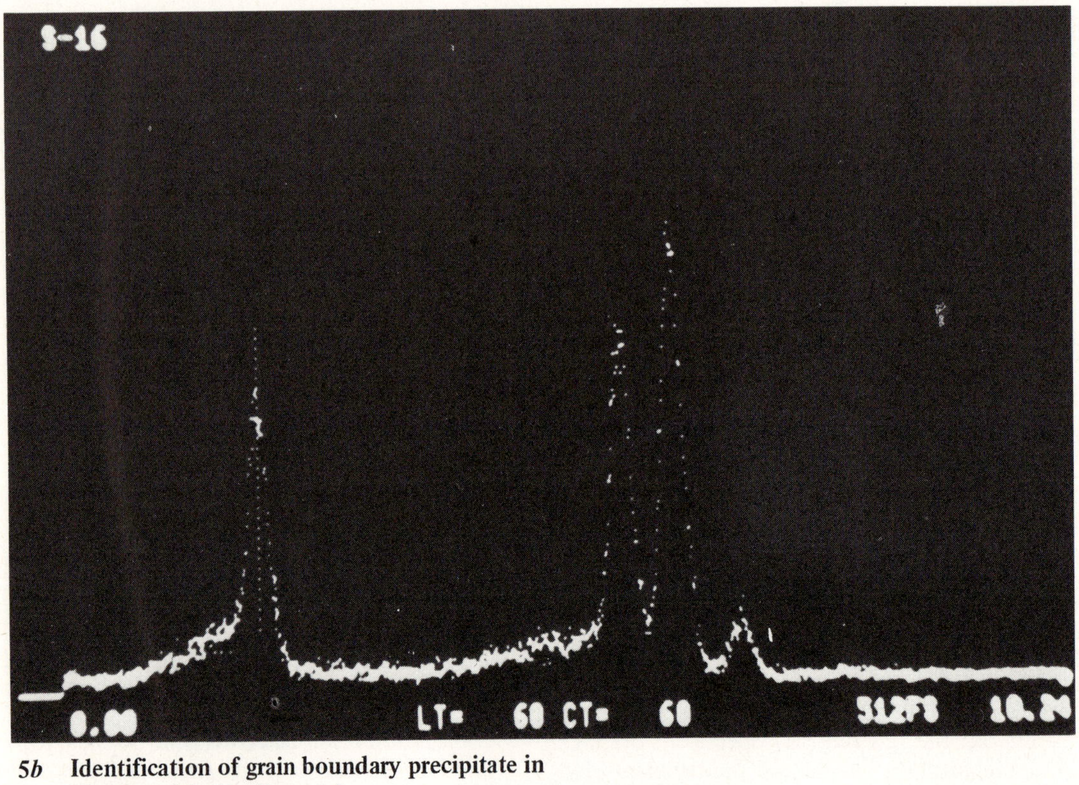

5b **Identification of grain boundary precipitate in Fig. 5a as MnS in Fe matrix**

Al-Zn-Mg and Cu-Ni-Nb alloys,[31] and the measurement of partitioning of silicon during pearlite growth in a eutectoid steel.[32]

(b) Grain boundary segregation

Grain boundary segregation is an important phenomenon in both ceramics and metals. The presence of thin films of a segregating species at a boundary can strongly influence the mechanical and physical properties of the material. Figure 4 shows the detection of grain boundary segregation of Cr_2O_3 in NiO.[33] The Cr_2O_3 is added to aid densification of the NiO powder during hot pressing. Another example includes an investigation of Sc, Fe and Cr segregation in polycrystalline MgO.[34] Similar examples have been reported in metals, for example studies of Sn and P segregation at grain boundaries in Fe-3 wt-%Ni[35] and Cr depletion at grain boundaries in 304 stainless steel.[36]

(c) Phase identification

Figures 5*a* and *b* show the chemical identification of a submicron MnS particle in plain-carbon steel weld metal.[37] Weldment microstructures are invariably complex and difficult to analyse by conventional electron optical techniques. *In-situ* chemical analysis, even at the qualitative level, permits relatively easy characterization of complicated microstructures. Other examples of phase identification include studies of submicron particles in glassy constituents of lunar rocks,[38] major phases in neutron irradiated stainless steels,[39] and segregates in $Ca_3Ga_2Ge_3O_{12}$ single crystals.[40]

AREAS OF ACTIVE RESEARCH IN TECHNIQUES FOR THIN FILM X-RAY MICROANALYSIS

1 Data handling: statistics

Only a few papers have discussed methods for determining analytical sensitivity, detectability limits and the total error for a given analysis; or have measured such important quantities.[3,27,28,41,42] In X-ray microanalysis of thin specimens, calculations of these parameters are critically important. The count rates and the peak to background levels in EDS analysis are typically low, yielding much poorer sensitivities and larger errors than in EMP analysis.

If X-ray counting statistics obey Gaussian behaviour, at the 2σ confidence level the relative error in the number of counts I is $2(I)^{-1/2} \times 100\%$. Since the Cliff-Lorimer relation (equation (1)) uses the X-ray ratio I_A/I_B, the relative error for this ratio is the sum of the relative errors for I_A and I_B. For example, one can consider an analysis from a point in the α phase of an Fe-Ni alloy which contains 6.6 wt-%Ni. For a typical analysis $I_{Fe} = 10\,968$ and $I_{Ni} = 870$.[28] At the 2σ confidence level the relative errors for I_{Fe} and I_{Ni} are ~2% and ~7%, respectively. The relative error for I_{Fe}/I_{Ni} is therefore 9%. This is a very high error which is caused mainly by the low count rate for Ni. Higher brightness guns and the ability to count for longer times in contamination-free systems (already referred to above under 'Improvements in instrumentation') will minimize these errors. However, even total accumulated counts of 10 000 will still introduce a minimum error of ~2%.

Unfortunately the total relative error in C_A/C_B for any one measurement is the sum of the error for I_A, I_B and k_{AB} (see equation (1)). For the Fe-Ni system, Romig and Goldstein[3] have shown that $k_{FeNi} = 0.88 \pm 0.04$ at the 95% confidence limit. Therefore the relative error for the composition ratio C_{Fe}/C_{Ni} in the α phase of the Fe-Ni alloy discussed above is 4.5 + 9% = 13.5%.

Relative errors for k_{ASi} given by Cliff and Lorimer[1] are of the order of 5%. Therefore any analysis using these k factors has a minimum error of 5%. New k_{AB} factors with a minimum relative error of ≤2% have been measured[4,48] and are necessary if more exact analyses are to be obtained. In summary, to reduce the relative error for the composition ratio C_A/C_B from the 10% to the 2-3% level, more accurate k_{AB} factors are needed and much higher total accumulated counts for I_A and I_B are necessary.

The detectability limit of element A, C_A(DL), in a matrix of element B, is given by:[3]

$$C_A(\mathrm{DL}) = \frac{3(2I_b^A)^{1/2}}{I_B} \cdot k_{AB} \cdot C_B \qquad (2)$$

where I_b^A is the continuum background for element A. This equation shows that C_A(DL) will decrease as the count rate increases. To increase the count rate one can use larger probe sizes and moderately thick foils if high spatial resolution is not required. For typical ion-thinned foils of Fe about 150 nm thick, a detectability limit of ~0.5 wt-%Ni has been reported in a Philips EM300 AEM operating at 100 kV and using a tungsten filament.[3] More recent work in our laboratory, using a Philips EM400T AEM where count times are not limited by carbon contamination, has demonstrated that ~0.3 wt-%Mn can be detected in Cu by counting for several minutes with a 10 nm probe to accumulate >10^3 counts. Extrapolation of these data using equation (2) indicates that, given sufficient counts, C_A(DL) can approach ~0.1 wt-% using a tungsten filament. High brightness guns (LaB_6 or field emission) should improve the detectability limits since they increase the X-ray intensity. Oppolzer and Knauer[27] used a field emission STEM to analyse a metallic glass and obtained detectability limits of approximately 0.15 wt-%. The minimum detectability limits for practical X-ray microanalysis of thin foils is not yet well established because of the lack of experimental data.

2 Ratio method

(a) k_{AB} *factors for light elements*

Figure 1 shows the large disagreement between measured[1] and calculated[6] k_{ASi} factors for Na, Mg and Al. Table 1 summarizes these discrepancies. Mehta *et al.*[32] remeasured k_{NaSi}, k_{MgSi} and k_{AlSi} using a Philips EM300 STEM at 100 kV. The measured values are shown in Table 1. There is invariably an uncertainty in the Be window thickness. Calculations of new k factors assuming (1) an EDS detector with a 12 μm thick Be window[26] and (2) an EDS detector with a 9.2 μm thick Be window, an 86 nm Si dead layer and an 8 nm Au layer,[44] are also listed in Table 1. The discrepancies between measured and calculated k factors

TABLE 1 Comparison of calculated and measured values of k_{ASi} for the light elements at 100kV

	Calculated k value			Measured k value		
Element	GCLR[6]	MGWR[26]*	C[44]	CL[1]	LAC[2]	MGWR[26]
Na	1.66	2.17	2.22	5.77	3.2	3.4
Mg	1.25	1.42	1.45	2.07	1.6	1.48
Al	1.12	1.18	1.19	1.42	1.2	1.25

* 12 μm Be window

are much less for Mg and Al although the Na values are quite different.

Zaluzec and Fraser[45] have shown that absorption of generated X-rays due to carbon contamination on the specimen can have a significant effect on the measured characteristic X-ray intensity ratio. This effect could also explain the differences between the calculated and measured k values. Mehta *et al.*[26] attempted to rationalize their measured k values in terms of this effect. They showed that the presence of a layer of carbon 2-3 μm thick on the Be window of the EDS could explain the observed differences between measured and calculated k values. It is clear from the results of their experiments that k factors for the light elements should be remeasured in a carbon contamination-free environment.

Detection of the light elements (C, N, O, Na, Mg, Al) can be enhanced by the use of a windowless or an ultra-thin window EDS.[46] Preliminary k factor measurements by Thomas[46] using an ultra-thin window detector on a 100 kV JEOL 100CX look promising. In addition measurements of O K_α X-rays with a TAP crystal in a wavelength-dispersive spectrometer (WDS) mounted on a Siemens Elmiskop 103 TEM have also been reported.[47] It appears that the use of EDS and WDS units mounted in contamination-free instruments will lead to the measurement of k factors not only for Na, Mg, and Al but also for C, N, and O.

(b) Absorption correction

Criteria are available which indicate when an absorption correction must be made.[6,48] If the calculation is necessary, equation (1) can be modified to the following form:

$$\frac{C_A}{C_B} = k_{AB}\frac{I_A}{I_B} \times [(\mu/\rho)^A_{spec}/(\mu/\rho)^B_{spec}] \times \left\{\frac{1-\exp-[(\mu/\rho)^B_{spec}\ \mathrm{cosec}\,\alpha(\rho t)]}{1-\exp-[(\mu/\rho)^A_{spec}\ \mathrm{cosec}\,\alpha(\rho t)]}\right\} \quad (3)$$

where $(\mu/\rho)_{spec}$ is the mass absorption coefficient of A or B in the specimen, ρ is the density of the thin film of thickness t, and α is the take-off angle, the included angle between the plane of the specimen and the EDS. The effect of absorption can be calculated if $\mu/\rho, \rho, \alpha$ and t are known. Unfortunately μ/ρ and ρ are a function of C_A and C_B and an iterative technique must be employed. If non-normal electron beam incidence is used a modification of equation (3) may be necessary.[7]

In some situations the distribution of X-ray generation $\phi(\rho t)$ may vary with depth. This will cause a further modification of equation (3).[6] Kyser[49] has calculated, using Monte Carlo techniques, the variation of $\phi(\rho t)$ for various films. For Al K_α in a 400 nm Al film there is no variation in $\phi(\rho t)$. However, for Cu K_α in a 400 nm Cu film and Au L_α in a 400 nm Au film, $\phi(\rho t)$ is predicted to vary by 5 and 20% respectively. Experimental measurements of $\phi(\rho t)$ in thin films at $\geqslant$50 kV have not been made except for those given in the paper presented by Stenton *et al.*[50] in this volume. In some specific cases, equation (3) may have to be modified.[6]

The necessary mass absorption coefficients for the absorption correction are generally well known. However, it is important to note that the term $(\mu/\rho)^A_{spec}$ includes the effect of absorption of all the elements present in the sample even if they are not detectable by EDS. Bender *et al.*,[33] in measuring the Ni and Mg contents of a MgO–10 mol-%NiO ceramic, point out the absolute necessity of including the mass absorption coefficient of Mg K_α radiation by O as well as the absorption by Ni.

In many instruments it is not possible to measure the take-off angle α accurately. Although an error of $\pm 2°$ in a $20°$ take-off angle may not appear to be significant, it is important for a highly absorbing system. For the intermetallic compound NiAl, $(\mu/\rho)^{Al}_{NiAl} = 3435$ cm^2/g. In this case the variation in the absorption correction for an uncertainty in take-off angle α of $\pm 2°$ at $\alpha \sim 20°$ is over 6% relative.

The major factor which limits one's ability to use the absorption correction is the accuracy of the measurement of the path length in the thin film over which absorption occurs. This path length is assumed to be t cosec α in equation (3). If the thin film specimen is of constant thickness in the analysis region, the value of t can be measured by the contamination spot method, relative electron transmission method, convergent beam diffraction method, etc. Love *et al.*[51] have pointed out that the contamination method determines the total foil thickness including any surface film (crystalline or amorphous) while the convergent beam diffraction method determines only the thickness of foil which diffracts. In some cases the differences between the measurement of t from the two methods can be as large as 50%. Another method to measure film thickness involves the use of characteristic X-rays to calibrate foil thickness. However, conversion to absolute values of t necessitates calibrating the X-ray intensity measurement as a function of foil thickness, measured by a separate technique.[51]

Nockolds *et al.*[52] have proposed an expression for estimating specimen thickness directly from the characteristic X-ray intensity data obtained during the analysis. Cliff and Lorimer[53] have modified the expression to provide a measure of ρt directly. This method involves ZAF corrections of bulk specimen data taken at high voltages. Such ZAF corrections have their own inaccuracies, leading to uncertainties in this method. Two intensity-ratio methods have been described by Morris *et al.*[54] The first uses the observed ratio of K to L intensity for a given element to deduce the absorption path length and is suitable provided both peaks are visible in the spectrum (usually for $Z > 27$). In practice the method is difficult to use because an iterative procedure (composition vs. ρt) must be used and the L line intensity is often quite low and difficult to separate from the continuum background. The second method is more general and requires several spectra to be recorded from the same area of the specimen at different tilt angles. The composition of the analysis point remains the same and all that is changed is the relative X-ray absorption path lengths. An iterative procedure is used adjusting ρt and composition values until the corrected analyses at each tilt approach the same value. This method is somewhat inconvenient as several spectra (>3) must be obtained from the same area at several tilt angles. Successful measurements using this method are given by Statham and Ball.[55] The major source of error for this method is the uncertainty of the sample geometry.

If the thin foil is not of constant thickness, as in wedge-shaped foils (ion beam or electropolished disc samples), the path length is t cosec α only if the detector is oriented such that the analysed X-rays have a constant thickness exit path. The more common situation is shown in Fig. 6. The X-ray absorption distance for an analysis point is clearly a function of the foil orientation with respect to the EDS at a given take-off angle α.

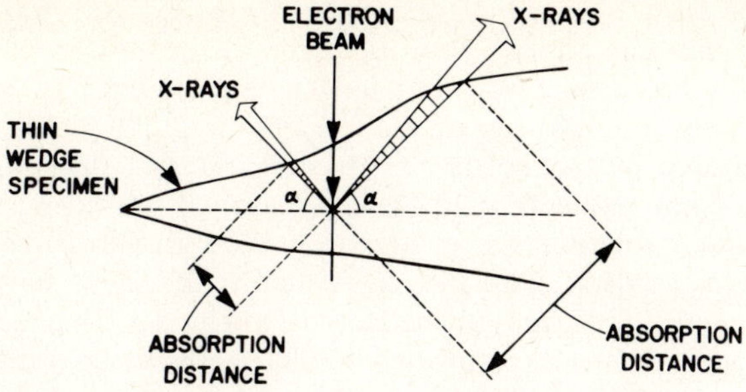

6 Geometry of thin wedge specimens for X-ray microanalysis

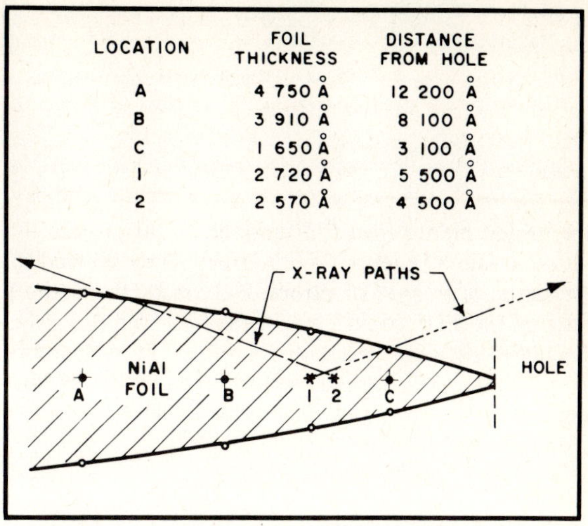

7 Geometry of NiAl thin foil cross-section[56]

The effect of the variation of absorption path length with detector orientation was studied by Glitz *et al.*[56] A highly absorbing ion beam thinned foil of NiAl was employed. Initially, an analysis was taken with the NiAl foil oriented such that the detected X-rays passed through the thinner portion of the wedge-shaped cross-section. The foil was then physically rotated 180° and the analysis repeated. Foil thicknesses were measured by the contamination spot method.

The geometry of the NiAl thin foil cross-section used in this demonstration is illustrated in Fig. 7. Analysis *1* was taken at a foil thickness of ~2700 Å and X-rays from point *1* passed through the thinner portion of the wedge en route to the detector. Analysis *2* was taken at a foil thickness of ~2600 Å and the X-rays from point *2* passed through the thicker portion of the wedge (Fig. 1). The ratio of Ni-K_α to Al-K_α peak intensities, foil thicknesses, and X-ray exit path lengths are listed in Table 2. One path length is calculated as t cosec α, and the other is the actual path length determined from the foil geometry.

In spite of the fact that analysis *1* was taken in a slightly thicker portion of the foil, the measured Ni-K_α/Al-K_α intensity ratio from point *1* is 0.50 that of point *2*, indicating that the Al-K_α X-rays from point *1* were absorbed less than those from point *2*. Absorption calculations using equation (4) and based on the calculated path length (Table 2) show that the intensity ratio at point *1* should be 1.04 times that at point *2*. However, calculations using the actual path lengths which were based on foil geometry/detector orientation considerations show that the intensity ratio at point *1* should be 0.465 times that at point *2*. From this measurement it is obvious that absorption calculations based on foil thickness without consideration of foil geometry can lead to erroneous answers in highly absorbing systems. In addition, some of the methods for measuring t or ρt[54] depend on having a constant thickness foil in the analysis region. In summary, much more work will be necessary before the absorption correction can be used with confidence in the analysis of thin foils.

(c) Characteristic fluorescence correction

The only measured data on characteristic fluorescence in thin foils are for Cr K_α radiation.[2] Although the fluorescence effects in the thin foil are less than those in an equivalent bulk sample, they are still significant. Corrections for characteristic fluorescence in thin films have been developed by Philibert and Tixier[57] and Nockolds *et al.*[52] An important difference between the two models is the variation of the fluorescence correction with foil mass thickness, ρt. Philibert and Tixier predict a $(\rho t)^2$ dependence while Nockolds *et al.* predict a $\rho t(0.9 - \ln \rho t)$ dependence for the correction. The latter authors have shown that their fluorescence correction fits the experimental measurements for an Fe–10.5 wt-%Cr alloy. Data from other systems in which fluorescence is significant are necessary before deciding when the correction must be applied and which formulation of the correction is more accurate.

3 Beam spreading/spatial resolution

(a) Analytical and computer models

The spatial resolution for chemical analysis in a thin foil is a function of atomic number Z, specimen thickness t, and the accelerating voltage E_0. Goldstein *et al.*[6] have estimated the amount of beam spreading using a single scattering model where electron scattering takes place at the centre of the thin film. The equation which relates beam broadening of a point probe as a function of Z, t and E_0 is:

$$b = 625 \left(\frac{Z}{E_0}\right)\left(\frac{\rho}{A}\right)^{0.5} t^{1.5} \qquad (4)$$

where b is the broadening (cm), A is the atomic weight, E_0 is the kV, ρ is the density of the film (g/cm^3), and t is in cm. The beam broadening equation is derived by assuming that the source size can be defined as that within which 90% of the electron trajectories lie.

Monte Carlo calculations of beam broadening in thin film X-ray microanalysis have been performed by numerous investigators.[49, 58–61] Paths of individual electrons are calculated taking into account both elastic and inelastic scattering. The advantage of this calculation procedure is that the effect of initial beam size and specimen-interface geometries can be taken into account. Geiss and Kyser[58] and

TABLE 2 X-ray microanalysis measurements from a wedge-shaped NiAl foil[56]

Analysis	$I_{Ni\,K\alpha}/I_{Al\,K\alpha}$	Foil thickness, nm	Path length, nm	
			calculated	actual
No. *1*	2.64	270	398	270
No. *2*	5.18	260	376	670

Newbury and Myklebust[59] have calculated the beam spreading for the 90% condition and the point probe used by Goldstein *et al.*[6] When comparisons of b are made, the deviations are no greater than 40% for any of the calculations.

Faulkner *et al.*[60,61] have used a more simplified Monte Carlo approach than some of the more recent studies.[58,59] Two factors arise which limit the validity of their model for thin films: (1) the first interaction is assumed to take place right at the surface and the step length between interactions is large (~43 nm in Fe at 100 kV); and (2) the minimum scattering angle for each interaction is high (15°). The former factor results in very few interactions in a typical thin foil which gives a very coarse approximation to the beam spreading and the latter gives rise to an anomalously high backscatter coefficient (~9% in a 150 nm thick foil of Fe at 100 kV). At 100 kV for an Fe foil 150 nm thick Faulkner *et al.* predict that 75% of the electrons are within a diameter $b = 100$ nm. At 200 kV, b is 50 nm.[60] The single scattering equation for 90% of the electrons yields $b = 35.4$ nm at 100 kV and 17.7 nm at 200 kV. Other Monte Carlo calculations yield slightly smaller values of b than the single scattering model.[58,59] The Faulkner *et al.*[60] beam broadening values are more than three times those given by the more sophisticated Monte Carlo techniques.

Doig *et al.*[62] have proposed that the intensity distribution of electrons within a thin foil is analogous to the solution of the equation for the radial diffusion of heat from a point source. They argued that the incident beam intensity given by a normal distribution decays to a broader normal distribution within the foil depth t described by:

$$I_{r,t} = \frac{I_e}{4\pi D(t_0+t)} \exp\left[\frac{-r^2}{4D(t_0+t)}\right] \qquad (5)$$

where D and t_0 are constants which define the incident electron beam diameter; $I_{r,t}$ is the electron current density at a point defined by the cylindrical coordinates r and t; and I_e is the total incident electron current. It is not clear whether this analogy is a valid one.[63] However, if an analytical description of the electron scattering can be developed, it will be most useful for solving problems of X-ray generation at phase interfaces and grain boundaries.

Calculations of beam spreading across planar interfaces have recently been performed.[62,64] Newbury and Myklebust,[64] using Monte Carlo simulations, predict a boundary width b of only 16.5 nm for an iron/nickel interface in a foil thickness of 200 nm. The beam size is 10 nm and the accelerating voltage is 100 kV. Doig *et al.*,[62] using an analytical description of electron scattering, predict a boundary width b of over 60 nm for an interface in an iron foil 100 nm thick. The beam size is 20 nm and the accelerating voltage is 100 kV. The lack of agreement on the amount of beam spreading is due in part to the choice of the Faulkner *et al.*[60] beam spreading values to calculate D in equation (5) for the analytical model. The use of these Faulkner *et al.* values will cause an overestimate of the amount of beam spreading across planar interfaces.

(b) Measurements of beam spreading

Hutchings *et al.*[65] have measured the spatial resolution for X-ray microanalysis in Si thin films. Narrow strips of gold 30 nm thick were placed on one surface. The electron probe was moved across the Au strips and the variation of Au intensity measured. Determinations of the probe size were made when the Au strip was oriented so that it was on the top and on the bottom of the Si film in the AEM. Beam broadening was obtained by subtracting the 90% probe size at the top of the foil from the 90% probe size measured at the bottom of the foil. For foil thicknesses ~500 nm normally used in defect analysis in silicon, the spatial resolution was ~70 nm at 100 kV. This value of b compares well with that calculated by the single scattering model[6] (~90 nm) or the Monte Carlo model[59] (~65 nm). In addition the experimental results for reasonable thicknesses of Si are consistent with the $t^{1.5}$ and $1/E_0$ dependence predicted by the single scattering formula.

Faulkner *et al.*[60] reported values of the spatial resolution by an indirect method. They measured the X-ray output from spherical TiC precipitates of known diameter in a foil of a ferrous alloy of known thickness (150 nm). The assumptions that the X-ray emission volume is cylindrical and that the X-ray generation of Fe and Ti are the same are incorrect and lead to an overestimate of the spatial resolution.

Most other experimental studies of beam spreading have used a vertical interphase boundary with a step function in composition to measure the 90% probe size. Figure 3*a* shows a composition profile of Ni taken every 25 nm across an α/γ interface in a 150 nm thin foil of an Fe-30.7 wt-%Ni alloy. The operating conditions are 100 kV with a 20 nm diameter electron probe and the value of b is about 50 nm. Results of a number of interface boundary measurements have been summarized by Goldstein[42] and by Romig and Goldstein.[3] The general validity of equation (4) and the Monte Carlo calculations of Geiss and Kyser[58] and Newbury and Myklebust[59] have been demonstrated by the experimental data. However, the measured beam spreading values do not correlate well with the calculated values of Newbury and Myklebust[64] and Doig *et al.*[62] for beam spreading at vertical interface boundaries. Clearly more careful experimental studies and more detailed calculations must be performed in order to be able to predict the X-ray resolution available for a given set of experimental conditions.

(c) Beam spreading effect on measurements at grain boundaries

Segregation of impurities to grain boundaries in metals and ceramics often controls mechanical properties and the behaviour of these materials during heat treatment. The segregants are often localized within a few nm of the boundary, perhaps even in a monolayer along the boundary. The spatial resolution of the AEM due to beam broadening is invariably not adequate to measure directly the concentration of the segregant at the grain boundary. The amount of X-rays generated by the segregant is only a small fraction of the X-rays generated from the interaction volume. Doig and Flewitt[66] have developed a model to take this spatial resolution problem into account. Based on Auger measurements the model assumes a slab of a thickness δ along the boundary and a cylindrical X-ray source with a spatial resolution b. In this or any other model, δ will be much smaller than b. The measured concentration using the Cliff–Lorimer[1] approach can be used to calculate the concentration of the grain boundary segregant.

The Doig and Flewitt[66] model has been improved by incorporation of the single scattering model[67] and by the use of the Monte Carlo approach.[68] Monte Carlo calculations can predict the X-ray intensity profiles as a function of distance from the grain boundary. Figure 8 shows a predicted profile for a 2.5 nm beam scanned across a 10 nm

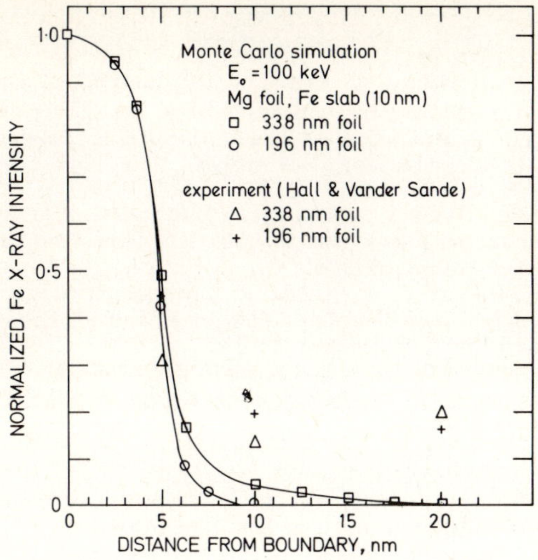

8 **Predicted Fe profile for 2.5 nm beam scanned across 10 nm thick slab of iron at grain boundary of Mg thin foil**[68]

thick slab of iron at the grain boundary of a Mg thin foil.[68] Measured data by Hall and Vander Sande[69] are also displayed. The apparent beam spreading is ~10 nm while the calculated value of beam spreading from equation (4) is ~30 to 50 nm depending on t. This apparent disagreement in the amount of beam spreading is not real. The slab has a width much smaller than the diameter of the X-ray source and the electron trajectory density is not a constant in the interaction volume in thin foils.[58,59] Therefore the concentration profile developed in this case is very different from that obtained across a planar interface between two phases of differing composition and does not yield a measurement of beam broadening. It is also interesting to note that the calculated profiles given in Fig. 8 do not differ very much with increasing foil thickness, as would be predicted for profiles across planar interfaces. The effect of beam broadening variations with foil thickness is to lower the maximum concentration at the boundary. Newbury and Myklebust[68] have also pointed out that concentration calculations cannot be made accurately using the Cliff–Lorimer technique since the concentration varies greatly within the analysis volume. This point also argues that any analytical model which describes the electron distribution as symmetrical[62] will not be accurate when calculating X-ray generation at grain boundaries containing segregants, or at interphase interfaces.

Vander Sande *et al.*[70] have also calculated the effect of beam spreading assuming an incident probe with an electron density described by a Gaussian distribution function and a broadened distribution with depth described by equation (4). Differing from Newbury and Myklebust,[68] they assume that the solute is distributed with distance from the boundary described by a Gaussian function rather than in the form of a slab. Their calculations also predict that the concentration profile is insensitive to foil thickness and that the peak concentration falls with increasing foil thickness/beam broadening. Vander Sande *et al.*[70] also point out that for an electron probe size of 10 nm typical of AEMs, rather than the 2.5 nm typical of dedicated STEMs, the concentration profile will be flattened out. A much larger apparent width will be obtained, perhaps leading to an undetectable X-ray signal.

Doig *et al.*[62] have also developed an analytical approach for handling a segregation profile at a planar boundary. They describe the segregation profile as

$$C_{x,t} = C_1 \exp - \left(\frac{|x|}{\alpha}\right) \qquad (6)$$

where $C_{x,t}$ is the elemental composition at a distance x from the boundary, C_1 is the concentration at the boundary ($x = 0$), and α is a constant which defines the extent of the segregation profile. The incident beam intensity is given by equation (5). Calculations of X-ray profiles are obtained and can be correlated with measured concentrations.

In summary there appear to be a number of models which can explain the very few experimental results on grain boundary segregation which are available. Improvement in the Monte Carlo and analytical models will occur as soon as more experimental data become available.

ACKNOWLEDGMENT

The authors wish to thank the National Aeronautics and Space Administration through grant NGR 39-007-043 for long-term support of their work on X-ray microanalysis in the AEM.

REFERENCES

1 G. CLIFF and G. W. LORIMER: *J. Microsc.*, 1975, **103**, 203
2 G. W. LORIMER, S. A. AL-SALMAN and G. CLIFF: 'Developments in electron microscopy and analysis', Inst. Phys. Conf. Ser. No. **36**, Inst. Physics, 1977, 369
3 A. D. ROMIG and J. I. GOLDSTEIN: 'Microbeam analysis/1979', ed. D. E. Newbury, San Francisco Press, 1979, 124
4 A. F. MARSHALL and C. ZERCHER: Proc. 38th EMSA Meeting, Claitors Press, 1980, 134
5 J. C. RUSS: Proc. 8th National Conference on Electron Probe Analysis, 1973, 30
6 J. I. GOLDSTEIN, J. L. COSTLEY, G. W. LORIMER, and S. J. B. REED: SEM/1977, IITRI Chicago, Ill., 1977, 315
7 N. J. ZALUZEC: 'Introduction to analytical electron microscopy', ed. J. J. Hren, J. I. Goldstein, and D. C. Joy, Plenum Press, 1979, 121
8 J. C. RUSS: SEM/1979, **II**, SEM Inc., AMF O'Hare, Ill., 1979, 673
9 A. G. S. JÁNOSSY, K. KOVACS, and I. TOTH: *Anal. Chem.*, 1979, **51**, 491
10 J. BENTLEY, N. J. ZALUZEC, E. A. KENIK, and R. W. CARPENTER: SEM/1979, **II**, SEM Inc., AMF O'Hare, Ill., 1979, 581
11 J. I. GOLDSTEIN and D. B. WILLIAMS: SEM/1978, **I**, SEM Inc., AMF O'Hare, Ill., 1978, 427
12 D. B. WILLIAMS and J. I. GOLDSTEIN: 'Artifacts encountered in energy dispersive X-ray spectrometry in the AEM', in NBS Conf. Energy Dispersive Spectrometry, Spring 1979, to be published 1981
13 H. SHUMAN, A. V. SOMLYO, and A. P. SOMLYO: *Ultramicroscopy*, 1976, **1**, 317
14 N. J. ZALUZEC and H. L. FRASER: Proc. 34th EMSA Meeting, Claitor's Publ., 1976, 420
15 D. C. JOY and D. M. MAHER: Proc. Specialist Workshop on AEM, Cornell University, 1976, 111
16 D. B. WILLIAMS and J. I. GOLDSTEIN: Proc. Specialist Workshop on AEM, Cornell University, 1978, 174
17 W. A. P. NICHOLSON, B. W. ROBERTSON, and J. N. CHAPMAN: 'Developments in electron microscopy and analysis', Inst. Phys. Conf. Ser. No. **36**, Inst. Physics, 1977, 373
18 D. B. WILLIAMS: Proc. 38th EMSA Meeting, Claitor's Publ., 1980, 82
19 C. E. FIORI and D. E. NEWBURY: SEM/1978, I, SEM Inc., AMF O'Hare, Ill., 1978, 51
20 J. I. GOLDSTEIN, D. E. NEWBURY, P. ECHLIN, C. E. FIORI, D. C. JOY, and E. LIFSHIN: 'Scanning electron microscopy and X-ray microanalysis', Plenum Press, 1981 (in press)
21 P. J. STATHAM: Electron Microscopy 1980, **3**, 7th European Congress on Electron Microscopy Foundation, Leiden, The Netherlands, 1980, 30
22 C. E. FIORI, R. L. MYKLEBUST, K. F. J. HEINRICH, and H. YAKOWITZ: *Anal. Chem.*, 1976, **48**, 172

23 P. J. STATHAM: *Anal. Chem.*, 1977, **49**, 2149
24 A. J. BOURDILLON, P. G. SELF, and W. M. STOBBS: 'Electron microscopy and analysis', Inst. Phys. Conf. Ser. No. **52**, Inst. Physics, 1979, 429
25 J. J. HREN: 'Introduction to analytical electron microscopy', ed. J. J. Hren, J. I. Goldstein, and D. C. Joy, Plenum Press, 1979, 481
26 S. MEHTA, J. I. GOLDSTEIN, D. B. WILLIAMS, and A. D. ROMIG: 'Microbeam analysis/1979', ed. D. E. Newbury, San Francisco Press, 1979, 119
27 H. OPPOLZER and U. KNAUER: SEM/1979, **I**, SEM Inc., AMF O'Hare, Ill., 1979, 111
28 A. D. ROMIG and J. I. GOLDSTEIN: *Metall. Trans.*, 1980, **11A**, 1151
29 R. A. RICKS: 'Electron microscopy 1980', **3**, 7th European Congress on Electron Microscopy Foundation, Leiden, The Netherlands, 1980, 184
30 M. L. MECARTNEY and R. SINCLAIR: Proc. 38th EMSA Meeting, Claitor's Publ., 1980, 364
31 M. RAGHAVAN: *Metall. Trans.*, 1980, **11A**, 993
32 S. A. AL-SALMAN, G. W. LORIMER, and N. RIDLEY: *Acta Metall.*, 1979, **27**, 1391
33 B. BENDER, D. B. WILLIAMS, and M. R. NOTIS: *J. Am. Ceram. Soc.*, 1980, **63**, 149
34 J. B. VANDER SANDE and A. J. GARRATT-REED: Proc. 38th EMSA Meeting, Claitors Press, 1980, 102
35 P. DOIG and P. E. J. FLEWITT: *Acta Metall.*, 1978, **26**, 1283
36 C. S. PANDE and R. L. SABATINI: 'Microbeam analysis/1980', ed. D. B. Wittry, San Francisco Press, 1980, 169
37 D. B. WILLIAMS: Proc. 37th EMSA Meeting, Claitors Press, 1979, 476
38 S. MEHTA and J. I. GOLDSTEIN: Proc. Lunar Planet. Sci. Conf. 10th, 1979, **10**, 1507
39 E. H. LEE and A. F. ROWCLIFFE: Proc. 38th EMSA Meeting, Claitors Press, 1980, 382
40 R. K. MISHRA: Proc. 38th EMSA Meeting, Claitors Press, 1980, 106
41 D. C. JOY and D. M. MAHER: SEM/1977, **I**, IITRI, Chicago, Ill., 1977, 325
42 J. I. GOLDSTEIN: 'Introduction to analytical electron microscopy', ed. J. J. Hren, J. I. Goldstein, and D. C. Joy, Plenum Press, 1979, 83
43 J. WOOD, D. B. WILLIAMS, and J. I. GOLDSTEIN: these Proceedings
44 G. CLIFF, Univ. Manchester: private communication (1980)
45 N. J. ZALUZEC and H. L. FRASER: Proc. Specialist Workshop on AEM, Cornell University, 1978, 122
46 L. E. THOMAS: Proc. 38th EMSA Meeting, Claitors Press, 1980, 90
47 A. ARMIGLIATO and A. GARULLI: 'Electron microscopy 1980', **3**, 7th European Congress on Electron Microscopy Foundation, Leiden, The Netherlands, 1980, 144
48 R. TIXIER and J. PHILIBERT: Proc. 5th Int. Cong. X-ray Optics and Microanalysis, eds. G. Mollenstedt and K. H. Gaukler, Springer Verlag, Berlin, 1969, 180
49 D. F. KYSER: 'Introduction to analytical electron microscopy', ed. J. J. Hren, J. I. Goldstein, and D. C. Joy, Plenum Press, 1979, 199
50 N. STENTON, M. R. NOTIS, J. I. GOLDSTEIN, and D. B. WILLIAMS: these Proceedings
51 G. LOVE, M. G. C. COX, and V. D. SCOTT: 'Developments in electron microscopy and analysis', Inst. Phys. Conf. Ser. No. **36**, Inst. Physics, 1977, 347
52 C. NOCKOLDS, M. J. NASIR, G. CLIFF, and G. W. LORIMER: 'Electron microscopy and analysis', Inst. Phys. Conf. Ser. No. **52**, Inst. Physics, 1979, 417
53 G. CLIFF and G. W. LORIMER: 'Electron microscopy 1980', **3**, 7th European Congress on Electron Microscopy Foundation, Leiden, The Netherlands, 1980, 182
54 P. L. MORRIS, M. D. BALL, and P. J. STATHAM: 'Electron microscopy and analysis', Inst. Phys. Conf. Ser. No. **52**, Inst. Physics, 1979, 413
55 P. J. STATHAM and M. D. BALL: 'Microbeam analysis/1980', ed. D. B. Wittry, San Francisco Press, 1980, 165
56 R. W. GLITZ, M. R. NOTIS, D. B. WILLIAMS, and J. I. GOLDSTEIN: to be published 'Microbeam analysis/1981', San Francisco Press, 1981
57 J. PHILIBERT and R. TIXIER: 'Physical aspects of electron microscopy and microbeam analysis', ed. B. M. Siegel and D. R. Beaman, J. Wiley, NY, 1975, 333
58 R. H. GEISS and D. F. KYSER: *Ultramicroscopy*, 1979, **3**, 397
59 D. E. NEWBURY and R. L. MYKLEBUST: *Ultramicroscopy*, 1979, **3**, 391
60 R. G. FAULKNER, T. C. HOPKINS, and K. NORRGÅRD: *X-ray Spectrometry*, 1977, **6**, 73
61 R. G. FAULKNER and K. NORRGÅRD: *X-ray Spectrometry*, 1978, **7**, 184
62 P. DOIG, D. LONSDALE, and P. E. J. FLEWITT: *Philos. Mag. A.*, 1980, **41**, 761
63 I. P. JONES and M. H. LORETTO: to be published in Proceedings of MICRO '80, *J. Microsc.*, 1981
64 D. E. NEWBURY and R. L. MYKLEBUST: 'Electron microscopy 1980', **3**, 7th European Congress on Electron Microscopy Foundation, Leiden, The Netherlands, 1980, 178
65 R. HUTCHINGS, M. H. LORETTO, I. P. JONES, and R. E. SMALLMAN: *Ultramicroscopy*, 1979, **3**, 401
66 P. DOIG and P. E. J. FLEWITT: *J. Microsc.*, 1978, **112**, 257
67 B. BENDER, D. B. WILLIAMS, and M. R. NOTIS: *J. Am. Ceram. Soc.*, 1980, **63**, 542
68 D. E. NEWBURY and R. L. MYKLEBUST: 'Microbeam analysis/1980', ed. D. B. Wittry, San Francisco Press, 1980, 173
69 E. L. HALL and J. B. VANDER SANDE: Proc. 37th EMSA Meeting, Claitors Press, 1979, 474
70 J. B. VANDER SANDE, T. F. KELLEY and D. IMESON: Proc. 38th EMSA Meeting, Claitors Press, 1980, 356

DISCUSSION of paper by Goldstein and Williams

A. J. GARRATT-REED:
In practical situations, an additional problem which we find to be often the limiting one is that of indeterminate surface layers formed by re-deposition, preferential polishing or sputtering, which results in an unknown (fixed) addition to the signal from the specimen.

J. I. GOLDSTEIN:
The presence of indeterminate surface layers is well known in microanalysis. If quantitative analysis is to be performed the layer(s) must be either removed or characterized as to composition and thickness. If this is not done, microanalysis is probably impossible.

P. J. STATHAM:
With the exception of Morris's L_α/K_α ratio method of determining absorption path length (Ref. 54 of your paper) it would seem that all methods of absorption correction are limited by the inability to measure local surface tilt accurately, particularly at low take-off angles. Since buckled and wedge-type specimens are commonly used, do you know of any method whereby local specimen geometry could be reliably determined?

J. I. GOLDSTEIN:
No, but I believe the section I showed in the presentation was an extreme case.

R. G. FAULKNER:
Can you describe any developments in the absorption correction for non-normal incident beam geometries?

J. I. GOLDSTEIN:
Yes. Please refer to the paper by N. Zaluzec in 'Introduction to analytical electron microscopy' (eds. J. Hren, J. Goldstein and D. C. Joy, 1979, Plenum Press).

Principles, limits and statistical evaluation of quantitative X-ray analysis

R Tixier, B Thomas, and J Bourgeot

The principles of the corrections for quantitative X-ray analysis of thin specimens are discussed and the theoretical and practical limits delineated, without entering into the mathematical details of the derivations. The basis of the statistical calculations to be applied in these forms of microanalysis are given, especially for the problems of measurement comparisons and limits of detection. These considerations are illustrated by various applications.

The authors are with IRSID,
St Germain en Laye, France.
[Dr Tixier subsequently joined ETCA, Arceuil, France]

Microprobe analysis of electron microscopy specimens has been considered since the early beginning of microprobe analysis. Castaing[1] evaluated the temperature rise in such specimens. Fisher[2] included exercises on extraction replica microanalysis in a summer school. Developments in quantitative analysis were published by Duncumb,[3] Hall,[4] Philibert *et al.*[5,6] The EMMA machine's possibilities opened a new era of applications in materials science and biology, as shown by Jacobs *et al.*[7] The present development of Si(Li) spectrometers for X-ray detection and of probe-forming lenses for small area selection has given a strong impetus to this technique with a side effect already pointed out by Beaman,[8] who noted that many microscopists are faced with problems that microprobe users have grappled with for several years. Instrumental and experimental difficulties have been discussed by Goldstein and Williams[9] and Zaluzec and Frazer[10] and are reviewed in other chapters in this volume.

In this contribution we shall discuss the principles and limits of quantitation. The detailed theory and the derivation of formulae will not be given. These can be found in Philibert *et al.*,[11] Tixier,[12] Goldstein,[13] and Zaluzec.[14] The quality of the results depends not only on *accuracy* but also on *precision* in the statistical sense of these terms. Some considerations on statistics and their application to metallurgical problems will be presented.

PRINCIPLES OF QUANTITATIVE ANALYSIS

In his thesis work Castaing[1] not only developed an instrument but also introduced a theoretical approach to quantitation based on the physics of electron/matter interaction. The same theory is extended consistently to the thin specimen analysis case. If one considers a very small element of volume dv of material irradiated *in vacuo* with a constant intensity I_0 by electrons of energy E_0, the number of characteristic X-ray photons emitted for an element A is proportional to the number of A atoms in the element of volume. As for a bulk specimen, the so-called correction theory makes it possible to derive from this concept the physical quantities of actual measurements (Henoc and Maurice[15]). The problems are conventionally classified into three classes:

(i) those due to differences in behaviour in electron excitation and ionization levels from atoms and lines of one element of atomic number Z to atoms and lines of another one (Z correction);
(ii) those due to differences in X-ray photon absorption by the atoms of the target, for the various characteristic lines considered (A correction); and
(iii) those due to indirect, or fluorescence, excitation of the X-ray lines considered by X-rays generated inside the specimen, either from the characteristics or the continuous (*bremsstrahlung*) spectra (F correction).

All these effects are taken into account for a target of known composition to get the number of ionizations. To obtain the intensities detected some factors are needed which can be calibrated on standards or, when possible, calculated.

A real specimen has a finite thickness; to apply a correction theory one needs to define some thickness range where it is relevant. We shall see that there is no single thickness criterion but various criteria depending on the class of phenomena considered.

CORRECTIONS AND LIMITS

Primary ionization (Z)

(a) Backscattering

Backscattering correction comes from the fact that a fraction η of the primary electrons is backscattered from the target with an energy which can be higher than the excitation threshold E of the level for the characteristic line of the element. This can be considered as equivalent to a loss of intensity in the primary beam and is expressed as a backscattering factor R. Electron retrodiffusion on thin specimens has been studied by Cosslett and Thomas[16] and Badde *et al.*[17] These authors have published curves giving, for various elements, the backscattering coefficient η and the energy distribution of backscattered electrons. Referring to these curves, one can derive that, generally, for the usual specimen mass thicknesses, $d\rho t$, and primary electron energy, E_0, $R = 1$. The range of validity of this approximation (first thickness criterion) is given by those parts of the curves where η is only a few percent.

(b) Ionization

The efficiency with which an electron of energy E can ionize the K level (for instance) of an element is characterized by the ionization cross-section, Q. If ρs is the electron path the average number of ionizations is given by

$$dn_K = \left[C_A \cdot \frac{N^0}{A}\right] \cdot [Q_K^A(E)\, d\rho s] \tag{1}$$

The first term is proportional to the number of atoms (C_A = mass concentration, N^0 = Avogadro's number, ρ = density, A = atomic mass); the second term is proportional to an effective volume. Various expressions have been proposed for $Q_K^A(E)$, some of which are discussed by Powell.[18] Generally if $U = E/E_I$, Q varies as $\ln U/U$; thus if $E_0 \gg E_I$ and if the most probable energy loss for the electrons transmitted by a thin specimen, ΔE, is low, Q can be calculated with $E = E_0$. This approximation can be tested using for ΔE the expression given by Landau[19] (second thickness criterion). The transmission coefficients have been calculated by Badde *et al.*[17] As for the backscattering coefficients one can refer to their curves to delineate the region where the transmission coefficients are close to 1.

Absorption (A)

X-ray photons are generated inside the specimen and some fraction is absorbed by the specimen material on the path to the detector. The absorption coefficient depends on the wavelength of the characteristic line and on the composition of the specimen. The correction takes the form:[20]

$$f(\chi) = 1 - \chi \frac{d\rho t}{2!} + \chi^2 \frac{(d\rho t)^2}{3!} - \ldots \tag{2}$$

which can also be written as an exponential.[9] In the developed form it is clear that $f(\chi) = 1$ if $\chi \cdot d\rho t < 0.1$. If this third thickness criterion is not fulfilled the absorption correction should be calculated. For such calculation one needs to determine the specimen mass thickness, $d\rho t$ (and not only the thickness). For instance Jacobs and Baborovska,[7] measuring copper-gold alloy compositions, found that using Au L_α the absorption was negligible up to 0.3 μm thickness, but was large using Au M_α even in the thinnest parts of the foil. Pozsgai *et al.*[21] and Zaluzec[14] have used an iterative approach to estimate ρ and the concentrations in specimens. In this case all elements have to be analysed (or all elements but one) as in bulk specimen analysis. Morris *et al.*[22] have proposed internal calibration techniques.

Fluorescence corrections (F)

Some X-rays, I_f, can be indirectly generated due to excitation of the A atoms of the specimen material by X-rays of higher energy coming from other atoms B excited by the electron probe in the specimen. This is the characteristic fluorescence effect. A formula has been given to correct for this secondary excitation by Philibert *et al.*,[11] assuming that the intensity is homogeneously generated in the target ($(\phi(\rho z))$ = const). This hypothesis has been confirmed by Monte Carlo calculations by Kyser.[23] The resulting formula shows that the correction is negligible if $\mu_B^{AB}\rho \cdot t \ll 1$ (fluorescence thickness criterion). Otherwise it can be calculated using the following formula:

$$\frac{I_f^A}{I_A} = 2\omega_B C_B \frac{r_A - 1}{r_A} \cdot \frac{A}{B} \cdot \mu_B^A \cdot \mu_B^{AB} \cdot \frac{\nu_A}{\nu_B} (d\rho t)^2 \tag{3}$$

where μ_B^{AB} is the mass absorption coefficient of line B in an AB alloy, ν_A is the characteristic frequency of the discontinuity for the line from A, r_A is the absorption jump ratio, and ω_B the fluorescence yield.

Another formula proposed recently gives different results (Nockolds *et al.*[24]). Armstrong,[25] using a Monte Carlo calculation, found results compatible with formula (3). The controversy is difficult to resolve on an experimental basis, due to other fluorescence effects coming from extraneous X-rays generated in the column, on the lenses, apertures and so on, which can be important but are not relevant to the problem of correction theory *per se* which is in no way concerned with the problem of correction of errors associated with instrumental imperfections.

Theoretically, secondary excitation by the continuous (or *bremsstrahlung*) spectra is also possible. This correction has been calculated by Philibert *et al.* and is found to be generally negligible.

MEASUREMENT

Due to variations in mass thickness from one point to another, it is common practice to measure the ratio of one element to another.[26] If the absorption and fluorescence corrections can be neglected, formula (1) shows that the mass thickness is thus eliminated and the number of ionizations has only to be corrected for X-ray cross-section variations to give the mass concentration ratio. The correction factor is thus independent of the concentration, which is an important simplification compared to the bulk specimen analysis case.[6] This technique has been recently designated the 'ratio method'.

STANDARDS

To obtain the photon intensity, dI_K^A, detected from the number of ionizations of atoms calculated in (1), dn_K^A, one has to take into account certain physical and experimental constants:

$$dI_K^A = dn_K^A \cdot \omega_K^A Z_A(K_\alpha) \cdot \frac{d\Omega}{4\pi} \cdot D_E(K_\alpha) \tag{4}$$

where ω_K^A is the fluorescence yield at the K level (for instance) of element A, $Z_A(K_\alpha)$ is the relative intensity of the line in the series, $d\Omega/4\pi$ is the solid angle of detection, and $D_E(K_\alpha)$ is the spectrometer efficiency for the line. Many of these constants are not known very accurately, as discussed by Henoc and Maurice.[15] For a quick analysis they can be calculated approximately, which makes it possible to carry out so-called standardless analysis.[27,9] Otherwise these factors are eliminated by using standards which can be either thin standards (pure or compound) or bulk standards corrected with the usual ZAF procedures. Compound thin standards give the simplest calculations for the correction factors and mass thickness cancels out. The difficulty is to obtain reliable standards on such a small scale. Known phases have, however, been successfully used.[28,6]

One should not forget that standards are not only useful for calibration but also provide one of the most efficient ways of controlling the instrument and the operator.

ACCURACY

When compared with bulk sample analysis, thin sample analysis derived in a consistent way should theoretically have a similar accuracy. In fact it does not present the advantage of numerous compensating phenomena in the approximations used in the theory for bulk sample analysis. Although there is already some controversy on the formulae to be used for ionization cross-section, fluorescence correction, etc. it seems quite obvious that in the present state the actual X-ray intensity measurements are still rather inaccurate due to inadequacies in the preliminary treatment of the raw data. The main sources of inaccuracy (leaving aside important instrumental problems which are treated elsewhere in this volume or in Zaluzec[14] and Goldstein and Williams[9]) are in the techniques of peak extraction from more or less complex spectra and in the way the background subtraction is calculated for Si(Li) detectors especially at low X-ray energies. Although there has been some progress recently,[29, 30] many improvements are still needed.

PRECISION

X-ray counts obtained in a fixed time mode follow a Poisson distribution. In thin specimen analysis the count rates are frequently low and it is important to consider the problems of precision and statistical evaluation of measurements.

Instrument control and rogue value detection

Ancey *et al.*[31] have shown that it is useful to control the instrument using a chi-squared (χ^2) test. Rogue values in the counts measured can be detected and rejected using statistical tests. These procedures are specially important if the instrument presents sources of instability in electron intensity – which is frequently the case in transmission electron microscopes.

Ratio confidence interval

The measured ratio is a function of random variables. Its value is random and the precision of measurement has to be defined using a statistical approach. The confidence interval of such a ratio has been studied. If P_A and B_A are the total number of counts measured respectively on the peak and on the background for element A during the integration time, using a Si(Li) detector, P_B and B_B are the counts for element B, and if R is the measured ratio element A/element B, the confidence interval for R at the significance level α is given by the roots of a second degree trinomial in R:

$$R^2[(P_B+B_B)\chi^2_{1-\alpha}-(P_B-B_B)^2]+2R(P_A-B_A)(P_B-B_B) + (P_A+B_A)\chi^2_{1-\alpha}-(P_A-B_A)^2=0 \quad (5)$$

where $\chi^2_{1-\alpha}$ is the critical value of chi-squared at the probability level α for one degree of freedom.

Such a confidence interval depends on the significance level (α) required. Its spread increases quickly when the number of counts integrated decreases, i.e. in the thinnest parts of a thin foil for instance.

It is an important characteristic of thin specimen analysis that the precision of measurement can be extremely variable from one point to another on the specimen. This should not be forgotten when comparing various measurements.

Comparison of measurements

Statistical tests have been derived by Ancey *et al.*[31] for this purpose. If r_i denotes the ith value of the R ratio among k measurements, it can be shown that the equality of the mathematical expectancies of the r_i's can be tested using the expression:

$$S=\sum_{i=1,k}\frac{(r_i-\bar{r})^2}{V(r_i)} \quad (6)$$

in which $V(r_i)$ is an estimator of the variance of r_i and $\bar{r}$ is a mean defined by:

$$\bar{r}=\frac{\sum_{i=1,k} r_i/V(r_i)}{\sum_{i=1,k} 1/V(r_i)} \quad (7)$$

If r_i' and r_i'' are the limits of the confidence interval of the ith measurement with the significance level α_i (they are the roots of (5) for the ith measurement) one can calculate $V(r_i)$ by

$$V(r_i)=\frac{(r_i''-r_i')^2}{4\chi^2_{1-\alpha_i}} \quad (8)$$

The value of $\bar{r}$ calculated from (7) using these factors is the best estimate of R compatible with the r_i's and their precision. The value of S given by (6) must be less than the critical $\chi^2_{1-\alpha}$ from the table of χ^2 for $k-1$ degrees of freedom. Otherwise the hypothesis that the r_i's are homogeneous should be rejected at the significance level α.

Thus it is possible to compare various ratio measurements even if they are measured with variable precision. In addition one obtains an estimate $\bar{r}$ of the intensity ratio which takes into account the weights of the individual measurements when they are homogeneous.

The test presented here is an approximate or asymptotic solution. Ancey *et al.*[31] give also a rigorous test which is more difficult to compute. The asymptotic solution is generally sufficient.

Detection threshold

The problem of detection threshold should also be treated using a statistical approach.[31] For thin specimen analysis it can be presented as follows.[32]

If I_P and I_B are the mathematical expectancies of the intensities measured for time t_A on peak and background to determine whether the element A is present, then asymptotically

$$\lambda(\alpha,\beta)=\frac{(I_P-I_B)^2\cdot t_A}{I_P+I_B} \quad (9)$$

where $\lambda(\alpha,\beta)$ is the parameter of the so-called noncentral χ^2; α and β are the two risks of a false conclusion, namely α (first kind risk) is the probability of concluding that there is a significant amount of the element when in fact its concentration $C=0$; while β (second kind risk) is the probability that there is no significant amount of the element when in fact $C\neq 0$. Of course both risks are to be taken at a reasonably low level. The values of $\lambda(\alpha,\beta)$ are tabulated.[33,34]

Equation (9) can be used to determine the minimum observation time for detection, given the intensities on peak and background. If the time t_A has been set to a predetermined value, the minimum detectable concentration can be derived:[34]

$$C_{min}=\left[\frac{F\cdot C_t}{(I_P-I_B)_t}\right]\left\{\frac{\lambda(\alpha,\beta)}{2t_A}\left[1+\frac{\sqrt{(1+8I_Bt_A)}}{\lambda(\alpha,\beta)}\right]\right\} \quad (10)$$

The first factor introduces an intensity-concentration cali-

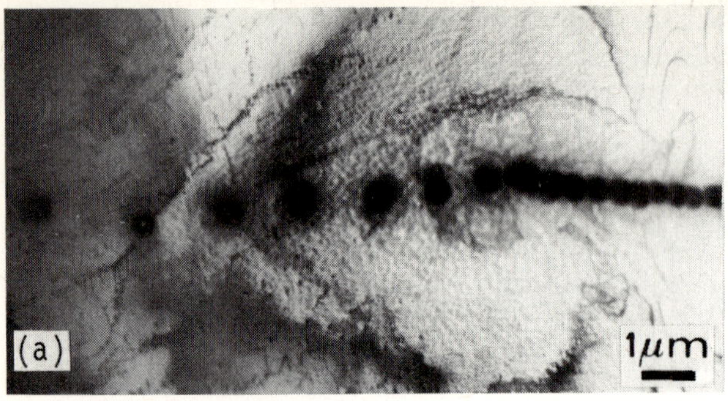

1 **Effect of foil thickness on measured intensity ratios in an iron–chromium alloy (constant counting time); (a) electron micrograph of observed zone; (b) variation of measurements of Cr/Fe intensity ratio and chromium concentration as function of foil thickness**
Open circles indicate rogue values

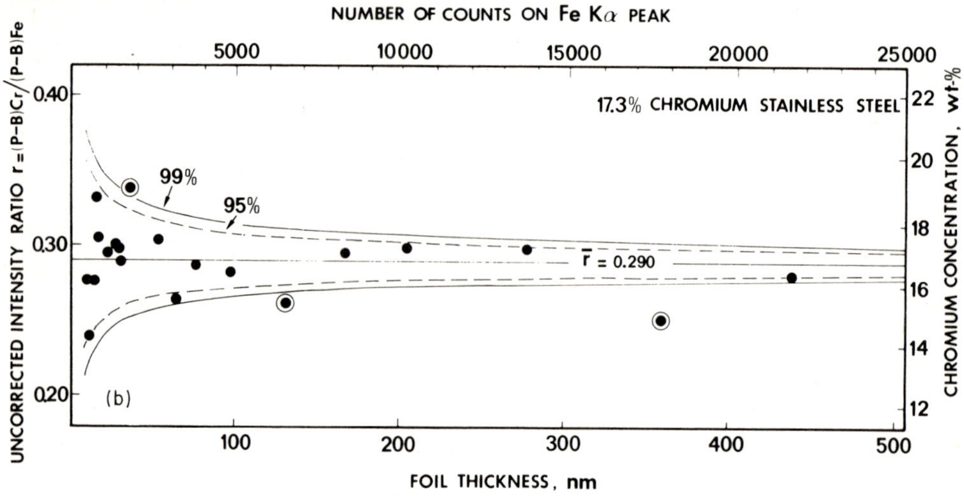

bration based on measurements of I_P and I_B on a standard in which the concentration of the given element is C_t and F is the appropriate correction factor. For thin specimens, we can put

$$I_{P_{min}} = k \cdot \rho t \cdot C_{min} \tag{11}$$

as the characteristic intensity will be proportional to the specimen mass thickness ρt at the point analysed and to the concentration of element A.

$I_{P_{min}}$ can be calculated as a function of I_B from (9). For a given $I_{P_{min}}$, if the mass thickness ρt decreases, C_{min} increases up to one, which corresponds to the minimum thickness of pure A which can be detected with the probability thresholds α and β. The corresponding number of A atoms should be approximately equal to $C_{min}(\rho N^0/A)$, where C_{min} can be calculated from (10) using a measured value for I_B. The practical limits of such a calculation are due to the fact that the real background intensity is composed of two parts: the *bremsstrahlung* in the sample, which is proportional to the mass thickness for a given composition, and an instrumental term which might have a more complex behaviour.

In practice the application of these formulae leads to the conclusion[32, 35] that given a gun of sufficient brightness (LaB_6 for instance), ultra-microanalysis makes possible the identification of a layer of atoms segregated to an interface, or the detection of precipitates of 10 nm diameter or less. Applications to segregations have been published by Doig and Flewitt,[36] who compared their measurements with Auger electron analysis[37] and discussed the limits due to beam broadening in the foil.[38] Segregation of a few layers at the surface of the specimen is also detectable and can give rise to significant errors when due to electropolishing of the sample as shown in Refs. 39 and 40 among others.

Many authors use for the minimum detectable concentration formulae derived from an incorrect approximation of the standard deviation of the measurement. This leads to over-optimistic values, as shown for instance in a discussion of iron analysis in blood cells by Quintana.[41]

APPLICATIONS

To illustrate some of the preceding considerations one or two examples of work on metallurgical specimens are presented below.

Effect of foil thickness in some iron-base alloys

A number of experiments have been carried out in our laboratory to study the effects of foil thickness on measured intensity ratios. Two of these are described below.

(1) Fe–Cr ferritic stainless steel

The specimen in this case was a thin foil prepared from a rolled sheet of ferritic stainless steel containing 17.3 wt-% chromium and less than 0.2 wt-% of residuals. Electron microprobe analysis on bulk specimens was initially carried out to control the chemical homogeneity and it was found that the chromium content varied between 16.6 and 18% over distances of the order of 100 μm perpendicular to the rolling direction. It was considered, however, that the composition would not vary significantly over distances of the order of a few μm in the thin foil although measurable variations could be expected from one part of the foil to another. The surface composition of samples after electrolytic polishing was also investigated by optical emission spectroscopy using a glow discharge lamp.[42] These experiments showed that the specimen surface is enriched in chromium due to the formation of chromium oxides and hydroxides, with a chromium-depleted layer in the metal

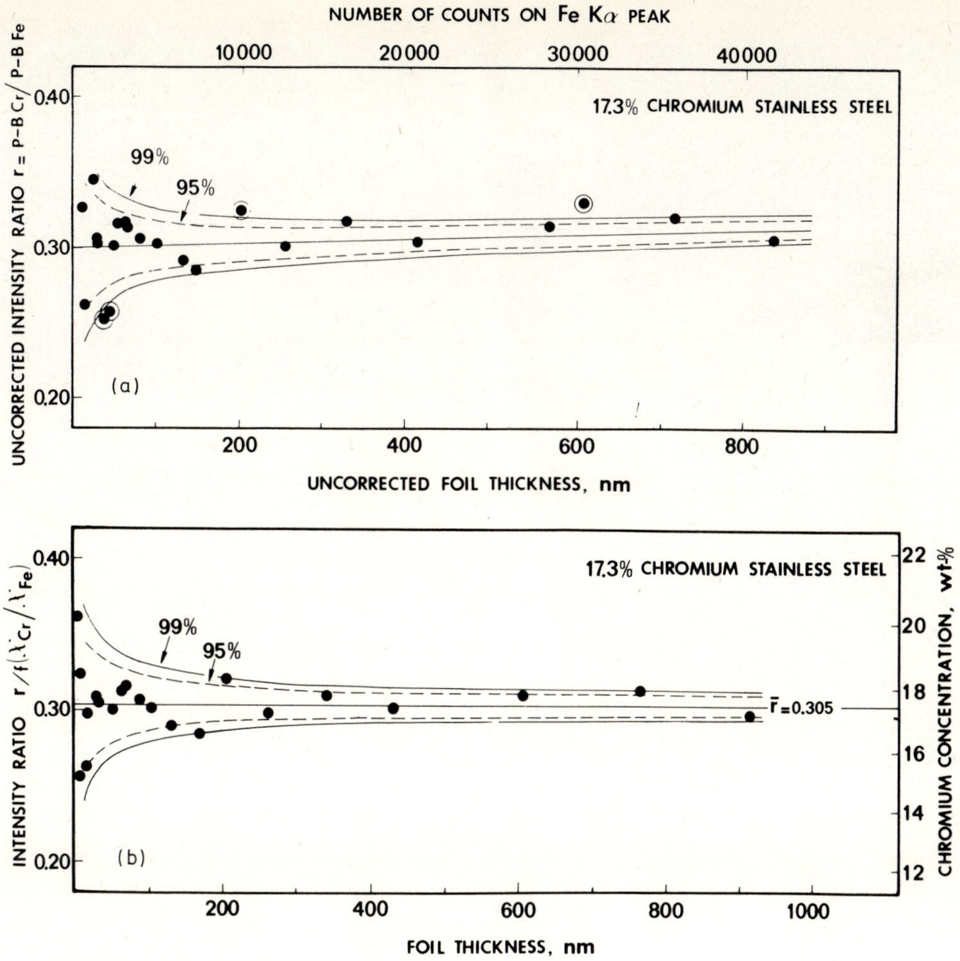

Open circles indicate rogue values

2 **Effect of foil thickness on intensity ratios in an iron-chromium alloy (constant counting time): (a) variation of observed Cr/Fe intensity ratio; (b) variation of Cr/Fe ratio corrected for absorption (Heinrich[44] absorption coefficients)**

underneath the oxide film. Measurements made on similar specimens using an X-ray interference technique[43] indicate that the overall thickness of the perturbed layers is of the order of 3.5 nm. These observations raise some doubts about the validity of measurements on areas where the foil thickness is less than 50 nm. It is possible, however, that the total iron and chromium contents remain approximately constant in the thin foil.

Analytical conditions The measurements in the foils were made at 100 kV using an EMMA 4 instrument with crystal spectrometers which had been very carefully aligned and adjusted to minimize spurious X-ray detection and to optimize the sensitivity. The emergence angle was 45° and the beam diameter was about 300 nm.

Results and discussion The results of a series of constant counting time measurements carried out in a thickness range up to 450 nm are shown in Fig. 1*b* in which the Cr K_α/Fe K_α intensity ratios are plotted as a function of specimen thickness. This thickness was measured at three points in the range from 50 to 130 nm using convergent beam diffraction patterns[45] and monitored at other points using the intensity of the Fe K_α line. The chromium concentrations were calculated (right vertical axis) using bulk standards of the pure metals measured at 40 kV.

The precision of the measurements at the 95 and 99% levels around the mean (equations (5) and (7)) are indicated on the graph. These results illustrate clearly the influence of the number of counts and thus the influence of thickness in constant counting time measurements. After elimination of rogue values the comparison test (see previous Section 'Ratio confidence interval') of all the values measured at different thicknesses showed that they are statistically consistent and the best estimate of the mean concentration ratio is 0.290 which after correction (neglecting absorption and fluorescence) corresponds to a chromium content of 16.6%, which is in agreement with the bulk specimen analysis. These results indicate that *within the limits of the precision of the present measurements* the effects of absorption and fluorescence are negligible in this alloy in the thickness range up to 450 nm, since if the effects were significant the statistical test would have indicated that the *r* values were incompatible. This conclusion is furthermore consistent with estimations of the effects of absorption and fluorescence based on equations (2) and (3).

At higher levels of precision and in thicker foils, however, absorption corrections become necessary as shown in Fig. 2

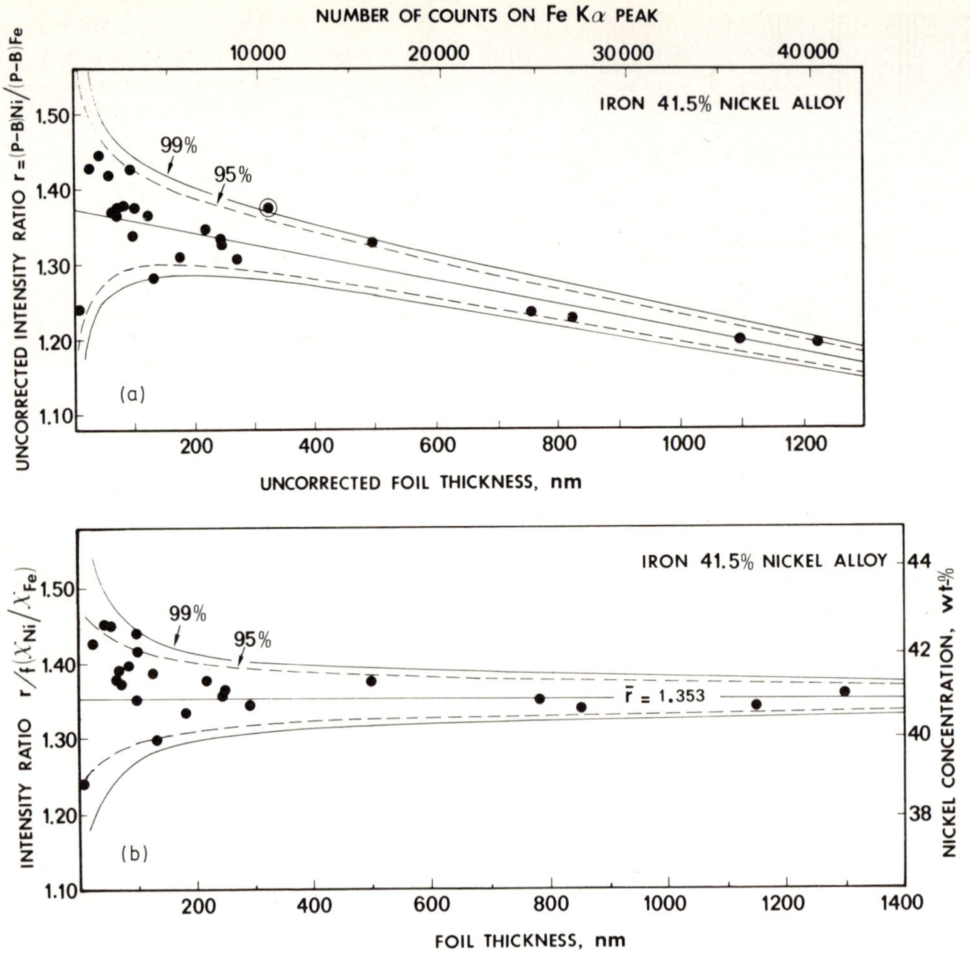

Open circle indicates rogue value

3 **Effect of foil thickness on measured intensity ratios in an iron–nickel alloy (constant counting time); (a) variation of observed Ni/Fe intensity ratios; (b) variation of Ni/Fe ratio corrected for absorption**

representing another set of constant counting time measurements carried out at a slightly higher beam current and in a thickness range extended up to 900 nm. The statistical comparison test shows that the uncorrected intensity ratios (plotted in Fig. 2*a*) are inconsistent and a trend of increasing Cr/Fe intensity ratios with increasing foil thickness is detected by a least-squares regression analysis using the appropriate weightings defined by equation (8). Recalculation of the 95 and 99% confidence intervals about the regression line then permits the elimination of rogue values.

The observed trend in the Cr/Fe intensity values can be accounted for by applying a calculated first-order absorption correction:

$$f(\chi^{\mathrm{Cr}}/\chi^{\mathrm{Fe}}) = 1 - \frac{(\chi^{\mathrm{Cr}} - \chi^{\mathrm{Fe}})\,\rho t}{2!} \qquad (12)$$

to the measured intensity ratios and correcting the foil thickness estimates for absorption of the Fe K_α X-rays.[46]

The results of these corrections are illustrated in Fig. 2*b*. The comparison test shows that the corrected ratios are compatible and the points all lie within the 99% confidence levels about the mean intensity ratio of 0.305 corresponding to a chromium content of 17.3%. This latter figure is again within the bulk specimen analysis spread. These results indicate that within the precision limits of the experiment the effects of fluorescence of the chromium K_α line by the iron K radiation are negligible.

Finally it can also be said that the perturbation of the surface layer composition of the foil due to electropolishing does not significantly affect the microanalysis results in the thinnest parts of the foil within the corresponding precision limits (±1% absolute about the mean chromium concentration at t = 20 nm).

(2) Fe–41.5 wt-%Ni alloy

A rather more striking example of the effects of foil thickness on the observed intensity ratios occurring in an iron–nickel alloy is shown in Fig. 3. The Ni K_α/Fe K_α intensity ratio is plotted as a function of Fe K_α intensity in Fig. 3*a*. Calculations similar to those described above show that the decrease in the Ni K_α/Fe K_α intensity ratio with increasing foil thickness can be fully accounted for by applying the appropriate absorption correction, the results of which are shown in Fig. 3*b*. It is again concluded that fluorescence of Fe K_α by Ni K is negligible within the limits of precision of the experimental measurements. Similar conclusions have been drawn for Fe–Cr–Ni alloys.[46]

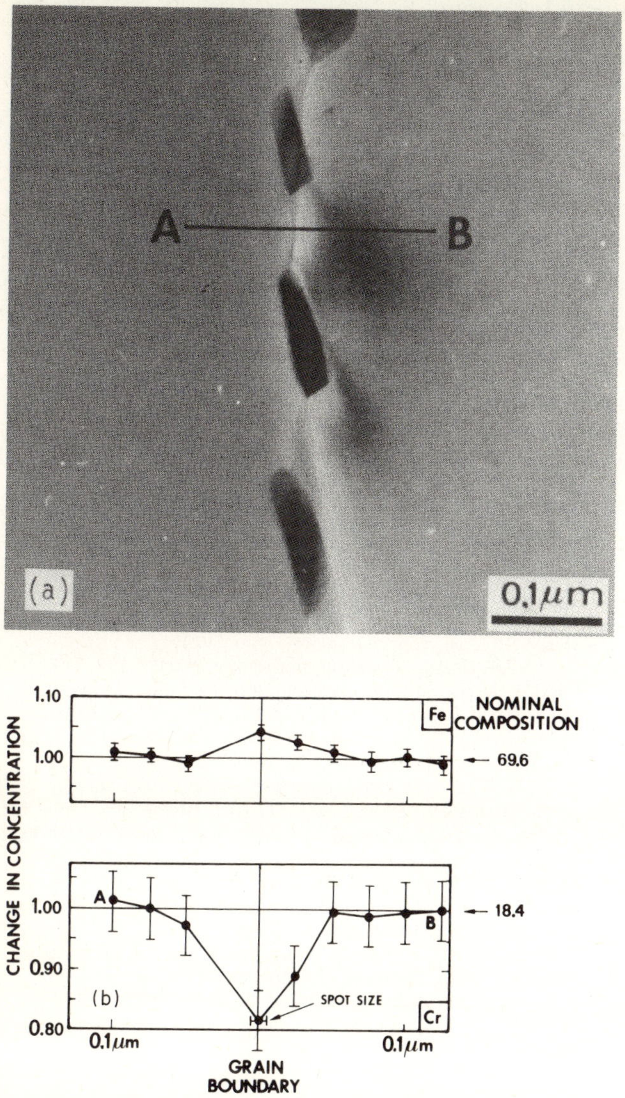

Error bars indicate so-called probable error in absolute concentrations based on 95% confidence limits of individual intensity ratios (Cr/Fe, Ni/Fe, Mn/Fe, Si/Fe)

4 Chromium depletion at grain boundaries in an austenitic stainless steel: (a) micrograph showing observed zone; (b) variation of iron and chromium concentration across grain boundary

Chromium depletion near grain boundaries due to carbide precipitation in austenitic stainless steel

It is well known that in these steels grain boundary precipitation of chromium carbides during cooling or aging can lead to increased grain boundary corrosion sensitivity, which was attributed by Bain *et al.*[47] to chromium depletion near the grain boundaries. The first direct evidence of this depletion was given by Philibert and Penot[48] using microprobe analysis at low incident electron energies on the intergranular fracture surfaces of bulk specimens. Measurements on thin foils have been made more recently by Pande *et al.*[49] and by Rao and Lifshin.[50]

In the present experiments the thin foils were prepared from specimens of a type 304 austenitic stainless steel (18.4 wt-%Cr-9.7Ni-1.8Mn-0.377Si-0.058C) cooled at various cooling rates after solution treatment.

Analytical conditions The measurements were made at 120 kV using a Philips EM400T microscope equipped with scanning transmission facilities and a Si(Li) detector. The spot size used was 10 nm and the emergence angle 20°. The thickness limits for negligible absorption and fluorescence effects vary from one element to another and because of the low emergence angle in this case some of these values are within the electron transparency limit. Theoretically one should find some advantage in tilting the specimen towards the detector in order to increase the emergence angle but this may also degrade the background noise. Experiments are currently being carried out to investigate this problem. In the present work foil thicknesses greater than 100 nm were avoided.

The intensity ratios were corrected using bulk standards measured at 20 kV and the results were analysed statistically to test the significance of the observed variations from point to point across the grain boundary.

Results and discussion An example of the results concerning the variation of the chromium and iron concentrations across a grain boundary are shown in Fig. 4. Statistical treatment of the confidence limits on the measurements of the individual intensity ratios (Cr/Fe, Ni/Fe, Mn/Fe, and Si/Fe) showed that only the iron and chromium contents vary significantly. The chromium and iron concentration variations shown in Fig. 4 are coherent and similar to those already published.[48, 49] It should be noted that the width of the chromium-depleted zone measured from the grain boundary is less than 100 nm. This work forms part of a more general study on grain boundary corrosion resistance.

CONCLUDING REMARKS

The quantitative X-ray analysis techniques described above have now been used successfully for a number of years on a variety of thin specimens and with several different kinds of instrument (electron-probe analyser with a transmission electron attachment, EMMA-4 with crystal spectrometers, scanning electron microscopes, and combined transmission/scanning transmission microscopes with Si(Li) detectors). The thin specimen technique has been applied not only to metallurgical and mineralogical specimens but also to biological specimens as, for example, in the work of Quintana[41] or Ballan-Dufrancais *et al.*[54] who proved recently the existence of metallathionein-like proteins in insect lysosomes.

It is to be emphasized that the statistical evaluation of the results is very important in thin specimen analysis in view of the wide scatter which can result from the low count rates often encountered. Overlooking this fact can lead to grossly erroneous interpretations which unfortunately have already appeared in the literature.

In our own laboratory, thin specimen microanalysis has been applied to a wide variety of specific studies including the quantitative analysis of the metallic elements in carbides in steels, the variation in composition of these carbides and other phases with aging and heat treatment conditions, and so on. However, it is not only in these specific applications that microanalysis techniques are useful but also in everyday microscopy in a qualitative mode as a rapid means of phase identification and as a complementary technique in the interpretation of microstructures. The routine use of this technique is, however, full of pitfalls which can be avoided only if the operators are well acquainted with the physics of X-ray analysis and with the limitations of their instruments. The difficulties of qualitative and quantitative microanalysis in standard electron microscopes have been well reviewed by Mulvey,[51] by Isaacson and Silcox,[52] and in the conclusions of an analytic electron microscopy workshop.[53]

It is our opinion that further work is required to improve quantitation in this field, not so much to refine the existing correction theory but rather to improve the instrumental imperfections such as noise due to incomplete optimization of the electron optics and of the spectrometer and collimator geometry. Further endeavour to reduce incomplete charge collection effects and background noise in energy-dispersive detectors in the low X-ray energy range is also to be recommended. Moreover, a clear need is felt for improvement in the techniques of background stripping and peak deconvolution. Last but not least, attention to the problems of specimen preparation and contamination is required.

In spite of these problems X-ray microanalysis in the electron microscope can now be considered to be an integral part of the possibilities of microstructural characterizations in electron microscopy.

ACKNOWLEDGMENTS

The authors are grateful to Professor R. Castaing, Professor J. Philibert, and Dr P. Duncumb for many stimulating discussions on quantitative X-ray microanalysis. They would also like to express their thanks particularly to Dr F. Bastenaire for his fundamental contribution to the statistical evaluation of microanalysis measurements.

REFERENCES

1 R. CASTAING: thesis Univ. Paris, serie A no. 2423, no. d'ordre 3295, 27–31, 1952, ONERA Publ. 55, Chatillon

2 R. M. FISHER: *J. Appl. Phys.*, 1957, **28**, 1379

3 P. DUNCUMB: *J. de Microscopie*, 1968, **7**, 581

4 T. A. HALL: 'Quantitative electron probe microanalysis', 269–299, 1968, NBS Spec. Pub. 298

5 J. PHILIBERT, G. HENRY, M. ROBERT, and J. PLATEAU: *CR Acad. Sci.*, 1961, **252**, 1320

6 J. PHILIBERT and R. TIXIER: *Brit. J. Appl. Phys.*, 1968, **1**, 685

7 H. JACOBS and J. BABOROVSKA: Proc. 5th Europ. Cong. on Electron Microscopy, 136–137, 1972

8 D. R. BEAMAN and J. A. ISASI: 'Electron beam microanalysis', 1–3, 1972, ASTM STP 506

9 J. I. GOLDSTEIN and D. B. WILLIAMS: Scanning electron microscopy, 1977, **10**, 651–662, Chicago IIT

10 N. J. ZALUZEC and H. L. FRASER: Proc. 11th MAS Conf., 1976, 14A–14B, San Francisco Press

11 J. PHILIBERT and R. TIXIER: 'Physical aspects of electron microscopy and microbeam analysis', 333–354, 1975, New York, John Wiley

12 R. TIXIER: thesis Univ. Paris, Orsay serie A no. 196, 1–166, 1973

13 J. I. GOLDSTEIN: 'Introduction to analytical electron microscopy', 83–120, 1979, New York, Plenum Press

14 N. J. ZALUZEC: 'Introduction to analytical electron microscopy', 121–167, 1979, New York, Plenum Press

15 J. HENOC and F. MAURICE: 'Microanalysis and scanning electron microscopy', 281–317, 1980, Orsay, Les Editions de Physique

16 V. E. COSSLETT and R. N. THOMAS: *Brit. J. Appl. Phys.*, 1964, **15**, 235, 883, 1283; 1965, **16**, 779

17 H. G. BADDE, H. DRESCHER, E. R. KREFTING, L. REIMER, H. SEIDEL, and W. BÜHRING: 'Electron microscopy and analysis', 74–77, 1971, London, Inst. Physics, Conference Series 10

18 C. J. POWELL: 'Use of Monte Carlo calculations in electron probe microanalysis and scanning electron microscopy', 97–104, 1976, Washington, NBS Sp. Pub. 460

19 L. LANDAU: *J. Phys. USSR*, 1944, 8, 201

20 R. TIXIER and J. PHILIBERT: Proc. 5th Int. Conf. X-Ray Optics Microanal., 180–186, 1969, Berlin, Springer

21 I. POZSGAI and N. P. ILYIN: Proc. 9th Int. Cong. on Electron Microscopy, **1**, 500–501, 1978, Toronto

22 P. L. MORRIS, M. D. BALL, and P. J. STATHAM: 'Electron microscopy and analysis', 1979, 413–416, 1980, London, Inst. Physics, Conference Series 52

23 D. F. KYSER: 'Introduction to analytical electron microscopy', 199–221, 1979, New York, Plenum Press

24 C. NOCKOLDS, M. J. NASIR, G. CLIFF, and G. W. LORIMER: 'Electron microscopy and analysis', 1979, 417–420, 1980, London, Inst. Physics, Conference Series 52

25 J. T. ARMSTRONG: private communication

26 M. ANCEY, G. HENRY, J. PHILIBERT, and R. TIXIER: Proc. 5th Int. Conf. X-ray Optics Microanal., 509–518, 1969, Berlin, Springer

27 J. C. RUSS: Proc. 8th National Conf. Electron Probe Microanalysis, 30A–30C, 1973, San Francisco Press

28 T. A. HALL and P. R. WERBA: 'Electron microscopy and analysis', 146–149, 1971, London, Inst. Physics, Conference Series 10

29 W. M. SHERRY and J. B. VANDER SANDE: *X-ray Spectrometry*, 1977, **6**, 154

30 P. J. STATHAM: 'Electron microscopy analysis 1980', 30–31, 1980, Leiden, 7th European Cong. Electron Microscopy

31 M. ANCEY, F. BASTENAIRE, and R. TIXIER: 'Microanalysis and scanning electron microscopy', 319–343, 1980, Orsay, Les Editions de Physique

32 R. TIXIER: *J. de Microsc. Spectr. Electron.*, 1979, **4**, 295

33 E. S. PEARSON and H. O. HARTLEY: 'Biometrika Tables for Statisticians', **2**, 53–57 and 240–241, 1972, 3rd ed., Cambridge Univ. Press

34 M. ANCEY, F. BASTENAIRE, and R. TIXIER: Proc. 7th Int. Cong. on X-ray Optics and Microanalysis, 49–56, 1980, Midland, Pendell Publishing Co.

35 R. TIXIER: 'Microanalysis and scanning electron microscopy', 429–446, 1980, Orsay, Les Editions de Physique

36 P. DOIG and P. E. J. FLEWITT: *J. Microsc.*, 1978, **112**, 257

37 P. DOIG, P. E. J. FLEWITT, and R. K. WILD: *Philos. Mag.*, 1978, **37**, 759

38 P. DOIG, D. LONSDALE, and P. E. J. FLEWITT: *Philos. Mag.*, 1980, **41**, 761

39 P. DOIG and P. E. J. FLEWITT: *J. Microsc.*, 1977, **110**, 107

40 P. L. MORRIS and H. J. LAMB: *J. Phys. D.*, 1978, **11**, L. 73

41 C. QUINTANA: *Biologie Cellulaire*, 1980, **39**, 43

42 R. BERNERON, J. C. CHARBONNIER, R. NAMDAR, and R. TIXIER: *Revue de Métallurgie – C.I.T.*, 1981, **2**, 73

43 P. CROCE and L. NEVOT: *Revue Phys. Appl.*, 1976, **11**, 113

44 K. F. J. HEINRICH: 'The electron microprobe', 296–377, 1966, New York, John Wiley

45 P. M. KELLY, A. JOSTONS, R. G. BLAKE, and J. G. NAPIER: *Phys. Status Solidi (a)*, 1975, **31**, 771

46 R. TIXIER, B. J. THOMAS, and J. BOURGEOT: to be published

47 E. C. BAIN, R. H. ABORN, and J. J. B. RUTHERFORD: *Trans. Amer. Soc. Steel Treating*, 1933, **21**, 481

48 J. PHILIBERT and D. PENOT: Proc. 4th Int. Cong. X-Ray Optics and Microanalysis, 365–373, 1966, Paris, Hermann

49 C. S. PANDE, M. SUENAGA, B. VYAS, and H. S. ISAACS: *Scr. Metall.*, 1977, **11**, 681

50 P. RAO and E. LIFSHIN: Proc. 12th Annual Conf. Microbeam Analysis Soc., 118A–118F, 1977, San Francisco Press

51 T. MULVEY: *Ultramicroscopy*, 1976, **2**, 109

52 M. S. ISAACSON and J. SILCOX: *Ultramicroscopy*, 1976, **2**, 89

53 E. ZEITLER and J. SILCOX: *Ultramicroscopy*, 1978, **3**, 449

54 C. BALLAN-DUFRANCAIS, J. RUSTE, and P. Y. JEANTET: *Biologie Cellulaire*, 1980, **39**, 317 and 325

DISCUSSION of paper by Tixier, Thomas, and Bourgeot

P. J. STATHAM:
When concentration values are determined by normalization to a total of say 100%, the statistical error is not as straightforward to define as for the ratio of two concentrations. What method would you suggest for estimating errors for normalized results?

R. TIXIER:
In normalizing the results one introduces one more relationship between the variables. From a statistical point of view this could be taken into account, in the calculation of the precision, as a reduction of the number of degrees of freedom. I would suggest that the problem in normalization is that it is prone to introduce systematic (non-random) errors, thus reducing the accuracy.

G. W. LORIMER:
As Dr Tixier points out, it is important not to use too simple an approach to the quantitative analysis of thin foils. However, it is also important not to complicate the problem in such a way that so many calculations and corrections have to be made that there is no time left in which to carry out the analysis!

R. TIXIER:
Of course, as you suggest, one can select arbitrarily very fast ways of getting the wrong result. On the other hand, physical calculations and corrections are usefully made at least one time by hand, to understand how it works. Subsequently these can be done by your spectrometer minicomputer in a matter of a few seconds.

J. I. GOLDSTEIN:
I am surprised that you did not observe fluorescence in your Fe–Cr alloy. Can you explain why Lorimer and his colleagues observe such a fluorescence effect while you do not?

R. TIXIER:
I do not know how Lorimer *et al.* came to observe such a high fluorescence in their experiments. In our measurements, and considering the attainable precision, fluorescence was found to be a negligible effect in the range of specimen mass thicknesses observable. This is consistent with theory and also with observations by others.

Determination of Cliff–Lorimer *k* factors for Philips EM400T

J Wood, D B Williams, and J I Goldstein

The Cliff–Lorimer standardless ratio technique for thin foil X-ray microanalysis requires the knowledge of k *factors that relate the measured characteristic X-ray intensities to the composition of the specimen. This paper reports the determination of a comprehensive set of* k *factors at 120 kV for a Philips EM400T analytical electron microscope, using well characterized mineral standards and homogeneous single-phase alloys. The experimental data are compared with previous experimental results and also with calculations from first principles of the theoretical* k *factor values.*

The authors are in the Department of Metallurgy and Materials Engineering at Lehigh University, Bethlehem, Pa, USA.

The Cliff–Lorimer ratio approach[1] to *in-situ* quantification of thin film X-ray microanalysis data has gained popularity due to its simplicity. It is a standardless technique that for a binary system *A–B* relates the characteristic peak intensities above background I_A and I_B to the actual concentration of the species in wt-%, C_A and C_B, according to the relationship:

$$\frac{C_A}{C_B} = k_{AB} \frac{I_A}{I_B} \qquad (1)$$

where k_{AB} is termed the Cliff–Lorimer factor. This factor is independent of variations in specimen thickness if the 'thin film criterion' is fulfilled. This criterion assumes that X-rays are neither absorbed nor fluoresced in the specimen.

By convention, k_{AB} factors are ratioed to Si, i.e. tabulation usually occurs as k_{ASi}, k_{BSi}, etc., and the values are called *k* factors. When Si is not present in the specimen the relative k_{AB} factor can be obtained by dividing the appropriate *k* factors, for example

$$k_{AB} = \frac{k_{ASi}}{k_{BSi}}$$

Since I_A and I_B are measured simultaneously their ratio is independent of fluctuations in beam current although, as shown by Goldstein *et al.*,[2] the *k* factors decrease with increasing operating voltage. To date, the only comprehensive range of *k* factors for K_α lines determined experimentally are those reported by Cliff and Lorimer[1] and Lorimer *et al.*[3] which cover a selection of elements with characteristic peaks in the range 1–10 keV. These results were obtained using an EMMA-4 analytical instrument which utilizes a mini-lens to form a probe down to $\sim 10^2$ nm. Limited work has been carried out on modern analytical instruments[4,5] but most reported results refer to a single k_{AB} determination for a system of interest to the investigator. The aim of the present work is to calibrate our Philips EM400T system by determining, experimentally, a comprehensive range of *k* values, thereby allowing future quantitative analyses to be carried out without reference to standards at the time of the analysis. The experimental values are compared with the experimental results of Lorimer and coworkers[1,3] and calculated values using the method of Goldstein *et al.*[2]

EXPERIMENTAL PROCEDURE

The *k* values were determined from a set of well characterized mineral standards and single-phase homogenized alloys summarized in Table 1. The minerals marked *CL* in Table 1 are the original standards used by Cliff and Lorimer and were prepared and chemically analysed in their laboratory. The mineral samples prepared in our laboratory were crushed under methanol and the slurry dispersed ultrasonically to separate out the finest particles. These samples were floated onto formvar and placed on carbon-coated Be microscope grids.

A series of single-phase alloys (also listed in Table 1) were induction-cast and homogenized. Composition and homogeneity were determined by electron microprobe analysis. Thin foils were prepared by diamond wafering thin slices from bulk samples, then electrodischarge machining 3 mm discs. These discs were ground to $\sim 250\ \mu m$ on SiC 600 mesh paper. Final thinning to electron transparency was accomplished by jet polishing followed by ion beam thinning to remove any residues deposited on the foil from jet polishing.

Measurements of *k* factors were carried out using a Philips EM400T analytical electron microscope fitted with an EDAX energy-dispersive Si(Li) detector. The detector resolution was nominally 153 eV FWHM at Mn K_α. Specimens were held in a Be low-background holder to minimize background X-ray emission from specimen-generated *bremsstrahlung*. All analyses were carried out at 120 kV with an emission current of 25–40 μA and a minimum of 10 000 counts were collected for each peak in the spectrum.

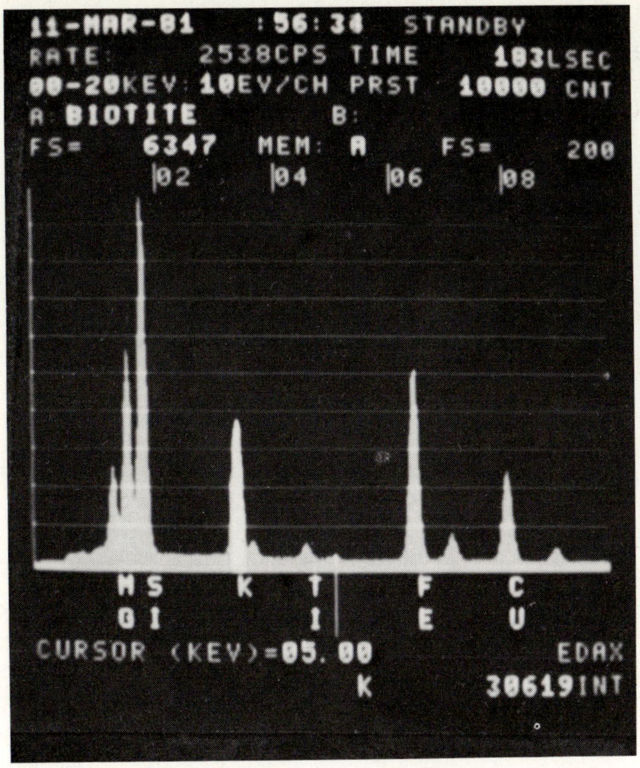

1 **X-ray spectrum of Biotite; typical, well characterized mineral used in *k* factor determination**

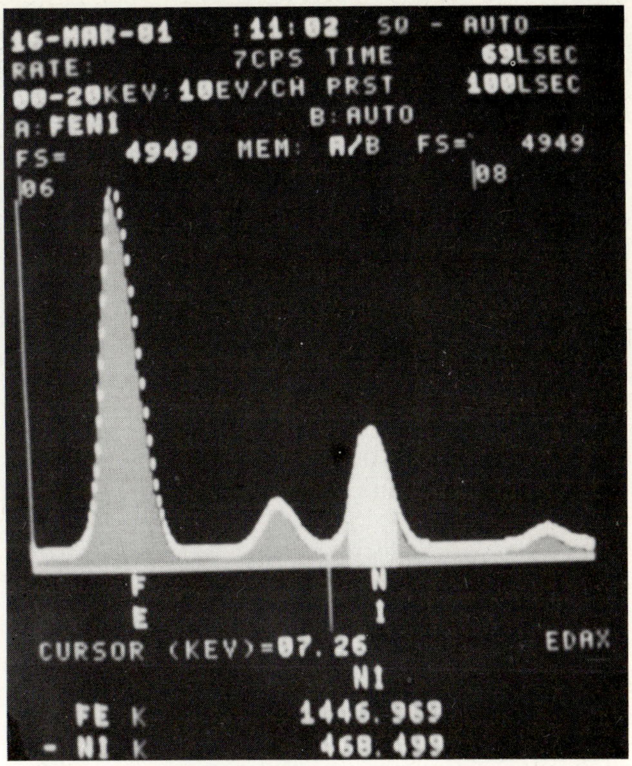

3 **EDS output in Fig. 2 showing peak fitting routine performed by EDAX SWTHIN program to determine integrated X-ray intensities**

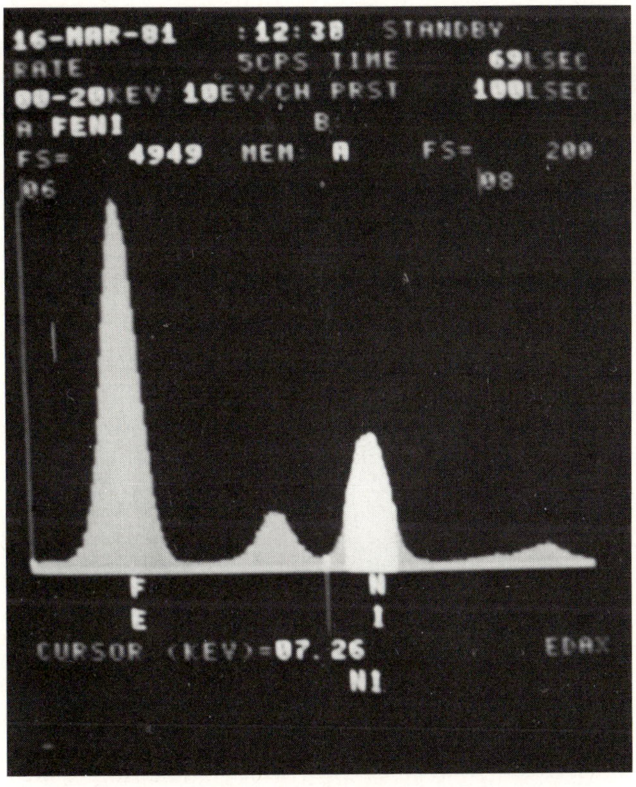

2 **X-ray spectrum from Fe–29.5%Ni; cast homogeneous alloy typical of those used in *k* factor determination**

To ascertain the limits of applicability of the thin film criterion for a given system, the effect of absorption on the k_{AB} factor for each system was determined using the equation:[2]

$$CF = \left[\frac{(\mu/\rho)^{A}_{spec}}{(\mu/\rho)^{B}_{spec}}\right] \times \left\{\frac{1-\exp-[(\mu/\rho)^{B}_{spec}\operatorname{cosec}\alpha(\rho t)]}{1-\exp-[(\mu/\rho)^{A}_{spec}\operatorname{cosec}\alpha(\rho t)]}\right\} \quad (2)$$

where CF is the correction factor applied to the value of k_{AB}; the terms $(\mu/\rho)^{A}_{spec}$ and $(\mu/\rho)^{B}_{spec}$ are the mass absorption coefficients of A and B in the specimen, ρ is specimen density, α is the take-off angle ($\sim 20°$ in a Philips EM400T), and t is the measured film thickness in cm.

Wherever possible, systems that required an absorption correction of $<10\%$ were used. In the results discussed below, only the values of k_{SSi} and k_{PSi} were corrected. If the system satisfied the thin film criterion for foil thicknesses up to 500 nm, or the material appeared to be beam-sensitive, data were collected by rastering the beam over a thin area of 250 nm^2. For systems showing significant absorption below a foil thickness of 500 nm, data were collected from a spot 5–10 nm in diameter. The thickness of the foil in the analysis area was determined by the contamination spot separation method.[6] Although this technique is known to overestimate the actual thickness, the absorption correction was of the order of the counting statistics error.

Thin edges of crushed minerals and foil regions transparent at 120 kV were chosen for analysis. Counting time was varied to give at least 10 000 accumulated counts from each element at each data point. Integration of peak counts and background subtraction were effected by the EDAX SWTHIN program using full shell integrated counts and adjusted to determine K_{α} intensities. Typical spectra collected are shown in Figs. 1 and 2, while Fig. 3 shows a typical peak and background fit to Fig. 2 given by the SWTHIN program.

THEORETICAL CALCULATION OF k FACTORS

Goldstein *et al.*[2] have developed a method for calculating k factors for K_α lines of any element X as a function of operating voltage. The equation for K_{XSi}[2] and for K_α lines is:

$$k_{XSi} = \frac{(Q_K \omega_K a_K / A)_{Si}}{(Q_K \omega_K a_K / A)_X} \cdot \frac{\epsilon_{Si}}{\epsilon_X} \quad (3)$$

where Q_K is the ionization cross-section for K lines, ω_K is the fluorescence yield for K lines, a is the ratio of K_α and K_β intensities expressed as $K_\alpha/(K_\alpha + K_\beta)$, A is the atomic weight, and ϵ is the EDS detector efficiency. The value of ϵ can be obtained by taking into account the absorption of X-rays as they enter the EDS detector in the Be window, gold surface layer and silicon dead layer:

$$\epsilon_X = \exp[-(\mu/\rho)^X_{Be} \rho_{Be} Y_{Be}] \cdot \exp[-(\mu/\rho)^X_{Au} \rho_{Au} Y_{Be}] \cdot \exp[-(\mu/\rho)^X_{Si} \rho_{Si} Y_{Si}] \quad (4)$$

where μ/ρ, ρ and Y are the appropriate mass absorption coefficients, densities and thicknesses of the Be window and the gold and silicon layers. In equation (3) only Q_K varies with operating voltage.

To calculate k_{XSi} the Green–Cosslett[7] cross-sections were chosen. Values of ω_K were taken from the fitted ω_K values of Bambynek *et al.*[8] using the Burhop equation.[9] The values of a and μ/ρ were obtained from Heinrich *et al.*[10] and Heinrich[11] respectively. The detector parameters, as suggested by Zaluzec,[12] used in this calculation are $Y_{Be} = 7.6\ \mu m$ (0.3 mil), $Y_{Au} = 0.02\ \mu m$ and $Y_{Si} = 0.1\ \mu m$.

TABLE 1 Specimens used in k factor determination

Element	Atomic no.	Specimen	Composition	wt-%
Na	11	Albite	Na	8.67
			Al	10.45
			Si	31.81
			O	balance
Mg	12	Antigorite (CL)	Mg	24.72
			Al	0.54
			Fe	1.26
			Si	20.37
			O	balance
Al	13	Biotite (CL)	Al	10.48
			Mg	5.37
			Si	17.33
			Fe	15.09
			K	6.37
			Ti	0.93
			O	balance
P	15	Fe_3P	Fe	85
			P	15
S	16	Pyrite	Fe	46.0
			S	53.0
			O	balance
K	19	Biotite	*as above*	
Ca	20	Tremolite (CL)	Ca	8.16
			Si	27.02
			Mg	14.63
			Mn	2.25
Ti	22	Ilmenite	Ti	27.41
			Mn	3.87
			Fe	36.36
			O	balance
Cr	24	Ni-Cr	Ni	80
			Cr	20
Mn	25	Ilmenite	*as above*	
Fe	26	Actinolite (CL)	Fe	9.1
			Si	24.31
			Ca	15.94
			Al	0.71
			Mg	6.68
			O	balance
Ni	28	Fe-Ni	Fe	70.5
			Ni	29.5
Cu	29	Cu-Mn	Cu	3.36
			Mn	96.64
Mo	42	Fe-Mo	Fe	90
			Mo	10
Ag	48	Ag–Al	Ag	9.39
			Al	90.61

RESULTS AND DISCUSSION

Figure 4 shows a selection of the k factor measurements plotted as a function of K_α characteristic X-ray energy. In Fig. 5 the experimental results are compared with the calculated k factors for 120 kV using the method outlined above. In Fig. 6 the results of Cliff and Lorimer[1] and Lorimer *et al.*[3] for 100 kV are compared with the results of this study. Table 2 gives a summary of all the data used in Figs. 4, 5 and 6. In addition experimentally measured k factors for P, Cr, Ni, Mo and Ag are listed. The error bars associated with the data in Figs. 4, 5 and 6 are discussed in the next section.

Comparison with calculated values

From Fig. 5 it is clear that above Si the calculated and experimental values follow the same trend: increasing with increasing K_α energy. It is not possible at this stage to explain any differences between the calculated and experimental values with certainty but the following factors should be considered:

(a) There are alternative expressions for ionization cross-sections Q_K which may result in closer agreement between theory and experiment. However, use of the cross-section expression recommended by Powell[13] yields an even greater discrepancy between calculated and measured k factors.

(b) The customary practice of ratioing each k factor to Si means that any error in the values for Si used in equation (2) will introduce a systematic error into the calculated data plotted in Fig. 5.

The experimental k value for Al coincides with the theoretical value but both Mg and Na show values significantly greater than the calculated figures. There is strong evidence, as reported by Mehta *et al.*,[14] that mass volatilization of Na occurs when Albite is exposed to the electron beam. The specimen was invariably observed to degrade in the EM400 during the time necessary to accumulate 10 000 counts, so this factor may contribute to the discrepancy in Na values. Yet this does not explain the high value for Mg. It is assumed that carbon contamination either on the specimen or on the Be window was not a limiting factor. This is considered reasonable in view of the minimal observed contamination in the ion-pumped stage of an EM400T. Another possible limitation on the quantification procedure is the current inadequacy of the EDAX SWTHIN program in terms of an accurate description of the continuum background at the low energy end of the spectrum where the *bremsstrahlung* intensity changes rapidly with

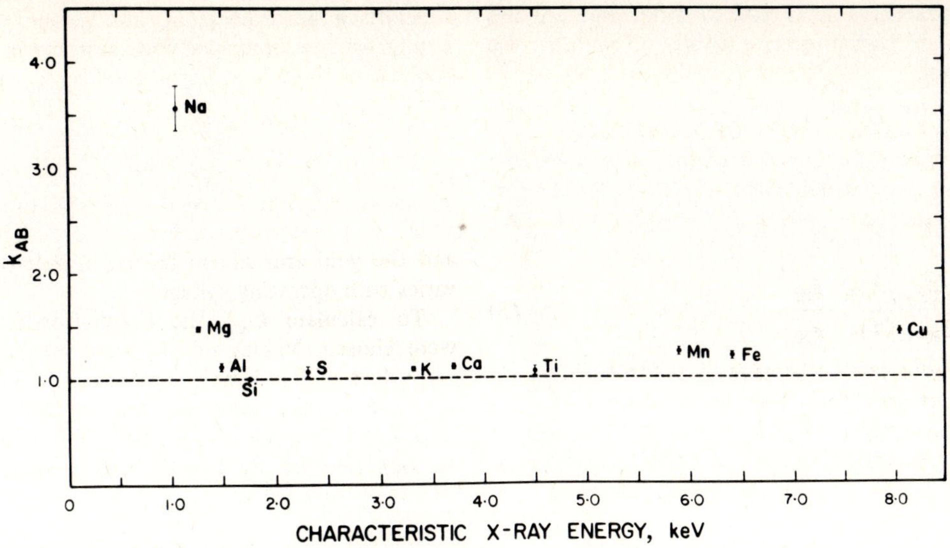

Error bars represent 95% confidence limits

4 Variation of experimentally determined *k* factors with characteristic X-ray energy for a selection of elements studied

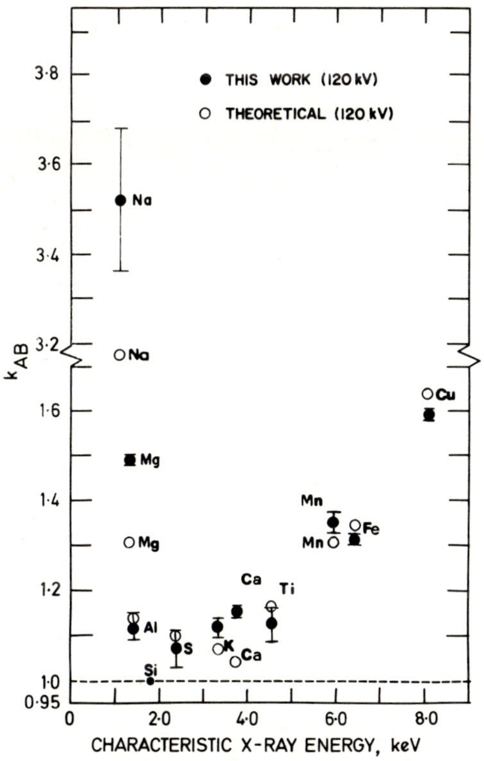

Error bars represent 95% confidence limits

5 Comparison of experimental results in Fig. 4 with calculated *k* factors at 120 kV

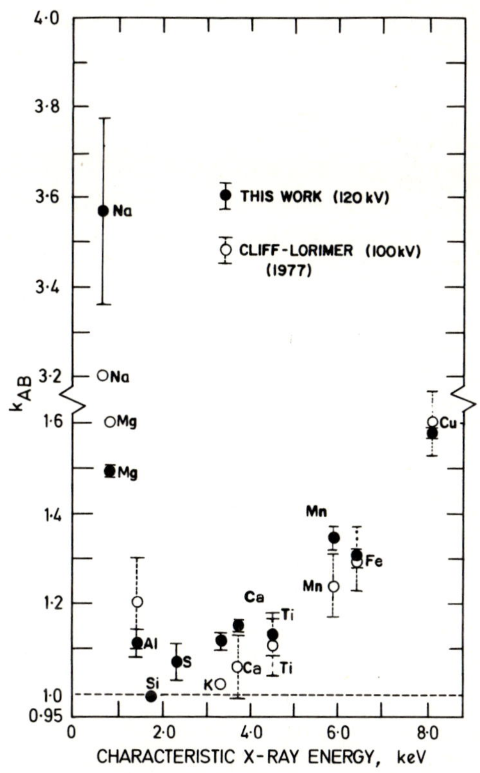

Open circles: reported by Lorimer *et al.*[3]
Dotted error bars (where shown) reported originally by Cliff and Lorimer[1]

6 Comparison of experimental results in Fig. 4 with previous experimental work of Lorimer and coworkers

TABLE 2 Comparison of experimental and calculated *k* factor data

k FACTORS						
This investigation (120 kV)						
		Relative error at 95% CL				
Element	*k*	Before interpolation	Final error	Cliff–Lorimer (100 kV) 1975[1]	1977[3]	Calculated (120 kV)
Na	3.57	0.21	0.21	5.77	3.2	1.83
Mg	1.49	0.007	0.007	2.07 ± 0.1	1.6	1.31
Al	1.12	0.03	0.03	1.42 ± 0.1	1.2	1.14
P	0.99	0.007	0.016	–	–	1.04
S	1.08	0.04	0.05	–	–	1.07
K	1.12	0.014	0.027	–	1.03	1.05
Ca	1.15	0.007	0.02	1.0 ± 0.07	1.06	1.04
Ti	1.12	0.04	0.046	1.08 ± 0.07	1.12	1.17
Cr	1.46	0.016	0.03	1.17 ± 0.07	1.18	1.24
Mn	1.34	0.025	0.04	1.22 ± 0.07	1.24	1.32
Fe	1.30	0.008	1.28	1.27 ± 0.07	1.3	1.36
Ni	1.67	0.04	0.06	1.47 ± 0.07	1.48	1.48
Cu	1.59	0.009	0.05	1.58 ± 0.07	1.60	1.65
Mo	4.95	0.12	0.17	4.3	–	4.68
Ag	12.4	0.09	0.63	8.49	–	7.08

energy. Any underestimation of the actual intensity of Na and Mg K_α X-rays would result in an increase in the values for k_{NaSi} and k_{MgSi}.

Comparison with previous experimental data

In Fig. 6 the current data are compared with the experimental results of Lorimer and coworkers.[1,3] The data points shown are those given in the latest paper by Lorimer *et al.*[3] but the error bars (where shown) are those reported in the original work of Cliff and Lorimer,[1] since the later work showed no error bars. In the majority of cases there is overlap between the experimental results, although invariably the errors in this current investigation are significantly lower. Discrepancies at the lower end of the energy range may be related to the problems of specimen degradation, relative differences in carbon contamination rates between the EMMA4 and the EM400T, and differences in the methods of data handling. But in conclusion it is clear that significant differences in the value of k_{AB} exist and the necessity for the empirical determination of *k* factors for a specific instrument are obvious.

QUALITY OF DATA

The error bars on the experimental data in Figs. 4, 5 and 6 represent the relative error at the 95% confidence limit for the *k* factors, determined using the Student *t* analysis for 30 individual readings taken from ~10 different areas on each sample. The calculated error in k_{AB} can be given by:

$$\% \text{ error} = \frac{t_{95}^{n-1}}{\sqrt{n}} \frac{S_c}{\bar{k}_{AB}} \times 100 \qquad (5)$$

where t_{95}^{n-1} is the Student t value for n readings and a 95% confidence level, S_c is the standard deviation for n readings, and $\bar{k}_{AB}$ is the mean k_{AB} value.

If the k_{AB} values are interpolated, individual errors must be added after each interpolation. As shown in Table 2, the final cumulative error in the k_{ASi} value may be considerably larger than the original k_{AB} value. Clearly errors in interpolated *k* factors will always be larger than experimentally determined factors. To obtain the most accurate experimental data it is always advisable to create the required standards so that interpolation errors need not be taken into account. However, it is occasionally impossible to create the required standards (for example if suitable single-phase alloys or stoichiometric intermetallic compounds cannot be found) and in this case interpolation of existing *k* factor data or calculated *k* factors is the only possible route for quantification.

ACKNOWLEDGMENTS

The authors acknowledge the financial support of NASA (NGR 39-007-043). They also wish to thank Dr G. W. Lorimer and Mr G. Cliff for provision of the mineral samples used in their original studies. Specimens were also supplied by C. H. Chen, J. R. Michael, S. Merchant and P. M. Novotny of Lehigh University.

REFERENCES

1 G. CLIFF and G. W. LORIMER: *J. Microsc.*, 1975, **103**, 203
2 J. I. GOLDSTEIN, J. L. COSTLEY, G. W. LORIMER, and S. J. B. REED: SEM/1977, ed. O. Johari, Chicago IITRI, 1977, 315
3 G. W. LORIMER, S. A. AL-SALMAN, and G. CLIFF: 'Developments in electron microscopy and analysis 1977', ed. D. L. Misell, Bristol and London, Inst. Physics, 1977, 369
4 A. D. ROMIG and J. I. GOLDSTEIN: 'Microbeam analysis 1979', ed. D. E. Newbury, San Francisco Press, 1979, 124
5 A. F. MARSHALL and C. ZERCHER: Proc. 38th EMSA Meeting, Baton Rouge, La, Claitors Press, 1980, 134
6 G. W. LORIMER, G. CLIFF, and J. N. CLARK: 'Developments in electron microscopy and analysis 1975', ed. J. A. Venables, London, Academic Press, 1977, 153
7 M. GREEN and V. E. COSSLETT: *Proc. Phys. Soc.*, 1961, **78**, 1206

8 W. BAMBYNEK, B. CRASEMANN, R. W. FINK, H. U. FREUND, H. MARK, C. D. SWIFT, R. E. PRICE, and P. V. RAO: *Rev. Mod. Phys.*, 1972, **44**, 716
9 E. H. S. BURHOP: *J. Phys. Radium*, 1955, **16**, 625
10 K. F. J. HEINRICH, C. E. FIORI, and R. L. MYKLEBUST: *J. Appl. Phys.*, 1979, **50**, 5589
11 K. F. J. HEINRICH: 'The electron microprobe', ed. T. D. McKinley, K. F. J. Heinrich, and D. B. Wittry, Wiley, New York, 1966, 296
12 N. J. ZALUZEC: 'Introduction to analytical electron microscopy', ed. J. J. Hren, J. I. Goldstein, and D. C. Joy, New York, Plenum Press, 1979, 121
13 C. J. POWELL: National Bureau of Standards, Special Publication 460, 1976, 97
14 S. MEHTA, J. I. GOLDSTEIN, D. B. WILLIAMS, and A. D. ROMIG: 'Microbeam analysis 1979', ed. D. E. Newbury, San Francisco Press, 1979, 119

DISCUSSION of paper by Wood, Williams, and Goldstein

P. DOIG:
Do you consider it necessary to use long-exposure STEM probe excitation of standard foils to establish accurate k values? We found that for our metal foils better X-ray counting statistics were obtained, with virtually no contribution from contamination deposition, by using a low beam current in a condensed TEM probe ~ 1 μm diameter. Our foils were of uniform thickness.

D. B. WILLIAMS:
No, we do not consider it necessary to use long exposures in STEM mode to obtain the necessary counts, except (a) in the cases where specimens are of a very non-uniform thickness (as was often the case for the crushed mineral samples) and (b) in cases where significant absorption is expected and therefore the accurate local thickness has to be determined.

P. J. STATHAM:
In your abstract you imply an instrumental dependence of k factors but omit to mention the efficiency of the Si(Li) X-ray spectrometer which probably provides the largest source of uncertainty in k values for elemental lines below 2 keV. Are these instrumental factors associated with the microscope that you consider to be important?

D. B. WILLIAMS:
The instrumental factors referred to in the abstract certainly include the variability associated with different EDS detectors below X-ray energies of 2 keV, but also we are concerned about the accuracy of the $+20°$ take-off angle that Philips report for their EM400T. If this figure is wrong by a few degrees then significant errors can be introduced into absorption correction factors, as discussed in the text of the review paper by Professor Goldstein (these proceedings).

G. CLIFF:
We have found that k values determined on EMMA-4 in 1974 for K_α radiation from Ti to Mo are still valid today on both EMMA-4 and an EM400T. The k values from Na to Ti are significantly different on the two instruments. This reinforces the suggestion that k values should be experimentally determined for elements $Z <$ Si.

D. B. WILLIAMS:
We too have found it necessary to redetermine k factors if the detector is changed, but otherwise we are confident that the values reported in this investigation will remain viable for the life of our EM400T.

Thin film *k*-value calibration for low atomic number elements using silicate mineral standards

R J McGill and F H Hubbard

A thin film k-*value calibration for low atomic number elements has been developed for use in the analysis of powdered silicate mineral samples using a modified AEI EM6G transmission electron microscope. Peak interference and background difficulties can be overcome by computer peak deconvolution and background correction procedures. Element loss and carbon contamination effects are reduced because of the large beam diameter (~3 μm) of the instrument. The loss characteristics of Na and K vary with the containing lattice. For this reason, element loss profile determination should precede the quantitative analysis of any unfamiliar mineral containing these elements.*

The authors are in the Department of Geology, University of Dundee.

Microprobe analytical systems have been successfully applied to the major element analysis of silicate and aluminosilicate minerals for many years. While high resolution thin film microanalysis using the transmission electron microscope has been applied to specific mineral systems (Champness *et al.*,[1] Lorimer and Cliff[2]) little has been published on the problems associated with the use of this technique for routine major element mineral analysis.

The special problems associated with the analysis of the low atomic number elements have been investigated. An element *k*-value calibration has been developed for the element range Na to Fe for routine analysis of finely powdered silicate minerals.

EQUIPMENT

The microanalytical equipment used comprised a modified AEI EM6G transmission electron microscope coupled to a Links Systems Model 290 analyser and Kevex energy-dispersive detector (resolution 146 eV, beryllium window thickness ~8 μm). The level of backscattered electrons entering the detector was significantly reduced by placing an Eclipse button magnet 50 mm in front of the beryllium window (Hall[3]) while extraneous and *bremsstrahlung* X-radiation from the stage and surroundings were effectively removed using a specially modified Swann double-tilting cold stage (Swann and Swann[4]) and careful carbon shielding of areas of the tilting tube, specimen holder and stage exposed to backscattered electrons. Full details of the modifications are given by Fitzgerald and Storey.[5]

When fully focused the EM6G has a beam diameter of approximately 3 μm which is significantly larger than in the majority of transmission microscopes specifically designed for microanalysis (e.g. EMMA: 0.1 μm). Although the large beam diameter has considerable disadvantages in terms of spatial resolution it has many advantages in low atomic number element microanalysis where element volatility and instrumental carbon contamination become increasingly important.

CALIBRATION STANDARDS

Independently analysed silicate minerals were used as standards for the determination of the low atomic number element *k*-value calibration for the instrument. A convenient compositional spread of the range of elements under investigation in a variety of lattice environments was thus obtained. The *k*-value independence of composition could therefore be tested and any crystal lattice effects assessed. The minerals used are listed in Table 1.

EXPERIMENTAL PROCEDURES

The mineral samples were finely fragmented and suspended in acetone. Single droplets of the dilute suspensate were deposited on carbon film superstrates on carbon-coated nylon microscope grids.

An accelerating voltage of 100 kV and a beam current of 200 μA were used. Mineral spectra were recorded for typical analytical live times of 200 s. Total spectral count rates were of the order of 1000 c/s for mineral grains in the size range 1–3 μm.

k-VALUE CALCULATION

Calculation of elemental *k* values followed the equation proposed by Cliff and Lorimer:[6]

$$k_{XSi} = \frac{C_X}{C_{Si}} \cdot \frac{I_{Si}}{I_X} \qquad (1)$$

where I_{Si} and I_X are the simultaneously recorded K_α characteristic X-ray intensities and C_{Si} and C_X the weight fractions of the elements Si and X in the thin film. Calibration of *k* values for low atomic number elements is complicated by several interrelated elemental and instrumental factors. With decreasing atomic number fluorescent yields decrease, resulting in poorer peak-to-background ratios. Low-energy Na and Mg characteristic X-rays are more readily absorbed within the thin film and in passing through the Be detector window, resulting in breakdown of the relationship of equation (1).

Energy-dispersive detectors record characteristic X-radiation as a Gaussian energy pulse where the area under the Gaussian peak, the peak integral intensity, reflects the

TABLE 1 Mineral standards used in *k*-value calibration

Mineral	Formula	Source
Anthophyllite	$(Mg, Fe)_7Si_8O_{22}(OH)_2$	Department of Geology, Univ. Dundee
Antigorite	$Mg_3Si_2O_5(OH)_4$	Department of Metallurgy, UMIST/ Univ. Manchester
Cummingtonite	$(Mg, Fe)_7Si_8O_{22}(OH)_2$	
Actinolite	$Ca_2(Mg, Fe)_5Si_8O_{22}(OH)_2$	
Hornblende	$Ca_2(Mg, Fe^{2+})_4(Al, Fe^{3+}, Ti)_3Si_8O_{22}(OH)_2$	
Amelia Albite	$NaAlSi_3O_8$	Department of Mineralogy and Petrology, Univ. Cambridge
Madagascar Orthoclase	$KAlSi_3O_8$	
Soda-feldspar	$NaAlSi_3O_8$	British Chemical Standards
Potash-feldspar	$KAlSi_3O_8$	

TABLE 2 Comparison of *k* values (100 kV)

	Theoretical*	I	II	III	IV
Na	2.02	2.85+	2.85+	3.20	3.40
Mg	1.39	1.67	1.67	1.60	1.48
Al	1.16	1.15	1.18	1.20	1.20
Si	1.00	1.00	1.02		
P	1.02		1.00		
S	1.08		1.01		
Cl	1.08		1.04		
K	1.08	1.10+	1.08+		
Ca	1.11	1.08	1.11		
Sc	1.13		1.15		
Ti	1.20		1.21		
V	1.25	1.28	1.27		
Cr	1.27		1.33		
Mn	1.33		1.39		
Fe	1.37	1.43	1.42		

* Calculations based on 8 μm Be window; 0.02 μm Au layer; 0.1 μm Si dead layer (Goldstein *et al.* 1977)
I Calculated from mineral standards (+50 s live time)
II Chebyshev refined values (+50 s live time)
III Lorimer *et al.* 1977 (EMMA-4 100 kV)
IV Mehta *et al.* 1979 (10 s live time, Philips EM300 100 kV)

TABLE 3 Statistical analysis k_{AlSi} values

Statistical function	I Range: all values 0.95–1.28	II Range: $\bar{X}^I \pm 3\sigma^I$ 1.00–1.30	III Range: $\bar{X}^{II} \pm \sigma^{II}$ 1.11–1.19
$\bar{X}$	1.15	1.15	1.15
M	1.15	1.16	1.15
σ	0.05	0.04	0.03
%CV	4.35	3.56	2.00
n	49	48	42

$\bar{X}$ mean
M median
σ standard deviation
%CV percentage coefficient of correlation
n number of values

concentration of the element in the specimen film. When dealing with 'thin' specimens (eqn (1)) peak integrals are directly proportional to element weight percent. Thus in order to calculate elemental k values characteristic peak integrals must be accurately determined from mineral standard spectra. Since the characteristic Gaussian peaks are superimposed on the *bremsstrahlung* profile corrections must be made for background counts when calculating peak integrals.

For elements of atomic number >17 the *bremsstrahlung* profile is essentially flat and linear and background-corrected peak integrals can be calculated to sufficient accuracy using the internally stored Links Systems program. For the element range Na to Si, however, there is significant peak overlap and the peaks are superimposed on the low-energy *bremsstrahlung* bulge. Use of the Links System procedure for these elements leads to significant errors (Fig. 1). Computer peak stripping, integration and background correction using the data from all channels over the range of interest overcomes this difficulty. Programs have been developed which calculate a least-squares, second-order, polynomial fit to the background profile and least-squares Gaussian estimates to a maximum of four overlapping peaks (McGill[7]). Peak intensities corrected for background and mutual interference are calculated using the mathematical estimates to the experimental spectrum.

LOW-Z ELEMENT MOBILITY AND CARBON CONTAMINATION

Element mobility and volatilization in microprobe specimens is well documented (Ribbe and Smith,[8] Fredriksson,[9] Lineweaver[10]). Feldspars are particularly prone to the loss of sodium and potassium. Similar observations have been reported (Lorimer and Champness,[11] Hodson and Marshall[12]) for transmission microanalytical systems. An investigation was undertaken to assess the extent of element mass volatilization in the EM6G system.

Any change in the concentration of a volatile element results in a change in the element k value. Focusing the electron beam on a single 'thin' mineral grain, the volatility of an element (X) was monitored by recording the values of k_{XSi} and k_{AlSi} after progressively longer periods of irradiation by the beam. Initially the standards Amelia Albite and Madagascar Orthoclase were selected for this investigation as microprobe studies had shown these minerals to be susceptible to element mass volatilization. Figure 2 illustrates the changes in k_{NaSi} and k_{AlSi} with irradiation time (s) for the Amelia Albite. After 100 s Na is progressively lost. The constant k_{AlSi} values imply no change in the relative elemental concentrations of Al and Si. Measurements of K loss from the Madagascar Orthoclase produced a similar K-loss profile with increasing irradiation time (McGill[7]). Na and K loss were significantly reduced if the beam was defocused but this had a detrimental effect on the overall spectral count rate and peak-to-background ratios. Having established Na and K loss for the feldspars, a wider investigation of element mobility was undertaken by monitoring changes in k_{NaSi}, k_{MgSi}, k_{AlSi} and k_{KSi} with irradiation time for a range of mineralogical specimens. The clay mineral Na montmorillonite and the calcium-sodium silicate pectolite showed no Na loss with irradiation time and a second clay mineral, K-bentonite, no K loss. The four UMIST standards (Table 1) showed no change in the Mg and Al k values with irradiation time.

This limited survey suggests that element mass volatilization is not solely dependent on atomic number and electron

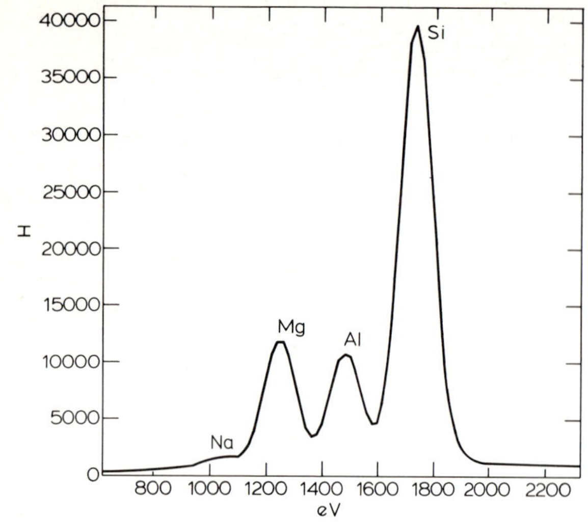

1 **Peak overlap and *bremsstrahlung* profile for low atomic number elements (100 kV)**

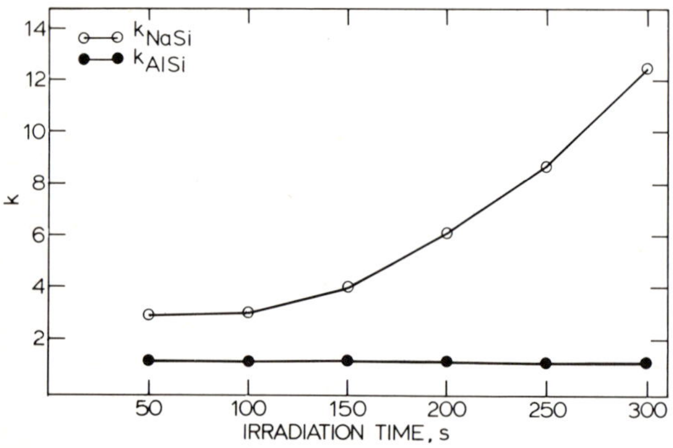

2 **Variation in k_{NaSi}, k_{AlSi} with irradiation time (s) for Amelia Albite**

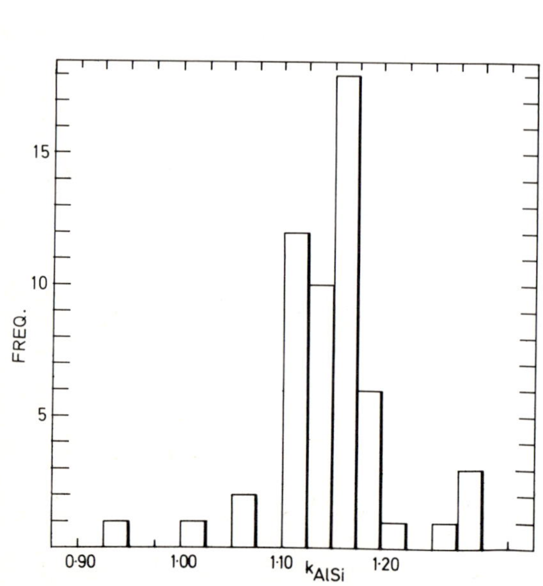

3 **Distribution of k_{AlSi} values**

beam intensity but that lattice controls may also influence the behaviour of specific elements in the electron beam.

Mehta *et al.*,[13] in an attempt to account for the differences between experimentally recorded k values for the elements Na, Mg and Al, suggested cracking of the microscope hydrocarbon oils and subsequent build-up of carbon contamination at the point of irradiation as a possible explanation for changes in k_{NaSi}, k_{AlSi} and k_{MgSi} with irradiation time. Investigations with the EM6G system have shown no change in the k values for Mg and Al, nor any change in Na k value in certain mineral groups, suggesting that carbon contamination is either not present or present at a level which does not affect the k values of low atomic number elements. These conclusions are supported by the work of Beaman and Walker[14] who studied the effects of beam intensity and analysis time on carbon contamination. They measured little change in I_{Mg}/I_{Si} with beam diameters of 3.7 μm (EM6G ~ 3 μm) over extended periods of irradiation.

EXPERIMENTAL *k*-VALUE CALIBRATION

Table 2 lists the derived set of experimental k values with the comparative values for Na, Mg and Al reported by Lorimer *et al.*[5] and Mehta.[13] All k values were calculated for live times of 200 s with the exceptions of Na and K which are quoted as 'no-loss' values for live times of 50 s, in accordance with the mass volatilization profiles (Fig. 2).

Figure 3 illustrates the distribution of k_{AlSi} values derived from four distinct mineral species. The distribution spread was controlled by several factors: (a) the chemical homogeneity of the standards, (b) the accuracy of the chemical analysis of the standards, and (c) variation in film thickness from the thin film criterion. The final experimental k values were calculated according to the statistical procedure presented in Table 3. All other major elements produced similar k-value distributions and were evaluated by the same statistical procedure.

Figure 4 is a plot of K_α energy versus experimental k value where the error bars indicate the standard deviation associated with the adopted k value. Because of the difficulty in obtaining suitable mineral standards, k values for the remaining elements in the range Na$\leqslant Z \leqslant$Fe were interpolated from the experimental k curve. In preference to interpolation from a hand-drawn curve, Chebyshev least-squares polynomial approximations were calculated to the experimental k curve and interpolations were based on the equations of the mathematical fit.

Figure 5 illustrates the 6th degree Chebyshev polynomial fit to the experimental k curve where K_α energies have been transformed to ln K_α values in order to accommodate the curve fitting procedures (Hayes[6]). Theoretical k values (Goldstein *et al.*[7]) are compared with interpolated values from the Chebyshev polynomial fit in Table 2.

Figure 6 combines the theoretical k values and the experimental Chebyshev k curve. Agreement between the theoretical and interpolated experimental k values is good for all but the elements Mg and Na. As the experimental Mg and Na k values are 'no-loss' figures the discrepancy cannot be attributed to mass volatilization. The invariance of k_{MgSi} (amphiboles) and k_{NaSi} (Na montmorillonite, pectolite) with irradiation time suggests that carbon contamination build-up during analysis is not a factor in the disparity between theoretical and experimental values in the EM6G system. Goldstein *et al.*[7] state that a 50% increase in Be window thickness will produce a 25% increase in the k values for Mg and Na. Window thickness is known to vary from detector to detector and variation

TABLE 4 Thin film analysis of amphibole

	SiO_2	Al_2O_3	MgO	Na_2O	CaO	K_2O	TiO_2	MnO	Fe_2O_3
EM6G	46.07	11.85	16.51	1.85	12.04	0.42	0.51	0.33	10.42
XRF	46.14	12.67	16.17	–	12.98	0.49	0.54	0.37	10.84
Microprobe	46.75	12.08	15.88	1.64	12.75	0.38	0.38	0.37	10.85

Analyses are quoted on a water-free basis

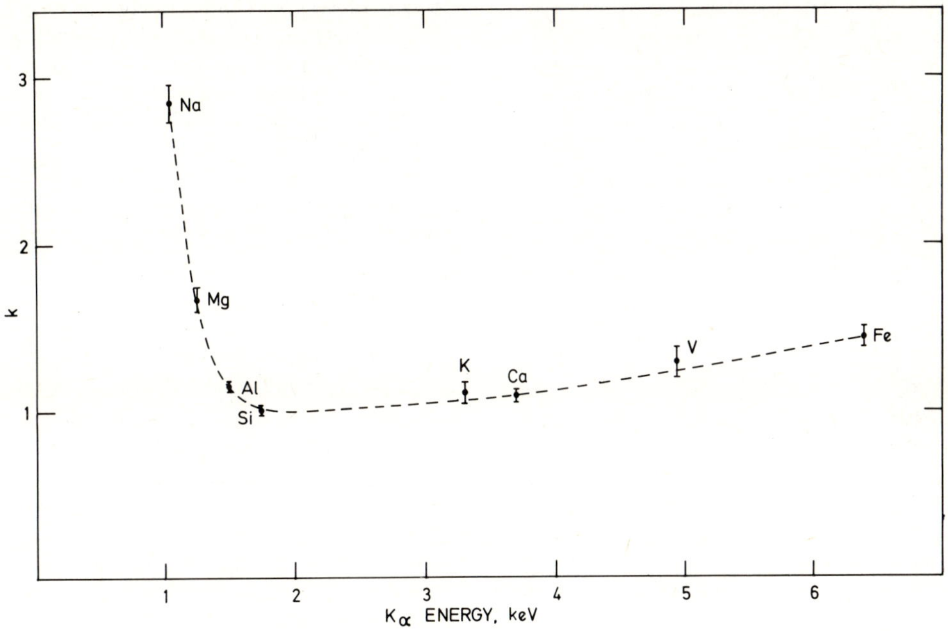

4 Experimental k values versus K_{α} energy (keV)

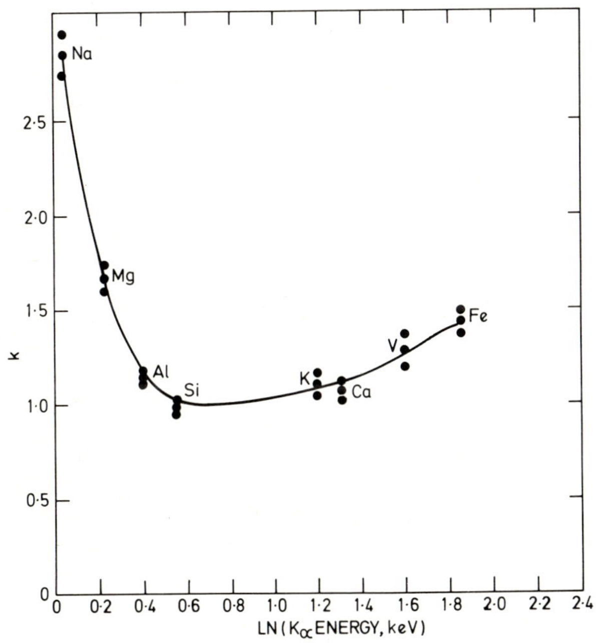

5 Experimental k values versus ln (K_{α} energy keV) with 6th degree Chebyshev polynomial curve fit

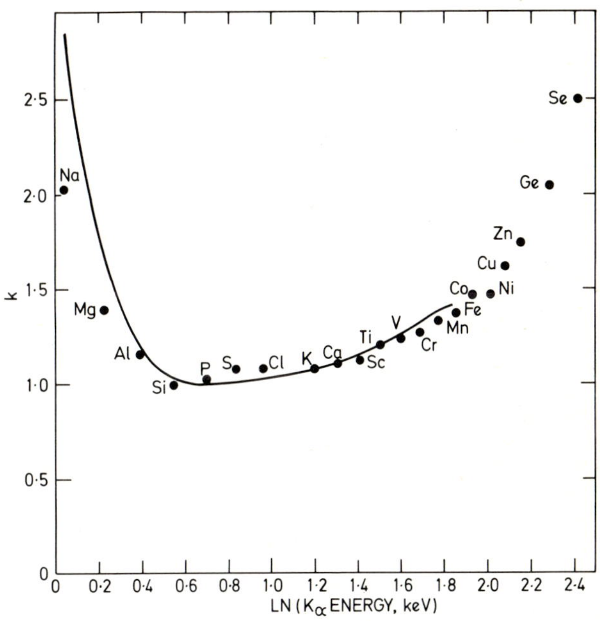

6 Theoretical k values and Chebyshev experimental k curve

from the 8 μm assumed in the calculation may partially account for the difference between the theoretical and experimental k values. Thickness increases of this order would raise the theoretical k for Mg from 1.37 to 1.74 (experimental 1.67), and k for Na from 2.02 to 2.53 (experimental 2.85). Thus an increase of less than 50% in Be window thickness could account for the discrepancy in k_{MgSi} but it would seem that some other parameter must be considered to resolve the deviation of the experimental k_{NaSi} from the theoretical value.

QUANTITATIVE MINERAL ANALYSIS

The interpolated k set (Table 2) has been applied to a range of mineral species with encouraging results. As an illustration, Table 4 compares the thin film microanalysis of an amphibole with results obtained by microprobe analysis on polished thin section (Department of Mineralogy and Petrology, University of Cambridge) and X-ray fluorescence analysis of a glass prepared from the mineral. 'No-loss' k values were adopted for Na and K following the establishment of no-loss profiles for the mineral after irradiation for periods in excess of the analytical live time (200 s). The same peak deconvolution and background correction procedures as used in the k-value determinations were employed and Ca K_α peak integrals were corrected for interference from the K K_β peak by use of a conversion factor obtained from determination of the K K_α/K K_β peak intensity ratio for calcium-free potassium-bearing minerals.

CONCLUSIONS

Where high-resolution (beam diameter <1 μm) microanalytical transmission electron microscope facilities are not available, relatively low-resolution instruments can be modified to enable thin film calibrations for the major element analysis of finely powdered silicate and aluminosilicate minerals. Poor analytical resolution is compensated for by a reduction in the level of both mass volatilization of low atomic number elements and carbon contaminant build-up during analysis.

Prior to qualitative analysis, mass volatilization profiles should be determined for each mineral to establish suitable analysis live times. Mass loss varies not only with element atomic number and beam intensity but also with lattice environment.

The difficulties of low atomic number element peak overlap and background correction can be overcome by computer peak deconvolution procedures.

ACKNOWLEDGMENTS

The cooperation of the Department of Physics, University of Dundee, the Department of Metallurgy, UMIST, and the Department of Mineralogy and Petrology, University of Cambridge, is gratefully acknowledged.

REFERENCES

1 P. E. CHAMPNESS, G. CLIFF, and G. W. LORIMER: *J. Microsc.*, 1976, **108**, 231
2 G. W. LORIMER and G. CLIFF: 'Electron microscopy in mineralogy', 506–519; 1976, Berlin, Heidelberg, New York, Springer-Verlag
3 T. A. HALL: *J. Microsc.*, 1977, **110**, 103
4 P. R. SWANN and G. R. SWANN: Proc. 28th Annual Meeting of EMSA, Houston, Texas, 372, 1970, Claitor's Publ. Div.
5 A. G. FITZGERALD and B. E. STOREY: *J. Microsc.*, 1979, **115**, 73
6 G. CLIFF and G. W. LORIMER: *J. Microsc.*, 1975, **103**, 203
7 R. J. McGILL: MSc thesis, Univ. Dundee, 1980
8 P. H. RIBBE and J. V. SMITH: *J. Geology*, 1966, **74**, 217
9 K. FREDRIKSSON: "Optique des rayons X et microanalyse', 1966, Paris, Hermann
10 J. L. LINEWEAVER: *J. Appl. Phys.*, 1963, **34**, 1786
11 G. W. LORIMER and P. E. CHAMPNESS: Proc. Third Int. HVEM Conf., 1974, Academic Press
12 S. HODSON and J. MARSHALL: *J. Microsc.*, 1971, **93**, 49
13 S. MEHTA, J. I. GOLDSTEIN, D. B. WILLIAMS, and A. D. ROMIG Jr: 'Microbeam analysis', 119–123; 1979, San Francisco, San Francisco Press Inc.
14 D. R. BEAMAN and H. J. WALKER: Proc. Workshop on Asbestos, NBS special publication No. 506, Washington DC, 1978, 249
15 G. W. LORIMER, S. A. AL-SALMAN, and G. CLIFF: 'Developments in electron microscopy and analysis', Inst. Phys. Conf. Ser. No. 36, Inst. Physics, 369, 1977
16 J. G. HAYES: 'Numerical approximation to functions and data', 1970, London, Athlone Press
17 J. I. GOLDSTEIN *et al.*: 'Quantitative X-ray analysis in the electron microscope', Proc. Workshop Anal. Electron Microscopy, ITTRI, Chicago, Ill., 1977, 315

DISCUSSION of paper by McGill and Hubbard

S. J. B. REED:
Was the background interpolation for the low-energy peaks (Na, Mg, Al, Si) carried out by fitting a curve to the minima between the peaks? This would not allow for the fact that these points generally include a contribution from the tails of the peaks. Also a second-order polynomial would probably not represent the true continuum shape very accurately.

R. J. McGILL:
No. The second-order polynomial fit is calculated from background levels recorded for the continuum on the high- and low-energy sides of the region of peak overlap. We accept that there is some error in the second-order polynomial fit to the continuum.

J. I. GOLDSTEIN:
What justification do you have for fitting a k versus E (keV) curve? I believe each k value is discrete and there is *no* theory which argues for a smooth variation between values.

R. J. McGILL:
A sixth-degree polynomial curve was found to agree closely with calculated theoretical k values. Agreement between the interpolated experimental k values and theoretical k values is good.

Determination of $\phi(\rho t)$ curves for thin foil microanalysis

N Stenton, M R Notis, J I Goldstein, and D B Williams

It has been generally assumed that, in electron-transparent thin foils, the depth distribution of X-ray production $\phi(\rho t)$ is constant with thickness t. This assumption is critical when absorption corrections are applied to the X-ray intensity data. The authors report the experimental determination of $\phi(\rho t)$ for thin nickel foils up to 380 nm thick and show that $\phi(\rho t)$ increases by ~30% over this thickness range. Comparison is made with Monte Carlo calculations describing $\phi(\rho t)$ which predict a similar trend but a smaller (~10%) increase.

The authors are in the Department of Metallurgy and Materials Engineering at Lehigh University, Bethlehem, Pa, USA

In quantitative microanalysis of thin specimens it is generally assumed that absorption and fluorescence of X-rays within the specimen can be ignored, in which case the Cliff–Lorimer[1] equation can be used to determine the compositions C_A and C_B of elements A and B. Thus:

$$\frac{C_A}{C_B} = k_{AB}\frac{I_A}{I_B} \tag{1}$$

where k_{AB} is constant for a fixed accelerating voltage and I_A and I_B are the characteristic peak intensities generated by elements A and B above the continuum background. Apart from isolated cases such as Fe–Cr, fluorescence can indeed be neglected, but it is becoming increasingly apparent[2–4] that absorption of X-rays within thin foils is very often significant, leading to errors in excess of ±10% relative in the calculated composition.

The Cliff–Lorimer equation can be adjusted to take into account the preferential absorption of one characteristic X-ray with respect to another in the following manner:[4]

$$\frac{C_A}{C_B} = k_{AB}\frac{I_A}{I_B}\left\{\frac{\int_0^t \phi_B(\rho t)\exp-[(\mu/\rho)^B_{spec}\operatorname{cosec}\alpha(\rho t)]\,dt}{\int_0^t \phi_A(\rho t)\exp-[(\mu/\rho)^A_{spec}\operatorname{cosec}\alpha(\rho t)]\,dt}\right\} \tag{2}$$

where $\phi(\rho t)$ is the depth distribution of X-ray production; $(\mu/\rho)^i_{spec}$ is the mass absorption coefficient for X-rays from element i in the specimen; ρ the density of the specimen; t the specimen thickness; and α the X-ray take-off angle.

If it is assumed that there is a constant rate of production of X-ray photons through the thickness t of the thin foil specimen, that is, the depth distribution of X-ray production $\phi_A(\rho t) = \phi_B(\rho t) = 1$, then equation (2) reduces to the expression:[4]

$$\frac{C_A}{C_B} = k_{AB}\frac{I_A}{I_B}\left[\frac{(\mu/\rho)^A_{spec}}{(\mu/\rho)^B_{spec}}\right] \cdot \left[\frac{1-\exp-[(\mu/\rho)^B_{spec}\operatorname{cosec}\alpha(\rho t)]}{1-\exp-[(\mu/\rho)^A_{spec}\operatorname{cosec}\alpha(\rho t)]}\right] \tag{3}$$

Generally, if absorption is sufficient to cause a change of >10% in the k_{AB} factor, it is considered necessary to use equation (3) rather than (1). This arbitrary guideline is already becoming more restrictive as the accuracy of quantification procedures increases and relative errors in composition measurements are reduced below ±10%.

The assumption that $\phi(\rho t)$ is a constant may not apply in all cases where an absorption correction is necessary. Kyser[5] has calculated, using Monte Carlo techniques, the variation of $\phi(\rho t)$ at high operating voltages for various thin films. For example, for Al K_α in a 400 nm Al film there is no variation in $\phi(\rho t)$. However, for Cu K_α in a 400 nm Cu film and Au L_α in a 400 nm Au film, $\phi(\rho t)$ is predicted to increase by 5% and 20% respectively. Furthermore there is evidence[6] that in relatively thick specimens in which there is a considerable amount of absorption, equation (3) is not an accurate description of the absorption correction. This may be due in part to the assumption that $\phi(\rho t)$ is constant. Accordingly we have attempted to determine experimentally the term $\phi(\rho t)$ for nickel X-rays at 120 kV. Comparisons are also made between the experimental data and theoretical calculations of $\phi(\rho t)$ at 120 kV using Monte Carlo computer techniques.

The use of $\phi(\rho t)$ curves is an integral part of bulk microanalysis in the electron microprobe or SEM where absorption is invariably a problem. In thick specimen microanalysis the term $\phi(\rho t)$ or $\phi(\rho z)$ is used where t or z is the measure of the depth of X-ray generation in the sample. Experimentally, it is necessary to measure the intensity of radiation emitted from tracer layers at known depths relative to a similar tracer layer isolated in space. To do this requires the use for the tracer layer of a material with slightly different characteristic X-ray energy but similar X-ray generation properties to the matrix. An example of a $\phi(\rho z)$ curve for Al K_α X-rays using a Mg tracer in a bulk specimen of Al is shown in Fig. 1. The major difference between a thin foil specimen and a bulk

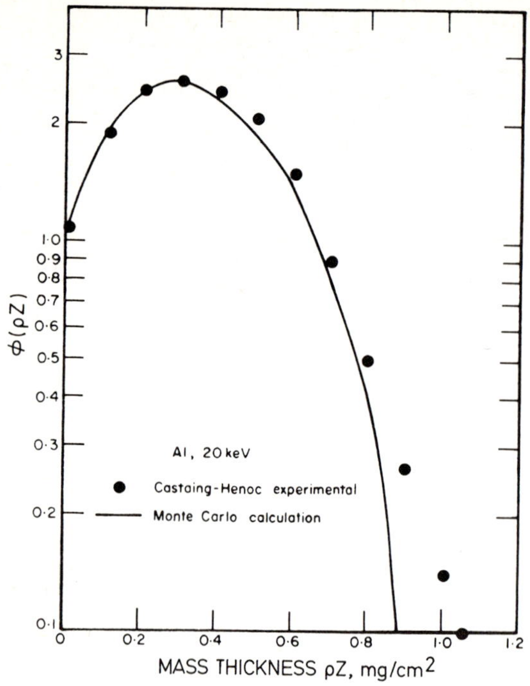

1 **Depth distribution of X-ray production using 20 kV electrons as function of mass thickness for bulk specimen of Al, showing comparison between experimental data and Monte Carlo calculations**[7]

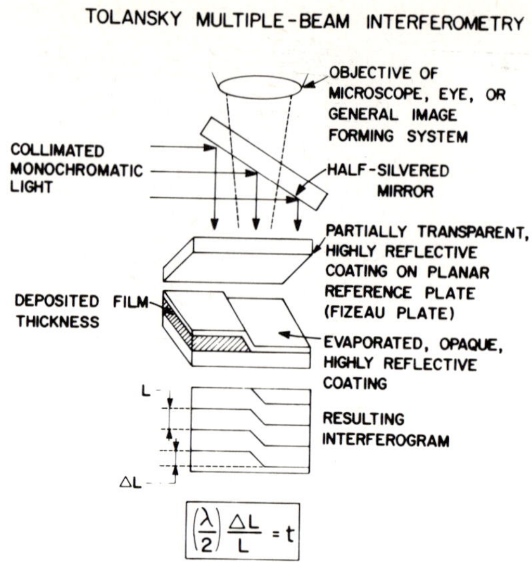

3*a* **Schematic: principles of Tolansky interferometry**

specimen is that very few electrons are scattered or lose significant energy before emerging from the thin film. This leads to a small fraction of backscattered electrons and a small X-ray source (good spatial resolution). Monte Carlo calculations are much more simplified for thin films.

EXPERIMENTAL PROCEDURES

1 Considerations of specimen geometry

For a thin film specimen, it is necessary to measure the intensity of radiation emitted from tracer layers at known depths relative to an identical tracer layer isolated in space. However, the total thickness t of the film (including the tracer) should be held constant and a $\phi(\rho t)$ curve obtained for that thickness. Experimentally, the production of several thin films of a given thickness each containing tracer layers at various depths is very difficult. However, if the backscatter coefficient for electrons in these films is very low, the effect of the material below the tracer layer on X-ray production in the tracer layer is very small. In this case $\phi(0) = 1$; that is, the X-ray emission from the tracer layer on the surface of the film will be equal to that of the tracer layer isolated in space. For the $\phi(\rho t)$ experiments in this study, negligible backscatter is assumed and

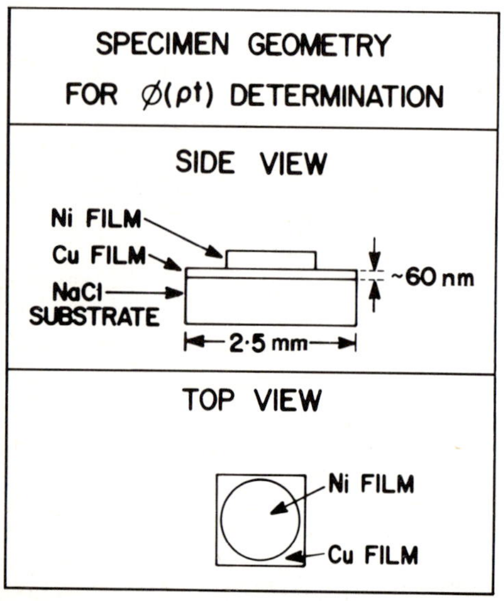

2 **Specimen geometry used in determination of $\phi(\rho t)$. Thicknesses of Ni varying from ~25 to 380 nm were deposited on top of Cu film leaving portions exposed**
To determine $\phi(\rho t)$ comparison is made between intensity of Cu X-rays coming through Ni (I_t) and intensity coming from exposed film (I_0)

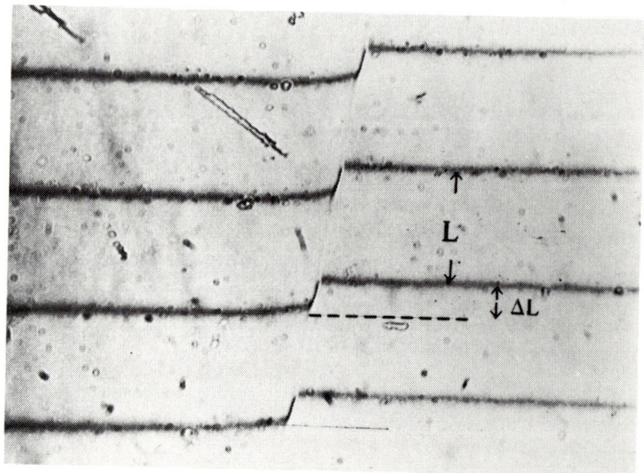

3*b* **Typical interferogram showing how L and ΔL are measured to enable specimen thickness t to be determined using equation in Fig. 3*a***

TABLE 1 Specimen thicknesses for $\phi(\rho t)$ determinations

Cu thickness, nm	Ni thickness, nm	ρt for Ni, mg cm^{-2}
	25.0 ± 5	0.022
	55.0 ± 5	0.049
	80.0 ± 5	0.072
	110.0 ± 5	0.099
59 ± 5	167.5 ± 5	0.150
	200.0 ± 5	0.179
	260.0 ± 5	0.233
	310.0 ± 5	0.278
	380.0 ± 5	0.341

TABLE 2 Comparison of techniques for thin foil thickness measurements

Quartz oscillator, nm	Talysurf, ± 10 nm	Tolansky interferometer, ± 5 nm
100.8	110.0	107.1
188.6	200.0	186.8
350.3	380.0	368.7

thin film specimens of varying thickness are used, as described below.

2 Sample preparation

A series of thin film samples were prepared for the determination of the $\phi(\rho t)$ curve for Ni, each having an initial layer of Cu approximately 60 nm thick acting as the tracer element. Different thicknesses of Ni were then deposited on the samples, leaving a section of the tracer layer exposed on each specimen. The specimen geometry is shown in Fig. 2. Both the tracer and matrix layers were prepared by vacuum evaporation and deposition onto a 2.5 mm square single crystal of sodium chloride. The Cu was deposited by thermal evaporation from a heated tungsten filament, and the Ni by electron beam evaporation.

Film thickness measurements were made by the Tolansky multi-beam interferometric technique[8] using glass slides which had been placed in the evaporation system at the same distance from the evaporation source as the specimens. A schematic of the Tolansky technique is shown in Fig. 3*a*. The glass slides were partially masked so that material was only deposited on half of the slide. After deposition the mask was removed and the entire surface covered by evaporating Al, which has high reflectivity. The Al layer follows exactly the contour of the initial film, giving rise to a step on the surface of the slide with a height equal to the thickness of the initial deposited film. When this step is observed in the optical microscope using multi-beam interferometry a series of stepped fringes are observed (Fig. 3*b*). The distance between successive fringes (L) corresponds to half the wavelength of the incident light (monochromatic radiation from a sodium lamp, 589.3 nm). The ratio of the step height ΔL to the fringe width L gives a measure of the film thickness. Hence:

$$t_{\text{film}} = \frac{\lambda_{\text{Na}}}{2} \cdot \frac{\Delta L}{L} \tag{4}$$

Thicknesses measured using this technique are accurate to within ±5 nm.

Details of the specimens used in the $\phi(\rho t)$ determination are listed in Table 1. When both Cu and Ni evaporations were complete the NaCl substrates were dissolved in water and the films floated onto 3 mm Be grids.

In addition to the interferometry measurements the Ni film thicknesses were also measured using a quartz crystal oscillator[9] (Fig. 4) and by Talysurf[10] measurements. A comparison of thickness measurements made using each technique is given in Table 2.

3 Generation of data

The $\phi(\rho t)$ curve was determined using the Philips EM400T AEM, in which the electron incidence angle is normal to the sample surface and the X-ray take-off angle $\alpha \simeq 20°$. Intensity measurements from the Cu tracer layer beneath the Ni matrix (I_t) were made at 120 kV using a probe size of 10 nm, and normalized by measuring the intensities from regions not covered by Ni (I_0). A total of 20 measurements were made in both the Cu/Ni and Cu regions to minimize the effect of minor fluctuations in emission current. The maximum count time at each point was 30 s and in addition groups of four readings were taken alternately from the two regions of the specimen. Each intensity reading was corrected for background and measurements made in the Cu/Ni region were corrected for absorption. Values of $\phi(\rho t)$ for each specimen were obtained by taking the ratio of the Cu intensity from the Cu/Ni region to that of the intensity from the pure Cu region.

4 Statistical analysis

The determination of $\phi(\rho t)$ curves requires taking the ratio of two values (the Cu X-ray intensity from the Cu/Ni region and that from the pure Cu region) which are very nearly equal to each other. In order to distinguish a true difference between the two values in question, the observed difference in X-ray counts must be greater than the error in the individual readings. This is known as the analytical sensitivity and is expressed mathematically by:

$$\bar{N} - \bar{N}_1 \geqslant \frac{\sqrt{2}(t^{1-\alpha}_{n-1} S_c)}{n^{1/2}} \tag{5}$$

where $\bar{N}$ and $\bar{N}_1$ are the mean X-ray counts from each

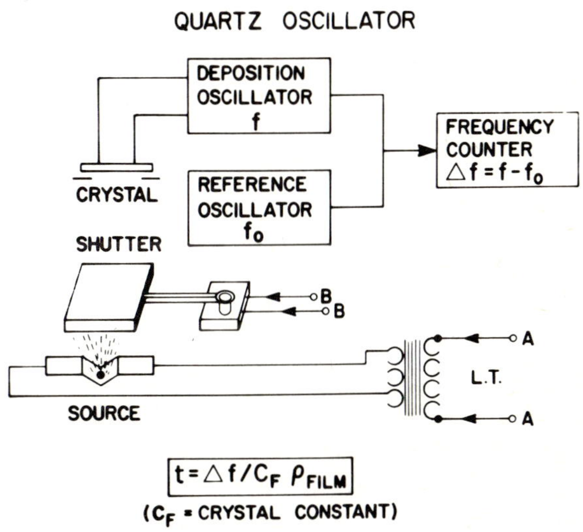

4 Schematic: how thickness of deposited film of density ρ may be measured by determining frequency shift (Δf) of quartz oscillator

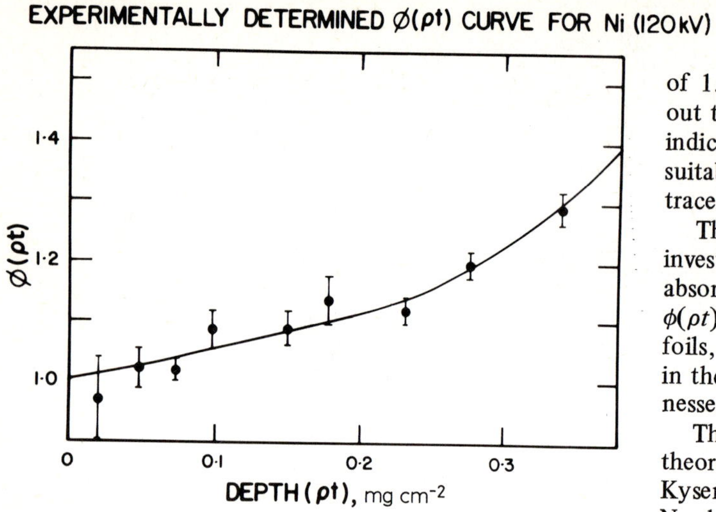

5 Experimental data showing increase in $\phi(\rho t)$ with mass thickness for thin films of Ni up to ~380 nm thick, using 120 kV electrons

region, S_c is the standard deviation (which is assumed to be the same for both numbers since they are almost equal), $t_{n-1}^{1-\alpha}$ is the Student t distribution value, and n is the number of counts taken for each set of numbers. Thus in terms of a percentage, the analytical sensitivity is given by:

$$\text{Sensitivity } (\%) = \frac{\sqrt{2}(t_{n-1}^{1-\alpha} S_c)}{n^{1/2}\bar{N}} \times 100 \qquad (6)$$

In this work the Student t value used was that for a 95% confidence limit using 16 measurements.

RESULTS AND DISCUSSION

The experimental $\phi(\rho t)$ curve for 380 nm of Ni is shown in Fig. 5. To determine the effect of electron backscatter, the 380 nm Ni film was re-examined with the Cu tracer layer placed on top of the film. In this case $\phi(0)$ is measured for the 380 nm thick film allowing for the maximum possible backscatter effect. The measured $\phi(0)$ value under these conditions is 1.03 ± 0.064. The analytical sensitivity at the 95% confidence level is too low to distinguish whether this value of $\phi(0)$ is significantly different from 1.0. Therefore the effect of electron backscatter is considered insignificant and the experimental thin foils used in this study are adequate for the measurement of $\phi(\rho t)$. Monte Carlo calculations by Newbury[11] for 120 kV electrons in 400 nm thin foils of Ni yield a very low backscatter coefficient of 0.0042. This backscatter coefficient is consistent with a measurement of 1.0 for $\phi(0)$. Because of this low backscatter coefficient, the experimental $\phi(\rho t)$ curve shown in Fig. 5 can be used for all Ni thin films $\leqslant$380 nm.

The error bars represent 95% confidence limits and only the first two data points (for Ni depths of 25 and 55 nm) fail to satisfy the analytical sensitivity criterion (equation (5)). This is not surprising since at small depths $\phi(\rho t)$ is very close to unity and the errors in thickness measurement and difficulties in obtaining a uniform film are the greatest.

The $\phi(\rho t)$ data vary by 30% over the thickness range 0–380 nm. The curve appears to be approximately linear up to a thickness of 200 nm but above this value $\phi(\rho t)$ increases more significantly with increasing depth.

The error bars imply that up to a thickness of 80 nm the value of $\phi(\rho t)$ does not vary significantly from a value of 1.0. Thus below 80 nm the X-ray production throughout the specimen may be considered uniform. This in turn indicates that the Cu tracer layer used (60 nm) was of suitable thickness so that the X-ray production within the tracer could be considered uniform.

The significant variation of $\phi(\rho t)$ observed in this investigation has important implications as far as thin film absorption corrections are concerned. The assumption that $\phi(\rho t)$ is always equal to 1, as is common practice for thin foils, is not necessarily a valid one. This work indicates that in the case of Ni this assumption is only valid for foil thicknesses of 80 nm or less, rarely achieved in practice.

The results of this investigation are compared with the theoretical $\phi(\rho t)$ curves determined by Newbury[11] and Kyser[5] based on Monte Carlo calculations[5,12] (Fig. 6). Newbury's model appears to agree more closely with the experimental data produced, even though it predicts only a 10% increase in $\phi(\rho t)$ over the thickness range investigated. The $\phi(\rho t)$ curve produced by Kyser for Cu X-rays at 100 kV shows only a 5% increase in $\phi(\rho t)$ over the same thickness range. It would be expected, therefore, that a corresponding curve for 120 kV would show an even smaller increase in $\phi(\rho t)$, since higher energy electrons give smaller variations in $\phi(\rho t)$ over the same thickness range. The reasons for the discrepancies between calculated and measured $\phi(\rho t)$ curves are not known but it should be noted that there are detailed differences in the assumptions and the theoretical expressions in the two Monte Carlo approaches.

FOIL THICKNESS DETERMINATION

The use of electron microscope foils the thickness of which is known to ~±5 nm using the Tolansky technique in combination with Talysurf and quartz oscillator frequency measurements has permitted comparison to be made with thickness measured by using more conventional techniques. Such thickness measurements are critical when absorption corrections have to be applied during quantification of X-ray data.

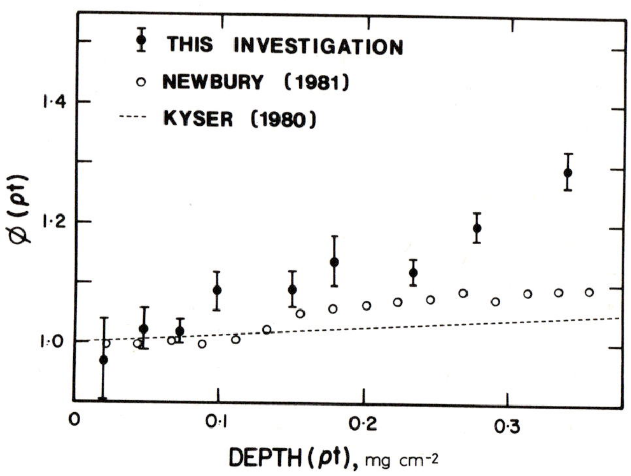

6 Comparison of experimental data from this current investigation with Monte Carlo calculations

Work of Newbury[11] is for Ni at 120 kV while data of Kyser[5] are for Cu at 100 kV

7 **Contamination spots created on upper and lower surfaces of evaporated Al foil (×100 000)**

There are various methods for foil thickness determination, not all of which are applicable to particular situations. Use of extinction distance (ξ_g) measurements under two-beam conditions requires tedious calculation or is restricted to systems where ξ_g has been tabulated. Measurement of the projected width of slip traces or any planar defects requires that they be present in the region of interest. The use of latex balls on either side of the foil has been suggested[13] but this requires dipping the specimen in alcohol which may encourage the formation of carbon contamination. The use of Kossel-Möllenstedt fringes in convergent beam patterns is accurate provided that such fringes can be observed, which in our experience has not been possible in many thin specimens of engineering materials. Usually, therefore, thickness is determined using the contamination spot separation method, since such spots can be easily generated on all specimens even in a Philips EM400T ion-pumped environment, by simply disengaging the cold finger. Although it is acknowledged that there are errors in such measurements (e.g. the work of Love, Cox and Scott[14]) the ease of the technique counters this drawback. Accordingly contamination spot separation measurements were made on a number of thin films of known thickness in order to determine the accuracy of the technique.

The films measured comprised two films of Al and a Cu film which were all produced by thermal evaporation from a heated tungsten filament onto NaCl substrates. Contamination spots were produced on each specimen using a 10 nm probe for several minutes with the specimen set at zero tilt. The thin films were mounted on formvar-coated grids to give sufficient support but contamination spots were only produced in regions where the support film was incomplete, so that the values obtained did not include the thickness of the formvar. The specimen was then tilted until the contamination spot was seen to separate into two sections (Fig. 7). The tilt angle, magnification and spot separation distance were then recorded and the film thickness determined geometrically.

At least ten measurements were made on different areas of each specimen, and the results are listed in Table 3, in which there is also a comparison of thickness measurements made by the Tolansky interferometry technique (which are accurate to within ±5 nm). The errors observed in contamination spot data are considerably greater for the thinner films both in terms of the variation of readings for the same film thickness and in the error over the Tolansky thickness. This is to be expected, since thin films are more likely to show slight thickness variations and also because the spot separation is considerably smaller for thin specimens and thus harder to measure accurately.

A source of the observed error must be attributed to the measurement of the spot separation distance after tilting, since such distances are relatively small. Also any error incurred in this measurement is greatly magnified when the actual thickness value is computed. The presence of an existing oxide film and/or contamination layer would also contribute to the observed difference as suggested by Love *et al.*[14] However, no diffraction evidence for significant surface oxidation was ever observed and it is not considered reasonable that a contamination layer of ~100 nm thick exists. Thus all that can be said with certainty is that the contamination spot separation measurement technique significantly overestimates the actual foil thickness and appropriate precautions should be taken when this is the only technique available to determine thickness.

TABLE 3 Comparison of thin foil thickness measurements

Film	Thickness (Tolansky), nm	Thickness (contamination spots), nm	Discrepancy, %
Al	117 ± 5	190 ± 31	62
Al	177 ± 5	270 ± 40	54
Cu	59 ± 5	170 ± 33	189

ACKNOWLEDGMENTS

The authors wish to thank Dr D. E. Newbury of the National Bureau of Standards, Gaithersburg, Md, for permission to publish the Monte Carlo calculations. They are also particularly grateful to Mr C. Kemmerer of Bell Laboratories, Allentown, Pa, for preparation of the nickel films. The suggestions of Dr R. Tixier, IRSID, France, on the specimen configuration for measurement of $\phi(\rho t)$ were very helpful. The support of the US Department of Energy is acknowledged under contract no. 76ER12408.

REFERENCES

1 G. CLIFF and G. W. LORIMER: *J. Microsc.*, 1975, **103**, 203

2 B. BENDER, D. B. WILLIAMS, and M. R. NOTIS: *J. Amer. Ceram. Soc.*, 1980, **63**, 149

3 S. MEHTA, J. I. GOLDSTEIN, D. B. WILLIAMS, and A. D. ROMIG: 'Microbeam analysis 1979', ed. D. E. Newbury, San Francisco Press, 1979, 119

4 J. I. GOLDSTEIN, J. L. COSTLEY, G. W. LORIMER, and S. J. B. REED: SEM/1977, ed. O. Johari, Chicago, IITRI, 1977, 315

5 D. KYSER: 'Introduction to analytical electron microscopy', ed. J. J. Hren, J. I. Goldstein, and D. C. Joy, New York, Plenum Press, 1979, 199

6 R. W. GLITZ, M. R. NOTIS, D. B. WILLIAMS, and J. I. GOLDSTEIN: to be published ('Microbeam analysis 1981', San Francisco Press, 1981)

7 R. L. MYKLEBUST, D. E. NEWBURY, and H. YAKOWITZ: 'Use of Monte Carlo calculations in electron probe microanalysis and scanning electron microscopy', National Bureau of Standards, 1976, 105

8 S. TOLANSKY: 'Multiple beam interference microscopy of metals', London, Academic Press, 1970

9 R. GLANG: 'Handbook of thin film technology', ed. L. I. Maissel and R. Glang, New York, McGraw-Hill, 1970, 1–107

10 K. L. CHOPRA: 'Thin film phenomena', New York, McGraw-Hill, 1969, 96

11 D. E. NEWBURY: private communication, 1981

12 D. E. NEWBURY and R. L. MYKLEBUST: *Ultramicroscopy*, 1979, **3**, 391

13 M. von HEIMENDAHL and V. WILLIG: *Norelco Reporter*, 1980, **27**, 22

14 G. LOVE, M. G. C. COX, and V. D. SCOTT: 'Developments in electron microscopy and analysis 1977', ed. D. L. Misell, Bristol and London, Inst. Physics, 1977, 347

DISCUSSION of paper by Stenton, Notis, Goldstein, and Williams

D. RAE:
You compared results from the Quartz Crystal Oscillation Method for determining the thickness of evaporated films with the multiple beam interference method. How was the quartz crystal oscillator calibrated? Is it not calibrated by multiple beam interferometry?

D. B. WILLIAMS:
I am not sure how the quartz oscillator was calibrated, but it may well have been by interferometry. However, I believe the absolute accuracy of the Tolansky interferometry as a measure of foil thickness. The reason for including oscillator and Tolansky measurements was that the Ni films were deposited in a different instrument from the Cu films. The Ni films were measured, as deposited, by quartz oscillator and Tolansky techniques and hence we had to be sure that there was agreement with the Cu films measured by Tolansky.

G. W. LORIMER:
I agree with Professor Williams that the contamination spot technique can give an overestimation of specimen thickness, particularly in the thinnest regions of a foil. This is due, at least in part, to the presence of a broad 'foothill' of carbon contamination which is beneath the contamination 'peak'. The distance between the contamination spots viewed in a tilted foil is the sum of the true specimen thickness plus the foothill.

D. B. WILLIAMS:
In light of the results reported by Rae *et al.* in these proceedings I agree that this is clearly the case and may well account for the enormous discrepancy between contamination spot measurements and interferometry. If there is some way to gain an accurate estimate of this systematic error, then the contamination spot measurement technique will be invaluable as a rapid method for determination of thickness.

Electron intensity distributions in STEM-EDS X-ray microanalysis of thin foils

P Doig, D Lonsdale, and P E J Flewitt

A simple analytical description of the electron intensity distribution within a thin foil during STEM-EDS X-ray microanalysis is derived. The results of this analysis are compared with published predictions of Monte Carlo calculations of electron probe spreading. The relative importance of electron probe diameter, electron accelerating voltage, foil material and thickness in controlling the spatial resolution of microanalysis are discussed.

The authors are in the Scientific Services Department of the CEGB South Eastern Region, Gravesend, Kent.

The scanning transmission electron microscope with small high-intensity electron probes at high accelerating voltages, when combined with efficient solid state energy-dispersive X-ray spectrometers (STEM-EDS), offers the possibility of undertaking chemical microanalysis on microstructural features within thin foils to a high spatial resolution which is limited by the scattering of the incident electron probe within the foil specimen.[1–3] This scattering gives rise to a distribution of electron intensity within the foil such that the X-ray intensity generated from any specific region within the foil volume will depend on its shape and position relative to the incident electron probe. This effect is particularly significant for microanalysis of features such as those shown schematically in Fig. 1 which do not constitute the total foil thickness or are not of sufficient extent to contain all scattered electrons. Hence it is necessary to consider the precise electron intensity distribution within the total irradiated volume of foil in order to establish quantitatively the proportion of the emitted X-ray intensity derived from the region of interest. In addition, for any given geometrical feature, it is desirable to predict the optimum experimental conditions to obtain the required chemical composition.

In order to undertake quantitative X-ray microanalysis of small features within thin foils it is necessary to obtain a description for the electron intensity distribution within the total foil volume. Electron scattering is determined by foil composition and density and incident electron probe accelerating voltage. Various theoretical approaches have been employed to establish electron scattering; the simplest of these, proposed by Goldstein *et al.*,[1] considers single Rutherford scattering which is arbitrarily selected to occur at the middle of the foil and thereby defines a measure of beam spreading at the bottom surface of a foil. This approach has the virtue of providing a simple analytical expression which gives an estimate of the spatial extent of scattering but no quantitative information on the volume distribution of electron intensity which is required for quantitative X-ray microanalysis. Kyser[2] has proposed a more sophisticated and potentially exact description of electron scattering based on the Monte Carlo method. Here, a large number of electron trajectories within the foil are calculated to give a statistical description of the volume electron intensity distribution. The method has been successfully developed with various degrees of sophistication and applied to quantitative X-ray microanalysis.[2,4] This approach, however, requires the use of relatively large computers and considerable skill in its application. With this in mind we proposed an analysis method based on a simple analytical function to describe the volume electron intensity distribution.[3] This provided a means of predicting the X-ray intensity generated from any geometrical feature within a foil as a function of material, electron accelerating voltage and, additionally, the incident electron probe size. This simple approach allows rapid evaluation of the X-ray signal either analytically or by use of simple numerical methods. The original function was developed from a simple interpretation of random elastic scattering but did not include a detailed consideration of the electron scattering behaviour. In view of the usefulness of this approach, we have extended the analysis to consider more accurately the scattering process and thereby improve the description of electron intensity distribution which is necessary to allow quantitative X-ray microanalysis.

This paper presents a theoretical description of the electron intensity distribution within a STEM electron probe in a thin foil. The predictions of this analysis are compared and shown to be in good agreement with published results of Monte Carlo analyses.

ELECTRON SCATTERING IN THIN FOILS

When a film is thick enough for an electron to suffer many elastic collisions in passing through it, the resulting angular distribution may be treated by error theory on the assumption that successive events are independent.[5] As a consequence, the central limit theorem, which forms the basis of Bothe's statistical theory,[6,7] results in a Gaussian angular

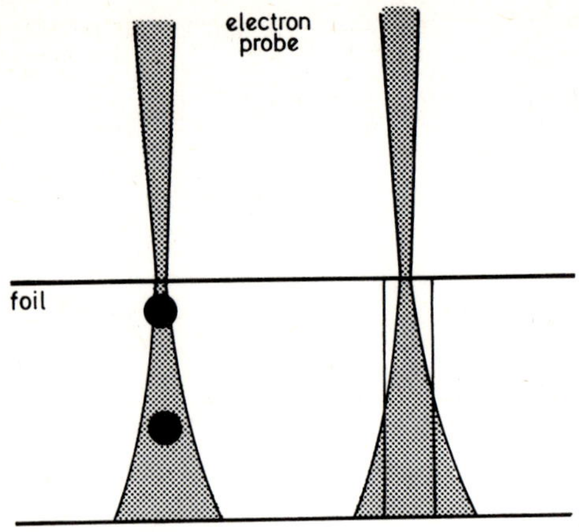

1 Schematic diagram showing transmitted intensity in STEM electron probe in relation to small microstructural features

distribution of electron intensity given by:

$$N(\theta) = (2\pi\lambda_B^2)^{-1} \exp(-\theta^2/2\lambda_B^2) \quad (1)$$

where $N(\theta)$ is the fraction of electrons scattered into a unit solid angle in the direction θ; $2\lambda_B^2$ is the mean square scattering angle of the Gaussian distribution; and λ_B is the most probable angle of total deflection. Experimental measurements of the transmission of fast beta particles through thin foils[8] verify this behaviour and provide a value for λ_B:

$$\lambda_B^2 = \left(\frac{4Z}{E_0}\right)^2 \cdot \frac{\rho t}{A} \cdot 10^3 \quad (2)$$

where E_0 is the electron accelerating voltage in volts; t is the foil thickness in nm; ρ is the specific gravity; Z is the atomic number; and A is the atomic weight of the foil material in grammes.

We can obtain the spatial distribution of electron intensity from its angular distribution by approximating the elastic scattering to occur at a mean depth $t/2$. For thin foils $\theta \simeq \tan\theta = 2t/r$ such that equation (1) becomes:

$$N(r) \propto \exp(-4r^2/2\lambda_B^2 t^2) \quad (3)$$

where r is the radial distance from the centre of the incident electron probe and t can now be considered as any depth in the foil. Substituting for λ_B^2 we obtain:

$$N(r) \propto \exp(-r^2/\beta t^3) \quad (4)$$

with

$$\beta = \left(\frac{4Z}{E_0}\right)^2 \cdot \frac{\rho}{A} \cdot \frac{10^3}{2} \quad (5)$$

Thus the radial distribution of electron intensity at a depth t in the foil is Gaussian and given by equation (4) for a point source of electrons at the foil surface. In general, the incident intensity in a STEM electron probe is Gaussian such that:

$$I(r) \propto \exp(-r^2/2\sigma^2) \quad (6)$$

where $I(r)$ is the intensity at a radial distance r and σ^2 is the variance of the Gaussian distribution and a measure of the probe size. Random elastic scattering of electrons passing through the foil increases this variance to give an electron intensity distribution at a depth t:

$$I(r, t) \propto \exp\,[-r^2/(2\sigma^2 + \beta t^3)] \quad (7)$$

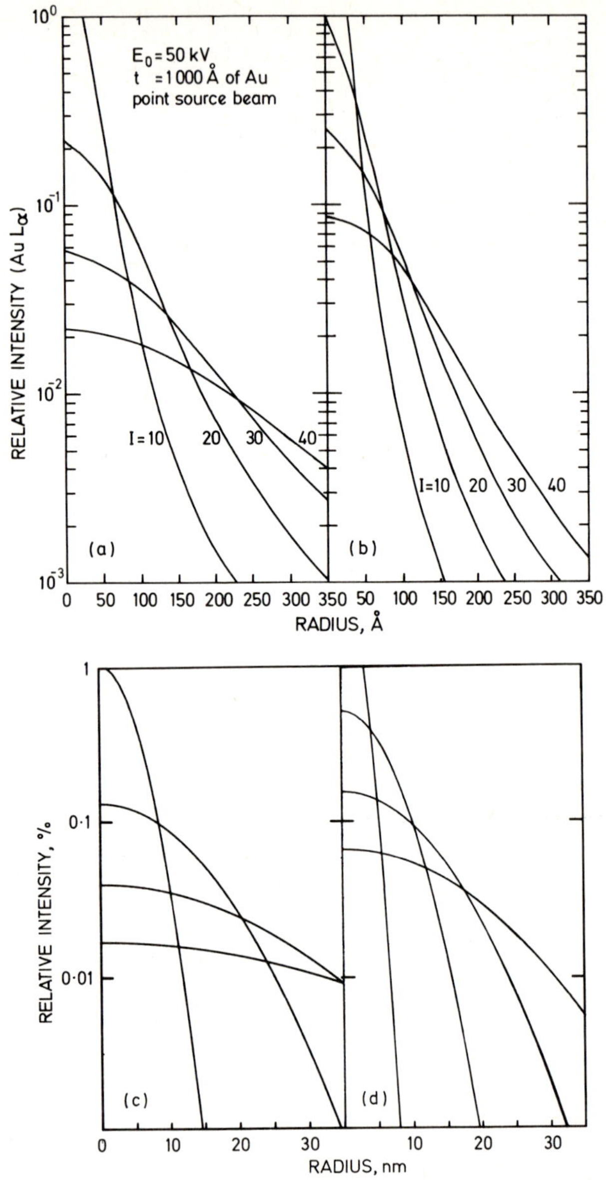

(a) and (b): foil depth in units of 2.5 nm (e.g. 10 ≡ 25 nm)

2 Radial electron intensity distributions in gold foils calculated using Monte Carlo methods for (a) 50 and (b) 100 keV electrons[2] and using the present analysis for (c) 50 and (d) 100 keV electrons

Normalization for a total incident electron intensity, I_e, gives:

$$I(r, t) = I_e\{\pi(2\sigma^2 + \beta t^3)\}^{-1} \cdot \exp\{-r^2/(2\sigma^2 + \beta t^3)\} \quad (8)$$

Thus we have an analytical expression for the electron intensity distribution within a foil as a function of incident electron probe size (σ), accelerating voltage and foil material (β). The incident probe size may be defined in terms of the full width at half maximum intensity (FWHM) such that from equation (6):

$$\text{probe size (FWHM)} = \sigma(8 \ln 2)^{1/2} \simeq 2.35\,\sigma \quad (9)$$

RESULTS OF ELECTRON SCATTERING ANALYSIS

Kyser[2] has recently published results of Monte Carlo analyses of electron probe scattering in thin foils and Figs. 2*a* and *b* reproduce his values for gold. Here, the relative

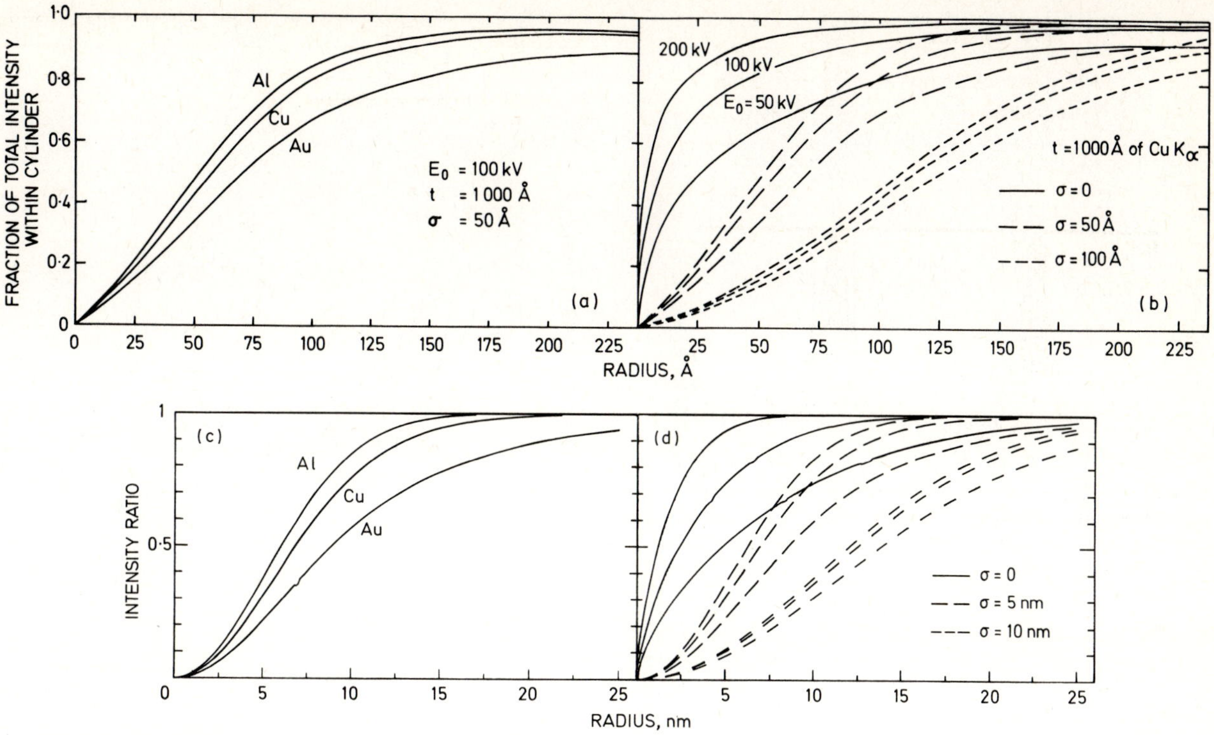

3 Total fraction of electron intensity contained within right circular cylinder as function of its radius calculated using:

(a) Monte Carlo method[2] for $\sigma = 5$ nm, $E_0 = 100$ kV
(b) Monte Carlo method[2] for range of σ and E_0
(c) present analysis as for (a)
(d) present analysis as for (b)

for foil thickness 100 nm

radial electron intensity distribution (%) is shown as a function of foil thickness in depth units of 2.5 nm (i.e. $10 \equiv 25$ nm) for a point electron source (i.e. $\sigma = 0$ in our notation) at electron accelerating voltages of (a) 50 kV and (b) 100 kV. Figures 2*c* and *d* show the corresponding radial electron intensity distributions calculated using the present analysis (equation (8)) for the same accelerating voltages and foil thicknesses. In general, we have excellent quantitative agreement except for higher scattering angles (i.e. higher *r*) where the present analysis predicts a lower electron intensity. The effect is small, however, and is exaggerated by semi-logarithmic plotting. Some error will result from the approximation $\theta \simeq \tan\theta$ (equations (1) and (2)) and the underestimate of electron path length for higher values of θ. The tails of the curves predicted by Kyser[2] are subject to an increasing statistical error resulting from the finite number of electron trajectories considered. This error is approximately $(N/10^3)^{1/2}$ for a reported relative intensity of N. Nevertheless, in the region of significant electron intensity there is good agreement between the results of the two analyses.

In the same paper, Kyser considers the effects of incident electron probe dimensions (σ), electron accelerating voltage (E_0), and foil material on electron scattering. The proportion of total electron intensity contained within a right circular cylinder with axis oriented parallel to the incident electron probe is considered as a function of cylinder radius. Figure 3*a* shows results for 100 nm thick foils of Al, Cu and Au at an electron accelerating voltage of 100 kV and an incident electron probe size defined by $\sigma = 5$ nm. Similarly Figure 3*b* reproduces results for a 100 nm thick Cu foil showing the effects of electron accelerating voltage 50, 100 and 200 kV for incident electron probe sizes defined by $\sigma = 0$, 5, and 10 nm. In our analysis the proportion of the incident electron probe contained within a radius r_1 at any foil thickness t_1 is given by the integral of equation (8):

$$I'(r_1 t_1) = \frac{1}{t_1}\int_0^{t_1}\int_0^{r_1} 2\pi r I(r,t)\,\mathrm{d}r\,\mathrm{d}t$$

$$= \frac{1}{t_1}\int_0^{t_1} \{1 - \exp(-r_1^2/(2\sigma^2 + \beta t^3)\}\,\mathrm{d}t \qquad (10)$$

This integral has been evaluated numerically (Figs. 3*c* and *d*) for conditions corresponding to those considered by Kyser in Figs. 3*a* and *b* respectively. These results are in excellent qualitative and quantitative agreement.

The spatial resolution for microanalysis may be defined as the diameter of the right circular cylinder which contains an arbitrary proportion of the scattered electron probe. We have previously adopted a value of 50%[3] where Kyser[2] has used a 90% value. The plots in Fig. 3 show that in this region $I'(r_1, t_1)$ is nearly independent of r_1 and therefore its precise value is subject to increasing statistical error when evaluated from the finite number of electron trajectories used in Monte Carlo evaluations. Adopting the criterion of 90%, we compare in Table 1 values for spatial resolution calculated from the present analysis with those due to Kyser. There is good overall agreement for all but the thicker foils (~400 nm) of elements with high atomic number. In practice, foil thicknesses less than ≃200 nm are used to optimize imaging and reduce X-ray fluorescence and absorption effects.

From the above criterion for defining spatial resolution it is useful to examine the relative influence of the experimental variables foil thickness, electron accelerating voltage and incident electron probe size. Table 2 shows our calculated results obtained for iron foils and 90% electron inten-

TABLE 1 Calculated spatial resolution for STEM X-ray microanalysis at 100 kV for incident electron probe sizes defined by σ = 0 nm (=0 nm FWHM) and 5 nm (=11.8 nm FWHM)

Element	Foil thickness, nm 40	100	200	400	σ, nm
C	– (–)	4 (5)	11 (13)	30 (25)	0
	22 (–)	22 (20)	24 (23)	35 (30)	5
Al	1 (2)	6 (9)	18 (20)	50 (46)	0
	22 (19.5)	22 (22)	27 (25)	53 (46)	5
Cu	4 (4)	17 (16)	47 (55)	132 (140)	0
	22 (20.5)	26 (25)	50 (55)	134 (140)	5
Au	10 (8)	38 (38)	107 (130)	302 (450)	0
	23 (22)	42 (40)	108 (130)	302 (450)	5

Values calculated using present analysis; those in brackets predicted by Monte Carlo method after Kyser[2]

sity. Clearly, decreasing foil thickness and increasing electron accelerating voltage from 100 to 200 kV significantly improves the spatial resolution for the range of beam sizes 2.5 to 10 nm which represents values generally available for STEM X-ray microanalysis. Decreasing incident probe size within this range, however, has a relatively less significant effect on the spatial resolution, particularly for thicker foils. These observations must be considered in conjunction with the problem of optimizing X-ray counting statistics. Decreasing incident electron probe size and foil thickness and increasing the accelerating voltage, although improving the defined spatial resolution, all serve to decrease the X-ray counts for a given electron flux. Thus any benefit of improved spatial resolution will be offset by decreased statistical accuracy of the recorded X-ray data.

Hutchings *et al.*[9] have reported results on the magnitude of the electron beam spread within silicon foils. Specimens of silicon were prepared with narrow strips of gold on one surface and analysed to measure the effective electron probe dimensions at the top and bottom surfaces of the foil. This was evaluated from the width of the measured gold X-ray intensity profile obtained by traversing the electron probe across the deposited gold strips. This experimental system may be considered analytically using the present scattering model.

The radial electron intensity distribution within the electron probe is shown schematically in Fig. 4*a*. The intensity I^* detected from a given microstructural feature using STEM-EDS X-ray microanalysis is given by:

$$I^* = \int_V K C_V I_V \, dv \tag{11}$$

where I_V is the electron intensity volume distribution; C_V is the particular elemental volume distribution; and K is a constant defining the efficiency of X-ray generation, emission and detection. Clearly, the present work provides a simple analytical form for I_V which when combined with any known, or assumed, C_V will provide a value for I^*. The relative X-ray intensity $I^*(x)$ generated by traversing this radial intensity distribution in the X direction across the planar gold layer is shown schematically in Fig. 4*b* and is given by:

$$I^*(x) = \int_{-\infty}^{x} \int_{-\infty}^{+\infty} I(r, t_1) \, dy \, dx \tag{12}$$

TABLE 2 Calculated spatial resolution for iron foils

Incident electron probe diameter, nm (FWHM)	Foil thickness, nm 50	100	150	200	Electron accelerating voltage, kV
2.5	12	18	29	43	100
	11	13	17	23	200
5	22	25	33	46	100
	22	23	25	29	200
10	43	45	50	58	100
	43	43	45	47	200

for a given foil thickness $t = t_1$ where $r^2 = x^2 + y^2$ and $I(r, t_1)$ is defined in equation (8). Substituting $2\gamma^2 = 2\sigma^2 + \beta t_1^3$ and $v = x/(\gamma\sqrt{2})$ reduces equation (12) to the error function:

$$I^*(x) = \frac{1}{\sqrt{\pi}} \int_{-\infty}^{x/(\gamma\sqrt{2})} \exp(-v^2) \, dv \tag{13}$$

Thus, following Hutchings *et al.*,[9] and defining the beam spread as the difference $x_1 - x_2$ for which $I^*(x_2) = 0.05$ and $I^*(x_1) = 0.95$ (Fig. 4*b*); from symmetry $x_1 = -x_2$ such that the beam spread $= 2x_2$:

$$I^*(x_2) = \tfrac{1}{2} \operatorname{erfc}(x_2/\gamma\sqrt{2}) \tag{14}$$

$$x_2 = \gamma\sqrt{2} \operatorname{erf}^{-1}(0.9) \simeq 1.64\gamma \tag{15}$$

At the top surface of the foil $2\gamma^2 = 2\sigma^2$ whereas at the bottom of the foil $2\gamma^2 = 2\sigma^2 + \beta t_1^3$. Thus, for a finite incident probe size, the beam spread as defined by Hutchings[9] is given by:

$$\text{beam spread} = 3.28\{[(2\sigma^2 + \beta t_1^3)/2]^{1/2} - \sigma\} \tag{16}$$

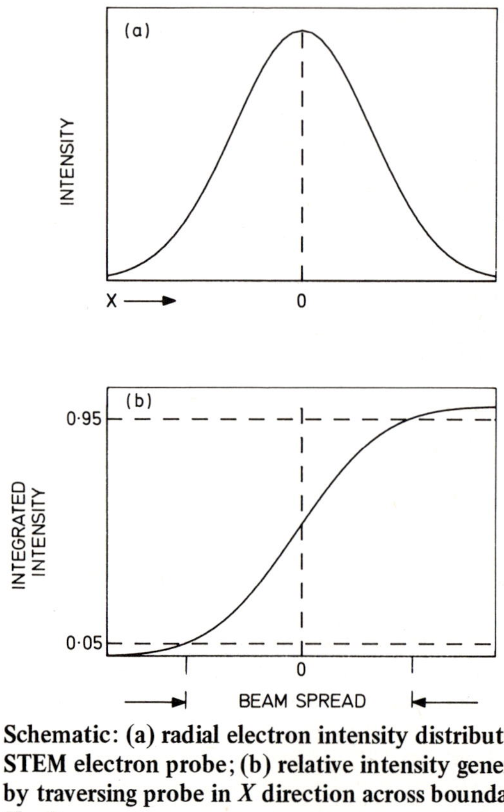

4 Schematic: (a) radial electron intensity distribution in STEM electron probe; (b) relative intensity generated by traversing probe in X direction across boundary at $X = 0$

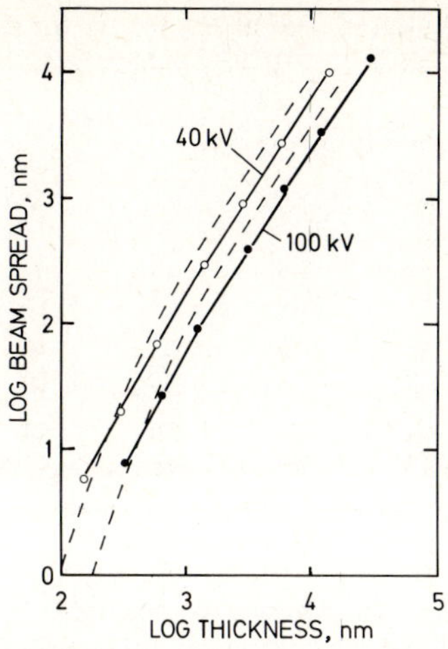

5 **Results of STEM electron probe scattering calculated using present analysis superimposed on results reported by Hutchings *et al.*[9]**

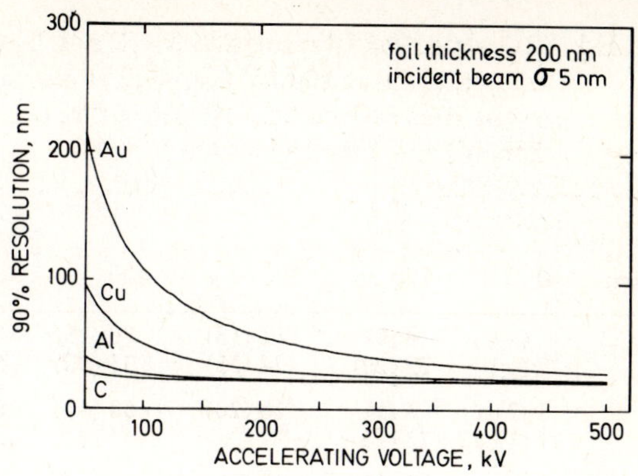

6 **Spatial resolution of STEM X-ray microanalysis in thin foils of range of elements as function of electron accelerating voltage for foil thickness 200 nm and incident electron probe diameter 5 nm**

We have evaluated this beam spread for silicon foils with an incident electron probe size of 2 nm (FWHM) as a function of foil thickness t_1 and electron accelerating voltages of 40 and 100 kV. These results, shown as broken lines in Fig. 5, are superimposed on Hutchings's calculated, values (corrected as recommended in his paper). There is excellent quantitative agreement between the two sets of data.

DISCUSSION

The present work derives a simple analytical expression to describe the electron intensity distribution within a thin foil during STEM X-ray microanalysis. The expression is developed from error theory on the assumption that successive scattering events occur independently and without energy loss. This treatment does not involve consideration of the detail of the particular individual scattering events but relies on the fact that for many events the angular and, for small angles, the radial distributions will always degrade to a Gaussian. In the present analysis we consider a finite size incident electron probe which has an initial Gaussian radial distribution; thus even the small number of scattering events which occur in very thin foils will preserve the Gaussian profile. Consequently, for real incident electron probe dimensions the present analysis should apply even under experimental conditions where single or plural scattering occurs within the foil.

As stated in the introduction, quantitative STEM-EDS X-ray microanalysis of geometrical features with dimensions which are less than the total volume of the sampling electron probe requires knowledge of the electron intensity distribution within the foil (Fig. 1). The simplest description of electron scattering, based on Rutherford single scattering, only provides a measure of the lateral spreading of electron intensity at the bottom surface of a foil. It provides no quantitative information to allow microanalysis of small geometric features within the foil volume. The present analytical function provides a volume description of electron intensity such that quantitative relationships between measured elemental compositions and true elemental compositions may be established for any assumed C_V.

The total scattering of the incident electron probe, characterized by the parameter β (equation (5)) includes a dependence on the electron accelerating voltage E_0 which allows consideration of its influence on the spatial resolution of STEM X-ray microanalysis as defined above under 'Electron scattering in thin foils'. The calculated spatial resolution for a range of elements as a function of electron accelerating voltage (Fig. 6) shows a decreasing improvement in spatial resolution with increasing accelerating voltage such that there is relatively little benefit in undertaking microanalysis at accelerating voltages greater than 300 kV. There is, however, a considerable improvement by increasing from 100 to 200 kV. Within this range it is noteworthy that there is a significant benefit in operating at 120 kV compared to 100 kV, particularly for high atomic number elements. This latter feature is relevant to those 100 kV STEM instruments which may be adapted for operation at these higher accelerating voltages.

CONCLUSIONS

The main conclusions which may be derived from the present analysis are:

1 The volume electron intensity distribution within a thin foil in STEM-EDS X-ray microanalysis may be adequately described by a simple analytical function.

2 The radial electron intensity distribution at a fixed depth within a thin foil may be approximated to a Gaussian.

3 The incident electron probe diameter can be of secondary importance in determining spatial resolution of X-ray microanalysis in STEM.

4 Increasing the electron accelerating voltage improves the spatial resolution of STEM X-ray microanalysis.

ACKNOWLEDGMENT

This paper is published by permission of the Director General, Central Electricity Generating Board, South Eastern Region.

REFERENCES

1 J. I. GOLDSTEIN, J. L. COSTLEY, G. W. LORIMER, and S. J. B. REED: SEM, 1977, **I**, IITRI, Chicago, 315

2 D. F. KYSER: 'Introduction to analytical electron microscopy', 1979, eds. J. J. Hren, J. I. Goldstein, and D. C. Joy, Plenum Publ. Corp., New York, 199–221

3 P. DOIG, D. LONSDALE, and P. E. J. FLEWITT: *Philos. Mag. A*, 1980, **41**, 761

4 'Use of Monte Carlo calculations in electron probe microanalysis and scanning electron microscopy', 1976, eds. K. F. J. Heinrich, D. E. Newbury, and H. Yakowitz, National Bureau of Standards Spec. Publ. 460

5 J. S. BENDAT and A. G. PIERSOL: 'Random data: analysis and measurement procedures', 1971, Wiley-Interscience, New York

6 W. BOTHE: *Z. Phys.*, 1921, **3**, 63

7 W. BOTHE: Handbook Phys. 1933, X 22/2, 1933, Springer, Berlin

8 J. A. CROWTHER: *Proc. R. Soc.*, 1910, **A84**, 225

9 R. HUTCHINGS, M. H. LORETTO, I. P. JONES, and R. E. SMALLMAN: *Ultramicroscopy*, 1979, **3**, 401

DISCUSSION of paper by Doig, Lonsdale, and Flewitt ('Electron intensity in STEM-EDS')

D. B. WILLIAMS:
Can your model handle the situation in which the interaction volume within a thin foil contains regions of widely differing composition or crystal structure, such as an interphase interface or a layer of segregant at a grain boundary?

P. DOIG:
The analysis presented here is based on sampling from a random scattering process where the population distribution is characterized by the parameter β. Where β is not constant, the sampling must be distorted and the resulting sample will be in error. The magnitude of this error will be a function of the difference in β and the spatial extent over which it occurs. For your example of a narrow layer (monolayer) at a grain boundary the present analysis should provide a reasonable solution.

G. CLIFF:
After how many elastic scattering events does the distribution become Gaussian and so validate the proposed modelling?

P. DOIG:
The angular electron intensity distribution becomes Gaussian, regardless of the angular intensity distribution (assuming this is continuous), after a number of scattering events. This number depends on the particular individual scattering behaviour, but for reasonable monotonic single scattering angular intensity distributions, sampling theory suggests that a number as low as 4 will give reasonable Gaussian behaviour with a number of 10 giving excellent Gaussian fit (see our Ref. 5). The purpose of the present paper, however, is to consider not the angular intensity distribution but rather the radial or spatial intensity distribution. Here, with a real incident electron probe size, the incident spatial electron intensity distribution is Gaussian and must remain sc regardless of the number of scattering events.

J. I. GOLDSTEIN:
Please explain the specific differences between your new model and the one you have proposed in your recent *Philos. Mag.* (1980) paper.

P. DOIG:
The philosophy of obtaining a simple analytical description for the volume electron intensity distribution within a thin foil in order to allow a quantitative analysis of the X-ray intensity generated from microstructural regions which are smaller than the total electron probe size was introduced in our earlier paper (*Philos. Mag. A*, 1980, **41**, 761–765). In common with this earlier work, we have considered the radial electron intensity distribution to be Gaussian which at the surface of the foil is characteristic of the incident electron probe and is broadened by many elastic scattering events as it penetrates the foil. The present analysis is based on the consequences of the central limit theorem and considers the radial electron intensity distribution to be obtainable from the angular electron intensity distribution which is characterized by the elastic Rutherford scattering. This has resulted in a different dependence of the radial intensity distribution on foil thickness. This is now given by a term in t^3 in the Gaussian variance in contrast to the original linear term in t. In addition we have now characterized the scattering behaviour by a parameter β which is derived from the Bothe scattering theory (Refs. 6 and 7) using the experimental data of Crowther (Ref. 8). The original analysis utilized a parameter D which was obtained from the only Monte Carlo results which were available to us at the time of that work. Since then it has been suggested that these Monte Carlo results have been improved.

Influence of plural electron scattering on X-ray spatial resolution in TEM thin foil microanalysis

G Cliff and G W Lorimer

The effects of plural scattering in thin specimens suitable for STEM X-ray analysis are outlined. A model based on plural scattering is derived to predict beam broadening and it is shown to agree exactly with the single scattering model of Goldstein et al. *(1977). A Monte Carlo programme has been developed using the basic equations of the plural scattering theory and compared with experimental measurements of beam broadening in an orthopyroxene,* $MgFeSi_2O_6$.

The authors are in the Department of Metallurgy, University of Manchester/UMIST, UK.

The extent of beam broadening during the microanalysis of thin specimens is an important parameter when the incident probe diameter is less than 100 Å. All of the electrons incident on a TEM thin foil potentially contribute to X-ray production and those scattered to the greatest angles define the effective width of the X-ray source size and hence the spatial resolution for analysis. Goldstein *et al.*[1] proposed a model for beam broadening in which it was assumed that each incident electron suffered a single, elastic, large-angle Rutherford scattering event at the centre of the foil (Fig. 1) and derived the expression

$$b = 6.25 \times 10^5 \frac{Z}{E_0} \left(\frac{\rho}{A}\right)^{1/2} t^{3/2} \quad \text{cm} \tag{1}$$

where E_0 is the incident beam energy, A the atomic weight, Z the atomic number, ρ the density and t the specimen thickness. The beam broadening parameter b was defined as containing 90% of the transmitted electrons.

Cosslett and Thomas[2] have shown that the number of scattering acts, p_e, suffered by an electron passing through a thin foil suitable for electron microscopy has a practical limit of $p_e \simeq 5$. (As pointed out by Halliday[4] and numerically confirmed by Smith and Burge[5] the plural scattering component, $p_e > 1$, cannot be neglected even for a film where $p_e = 1$.)

Although the model of Goldstein *et al.*[1] is based on a simple single scattering treatment, the values of beam broadening which it predicts agree remarkably well with the Monte-Carlo calculations of Kyser and Geiss[8] and with experimental measurements by Hutchins *et al.*[6] and Champness *et al.*[7] The object of this paper is to examine the effects of plural scattering, to derive an analytical expression for beam broadening based on plural scattering concepts, and to compare the predictions of the plural scattering model with the model of Goldstein *et al.* and other theoretical and experimental evaluations of beam broadening.

SINGLE SCATTERING

Electron scattering may occur either elastically or inelastically. The probability, and hence the cross-section, for inelastic scattering into large angles is smaller than that for elastic scattering and a first order treatment for $Z > 13$ need only take elastic collisions into account. Most scattering theories have been based on a model of the atom in which the atomic electrons exponentially screen the nuclear charge (Wentzel[9]). This gives the elastic differential scattering cross-section $d\sigma$ into a solid angle $d\Omega$ as

$$\frac{d\sigma}{d\Omega} = \left(\frac{Ze^2}{4E_0}\right)^2 (\sin^2 \tfrac{1}{2}\theta + \tfrac{1}{4}\theta_0^2)^{-2} \tag{2}$$

where e is the electron charge, θ is the angle of scattering and θ_0 is the screening parameter.

Lenz[10] made use of this cross-section when integrating transport equations, originally derived by Bothe,[11,12] to determine the angular distribution of electron scattering. The work of Hall and Inoue[3] at 50 and 100 kV on scattering from latex spheres provided early evidence supporting Lenz, and the measurements of Halliday[4] of the electron intensity distribution in electron diffraction patterns from gold at 117 keV supporting Lenz's theory up to $p_e = 6$. Cosslett and Thomas[13] applied Lenz's theory in their investigations of plural scattering in thin films at 20 kV. Their experimental results were in agreement with the model of Lenz and they proposed that they were also consistent with observations made at the higher voltages needed for transmission electron microscopy. Lenz's theory using the differential scattering cross-section of equation (2) has been used by Martinez *et al.** to interpret electron scattering data at energies from 1.2 to 3 MeV. They conclude that Lenz's theory gives a good representation of their measurements. Equation (2) appears to be valid over an energy range of 20 keV to 3 MeV.

---

* J. P. MARTINEZ, J. L. BALLADORE, and J. TRINQUIER: *Ultramicroscopy*, 1979, 4, 211–219

If the screening parameter θ_0 from the Wentzel model of the atom is combined with the Rutherford single scattering formula used by Goldstein *et al.*, it is possible to write the differential probability function $P_\theta \, d\theta$ for single scattering through an angle θ by multiplying (2) by the number of possible scattering sites. Thus, using the expression given by Willmott:[16]

$$P_\theta \, d\theta = \frac{N\rho t}{A}\left(\frac{Ze^2}{4E_0}\right)^2 (\sin^2 \tfrac{1}{2}\theta + \tfrac{1}{4}\theta_0^2)^{-2}\, 2\pi \sin\theta \, d\theta \tag{3}$$

where P_θ is the probability of a scattering event occurring through an angle θ and N is Avogadro's number.

To obtain the probability P_θ for single scattering through an angle θ (3) is integrated over θ to π and normalized over $\theta = 0$ to π:

$$P_\theta \quad \frac{\int_\theta^\pi P_\theta \, d\theta}{\int_0^\pi P_\theta \, d\theta} = \frac{\int_\theta^\pi (\sin^2 \tfrac{1}{2}\theta + \tfrac{1}{4}\theta_0^2)^{-2} \sin\theta \, d\theta}{\int_0^\pi (\sin^2 \tfrac{1}{2}\theta + \tfrac{1}{4}\theta_0^2)^{-2} \sin\theta \, d\theta}$$

$$P_\theta \quad \frac{(\sin^2 \tfrac{1}{2}\theta + \tfrac{1}{4}\theta_0^2)^{-1} - (1 + \tfrac{1}{4}\theta_0^2)^{-1}}{4\theta_0^{-2} - (1 + \tfrac{1}{4}\theta_0^2)^{-1}} \tag{4}$$

This expression equals unity for $\theta = 0$. It is critically dependent on the screening parameter θ_0 and is independent of t. (An approximate expression for θ_0 is given by $\theta_0 = 3.69 Z^{1/3} E^{-1/2}$ rad, if E is expressed in eV.)

To provide a more convenient equation it can be noted that at 100 kV $\theta_0 = 0.05$ for gold ($Z = 79$) and that for elements $Z < 79$ θ_0 is even smaller; thus (4) can be approximated by

$$\theta \simeq \frac{\theta_0}{(P_\theta)^{1/2}} \tag{5}$$

Equation (5) can be shown to be sufficiently accurate in both P_θ and θ for a first-order treatment.

For electrons incident on a specimen whose thickness compares with the mean free path for single scattering, (5) can be used to estimate the mean angle through which a fraction P_θ of the incident electrons are scattered.

PLURAL SCATTERING

For foils thicker than a mean free path it is necessary to know the scattering distribution for $p_e > 1$. When $p_e = 25$ the distribution becomes Gaussian, as observed by Cosslett and Thomas.[13] When the foil thickness is sufficient for Gaussian statistics to apply the transmitted cone of electrons containing a fraction $1 - P_\theta$ of the incident beam can be described by a cone of semi-angle

$$\theta_{P_\theta}^{p_e=25} = \theta_{P_\theta}^{p_e=1} \cdot (p_e)^{1/2} \tag{6}$$

where

$$\theta_{P_\theta}^{p_e=1} = \frac{\theta_0}{(P_\theta)^{1/2}}$$

When p_e is between 1 and 25 equation (6) should not apply as the distribution is not Gaussian. The problem of determining what the true distribution is has not yet been fully resolved. Evans[17] in his extensive discussions of nuclear scattering processes makes reference to plural scattering twice, firstly to define it and secondly to dismiss it as 'very complicated'. Further, Cosslett and Thomas point out that even Lenz's evaluation of Bothe's theory underestimates plural scattering into small angles <0.1 rad and conversely overestimates large-angle scattering in thin foils; thus even detailed theory is inadequate.

A simple procedure for expressing the scattering distribution is obtained if (6) is rewritten as follows

$$\theta_{P_\theta}^{p_e} = \frac{\theta_0}{(P_\theta)^{1/2}} \cdot (p_e)^{1/n} \tag{7}$$

where $n = 2$ when $p_e = 25$ and $n = \infty$ when $p_e = 1$. The parameter n can then be used to characterize the plural scattering distribution for p_e between 1 and 25.

To proceed it is now necessary to find an expression for p_e. When $p_e = 1$ the foil thickness t is equal to one mean free path and, in terms of a collision cross-section σ, p_e can be expressed as

$$p_e = \sigma \frac{N\rho}{A} t \tag{8}$$

But σ can be simply evaluated by integrating (2) over all angles, $\theta = 0 \to \pi$. For small θ_0

$$p_e = \pi \left(\frac{Ze^2}{E_0}\right)^2 \frac{N\rho t}{A} \frac{1}{\theta_0^2} \tag{9}$$

Combining (7) and (9):

$$\theta_{P_\theta}^{p_e} = \frac{\theta_0^{1-2/n}}{(P_\theta)^{1/2}} \left(\pi \frac{Z^2 e^4}{E_0^2} \frac{N\rho t}{A}\right)^{1/n} \tag{10}$$

To express a beam broadening parameter using (10) the single scattering beam broadening parameter b of Goldstein *et al.*[1] must be redefined to allow valid comparison with a simple plural scattering model.

In the single scattering model of Fig. 1 scattering is assumed to occur at the foil centre. It must be recognized that the event could occur anywhere through the thickness of the foil, along the projected line of the incident beam, and thus the parameter b is to be compared with the cone diameter at half the foil thickness (Fig. 2). This model defining b is discussed by Goodhew and Chescoe.[18]

If the model of Fig. 2 is used, for small angles, θ can be equated with $\theta_{P_\theta}^{p_e}$ of (10) and

$$b = \theta t$$

$$= \frac{\theta_0^{1-2/n}}{(P_\theta)^{1/2}} \left(\pi \frac{Z^2 e^4}{E_0^2} \frac{N\rho}{A}\right)^{1/n} t^{1+1/n} \tag{11}$$

When the specimen is sufficiently thick for the scattering distribution to become Gaussian the value of n is 2 ((6)

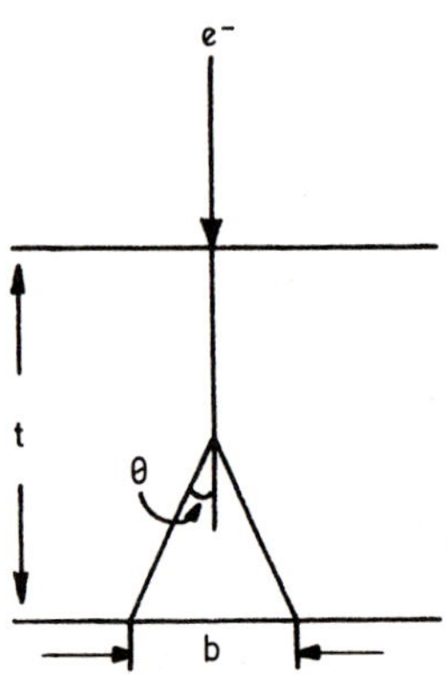

1 Single scattering model

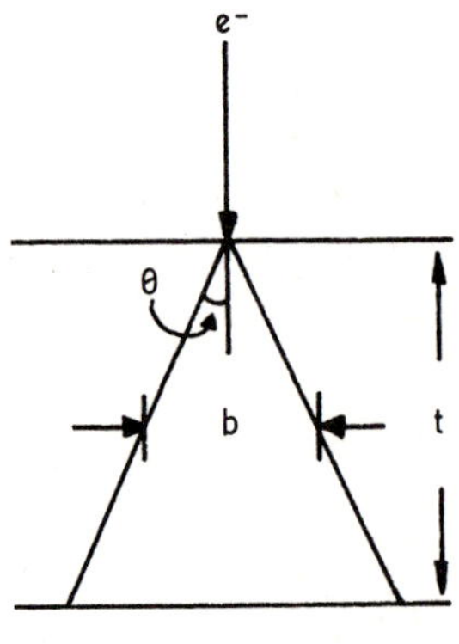

2 Plural scattering model

TABLE 1 Broadening parameter b(Å) for Cu at 100 kV

p_e	t, Å	n = 2	4	8	16	32	99
1	267	30	30	30	30	30	30
2	534	86	72	66	63	62	61
3	801	157	120	104	97	94	92
4	1068	242	171	144	132	126	123
10	2671	958	539	404	350	325	310
15	4006	1759	894	637	538	494	467
25	6077	3785					

p_e = no. of scattering acts, t = specimen thickness

TABLE 2 Variation in b as function of p_e for three elements at 100 kV with n = 16

	Al (Z = 13)		Cu (Z = 29)		Au (Z = 79)	
p_e	t, Å	b, Å	t, Å	b, Å	t, Å	b, Å
1	1057	92	267	30	101	16
2	2114	192	534	63	201	33
3	3171	295	801	97	302	51
4	4229	400	1068	132	402	69
10	10571	1059	2671	350	1005	184
15	15857	1630	4006	538	1508	283

p_e = no. of scattering acts, t = specimen thickness, b = broadening parameter

and (7)). Substitution of n = 2 in (11) will then yield

$$b = \frac{1}{(P_\theta)^{1/2}} \left(\frac{\pi Z^2 e^4 N \rho}{E_0^2 A} \right)^{1/2} t^{3/2} \tag{12}$$

For $e = 4.8 \times 10^{-10}$ e.s.u., $1\ \text{eV} = 1.6 \times 10^{-12}$ erg, $\pi = 3.14$, $N = 6 \times 10^{23}$ and $P_\theta = 0.1$:

$$b = 6.25 \times 10^5 \frac{Z}{E_0} \left(\frac{\rho}{A} \right)^{1/2} t^{3/2} \quad \text{cm} \tag{13}$$

which is identical to the expression of Goldstein *et al.* (1)!

This agreement is surprising, but it can be recognized that the success of (1) is due to its correct dependence on the critical parameters ρ, A, Z, and E_0. The constant is identical only when approximate values are chosen for π and N.

The influence of n on b can be seen in Table 1 where the change in b is monitored for Cu at 100 kV for various values of p_e and n. Equations (1) and (13) can now be understood as setting an upper limit to beam broadening, and it would be expected that the parameter b will over-estimate the extent of beam broadening and thus the X-ray resolution of TEM thin foils.

In Table 2, for an arbitrary value of n = 16, the variation in b as a function of p_e is shown for the elements Al, Cu, and Au at 100 kV.

THE SCREENING PARAMETER

For the purposes of this paper the approximate expression for the screening parameter θ_0 is sufficiently accurate when dealing with large-angle scattering $\theta > \theta_0$, but Cosslett and Thomas[2] emphasized the need to know θ_0 to an accuracy of a few percent for small-angle events and therefore retained θ_0 for the whole angular range they investigated, $10^{-2} < \theta < 1$. To obtain agreement between their results and Lenz's theory it was necessary to modify the approximate expression for θ_0. It can be shown that their modification reduces to

$$\theta_0 = \frac{\hbar Z^{1/3}}{p a_H (k/6)^{1/2}} \tag{14}$$

where $\hbar = h/2\pi$, h is Planck's constant, p is the electron momentum, a_H is the Bohr atomic radius with k taking the values 7.3, 6.8 and 5.3 for Al, Cu and Au respectively.

An alternative expression for θ_0 has been proposed by Nigam *et al.*[14] and used by them for 15.6 MeV electron scattering in Be and Au. In a form due to Shimizu and Murata,[15] and used in their Monte-Carlo calculation at 30 kV, it was expressed as

$$\theta_0 = \frac{\hbar Z^{1/3}}{p a_H (0.885/1.12)} \tag{15}$$

The value of k deduced from (15) must be 3.75, a very different value to those proposed by Cosslett and Thomas.[2] It would appear that some doubts exist as to the appropriate values of k but because 100 kV is only a factor of 5 removed from 20 kV, whilst 15.6 MeV is 780 times as great, the values of Cosslett and Thomas are to be preferred and should be included in any Monte-Carlo modelling of electron scattering.

MONTE-CARLO MODELLING

A simplified analytical approach to scattering can only hope to estimate a beam broadening parameter but in developing (13) it has been necessary to use expressions describing the mean free path for single scattering, t in (8) when p_e = 1, and an expression for the probability of the subsequent scattering angle being θ, (4). These expressions together with that for the screening parameter θ_0, using the modifications proposed by Cosslett and Thomas[2] instead of Nigam *et al.*, form the basis for Monte-Carlo modelling of the scattering process. A program has been developed from

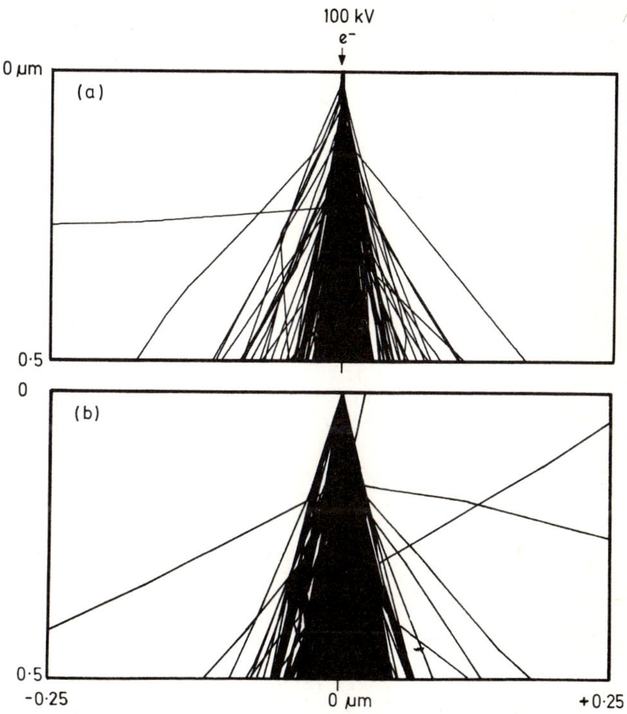

3 Monte-Carlo electron trajectories in ½ µm of Al at 100 kV: (a) this work; (b) calculated by Kyser[20]

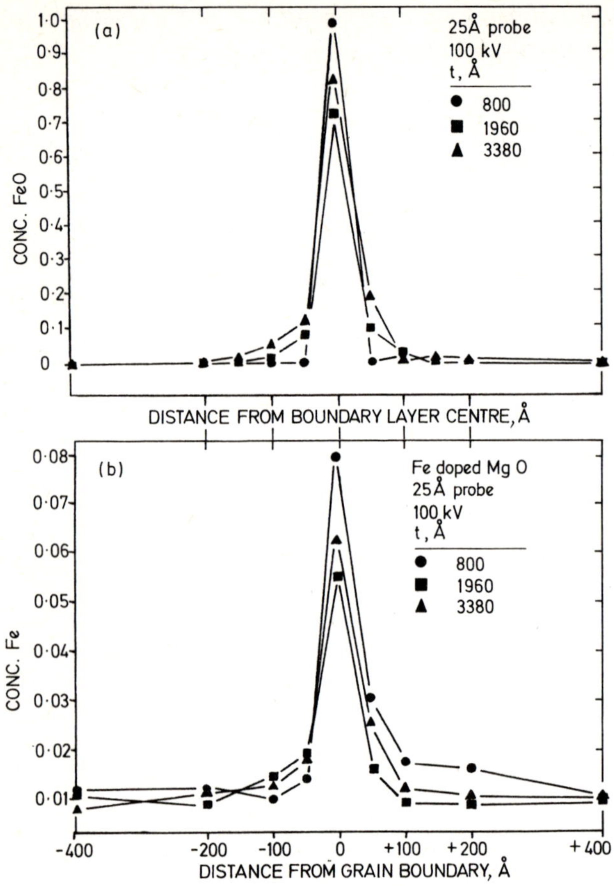

4 (a) Monte-Carlo calculation of FeO concentration across 60 Å boundary layer of FeO in matrix of MgO at 100 kV (this work); (b) experimental results of Vander Sande and Hall[21] of Fe concentration across grain boundary in Fe-doped MgO at 100 kV

these equations which also assumes an exponential probability distribution for the occurrence of the scattering event about the mean free path (Flowers and Mendoza[22]).

Results of this Monte-Carlo modelling are shown in Figs. 3*a* and 4*a*. In 3*a* 400 electrons incident on a foil of Al, $\frac{1}{2}$ μm thick, at 100 kV are plotted and compared with a similar result (3*b*) for Al at 100 kV, obtained by Kyser.[20] The two results are clearly very similar.

In Fig. 4*a* the Monte-Carlo predictions of FeO concentration profiles across an FeO grain boundary layer 60 Å wide in an MgO matrix are shown for various specimen thicknesses. These are shown with the experimental results taken from Vander Sande and Hall[21] (Fig. 4*b*) who measured Fe concentration profiles across grain boundaries in Fe-doped MgO using a 25 Å probe at 100 kV in specimen thicknesses of 800, 1960, and 3380 Å. The Monte-Carlo profiles were based on the same values of probe size and specimen thickness used by Vander Sande and Hall. They interpreted their results as indicating that there was no significant beam broadening in their specimen and that it was not possible to deconvolute the observed profiles using either Monte-Carlo techniques or the single scattering model of Goldstein *et al.* The Monte-Carlo calculation agrees moderately well with their experimental results at a thickness of 3380 Å, but indicates that if there was an abrupt boundary between the iron-rich layer and the MgO matrix it should be possible to detect it at a specimen thickness of 800 Å.

Newbury and Myklebust[23] have also carried out a Monte-Carlo analysis of the Vander Sande and Hall results, but they assumed a boundary slab of pure Fe. They reported that their Monte-Carlo simulations agreed with the Vander Sande and Hall experiments at thicknesses of 1960 and 3380 Å. They conclude that the Monte-Carlo simulation describes the Vander Sande and Hall experiment and,

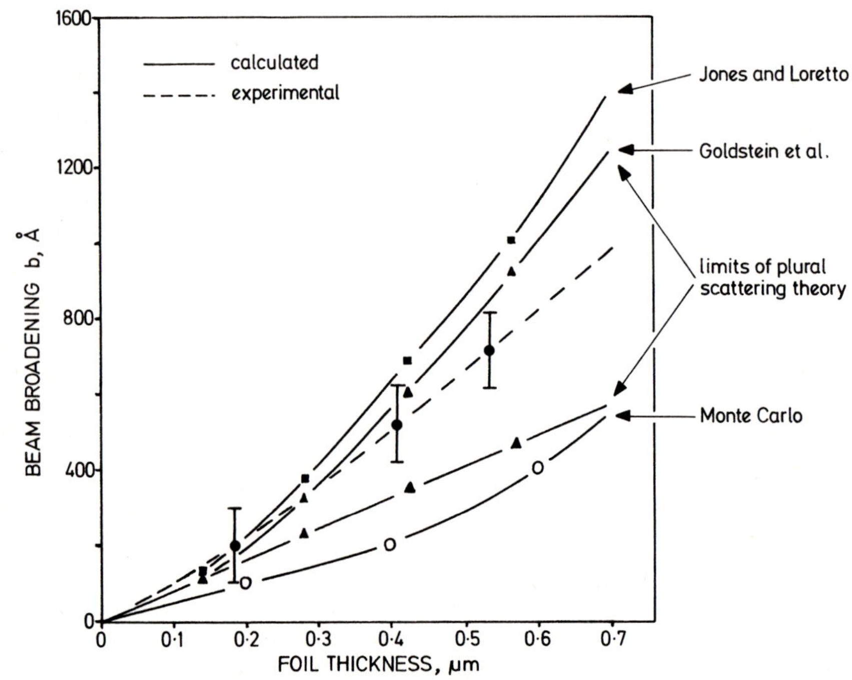

5 Comparison of experimental and theoretical beam broadening parameters *b* for orthopyroxene at 100 kV

by implication, that the results could be interpreted in terms of a grain boundary layer of constant composition and finite extent. Our calculation at 800 Å disagrees with this conclusion and we propose that the experimental results of Vander Sande and Hall imply that there is not a step function in the concentration profile which is obscured by beam broadening.

In Fig. 5 the experimentally observed beam broadening in orthopyroxene, $MgFeSi_2O_6$, is compared with: the predictions of the single scattering model of Goldstein *et al.*; a modified version due to Jones and Loretto[19] who introduced an additional factor of $2/\sqrt{3}$ to the original constant; the simple plural scattering theory of this paper; and the Monte-Carlo program developed from these discussions. The sample consisted of calcium-rich augite lamellae in an orthopyroxene matrix and could be orientated by diffraction techniques to ensure that the lamella/matrix interface was parallel to the incident electron beam ($\pm 1°$). Analyses were obtained at 100 kV using a probe size of about 75 Å, and the beam broadening was measured by monitoring the Ca/Fe ratio across the boundary as a function of the probe distance from the boundary. The measured separation of the 5% and 95% levels of the Ca/Fe ratio, corrected for probe size, were taken as the experimental estimate of beam broadening.

The results indicate that the single scattering model of Goldstein *et al.* (and the modified version of Jones and Loretto) are in moderately good agreement with the experimental results up to thicknesses of 0.4 μm but then diverge to give an overestimate of broadening. The results of simple plural scattering theory bracket the experimental results; the upper limit coincides with Goldstein *et al.* ($n = 2$ in (11) while the lower limit is for $n = \infty$). Our Monte-Carlo results, based on the concepts of the plural scattering theory, give similar results to the lower limit of plural scattering.

CONCLUSION

1 A simple model based on concepts of plural scattering has been shown to reduce to an expression identical with an earlier expression based on single scattering.

2 It has been shown that even for specimens for which the single scattering assumption is invalid, the expression of Goldstein *et al.* can be used to estimate the upper limit of the extent of beam broadening in a thin foil.

3 The importance of the screening parameter in expressions for elastic scattering cross-sections has been recognized and the modified screening parameter of Cosslett and Thomas has been used in Monte-Carlo calculations of electron scattering.

ACKNOWLEDGMENTS

The authors acknowledge the patient help and assistance of Dr S. J. B. Reed, Dr J. Calvert, Mr I. Brough and Mr P. Kenway. The experimental work was carried out on a Philips 400T (STEM) analytical electron microscope purchased with grants from SRC and NERC.

REFERENCES

1 J. I. GOLDSTEIN, J. L. COSTLEY, G. W. LORIMER, and S. J. B. REED: *SEM*, 1977, **1**, 315
2 V. E. COSSLETT and R. N. THOMAS: *Brit. J. Appl. Phys.*, 1964, **15**, 235
3 C. E. HALL and T. INOUE: *J. Appl. Phys.*, 1957, **28**, (11), 1346
4 J. S. HALLIDAY: *Brit. J. Appl. Phys.*, 1960, **11**, 259
5 G. H. SMITH and R. E. BURGE: *Proc. Phys. Soc.*, 1963, **81**, 612
6 R. HUTCHINS, M. H. LORETTO, I. P. JONES, and R. E. SMALLMAN: *Ultramicroscopy*, 1979, **3**, 401
7 P. E. CHAMPNESS, G. CLIFF, and G. W. LORIMER: in press, *Bull. de Min.*, 1980, Orleans
8 D. F. KYSER and R. H. GEISS: Proc. 12th Annual Conf. MAS, 1977, 110
9 G. WENTZEL: *Z. Phys.*, 1927, **40**, 590
10 F. LENZ: *Z. Naturforsch.*, 1954, **9a**, 185
11 W. BOTHE: *Z. Phys.*, 1921, **5**, 63
12 W. BOTHE: Handbuch d. Phyz., 1933, 22/2, 1, Springer, Berlin
13 V. E. COSSLETT and R. N. THOMAS (*Part I*): *Brit. J. Appl. Phys.*, 1964, **15**, 883
14 B. P. NIGAM, M. K. SUNDARESAN, and T. WU: *Phys. Rev.*, 1959, **115**, (3), 491
15 R. SHIMIZU and K. MURATA: *J. Appl. Phys.*, 1971, **42**, (1), 387
16 J. C. WILLMOTT: 'Atomic physics', 43; 1975, London, John Wiley and Sons Ltd
17 R. D. EVANS: 'The atomic nucleus', 597; 1955, New York, McGraw-Hill
18 P. J. GOODHEW and D. CHESCOE: *Micron*, 1980, **11**, 153
19 I. P. JONES and M. H. LORETTO: invited paper at Micro '80, 1980, Brighton, England
20 D. F. KYSER: in 'Introduction to analytical electron microscopy', 199; 1979, New York and London, Plenum Press
21 J. B. VANDER SANDE and E. L. HALL: Inst. Phys. Conf. Ser. No. 52, 1979, 183
22 B. H. FLOWERS and E. MENDOZA: 'Properties of matter', 153; 1970, London, John Wiley and Sons Ltd
23 D. E. NEWBURY and R. L. MYKLEBUST: 'Microbeam analysis 1980', ed. D. B. Wittry, 173; 1980, San Francisco Press

DISCUSSION of paper by Cliff and Lorimer

P. HAGEMANN:

Why does the Monte Carlo method work without taking into account that you are dealing with a crystalline specimen?

G. CLIFF:

The answer to the question is simple: it does not work! The *simple* Monte Carlo cannot predict the diffraction maxima in a diffraction pattern that result from Bragg diffraction with the crystal. If, however, the modelling is improved to take account of the sample crystallinity beam/specimen orientation can be shown to produce scattering maxima (Niedrig, 1978*). Monte Carlo modelling can only model the processes included in the model. It is the personal opinion of the authors that to include all possible processes would result in an almost unmanageable program which would only produce semiquantitative results because of our incomplete knowledge of some of the processes involved.

P. DOIG:
Is it possible to obtain a volume electron intensity distribution within the foil using your analysis?

G. CLIFF:
Yes. One could simply plot contours of electron intensity as a function of distance from the foil surface.

C. Y. F. COLLIEX:
In these various models concerning the beam broadening of electrons through a moderately thick foil (1000–5000 Å) there is a relationship between the angle of scattering and the lateral extension of the beam (i.e. $\theta \propto (b/t)$). Consequently it is important to start from a satisfactory knowledge of the angular distribution $N(\theta)$, such as a Gaussian profile. How do you justify such a model when you make the measurement on a specimen area which displays a very nice diffraction pattern with many beams?

G. CLIFF:
The question really deals with the specific problem of the way in which any Monte-Carlo-based approach models the complex interaction of electrons with solids. The generally accepted approach involves scattering of the electron in the Coulomb field of the atomic nucleus. To a first order this effect dominates, and modelling of scattering involving additional scattering processes will serve to give detailed information about the way in which these modify the effects of the nuclear scattering process. To include all possible scattering processes would require an extensive Monte Carlo program. Sheldon and Ogilvie (1977)† report an attempt at comprehensive Monte Carlo modelling but conclude that only semiquantitative agreement with experiment was possible. Monte Carlo calculation and any simple analytical expression derived from the equations on which the modelling is based can only be seen as providing useful semiquantitative information.

* H. NIEDRIG: SEM, 1978, **1**, 841

† G. L. SHELDON and R. E. OGILVIE: Proc. MAS, 1977, **28A**

Beam spreading in thin foils

T A Stephenson, M H Loretto, and I P Jones

The influence of specimen thickness on electron beam broadening has been measured at 40 kV using Si specimens. It has been shown that Monte Carlo calculations of beam broadening due to elastic scattering accurately reflect the form of the relation between specimen thickness (up to the maximum thickness, 2 μm, examined) and beam broadening, but they tend to underestimate the magnitude slightly. Inclusion of inelastic scattering would, it is suggested, remove this slight disagreement.

The authors are in the Department of Physical Metallurgy and Science of Materials at the University of Birmingham, UK.

With the advent of STEM there has been significant interest in experimental and theoretical work aimed at determining the influence of experimental variables, such as specimen thickness and accelerating voltage, on beam spreading.[1–4] Hutchings *et al.*[4] determined the dependence of beam spreading on specimen thickness in Si samples. These samples had thin strips of gold on one surface and the intensity of the Au L_α X-ray lines was monitored as the electron probe was stepped across a gold strip when the gold was on either the electron entrance or the electron exit side of the Si simply by inverting the specimen between measurements. The difference in the 90% X-ray source size* with the gold on the top or on the bottom was taken as the 90% probe size and hence as a measure of beam spreading. Measurements carried out over a range of thicknesses at 100 and 40 kV enabled the thickness and voltage dependence of beam spreading to be investigated. The results obtained by Hutchings *et al.* for foils of medium thickness (~3000 Å) were in reasonable agreement with Monte Carlo calculations (using Doyle and Turner elastic scattering cross-sections[5]) and with the single scattering model of Goldstein *et al.*[1] For thin specimens (below 1500 Å) the experimental scatter was large compared with the predicted spreading and thus did not provide a good test of the calculations. For very thick specimens (⩾1.5 μm) at 40 kV, the experimental results lay below the calculated curve. All of the various approximations made in the calculations, in particular the neglect of inelastic scattering, were such as to underestimate the beam spreading and it was unexpected to find that the calculated values were actually larger than the experimental ones. The present experiments were carried out in an attempt to determine whether this discrepancy was due to the difficulty of carrying out such measurements in very thick foils or whether there was a fundamental disagreement between theory and experiment. In addition, measurements have been carried out over a range of thicknesses in order to assess the extent of any agreement between the measurements and calculations.

* The 90% X-ray source size is given by measuring the distance which the electron probe has to be moved to obtain 95% and 5% of the maximum Au X-ray signal, this maximum being generated when the whole of the probe interacts with the Au (see Ref. 4).

EXPERIMENTAL

Specimens (kindly supplied by P. Augustus of Plessey Ltd) of Si were prepared with thin (~300 Å) strips of Cr on one surface using an electro-resist process. These strips were more uniform in thickness than the Au strips used in the earlier experiments and observations using secondary electron imaging showed that the edges of the strips were reasonably sharp. In order to determine the magnitude of beam broadening precisely the same technique as that used by Hutchings *et al.* was used, but modifications to the EM400 STEM enabled the measurements on the thick specimens to be carried out more accurately than could be done by Hutchings *et al.* Most importantly a new objective-condenser lens (the twin lens) was fitted to the microscope and this was used in conjunction with the new STEM unit (which incorporated a scan-rotate facility). These modifications, together with the new 20° take-off angle X-ray detector, which can be used at zero specimen tilt, enabled reasonably accurate measurements of beam spreading to be carried out on samples which could not have been imaged in the earlier experiments. The total range of thicknesses on which measurements have been carried out was thus extended so that the functional dependence of beam spreading on thickness could be confidently assessed.

In order to measure the thickness of thin regions (up to about 4000 Å) convergent beam diffraction was used. For thicker regions some preliminary experiments were carried out using a combination of convergent beam, stacking fault images, and X-ray background intensity. These experiments showed that the X-ray background intensity (integrated between 3.65 and 4.70 kV and between 9.0 and 19.0 kV) varied approximately linearly with thickness. For the measurement of thickness during the experiments to determine beam broadening, the background intensity was measured for every position of the X-ray probe and the absolute thickness assessed by calibrating the background intensity for the particular experimental conditions by obtaining a convergent beam pattern and background intensity measurement from a suitably thin region. By careful positioning of the sample in the single tilt holder the orientation used for the X-ray measurements, positive

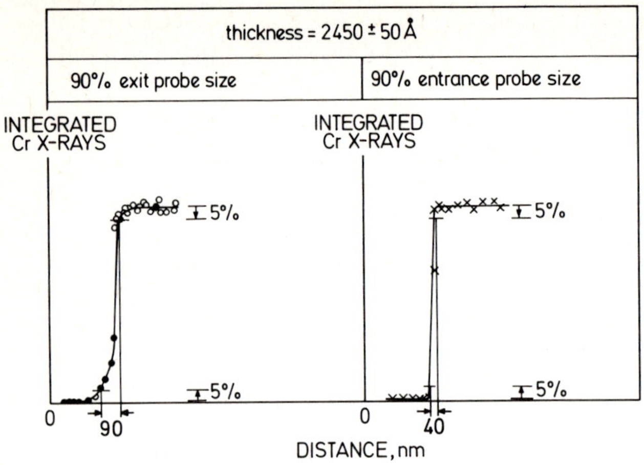

1 **Cr X-ray profiles obtained at 40 kV from Si specimen 2450 Å thick together with measurements of 90% X-ray source size obtained when Cr was on electron exit (○) or entrance (×) side of the Si (for full explanation see text)**

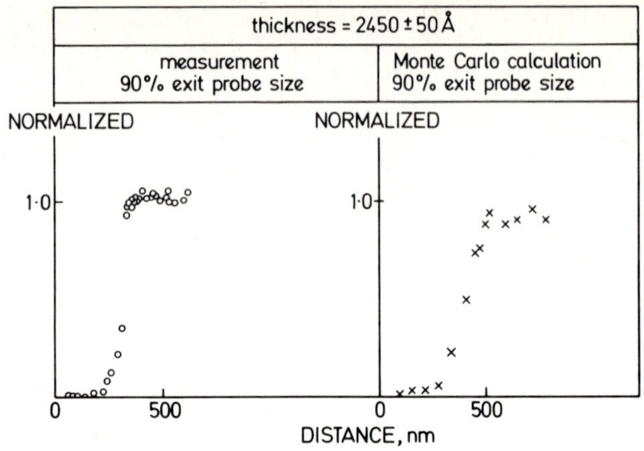

3 **Experimental and theoretical Cr X-ray profiles for Si specimens 2450 Å thick; theoretical curves were obtained using Monte Carlo calculations**

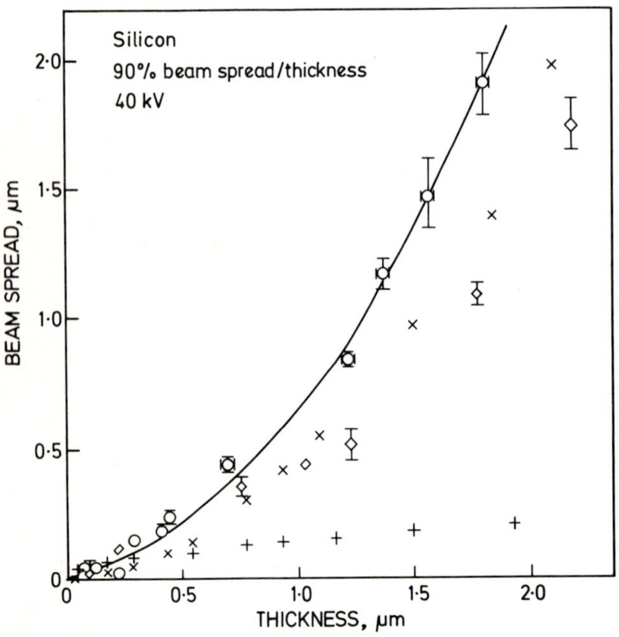

○ Stephenson *et al.* (present work)
◇ Hutchings *et al.*[4]
× Monte Carlo
\+ Doig *et al.*[3]

2 **Experimental relationship obtained at 40 kV between 90% X-ray source size and Si specimen thickness together with various calculated values for same thicknesses**

of a 440 reflection, could be changed to satisfying a 220 reflection for the convergent beam. This small tilt of ~1° produces a negligible change in foil thickness. For this technique of thickness measurement to be successful, and indeed for the determination of the beam profiles to be useful, it is essential that the beam current remain constant during the whole of the experiment. This was checked periodically by measuring the integrated Si count from some reference point on the specimen. If the count varied by more than $2\sqrt{N}$ (where N is the integrated counts) the run was discarded.

The thickness of the Cr layer (which needs to be known for the calculations of beam broadening) was determined by measuring the ratio of Si and Cr intensities from a region of known Si thickness. This ratio was converted to ratios of thickness by using intensity data obtained from a crushed Si–Cr intermetallic compound of known stoichiometry. In each case tilt angles were noted so that the measurements could be corrected for absorption.

RESULTS

Typical Cr X-ray profiles obtained from a specimen of Si 2450 Å thick are shown in Fig. 1 together with the corresponding measurements of 90% X-ray source size obtained when the Cr was on the electron entrance or on the electron exit side of the Si. A summary of all such measurements is shown in Fig. 2 together with corresponding Monte Carlo calculations and calculations using the model of Doig, Lonsdale, and Flewitt.[3] The errors marked in Fig. 2 are estimates only of the various uncertainties in the measurements. The errors are large at small thicknesses because the profiles obtained when the Cr was on the electron exit surface are only slightly larger than those obtained when the Cr was on the top. Figure 3 shows two examples of experimental profiles together with profiles calculated using Monte Carlo calculations. Profiles obtained when a 220 reflection was strongly excited were indistinguishable from those obtained for the condition used throughout this work, in which the crystal was tilted such that the deviation parameter for a 440 reflection was both large and positive.

DISCUSSION

The results shown in Figs. 2 and 3 demonstrate clearly that Monte Carlo calculations are entirely consistent with the experiments reported in this paper. The discrepancy noted in the earlier work by Hutchings *et al.*[4] was therefore associated with the difficulty of obtaining accurate measurements on very thick samples. The disagreement in Fig. 2 between the Monte Carlo calculations and the experimental observations is in the sense that would be expected – the calculation underestimating the beam spreading. Inclusion of inelastic scattering would presumably bring experiment and calculation into close agreement. The calculations using the model of Doig *et al.* are quite clearly inappropriate since the model predicts a $t^{1/2}$ rather than the $t^{3/2}$ dependence predicted by other models and observed in the present work.†

Somewhat disappointingly, the experimental measurements of beam broadening on specimens of a thickness appropriate for X-ray microanalysis were not sufficiently accurate to test the calculations. Further measurements are under way in which the initial probe size will be far smaller than that used in the present experiments. The 90% probe size, measured by stepping the probe across a Cr strip when the Cr was on the electron entrance surface of the Si, was found to be about 500 Å (cf Fig. 1). This unexpectedly large probe size arises both from the fact that the Cr is not vertically stepped and from the fact that the probe has significant electron tails. It appears that the size of these tails can be drastically reduced by lowering the specimen just below the eucentric height[7] and that the value of 500 Å could be reduced to about 100 Å under suitable conditions.

CONCLUSIONS AND GENERAL DISCUSSION

The influence of sample thickness on beam spreading has been measured at 40 kV in Si specimens. The experimental measurements are in agreement with Monte Carlo calculations using Doyle and Turner elastic scattering cross-sections. Any discrepancy between theory and experiment is in the sense that would be expected from the approximations made in the calculations, i.e., the neglect of inelastic scattering will tend to underestimate observed spreading (as is found).

The effects of crystallinity on beam broadening, which are not included in the calculations, have not yet been investigated experimentally in sufficient depth. In addition there is no quantitative information on the effect of varying specimen atomic number. From the Schrodinger equation, corrected for relativistic effects in a simple way, beam spreading is expected to be proportional to Z. There is, however, good reason for believing that the Dirac equation must be used under certain circumstances, particularly for heavier elements.[6] Preliminary calculations for scattering of 40 kV electrons by gold suggest that Mott cross-sections predict values of beam broadening very different from those predicted using Schrodinger cross-sections. If it proves possible to measure beam broadening in heavy materials to the same accuracy as in the present measurements in Si, these differences should be detectable.

Projected variations on the experiment are therefore as follows:

(1) beam broadening experiments on different Z materials, including both those expected to obey Schrodinger scattering factors and those where Mott cross-sections must be used;

(2) beam broadening experiments at different temperatures; and

(3) beam broadening experiments on Si before and after making it amorphous by bombarding with Si ions (the same areas can be examined before and after irradiation).

Variations (2) and (3) should demonstrate any effects of crystallinity.

† It should be noted that Doig *et al.* presented at this conference a new version of their model which yields a $t^{3/2}$ dependence.

ACKNOWLEDGMENTS

The authors gratefully acknowledge general support and encouragement, together with financial support for T. A. Stephenson, from Professor R. E. Smallman.

REFERENCES

1 J. I. GOLDSTEIN, J. L. COSTLEY, G. W. LORIMER, and S. J. B. REED: SEM vol. 1, IITRI, Chicago, 1977, 315
2 R. G. FAULKNER, T. C. HOPKINS, and K. NORRGARD: *X-ray Spectrometry,* 1977, 6, 73
3 P. DOIG, D. LONSDALE, and P. E. J. FLEWITT: *Philos. Mag. A,* 1980, 41, 761
4 R. HUTCHINGS, M. H. LORETTO, I. P. JONES, and R. E. SMALLMAN: *Ultramicroscopy,* 1979, 3, 401
5 P. DOYLE and P. S. TURNER: *Acta Crystallogr. A,* 1968, **24**, 390
6 H. G. BADDE, H. DRESCHER, E. R. KREFTING, L. REIMER, H. SEIDEL, and W. BÜHRING: Proc. 25th Anniversary Meeting of EMAG, Inst. Physics, 1971, 74–77
7 D. ROEBUCK: private communication

DISCUSSION of paper by Stephenson, Loretto, and Jones

P. J. GOODHEW:
Your results differ from those of Hutchings *et al.,** who apparently used a similar technique. Could you explain what experimental differences account for this disparity?

T. A. STEPHENSON:
The instrumental and specimen improvements lead (a fact which there was not time to mention in the presentation) to a significant increase of accuracy in thick regions. These improvements are listed in the paper.

D. A. RAE:
You calculated the foil thickness t of silicon using the projected width of a stacking fault. How did you avoid the

*R. HUTCHINGS, M. H. LORETTO, I. P. JONES, and R. E. SMALLMAN: *Ultramicroscopy,* 1979, **3**, 401 possibility of large errors occurring in t due to any non-coincidence of the foil normal and beam direction? Did you use bright-field, dark-field or weak beam microscopy to determine the projected fault width?

T. A. STEPHENSON:
Foil thicknesses were determined in known beam directions. The foil normal was accurately determined to be either ⟨111⟩ or ⟨100⟩. Bright-field and dark-field pairs were usually taken.

P. DOIG:
Is it possible to establish the electron intensity distribution within the STEM probe at the bottom surface of the foil from your measured Cr signal profile? Such a differentiation may allow you to establish whether single, plural or multiple scattering has occurred.

T. A. STEPHENSON:
It is possible in principle but poor statistics in the experiment impair the accuracy of the intensity distribution.

P. J. STATHAM:
Whereas the simple expression for b and the simple scattering formula of Doig *et al.* can be calculated very easily the Monte Carlo approach, although potentially more versatile, does not appear to afford a practical aid for on-line analysis. Is it reasonable to suggest that experimental and Monte Carlo results could be applied more constructively if they were used to develop and improve a simple parameterized formula for on-line experimental use?

T. A. STEPHENSON:
I agree the Monte Carlo approach does not seem practical for on-line analysis. Since it is at this time the most accurate method for calculating the electron beam profile it would certainly be worthwhile to use this technique to develop parameterized formulae. Whether complex geometries could still be handled in such a specific way is open to question.

Errors in foil thickness measurement using contamination spot method

D A Rae, V D Scott, and G Love

The contamination spot method for measuring foil thickness in transmission electron microscopy is investigated. It is shown by comparison with foil thickness determinations using crystallographic features in the specimen that the contamination spot technique may overestimate by as much as 80 nm. This discrepancy arises from a number of sources. With stainless steel foils layers of contamination such as oxide and/or carbonaceous material generally account for errors of ~10–20 nm. However, the major source of inaccuracy is shown to be associated with difficulties in locating the points on the image from which to measure; this is due to the fact that the contamination spots are surrounded by a disc of contamination which is not clearly visible. The measurement error is worse the larger the spot height; with stainless steel foils it is ~50 nm.

The authors are in the School of Materials Science at the University of Bath, UK.

The accurate measurement of specimen thickness is important in two areas of materials research which utilize transmission electron microscopy of thin foils: (i) quantitative estimates of the populations of dislocations, precipitates, etc.[1]; (ii) quantitative X-ray microanalysis of chemical compositions where it is necessary to know, firstly, when correction factors (X-ray absorption, etc.) need to be introduced[2,3] and, secondly, the size of the analysed volume.[2]

Several methods of thickness measurement which can be carried out in a conventional transmission electron microscope have been described, many of which make use of crystallographic features in the specimen such as slip plane traces, grain boundaries and stacking faults.[4] The idea here is to measure the projected width of a feature which intersects the top and bottom surfaces of the foil before and after tilting through a known angle; usually the parallax position, where the fault lies parallel to the electron beam, is chosen as one of the tilt orientations. Unfortunately, such crystallographic features will not generally be present in materials and to overcome this limitation Heimendahl[5] applied latex spheres to a foil. Parallax techniques can then be applied by measuring the projected distance between spheres on the top and bottom of the foil and tilting the specimen as described previously. However, all of these techniques of foil thickness measurement are prone to error.[6–8]

With the addition of scanning transmission (STEM) and X-ray analysis (EDS) attachments to the transmission electron microscope, several alternative techniques are now available. These include the contamination spot method,[9] convergent beam diffraction,[10] and measurement of the intensity of X-ray emission.[11,12] Such methods have been compared and discussed in an earlier paper[13] and it was shown that each gave a different answer for the measured foil thickness. These apparent inconsistencies were explained by the fact that every technique measures something different. For example, convergent beam diffraction measures the thickness of the crystalline layer whereas the contamination spot method gives the total thickness of specimen including any films of oxide or contamination present on the foil surface.[11,13]

In the present paper the inaccuracies associated with the contamination spot method are examined in some detail and it is shown that only part of the error may be attributed to surface films of oxide and contamination.

MATERIALS AND SPECIMEN PREPARATION

Thin foils suitable for transmission electron microscopy were prepared from stainless steel and from aluminium. The stainless steel was obtained as commercial sheet containing 18% chromium and 8% nickel, while the aluminium was in rod form of 99.999% purity. Discs 3 mm in diameter were cut from the respective materials, inserted into PTFE holders and thinned in a Tenupol electropolishing unit. In the case of stainless steel a solution of one part perchloric to 20 parts acetic acid at room temperature was used with a current of 150 mA. For aluminium the electropolishing solution consisted of 20% nitric acid in methanol held at -20°C, the same current of 150 mA being applied. When perforation of a specimen occurred it was quickly removed, thoroughly washed in methanol, and transferred immediately into the transmission electron microscope for examination.

In one series of experiments thinned foils of stainless steel were coated with a thin layer of aluminium using thermal evaporation in a high vacuum system. The thickness of aluminium deposit was determined using the following approach: a glass slide was thoroughly cleaned and part of its surface masked; it was positioned close to the specimen in the vacuum chamber and simultaneously coated with aluminium; removal of the mask left a step on its

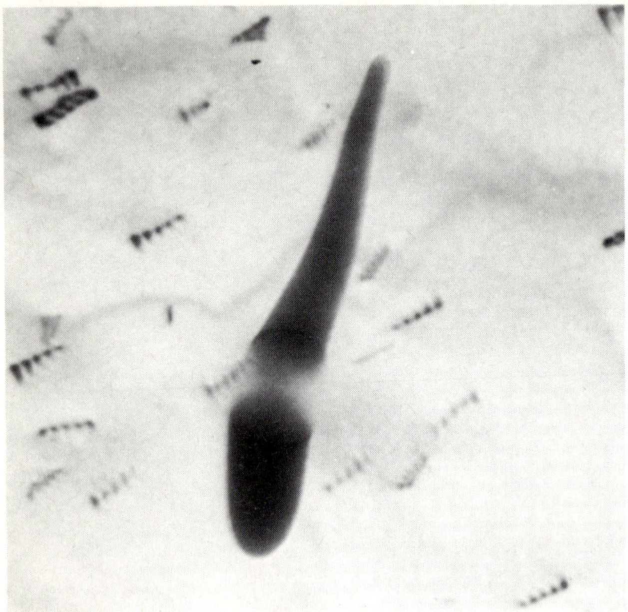

1 **Contamination spot deposited on stainless steel at normal incidence; foil tilted 45° (× 100 000)**

surface and the whole slide was thinly coated with a layer of silver to provide a highly reflecting surface. The step height was then measured using multiple beam interferometry.

MEASUREMENT OF FOIL THICKNESS

The electron microscope used in this work was a JEOL 100CX instrument fitted with scanning transmission facilities. Measurement of foil thickness using the contamination spot technique involves irradiation of the specimen with a focused beam of electrons to provide a 'cone' of contamination on the upper and lower surfaces of the foil; when the foil is tilted through a known angle (Fig. 1) the foil thickness may be determined by measuring the spot separation. In most of the work described here the contamination spots were deposited with the goniometer stage set at 0°, and hence, using a tilt angle of 45°, the foil thickness in the beam direction t_B is given by $\sqrt{2}x$, where x is the spot separation.

An alternative way of using the contamination spot method was adopted in one series of experiments (see following section) whereby the spots were deposited with the foil inclined to the beam and the spot displacement measured after tilting back through a known angle θ. This approach was used specifically in a comparison with the slip plane trace method for estimating foil thickness, the aim being to avoid errors incurred if the foil is not accurately perpendicular to the beam when the goniometer stage is at the 0° setting. The procedure was as follows. First an area of metal foil containing a slip plane trace was located and the specimen rotated until the trace lay parallel to the tilt axis of the goniometer stage (Fig. 2*a*). Next the foil was tilted until the slip plane was parallel to the electron beam (2*b*). This setting was achieved with an accuracy of better than 1° by making measurements of the projected width of the slip trace w' at small angles α either side of the estimated parallax position. From a graph of w' versus α (linear for small angles of α) the true parallax position was determined from $w' = 0$. With the foil in the parallax position, a row of contamination spots was deposited along the line of the slip plane trace (Fig. 2*c*) and the foil tilted through an angle θ (2*d*) until it was approximately perpendicular to the electron beam. Hence the width of the slip plane trace w is given by $w' \operatorname{cosec}\theta$. At the same time the separation of the contamination spots was measured to give a second determination which should, of course, correspond to w.

RESULTS AND DISCUSSION

Conditions for forming contamination spots

Some preliminary experiments were carried out in order to establish suitable conditions for depositing contamination spots on the foil. The electron beam voltage used was 100 kV while the diameter of the focused beam was approximately 5 nm. The cold stage was not in operation when depositing contamination spots. This enabled visible deposits to be formed after some seconds of irradiation rather than minutes, so minimizing any problems associated with drift of either electron beam or specimen.

Contamination spots were formed on a stainless steel foil for irradiation periods of up to 90 s. As would be expected[14] the height and diameter of the contamination mark increase with time, the volume of deposit being approximately proportional to irradiation time. An irradiation period of 30 s was chosen for the majority of the present experiments since this was considered the minimum time necessary to produce clearly visible spots on foils covering a range of thickness.

Measurements of width of slip plane trace using contamination spot method

Figure 3 was taken from a stainless steel foil containing a slip plane trace. The pair of contamination spots were produced upon the top and bottom surfaces of the foil when in the parallax position, and in the electron micrograph the foil is tilted 35° from this position.

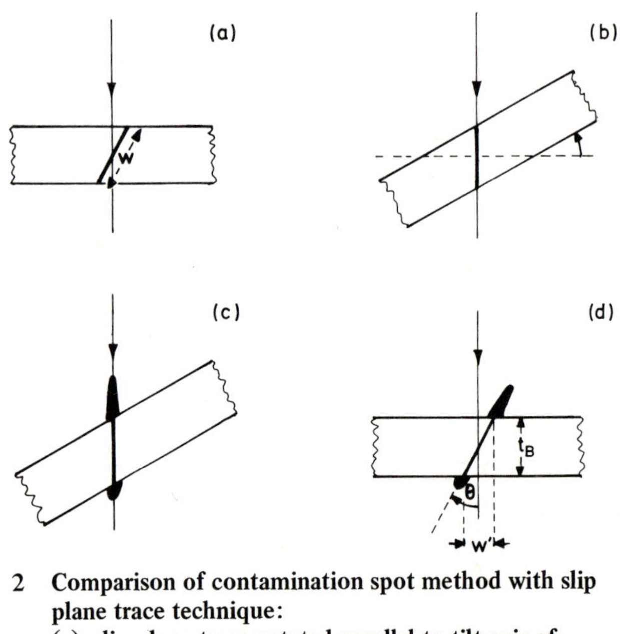

2 **Comparison of contamination spot method with slip plane trace technique:**
(a) slip plane trace rotated parallel to tilt axis of goniometer stage;
(b) foil tilted so that slip plane is parallel to electron beam;
(c) as (b) and contamination spots deposited;
(d) foil tilted back through measured angle θ.
Width of slip plane trace w given by measurement of w', i.e. $w = w' \operatorname{cosec}\theta$

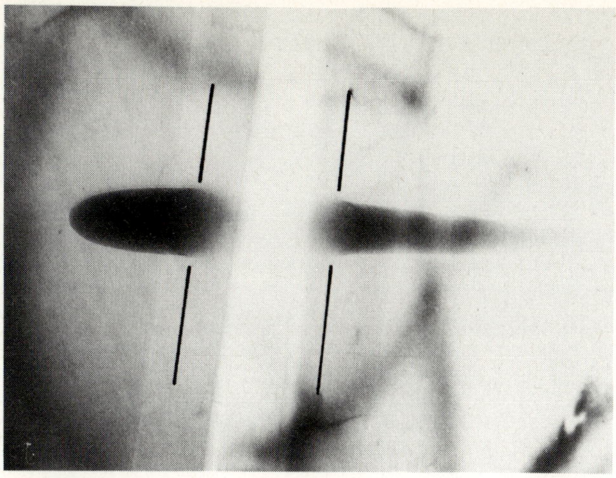

3 **Slip plane trace plus contamination spots on stainless steel foil (× 90 000)**

Since it is not immediately obvious from which points on the contamination spots the displacement distance is measured, we shall first examine the expected contrast profile of the contamination spots by means of the diagrams in Fig. 4, typical for 30 s irradiation on a 300 nm thick foil. Assuming the spots have the form shown in Fig. 4*a*, a plot of effective carbon thickness across the image would appear as in 4*b*. A plan view of the spots constructed from the carbon profile is shown in 4*c* from which it may be seen that the measurement required is the distance between the tips of the cusps, i.e. between the lines XX' and YY'. These have been drawn on the electron micrograph Fig. 3 and clearly they will give a value in excess of the projected width of the slip trace. Using the formula $w = w' \operatorname{cosec}\theta$ it was calculated that the contamination spot method gave a value of $w = 275$ nm, whereas measurement using the projected width of the slip trace gave $w = 175$ nm, the discrepancy Δw between the two values being 100 nm. If Δw is now converted into a discrepancy in measuring the effective foil thickness, t_B, we obtain $\Delta t_B = \Delta w \cos 35° = 82$ nm. A series of such measurements were carried out involving several different tilt angles and foil thicknesses from which a mean value of $\Delta t_B = 75 \pm 10$ nm was obtained.

Since it is difficult to believe that the slip plane trace is underestimating w it follows that the contamination spot method must be in error. Factors which could account for the observed discrepancy are:

(a) a surface oxide film
(b) a uniform layer of carbon contamination formed on the foil before insertion in the electron microscope
(c) a uniform layer of carbon contamination formed while the foil was being examined in the electron microscope
(d) an error in the practical application of the contamination spot method.

These factors are examined in sequence.

Oxygen and carbon measurements on electropolished thin foils using electron-probe microanalysis

Clearly an oxide film is present on the foil surfaces, otherwise slip plane traces would not be observed, and this experiment was designed to give an estimate of the upper limit of oxide thickness likely to be present. A number of stainless steel specimens were prepared and each was subjected to different periods of oxidation. The thicknesses of oxide films produced were estimated from the colour of the interference films[15] and ranged from ~45 to 70 nm. The specimens were inserted into the electron-probe microanalyser and the oxygen K emission of each was measured using 5 keV electrons and a specimen current of 100 nA. A calibration curve of oxygen count rate versus oxide thickness was constructed from these data. Thin foils of stainless steel were next analysed for oxygen and the results compared with the calibration curve. It was concluded that any oxide film present on the foil was less than 2.5 nm per surface.

In a similar manner carbon measurements were carried out on electropolished thin foils using the electron-probe microanalyser. In this instance the calibration curve of thickness of carbon count rate versus carbon thickness was produced using stainless steel samples coated by vacuum evaporation with known thicknesses of carbon. The results showed negligible carbon (<2 nm) to be present on the electropolished foil.

Hence it may be concluded that surface layers of contamination produced during foil preparation cannot account for the discrepancy.

Aluminium layer produced in stainless steel foil

To examine the possibility that a layer of contamination may have formed on the surfaces of the specimen while it was being examined in the electron microscope, foil thicknesses were measured using the contamination spot technique before and after an aluminium film of known thickness had been deposited. The intention was to remove any error in the method of contamination spot measurement (factor '(d)' under 'Measurements of width . . .' above) by subtracting the two sets of readings.

A row of contamination spots spaced 2.5 μm apart was deposited on a stainless steel foil with its surface approximately perpendicular to the electron beam. It was then tilted through 45° and the spot separation measured. Next

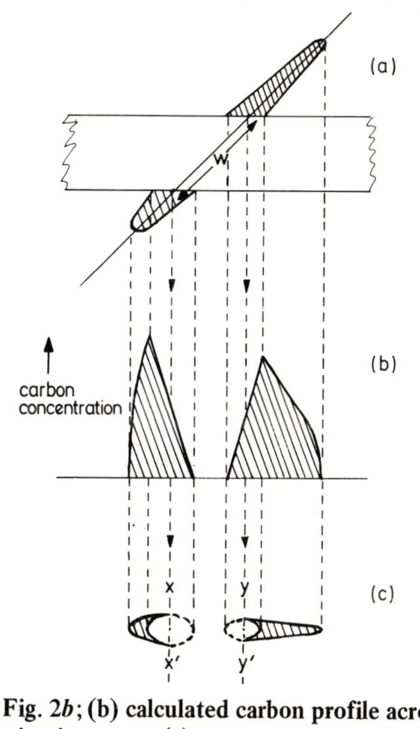

4 **(a) as Fig. 2*b*; (b) calculated carbon profile across contamination spots; (c) measurement required is distance between *xx′* and *yy′***

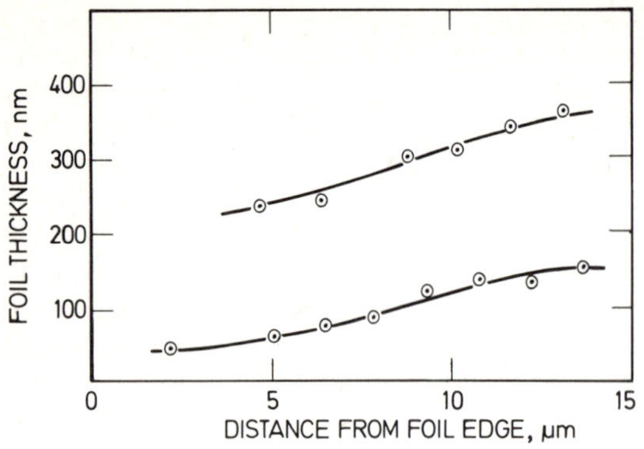

5 Thickness measurements on stainless steel foil before (*lower curve*) and after (*upper curve*) coating with layer of aluminium 170 ± 5 nm thick (contamination spot method)

the foil was coated with a layer of aluminium in a good quality vacuum system. The coating thickness, established from interferometric measurements on a simultaneously coated glass slide, was 170 ± 5 nm. The specimen was reinserted into the electron microscope and a second set of contamination spots placed between the first series. As before the spot separations were measured after tilting through 45°.

Both sets of data are plotted in Fig. 5 and the curve for the aluminium-coated specimen gives points which are consistently 180 ± 10 nm higher than those for the non-coated specimen (the undulation in both curves is due to a non-uniformly tapering section in the region of a metal grain boundary).

Comparison of the interferometric data with contamination spot measurements indicates therefore that if there is any contamination of the foil surface during examination its contribution to the discrepancy is at most 10 nm per surface, i.e. less than 20 nm overall. Hence the contamination spot method itself must contain the major source of error.

Effect of electron dose

In order to investigate whether the observed discrepancy was affected by the size of the contamination spot some measurements were carried out on stainless steel foils using 90 s spots as well as the standard 30 s spots. The former irradiation time produced spots that were approximately 1.5 times the height and diameter of the latter.

Firstly, a series of 90 s spots spaced 2.5 μm apart was deposited inwards from the edge of a stainless steel foil and the respective foil thickness calculated (see Fig. 6). Next a series of 30 s spots was placed between the first series and a second set of calculations of foil thickness carried out. Comparison of the data (Fig. 6) shows that the use of 90 s spots results in a greater overestimation of the true foil thickness, the difference between the two sets being consistently 20 nm.

The experiment was repeated, on this occasion forming the series of 30 s spots first. Again the use of 90 s spots gave a greater overestimate of the true foil thickness but this time the difference between the two sets of calculations was ~30 nm.

It may be concluded from these observations that:

(i) there is an error in the contamination spot method of measuring foil thickness since 90 s spots always gave larger overestimates of foil thickness than the use of 30 s spots; and

(ii) since the difference in the overestimates depends upon whether the 30 s or the 90 s spot was deposited first, part of the measurement error must be due to some overlap between neighbouring spots.

Hence it may be argued that there must be some broadening of the base of the core deposit. An indication of this was available from earlier observations[16] whereby a thin but visible annulus of carbon was formed around the base of the core. The annulus was, however, produced under unusual irrradiation conditions and such a feature was not visible in any of the electron micrographs taken in the present study.

In the following section we shall investigate the profile of a contamination spot in more detail.

Profile of contamination spot

Let us begin by constructing the expected carbon profile across a pair of contamination spots produced with the electron beam perpendicular to the foil surface. Assuming these have a similar conical form to that shown in Fig. 4, where the spot has been deposited on an inclined foil, the measurement required is again the distance between the tips of the cusps (see Fig. 7). However, if the spots do not have this idealized shape but are broadened at the base by a 'tailing' effect (Fig. 8*a*) then the carbon profile will be different. Figures 8*b* and *c* show the result of introducing such a tail. Firstly the carbon level does not drop to zero between the two maxima (8*b*) and secondly the edges of the spot images become generally more diffuse. Now, if the distance between the tips of cusps is measured (8*c*) the result after applying the tilt correction would correspond to t', which is greater than the foil thickness.

In order to check this hypothesis a photo-densitometer trace was obtained from the electron micrograph of a contamination spot. The spot had been formed on a thin foil of aluminium since this material gave regions of more uniform background contrast compared with stainless steel. Figure 9*a* shows spots formed on an area of foil ~300 nm thick. The corresponding densitometer trace, obtained using a finely collimated beam of light (50 μm slit), is given in Fig. 9*b*. Its close similarity with the carbon profile illustrated in Fig. 8*b* rather than the profile deduced from the idealized spot shape in Fig. 7*b* confirms the existence of a 'tail' on the contamination spots. The important point to note is that although the contamination spots appear to be well separated in the electron micrograph the densito-

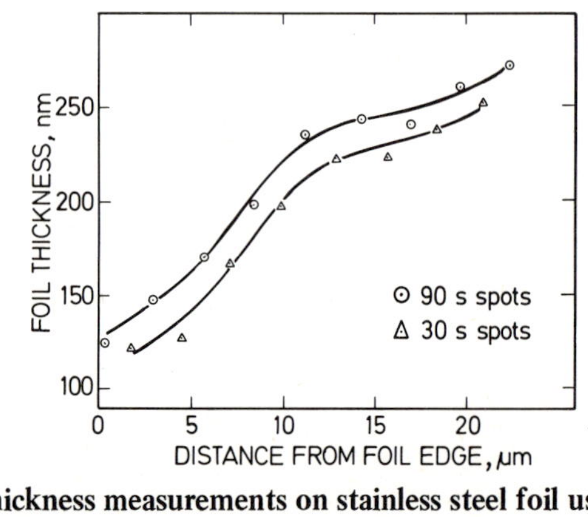

6 Thickness measurements on stainless steel foil using firstly 90 s spots and secondly 30 s spots

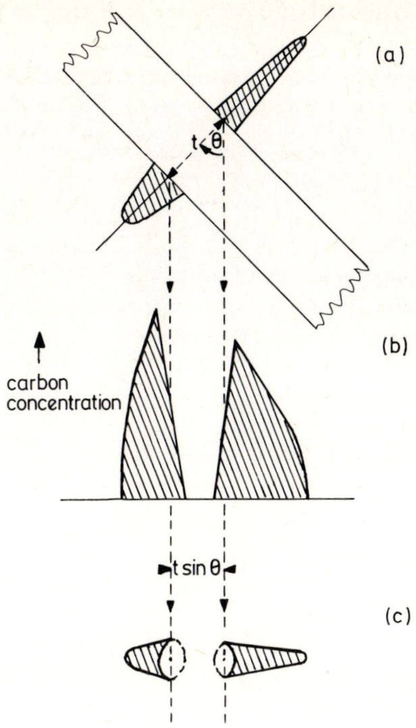

7 **Expected carbon profile and plan view of spots resulting from cones of contamination produced at normal incidence to foil**

meter trace shows a detectable level of carbon between them.

It follows that the bigger the spot for a given foil thickness the larger will be the 'tail' and the greater will be the overestimate of the true foil thickness, as was indeed found when carrying out thickness measurements using spots deposited for 90 s. It is possible to make some estimate from these data of the likely errors involved. For example, it was noted (see previous section) that a 90 s contamination spot was approximately 1.5 times the height produced after 30 s irradiation; furthermore, the difference between the measurements was ~20 nm when the 90 s spots were deposited first and ~30 nm when the 30 s spots were formed first. Using this information it was deduced that the measurement error involved with 30 s spots was an overestimate of 50 ± 5 nm. With 90 s spots the error was increased to ~75 nm while the overlap error (see item '(ii)' above) was ~5 nm.

CONCLUSIONS

1 It has been shown that the use of the contamination spot method for measuring foil thickness may be subject to large errors and that these errors arise from a number of sources.

2 Any layer of contamination covering the surfaces of the foil will, of course, be measured together with the actual foil thickness when using the contamination spot method. Such layers may include an oxide film produced during foil preparation and/or a carbonaceous layer formed either before insertion in the electron microscope or during examination. Even when extreme care is taken these factors can introduce errors of 10–20 nm.

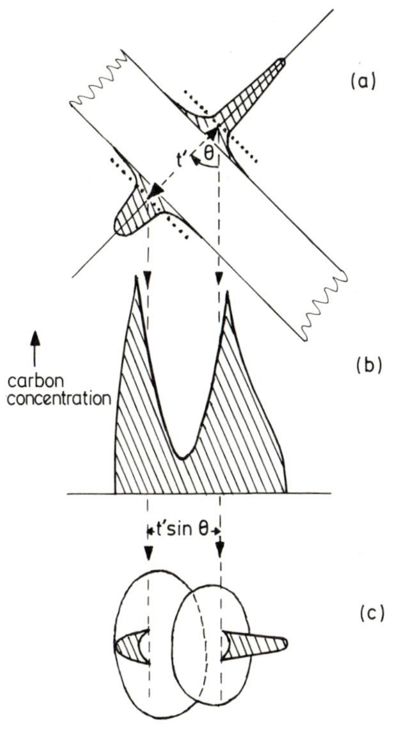

8 **As Fig. 7 but with disc of contamination surrounding cones; note that in (b) carbon level is not now zero between maxima**

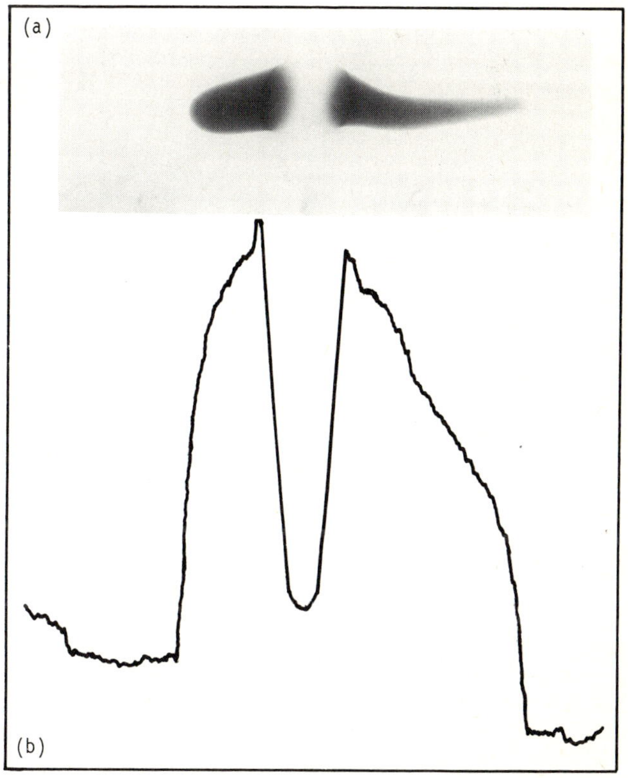

9 **(a) Contamination spot produced on Al foil at normal incidence (foil tilted through 35°); (b) corresponding photo-densitometer trace across spots; note profile is better represented by Fig. 8*b* than by Fig. 7*b***

3 The major source of error lies, however, with the actual spot measurements since the points on the image corresponding to the foil surface cannot be clearly defined; this is due to the presence of a disc of contamination which surrounds each spot. It has been shown that measurements inevitably overestimate the foil thickness and the larger the spot used the greater the error. In practice, the minimum spot height that can be usefully adopted is governed by the need to produce a visible feature against the background level in the image. For example, it was found with stainless steel foils that the spot heights required were approximately the same as the foil thickness, in which case measurement errors were ~50 nm on a 300 nm thick foil.

4 Care should be taken not to make spot measurements too close together since the disc of contamination surrounding the main spot may introduce additional errors. Although this disc will not generally be visible in the electron micrograph, errors of ~5–10 nm may be inadvertently introduced, depending upon the proximity of spots. The present work suggests that the minimum distance between neighbouring spots should be not less than 3 μm.

ACKNOWLEDGMENTS

Thanks are due to the SRC for funding the research. D.A.R. is supported by an SRC CASE award in collaboration with MOD, AUWE, Portland.

REFERENCES

1 G. LOVE, D. A. RAE, and V. D. SCOTT: 'Electron microscopy 1980', **1**, eds. P. Brederoo and G. Boom, Leiden, 7th European Cong. Electron Microscopy Foundation, 208

2 J. I. GOLDSTEIN, J. L. COSTLEY, G. W. LORIMER, and S. J. B. REED: Proc. Workshop on Analytical Microscopy, ITTRI, 1977, Chicago Press Corp., 153

3 J. BENTLEY and E. R. KENIK: *Scr. Metall.*, 1977, **11**, 261

4 P. B. HIRSCH, A. HOWIE, R. B. NICHOLSON, D. W. PASHLEY, and M. J. WHELAN: 'Electron microscopy of thin crystals', 1965, London, Butterworths

5 M. von HEIMENDAHL: *Micron*, 1973, **4**, 111

6 P. M. KELLY: *Phys. Status Solidi*, 1975, **32**, 529

7 E. L. HALL and J. B. VANDER SANDE: *Philos. Mag.*, 1975, **32**, 1289

8 L. M. CLAREBROUGH and A. K. HEAD: *Philos. Mag.*, 1976, **33**, 557

9 G. W. LORIMER and G. CLIFF: 'Developments in electron microscopy and analysis', 1976, ed. J. A. Venables, New York, Academic Press

10 P. M. KELLY, A. JOSTONS, R. G. BLAKE, and J. G. NAPIER: *Phys. Status Solidi*, 1975, **31**, 771

11 W. S. MILLER and V. D. SCOTT: *Met. Sci.*, 1978, **12**, 95

12 P. J. STATHAM and M. D. BALL: 'Microbeam analysis 1980', ed. D. B. Wittry, San Francisco, San Francisco Press Inc., 165

13 G. LOVE, M. G. C. COX, and V. D. SCOTT: 'Developments in electron microscopy and analysis', 1977, ed. D. L. Misell, Bristol, Inst. of Phys., 347

14 J. J. HREN: 'Introduction to analytical electron microscopy', 1979, eds. J. J. Hren, J. I. Goldstein, and D. C. Joy, New York, Plenum Press, 481

15 F. H. CONSTABLE: *Proc. R. Soc. A*, 1928, **117**, 376

16 G. LOVE, V. D. SCOTT, L. LAURENSON, and N. M. T. DENNIS: *Scanning*, 1981, **4** (1), 32

Part 2
Energy-dispersive X-ray analysis: applications

Measurement of grain boundary segregations using STEM-EDS X-ray microanalysis

P Doig, D Lonsdale, and P E J Flewitt

Results are described for STEM-EDS X-ray microanalysis of equilibrium and non-equilibrium solute segregations to prior austenite grain boundaries in a range of ferritic steels typically used in electrical power generating plant. Consideration is given to the requirements and limitations of the microanalysis technique. Examples have been selected which demonstrate the various solute profiles and distributions which may be developed and measured for various heat treatments and alloy compositions.

The authors are in the Scientific Services Department of the CEGB South Eastern Region, Gravesend, Kent.

The importance of chemical composition within the grain boundary region in metals and alloys has been widely recognized[1–3] and its influence on the creep,[4] stress corrosion and corrosion resistance[5–7] of electrical power plant components is well known. In ferritic steels changes in composition can result from either equilibrium or non-equilibrium segregation of alloying or impurity elements to the prior austenite (γ) grain boundaries. The spatial resolution of the STEM-EDS (scanning transmission electron microscope-energy-dispersive spectrometer) technique has been shown to be sufficient to measure such solute segregations to grain boundaries in thin foil specimens.[8,9] In addition the solute distributions can be related to the total microstructure conventionally resolvable in the transmission electron microscope.

The equilibrium segregation of Groups IV–VI impurity elements and major alloying additions to grain boundaries is effected by heat treatment within a critical temperature range.[2,3,7,10] The driving force for such segregation is the reduction of interfacial free energy. Solute segregations do not necessarily result in detectable changes in microstructure although their presence may be reflected in a change in the fracture mode.[3] Irrespective of the various theoretical models which describe such segregation on an atomic scale it is generally considered to be localized to within a few atomic layers of the grain boundary.[2,7,8] Its occurrence, however, is complicated by the presence of major alloying elements within the steel which can either promote or inhibit segregation of these elements.[3,10]

Alternatively, non-equilibrium segregations at grain boundaries have been observed in a range of alloy systems following either cooling at a critical rate[11–13] or isothermal heat treatment within a critical temperature range.[14] Non-equilibrium segregations developed by quenching have been considered to arise from the diffusion and decomposition of vacancy–solute complexes.[11,13] Changes in the equilibrium vacancy concentration with temperature provide the necessary driving force for diffusion to grain boundary sinks during cooling. In general, non-equilibrium segregations which occur as a result of heat treatment within a critical temperature range arise as a consequence of heterogeneous intergranular precipitation.[14]

In this contribution we describe the results of STEM-EDS microanalyses for equilibrium and non-equilibrium solute segregations in a range of ferritic steels typically encountered in electrical power generating plant. The examples selected demonstrate those solute profiles which may be developed and measured for various heat treatments and alloy compositions.

X-RAY MICROANALYSIS

Thin foil disc specimens (3.05 mm dia.) were examined in a JEOL JEM 100CX transmission electron microscope fitted with an ASID 4 scanning transmission attachment, a Kevex energy-dispersive X-ray spectrometer ($\sim$152 eV resolution) and a Link Model 860 analyser system. The operating vacuum adjacent to the specimen chamber was measured to be $<3 \times 10^{-5}$ Pa. This vacuum quality is reflected by the small contamination spots developed when in the STEM spot mode.

For X-ray microanalysis, discs were held in a beryllium 'low background' single-tilt specimen holder and the microscope illumination system was fitted with 1 mm thick platinum apertures which reduced spurious 'hole-count' X-ray signals to a level very much below the background recorded from thin foil samples.[15] Before any X-ray microanalysis was undertaken the whole specimen area was irradiated by the electron beam for a period of $\sim$300 s with the second condenser aperture removed. This 'fixed' any surface contamination introduced with the specimen and reduced the accumulation of contamination at positions of subsequent microanalysis. X-ray microanalysis was carried out with the specimen tilted between 25 and 35° towards the detector using a beam current of $\sim 2 \times 10^{-10}$ A, an accelerating voltage of 100 kV and a spot size of $\sim$5 nm.

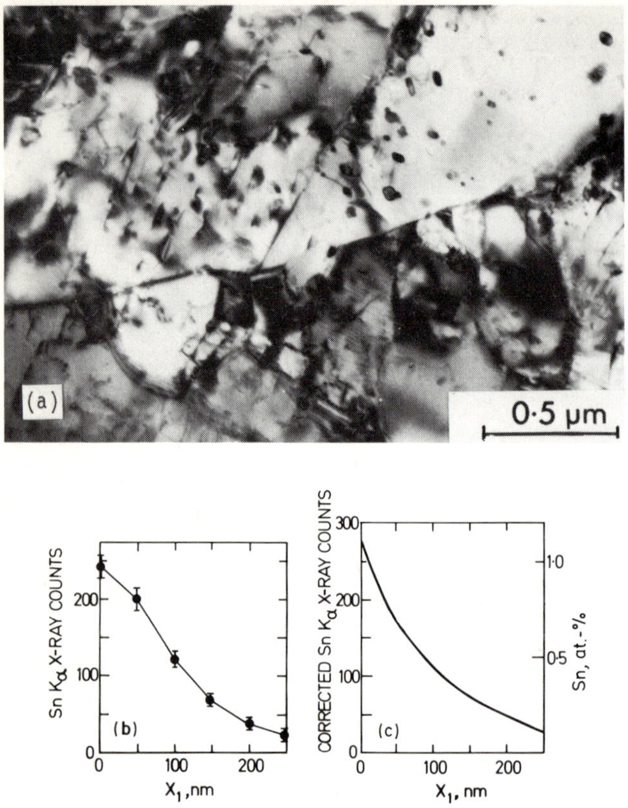

1 Transmission electron micrograph (a) showing microstructure within HAZ of $\frac{1}{2}$Cr–Mo–V steel weldment; measured (b) and corrected (c) Sn K_{α} X-ray intensity profiles across prior austenite grain boundary ($x_1 = 0$ is position of boundary)

Spectra were recorded for 100 s from regions of foil 100–200 nm thick in the direction of the incident electron probe with recorded X-ray count rates of between 100 and 400 counts s^{-1}. These were processed using the Link RTS 2 FLS thin section quantitative routine which includes a correction for X-ray absorption and foil orientation. The Cliff–Lorimer k factors[16] relating X-ray intensities to solute concentrations used in the program had been obtained previously using foils of known homogeneous binary alloys. Foil thicknesses were derived from projected widths of adjacent grain boundaries tilted to produce strong fringe contrast. Since we have a finite incident electron probe diameter and subsequent electron scattering in the foil there is a need to convert the experimentally measured concentration profiles to true compositions using the quantitative procedure described previously.[17,18] However, it is essential when undertaking measurements of these segregations that the boundary be oriented parallel (within ±3°) to the incident probe to ensure complete sampling. In addition, to enable quantitative evaluation of the composition it is necessary to accumulate sufficient X-ray counts to provide adequate statistical condidence in the finally determined composition.

RESULTS AND DISCUSSION

Equilibrium segregations

Three examples of equilibrium solute segregation in ferritic steels demonstrate the importance of both its spatial extent about a grain boundary and also the distribution within the total microstructure of a boundary. Two of these arise directly from problems associated with large turbogenerator plant, the other from a laboratory evaluation of a $2\frac{1}{4}$Cr–1Mo steel subjected to an embrittling heat treatment.

Segregation profiles about γ grain boundaries have been examined in a sample cut from a boiler steam chest weldment. The weld metal is a nominal $2\frac{1}{4}$Cr–1Mo steel and the base steam chest a $\frac{1}{2}$Cr–$\frac{1}{2}$Mo–$\frac{1}{4}$V steel. The microstructure within the weldment was complex but in the heat-affected zone (HAZ) where reheat cracking occurred it comprised ferrite laths with a distribution of coarse carbides (Fig. 1*a*). X-ray microanalysis of foils identified tin to be the segregating species and Fig. 1*b* shows the measured tin K_{α} X-ray intensity across a γ boundary in the HAZ microstructure. Here a significant concentration of tin extends for a distance (x_1) of up to 200 nm from the boundary. Using the procedure discussed by Doig, Lonsdale, and Flewitt[17] the true X-ray intensity profile is derived (Fig. 1*c*). Thus the concentration of tin on the γ grain boundary is 1.1 at.-% and such a segregation is of particular significance in controlling the susceptibility of this steel to weld reheat cracking.[3]

A similar examination has also been undertaken on samples from within the HAZ of a 1Cr–$\frac{1}{2}$Mo steel boiler header loop pipe butt weld which exhibited extensive creep cavitation along grain boundaries (Fig. 2*a*) after $\sim 10^8$ s

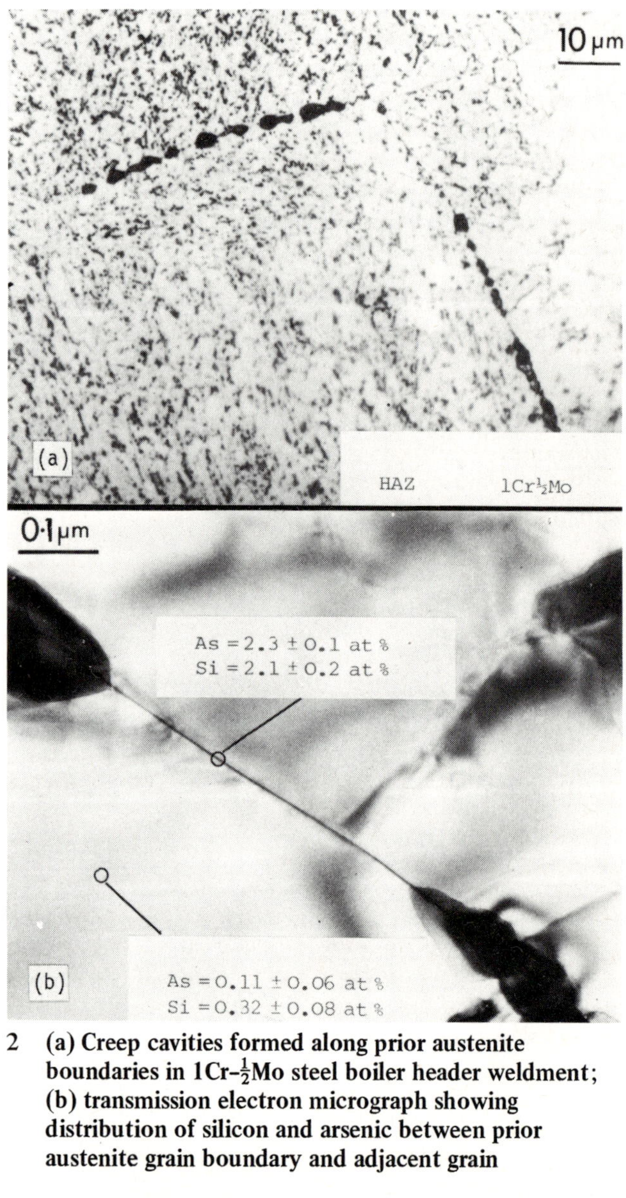

2 (a) Creep cavities formed along prior austenite boundaries in 1Cr–$\frac{1}{2}$Mo steel boiler header weldment; (b) transmission electron micrograph showing distribution of silicon and arsenic between prior austenite grain boundary and adjacent grain

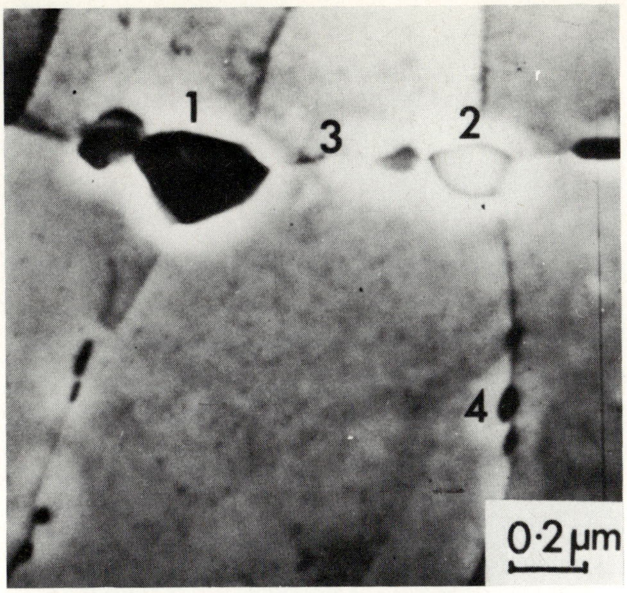

3 Scanning transmission electron micrograph showing tempered bainite microstructure of $2\frac{1}{4}$Cr–1Mo steel

operation at ~800 K. Microanalysis on foils at selected positions along γ grain boundaries and within the matrix show segregations of both silicon and arsenic at the boundaries (Fig. 2*b*). There was no evidence, however, of an extended concentration profile. Assuming that any boundary concentration is localized to a region ~2 nm wide[2] it was possible to establish that the silicon concentration at the boundary was 2.1 ± 0.2 at.-% compared with the matrix value of 0.32 ± 0.08 at.-%. The corresponding results for arsenic are 2.3 ± 0.1 at.-% at the boundary and 0.11 ± 0.06 at.-% in the matrix. Both arsenic and silicon are known to promote the nucleation of creep cavities in these low-alloy steels.[3,4,7]

Nominal composition $2\frac{1}{4}$Cr–1Mo steel specimens were austenitized at 1423 K for 7.2×10^3 s before furnace cooling to an intermediate temperature of 1223 K where they were held for 1.8×10^3 s before air cooling to room temperature and tempering at 983 K for 5.8×10^4 s to provide a stable distribution of carbides in a bainitic microstructure (Fig. 3). This was followed by classical embrittling heat treatments in the temperature range 793–873 K for times up to 3.6×10^5 s. Point microanalyses were undertaken on thin foil specimens at the γ and lath boundaries, on the ferrite laths and on the γ and lath boundary carbides. Figures 4*a* and *b* show the measured change in composition at the γ boundaries with heat treatment time. Clearly in all cases there is an increase in the concentration of silicon and also manganese which is known to co-segregate with many embrittling impurity elements. However, it is noteworthy that there was no measurable enrichment of the boundary in the embrittling element phosphorus. Similar results were obtained for the lath boundaries, although overall changes in concentration were smaller by at least a factor of four. Microanalysis undertaken at 20 nm intervals along a line normal to the γ boundaries showed no evidence of the enrichment except at the boundary position such that it may be concluded that any concentration profile about the boundary is narrow (<2 nm approximately). Furthermore a de-embrittling heat treatment, typically 923 K for 10^5 s, completely removed the enhanced concentration of silicon and manganese at these γ grain boundaries.

Three different composition carbides have been identified within the initial severely tempered microstructure of this steel. All carbides contained iron, chromium and molybdenum although the relative proportions differ so that, for convenience, we refer to these as molybdenum-rich, chromium-rich and iron-rich. Certain larger carbides were a combination of both a chromium-rich and a molybdenum-rich carbide.

X-ray microanalysis on these three carbide types showed their chemical composition to be $(Fe\text{-}Cr_{14-35}\text{-}Mo_5)_xC_y$, $(Fe\text{-}Mo_{14-40}\text{-}Cr_5)_xC_y$ and $(Fe\text{-}Cr_{5-12})_xC_y$, the latter representing the minor fraction of the total carbide population; the subscripts refer to the weight percentage measured and show a range of compositions since absolute chemical ratios were not established. In addition, the chromium-rich carbides also contained between 1 and 2% manganese which was absent from the molybdenum-rich carbides. The iron-rich carbides, in general, contained 0.5–1% manganese and 1–1.5% molybdenum. There is a small (~5%) but progressive increase in the proportion of molybdenum-rich carbides with embrittling time up to ~3.6×10^5 s at the expense of the chromium-rich carbides.

Following embrittlement heat treatment the molybdenum-rich carbides on γ grain boundaries contained a

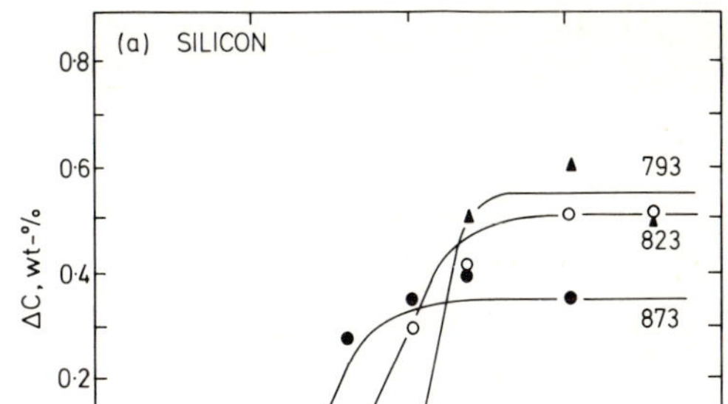

4 Variation of prior austenite grain boundary composition with embrittling heat treatment time for (a) silicon and (b) manganese (ΔC is difference between grain boundary composition and that of adjacent ferrite); (c) variation of number of molybdenum-rich grain boundary carbides containing segregated phosphorus (numbers on plots refer to temperature, K)

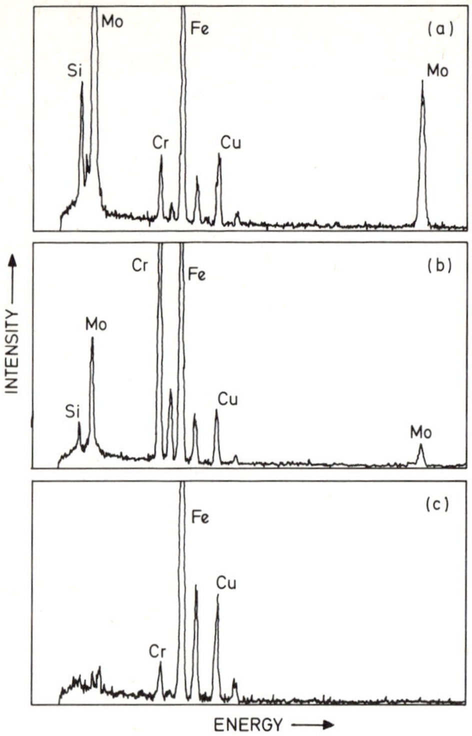

5 **X-ray spectra recorded from various types of extracted prior austenite boundary carbides: (a) molybdenum-rich; (b) chromium-rich; (c) iron-rich**

significant concentration of phosphorus and silicon. The proportion of carbides with associated phosphorus increased with time at the embrittling temperature (Fig. 4*c*) such that in the fully embrittled condition approximately 60% of the total molybdenum-rich carbide population was enriched with phosphorus and silicon. The concentration of phosphorus in individual carbide was up to ~2%. However, the confidence in this absolute value must be low due to the small number of X-ray counts accumulated. Thus, a simple criterion of phosphorus present or absent based on the associated statistical error was adopted. Phosphorus was always associated with carbides containing silicon although the converse was not necessarily the case. It must be emphasized that neither phosphorus nor silicon was associated with either the chromium- or iron-rich carbides. Within statistical error there was no significant change in phosphorus or silicon concentration with position within the carbide precipitate. Also included in Fig. 4*c* are the results of microanalyses for molybdenum-rich γ boundary carbides after a de-embrittling heat treatment at 923 K for 1.08×10^5 s. Clearly there has been a redistribution of both phosphorus and silicon since there is no longer preferential segregation to these carbides. As for the case of the γ boundary to lath boundary segregation the evidence of impurity segregation to lath boundary carbides was diminished.

To examine the distribution of phosphorus and silicon associated with the molybdenum-rich carbides on γ boundaries extraction replicas were prepared from fracture surfaces (77 K) of fully embrittled samples. Again three composition-type carbides were identified (Fig. 5*a–c*) with phosphorus and silicon associated with the molybdenum-rich carbides. Indeed there is excellent agreement between these data and those obtained from the thin foil samples. The similarity in measured compositions for impurity elements from these two types of microanalysis sample implies that segregation is not associated with the matrix side of the carbide/matrix interface but rather with the carbide itself. However, neither procedure provides information regarding the distribution of the segregated species within the molybdenum-rich carbide particles. It has been possible to establish from selected area electron diffraction from carbides extracted from fracture surfaces and the corresponding X-ray microanalysis coupled with the results of X-ray diffraction on electrolytically extracted carbides that the chromium-rich carbides are the $M_{23}C_6$ type, the molybdenum-rich carbides are the M_6C type and the iron-rich carbides are the Fe_3C type.[19]

Non-equilibrium segregations

Nominal $2\frac{1}{4}$Cr–1Mo steel specimens were austenitized at 1423 or 1323 K for 7.2×10^3 s and water-quenched to produce a lath- or plate-like martensitic microstructure as shown in Fig. 6. There was no evidence of precipitation either at γ and lath boundaries or within individual laths. Point microanalyses were undertaken on the thin foils at various positions along γ grain boundaries and at distances of up to 150 nm from the boundary (see Table 1). Generally, spectra recorded at the grain boundary positions showed an increase in measured chromium content relative to the intragranular positions for both quenched conditions. Two points emerge: (a) all segregation is contained within a distance not exceeding ~20 nm from the boundary; (b) the only segregating species is chromium. The scatter in the measured boundary analysis arose as a consequence of errors in positioning the electron probe centrally across the boundary. By assuming segregation profiles of various spatial extents (1, 10 and 20 nm) we evaluated the corrected grain boundary composition from the measured values[17] given in Table 1.

An analysis based on conjugate solute–vacancy migration theory has provided a basis for evaluating the magnitude and spatial extent of solute segregation profiles in such a low-alloy ferrite steel.[13] Figure 7*a* shows the variation of the grain boundary chromium content predicted by this analysis assuming a 10 nm wide profile, a value for the solute-vacancy binding energy of 0.5 eV, a grain size of 100 μm and a quench rate typical of a water quench. Also in Fig. 7*a* we compare this predicted distribution with that

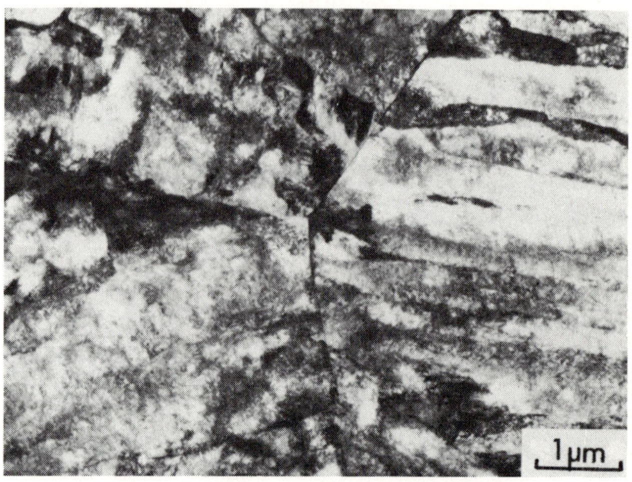

6 **Transmission electron micrograph showing typical martensite microstructure of water-quenched $2\frac{1}{4}$%Cr–1%Mo steel after austenitizing heat treatment at 1423 K**

derived on the basis of the experimental data in Table 1. By combining theoretically calculated segregation data with the time-temperature transformation characteristics of this steel it is possible to map the austenitizing temperature and cooling rate range for non-equilibrium segregation of chromium (Fig. 7*b*). The condition used for the present experimental examination fall within the predicted range for enrichment.

Samples of a commercial 12Cr-Mo-V stainless steel were austenitized at 1373 K for 1.8×10^3 s under vacuum (<1 mPa) argon gas quenched to room temperature and tempered under vacuum at temperatures in the range 823-1023 K for periods of up to 1.8×10^6 s. Figure 8 shows a typical microstructure of tempered martensite with coarse $M_{23}C_6$ type carbide precipitates distributed along the grain boundaries.

X-ray energy-dispersive spectra obtained from γ and interlath boundary $M_{23}C_6$ carbides showed them to be enriched in both chromium and molybdenum together with a high iron content. The measured compositions were not constant but were typically in the ratio 15:1:20 for chromium, molybdenum, and iron respectively. Microanalysis from the rod-like matrix (M_2X) precipitates was not unambiguous because of their small size relative to the spread of the incident electron probe within the foil.[5,17,18] However, they were enriched in both chromium and molybdenum in the ratio of ~8:1 with little or no iron content.

X-ray microanalysis of γ grain boundaries and martensite lath boundaries between the $M_{23}C_6$ carbides on more heavily tempered foils showed a significant depletion in chromium. Repeated analysis showed this depletion to be uniform along a given grain boundary. Analysis of results recorded from positions away from the boundaries showed depletion of chromium across all grain boundaries and those martensite lath boundaries in the more heavily tempered conditions where coarse carbide precipitates were present. A typical chromium concentration profile about a γ grain boundary is shown in Fig. 9*a*. The small number of X-ray

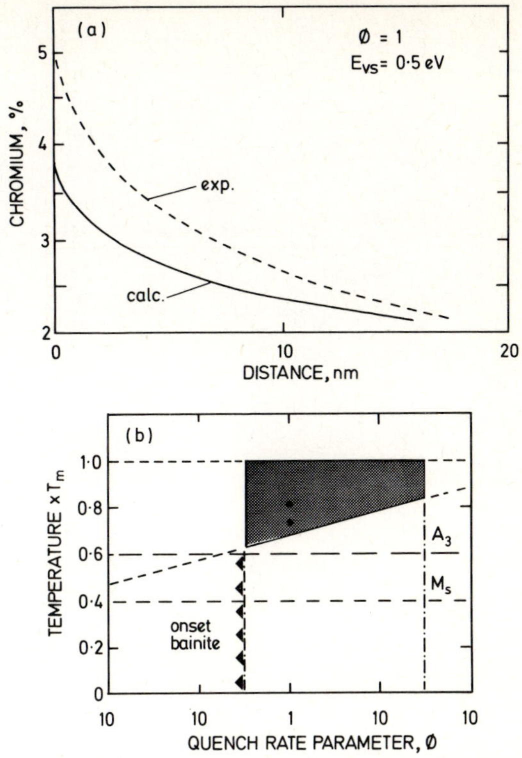

7 (a) Calculated chromium concentration profile ($\phi = 1$; $E_{vs} = 0.5$ eV) compared with that evaluated from experimental measurements (see Table 1). Quench rate parameter ϕ is defined by equation (4) in Ref. 13
(b) Map defining austenitizing temperature and cooling rate range (shaded area) over which non-equilibrium chromium segregation occurs in $2\frac{1}{4}$%Cr–1%Mo steel

TABLE 1 Typical values for microanalysis across prior austenite grain boundaries in water-quenched $2\frac{1}{4}$%Cr–1%Mo steel

Quenching tempera- ture, K	Analysis position: distance from grain boundary, nm	Measured chromium composi- tion, wt-%	Calculated prior γ g.b. Cr composition (wt-%) for assumed profile widths ... 1 nm	10 nm	20 nm
1423	0	3.01	9.2	4.4	3.8
	20	2.20			
	50	2.11			
	100	2.09			
	150	2.21			
1423	0	3.34	11.7	5.2	4.4
	100	2.07			
1423	area analysis	2.07			
1323	0	3.1	9.8	4.6	3.9
	20	2.09			
	50	2.11			
	100	2.20			
	150	2.10			
1323	0	3.21	9.9	4.7	4.1
	100	2.21			
1323	area analysis	2.11			
		2.00			

8 Precipitates of $M_{23}C_6$ tempered 3.6×10^3 s at 873 K at prior austenite grain boundary

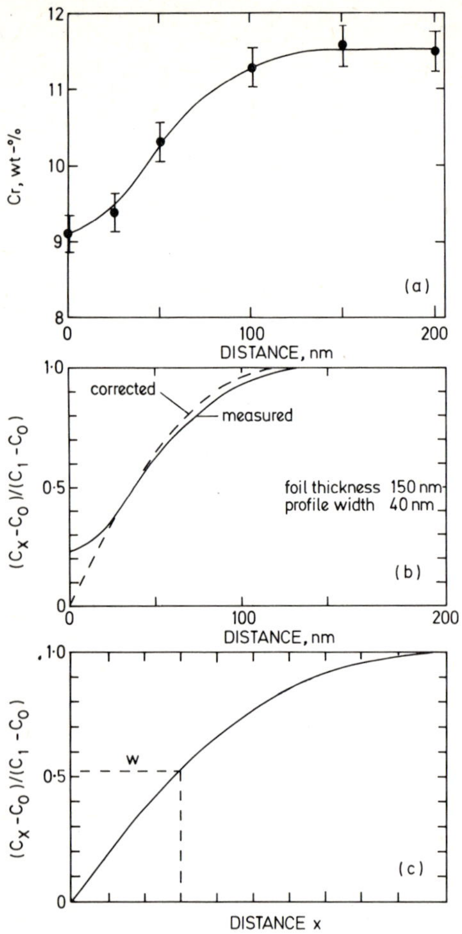

9 Experimental X-ray analysis of chromium concentration profile away from prior austenite grain boundary (position 0) in sample tempered for 3.6×10^3 s at 873 K (b) Schematic: measured and corrected chromium solute concentration profiles measured about prior austenite grain boundary for foil thickness 150 nm. (C_0 = grain boundary concentration, C_1 = bulk matrix concentration, C_x = concentration at distance x from grain boundary) (c) Schematic: construction used to define width of chromium solute concentration profile

counts and thus poor statistics for X-ray analysis for molybdenum did not allow any positive measurement of a molybdenum concentration profile although there was an indication of depletion. Applying the previous analysis[17] for obtaining a corrected solute concentration profile from a meaured profile, we derive the chromium profile for a typical microanalysis (Fig. 9*b*). This procedure was applied to all the measured profiles to derive a value for the grain boundary solute composition C_0 and the extent of the solute profile w as defined schematically in Fig. 9*c*. 'Corrected' compositions on γ boundaries were found to be constant within the overall accuracy of the experimental measurement and correction procedure at 8.4 ± 0.3 wt-% chromium. Profile widths w about the γ boundaries increased with both tempering time and temperature to a maximum value of ~50 nm. For tempering times and temperatures expected to produce values of $w > 50$ nm, no distinct solute profiles were obtained. Here the overall matrix chromium concentrations were less than the bulk value of 11.5 wt-%. These heat treatments coincided with extensive precipitation of martensite lath boundary carbides and coarsening of the M_2X intergranular precipitates. The measured chromium solute depletions at the γ boundaries may be related directly to changes in susceptibility of this martensitic stainless steel to stress corrosion cracking.[14]

CONCLUDING COMMENTS

The experimental examples described above illustrate the application of the STEM-EDS X-ray microanalysis technique to the detection and measurement of composition segregations about grain boundaries. Such composition fluctuations can occur over dimensions much less than the total volume of foil excited by the incident electron probe. However, these may be detected and, given a sufficient number of X-ray counts, quantitatively evaluated. In some of the cases presented, results can only be obtained and their importance evaluated using this technique since the segregation has to be related to the total microstructure with a spatial resolution characteristic of conventional transmission electron microscopy.

The consequences of equilibrium and non-equilibrium segregations to grain boundaries can be far-ranging particularly in the case of commercial steel components that are subject to a variety of heat treatments during manufacture, assembly, repair and operation. Certainly the STEM-EDS X-ray microanalysis technique provides a capability for quantitatively examining such solute distributions.

ACKNOWLEDGMENT

This paper is published with the permission of the Director-General, Central Electricity Generating Board, South Eastern Region.

REFERENCES

1 H. GLEITER and B. CHALMERS: 'Progress in materials science, eds. B. Chalmers, J. W. Christian, and T. B. Massalski, 1962, **16**, Pergamon (London)

2 E. D. HONDROS and M. P. SEAH: *Int. Met. Rev.*, 1977, **22**, 262

3 C. L. BRIANT and S. K. BANERJI: *Int. Met. Rev.*, 1978, **23**, 164

4 D. LONSDALE and P. E. J. FLEWITT: *Mater. Sci. Eng.*, 1979, **41**, 127

5 P. DOIG, D. LONSDALE, and P. E. J. FLEWITT: *Micron*, 1979, **10**, 277

6 P. DOIG and P. E. J. FLEWITT: *Mater. Sci. Eng.*, 1980, **43**, 77

7 M. P. SEAH and E. D. HONDROS: *Proc. R. Soc.*, 1973, **A335**, 191

8 P. DOIG, P. E. J. FLEWITT, and R. K. WILD: *Philos. Mag.*, 1978, **37**, 759

9 K. E. EASTERLING: *Int. Met. Rev.*, 1977, **22**, 1

10 C. J. McMAHON: 'Mechanics and physics of fracture', 1975, London, Metals Soc., 192

11 T. M. WILLIAMS, A. M. STONEHAM, and D. R. HARRIES: *Met. Sci.*, 1976, **10**, 14

12 M. GUTTMAN: 'Residuals, additives and materials properties', 1980, London, Royal Soc., 169

13 P. DOIG and P. E. J. FLEWITT: *Acta Metall.* (in press)

14 P. DOIG, D. J. CHASTELL, and P. E. J. FLEWITT: *Acta Metall.* (to be published)

15 J. I. GOLDSTEIN and D. B. WILLIAMS: 'Scanning microscopy', **1**, publ. SEM Inc., A. M. F. O'Hare, Illinois, 1978, 427

16 G. CLIFF and G. W. LORIMER: *J. Microsc.*, 1975, **103**, 203

17 P. DOIG, D. LONSDALE, and P. E. J. FLEWITT: *Metall. Trans. A* (in press)

18 J. GOLDSTEIN, J. L. COSTLEY, G. W. LORIMER, and S. J. B. REED: 'Scanning electron microscopy', **1**, (IITRI) Chicago, 1977, 315

19 P. DOIG, D. LONSDALE, and P. E. J. FLEWITT: to be published

DISCUSSION of paper by Doig, Lonsdale, and Flewitt ('G.b. segregations')

J. M. TITCHMARSH:
Have you correlated the quantitative results obtained by EDX with results obtained using Auger spectroscopy on samples fractured in-situ in UHV?

P. E. J. FLEWITT:
We have described the results of STEM X-ray microanalysis undertaken on grain boundaries and the correlation with Auger electron spectroscopy (Doig, Flewitt, and Wild: *Philos. Mag. A*, 1978, **37**, 759–768). In this work we considered two model alloys of Fe–3Ni–0.09Sn and a Fe–3Ni–0.06P and compared the measured grain boundary compositions after a full embrittling heat treatment. We were obviously confined to making this comparison in the embrittled condition in view of the requirements of the AES technique to analyse surfaces which have been prepared by intergranular fracture under a high vacuum (5×10^{-8} Pa). The boundary compositions determined by STEM X-ray microanalysis were derived using a simple approximation based on a knowledge of an 'effective' electron probe size. Thus for a foil of ~200 nm thickness we considered a right circular cylinder of diameter ~15 nm. This value for the diameter, although only approximating to a mean diameter of the excited analysing volume within the foil (Doig, Lonsdale, and Flewitt in previous section of these proceedings), was based on the best experimental data available at the time. Indeed it has been shown subsequently to be a realistic approximation.

J. M. TITCHMARSH:
Have you attempted to examine the spatial extent of the segregation by sputtering/Auger spectroscopy measurements?

P. E. J. FLEWITT:
Unfortunately, we were unable to undertake sputtering experiments to examine the spatial extent of these segregation profiles about the grain boundaries using AES. The STEM X-ray microanalysis indicated the profiles to be confined to the boundary layer and would probably extend a distance up to ~2 nm. It should be noted that caution has to be exercised when undertaking sequential sputtering experiments on samples containing restricted profiles since selective removal of a given atom species can introduce an artificially measured profile.

G. CLIFF:
Would the author comment on the procedure used to deconvolute the overlap of MoL_L on PK_{α_1,α_2} in spectra from Mo-rich carbide in $2\frac{1}{2}$Cr–1Mo steels: in pure Mo samples MoL_L interpreted as a PK yield a P content of ~1–3%.

P. E. J. FLEWITT:
We certainly agree that the phosphorus spectrum is difficult to deconvolute quantitatively in view of the close proximity of the two peaks, SiK_α and SK_α/MoL_α together with the superposition of the MoL_L. This problem is further enhanced by the relatively large X-ray 'background' counts accumulated at these low-energy positions. The Link RTS 2FLS quantitative program involves the use of a precalibration using standard specimens to identify the peak profiles for the K, L, K lines of all elements considered. On this basis a curve-fitting procedure is used to establish the correlation between a measured profile and that of the known standard. Thus a measure of confidence can be derived in deconvoluting the MoL_L and the PK_α peaks. However, in

view of the uncertainty associated with deriving quantitative values for phosphorus we have used a simple statistically based criterion to establish either presence or absence of phosphorus in a given carbide particle.

J. M. TITCHMARSH:
I have attempted to detect P within particles rich in Mo and have found considerable difficulty because of the close proximity of the PK and MoL_L peaks. One gets an apparent PK peak from a foil of pure Mo. The presence of the MoL_L peak must be carefully stripped to obtain accurate PK results.

P. E. J. FLEWITT:
We would accept the comments concerning the difficulties associated with the measurement of phosphorus on the Mo-enriched carbides but we did consider difference values between embrittled samples and those in the unembrittled and de-embrittled condition.

P. J. STATHAM:
With reference to the discussion on the PK_α/MoL_L interference, I would like to point out that separation *is* possible if the difference in shape between L and K series profiles is taken into account in a full least-squares analysis and provided particular care is taken to ensure accurate spectrum calibration. Statistical errors are obviously enhanced by the severe overlap but can be calculated and such an approach is used in the spectrum-processing program quoted by the authors.

R. G. FAULKNER:
What type of nonequilibrium segregation model was used to theoretically model the process? What binding energy and constants were used in this model?

P. E. J. FLEWITT:
The nonequilibrium segregation of solute (in this case chromium) to prior austenite grain boundaries of a $2\frac{1}{4}$Cr–1Mo steel is based on diffusion of vacancy–solute pairs to the boundary sink. The model considers the excess vacancy concentration generated under Newtonian cooling as providing the necessary driving force for segregation (P. Doig and P. E. J. Flewitt, *Acta Metall.*, in press). The 'degree' of solute segregation critically depends upon the solute–vacancy binding energy which is a parameter difficult to evaluate. We in fact derived a value based on an empirical relationship for fcc metals such that for chromium in γ-Fe this would be 0.36 eV. However, the model predicted the best correlation, for the quench rate parameter and grain size considered, with the magnitude and spatial extent of the profile for a chromium–vacancy binding energy of 0.5 eV.

R. G. FAULKNER:
The choice of an ideal boundary width to predict the actual concentration on the boundary may not be applicable for nonequilibrium segregation. The shape of the predicted profiles is much sharper for equilibrium sizes than for nonequilibrium segregation and so the actual microanalysis curves may be showing the rich effect much more efficiently for the nonequilibrium segregation results.

P. E. J. FLEWITT:
We would accept that there must always be some doubt regarding the deconvolution of a true concentration profile from a measured profile when the procedure is based on a profile of a given functional form $(C = C_1 \exp - |x/\alpha|)$ [Doig, Lonsdale, and Flewitt: *Metall. Trans. A*, May 1981]. However, a measure of confidence in this approximation is provided by the similarity with the form of the profile predicted by the theoretical analysis as given by the comparison in Fig. 7*a* of our paper [*page 69*].

Transition metal carbides in low-alloy steels

T N Baker and S P Duckworth

The contribution made to the strength of structural steels by transition metal carbides, nitrides and carbonitrides has recently been calculated using the Hall–Petch equation, and several attempts have been made to measure the volume fraction of the particles by electron microscopy; correlation with calculations based on the stoichiometric composition of these particles usually shows an under-estimated value. The present work has been carried out to consider possible deviations in composition from the stoichiometric ratio approach employed in earlier studies. Precipitation hardening by the carbides of niobium, titanium and vanadium has been observed, and the chemical composition of particles extracted on carbon replicas has been examined using a Philips EM400T electron microscope fitted with an EDAX 9100 energy-dispersive system. Preliminary results show that deviations from the stoichiometric composition of the carbides exist and that, in many particles continuously cooled or rolled through the γ–α phase transformation, considerable quantities of iron are present. In tempered alloy steels, only traces of iron have been found in the particles. Work is in progress to find if there is any correlation between particle size and composition or any variation in composition between matrix precipitates and those at grain boundaries; and to consider the significance of deviations from stoichiometry on particle volume fraction and the resultant precipitation hardening component of the yield strength in low-alloy steels.

The authors are in the Department of Metallurgy, University of Strathclyde, Glasgow.

Hobstetter[1] was among the first to appreciate that the initial precipitation occurring by classical nucleation and growth need not have the equilibrium composition. He also pointed out that this may be particularly true for coherent precipitates.

Confirmation of a variable nucleus composition has been possible using the FIM atom probe. Goodman, Brenner, and Low[2] have studied an iron–1.4 at.% copper alloy and chemically analysed precipitates under 1.0 nm diameter. They showed that the precipitates contained less than 50% copper, and concluded that the equilibrium composition of the precipitates was only reached in the overaged state.

Transition metal carbides, nitrides and carbonitrides have been used as a means of strengthening steels for many years. Vanadium, titanium and niobium have all been added singly or in combination to steels, which have been processed by quenching and tempering, or by hot rolling to a wide variety of shapes and sizes. Only infrequently do the cooling rates, during or following the processing, allow equilibrium conditions to be established.

Precipitation hardening by small particles of carbides and nitrides of V, Ti and Nb is well established. Comparison of measured and calculated strengthening of steels by precipitation hardening has been reported.[12] Discrepancies are usually attributed to the difficulties in obtaining reliable particle volume fraction data. Many studies have overcome this problem by calculating the particle volume fraction assuming an equilibrium stoichiometric relationship between the transition metal and carbon and/or nitrogen. However, deviations from stoichiometry during the early stages of precipitation have been reported by Goodman *et al.*[2]

Atom-probe field ion microscopy on a low-alloy steel containing both vanadium and titanium[3] has shown that the MC carbides which precipitate during direct transformation have a mixed composition of the type $(V, Ti)C_{0.84}$ after long term aging. After short isothermal aging times of 5 to 10 min, however, iron was detected in 6 nm diameter particles, but only in small quantities, i.e. <10%. Analysis carried out following 27 h at temperature indicated no significant levels of iron in 40 nm diameter particles.

An alternative method for obtaining the chemical analysis of small particles is to use a transmission electron microscope capable of producing a small probe diameter coupled with an X-ray energy-dispersive analyser. STEM units can give a focused spot diameter down to 2 nm and a twin objective lens system such as that on the Philips EM400T used in the nanoprobe mode will produce a 4 nm spot diameter. Increasingly this approach is being used for particle analysis and promises to be widespread during the next few years for obtaining information about elements with atomic number $Z = 11$ and above.

Recently, Baker,[9] using STEM with an EDA system, has found considerable quantities of iron in both fibrous and interphase precipitates present in 0.024C–0.10V–0.012N iron alloys after air cooling following hot rolling. Similar results have been reported in niobium steels by Hirsch and Parker,[10] who also note that the Fe/Nb ratio decreased with increasing particle size.

The present study was undertaken to substantiate the observation that deviations from stoichiometry through the replacement of vanadium, titanium or niobium by iron occurred in alloys over a wide range of composition and heat treatment, and to examine the variation in the Fe/M ratio with particle size.

EXPERIMENTAL METHOD

The compositional analysis and thermomechanical treatment of the vacuum-melted and cast steels studied in the work are given in Table 1.

TABLE 1 Chemical composition (wt-%) and thermochemical treatment of steels

Code	Cast No.	C	Mn	Si	S	P	V	Nb	Ti	O	N	Treatment
V1	K3006	0.12	–	<0.035	<0.01	<0.005	0.48	–	–	0.031	0.0018	Solution-treated, aged, WQ
V2	K3007	0.13	1.50	<0.035	<0.01	<0.005	0.49	–	–	0.029	0.0019	Solution-treated, aged, WQ
V3	V26	0.024	0.015	0.007	0.004	0.009	0.095	–	–	0.008	0.012	Solution-treated, hot-rolled
V4	V32	0.007	1.0	0.020	0.004	0.008	0.10	–	–	0.020	0.011	Solution-treated, hot-rolled
T1	T08	0.11	1.26	0.05	0.005	0.005	–	–	0.53	0.032	0.0024	Solution-treated, WQ, tempered, WQ
T2	T08	0.11	1.26	0.05	0.005	0.005	–	–	0.53	0.032	0.0024	Solution-treated, WQ, hot-rolled
T3	T11	0.11	1.3	0.015	0.005	0.005	0.11	–	0.36	0.057	0.003	Solution-treated, WQ, hot-rolled
N1	S2945	0.041	1.61	0.025	0.008	0.004	<0.01	0.065	–	–	0.006	Solution-treated, hot-rolled
N2	K3005	0.02	<0.01	–	<0.01	<0.005	–	0.14	–	–	–	Solution-treated, aged, WQ

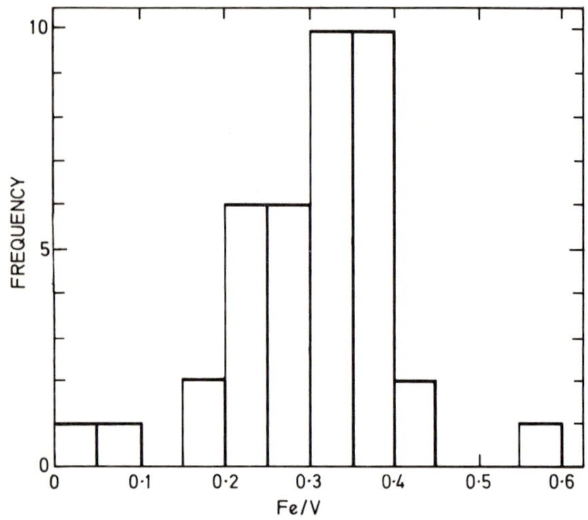

1 **Histogram of Fe/V ratio vs. number of particles for steel V1 aged at 873 K for 85 s**

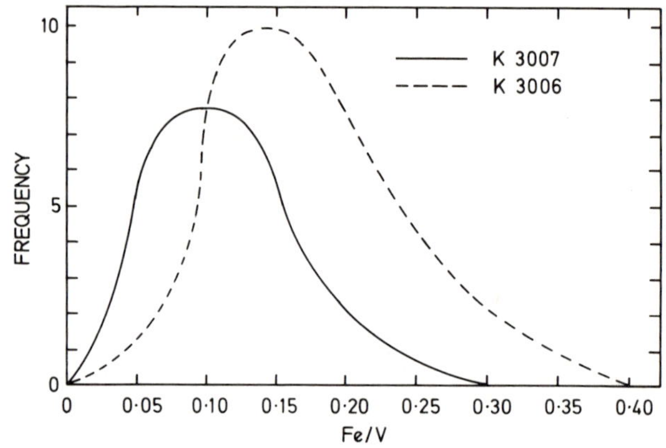

3 **Schematic: influence of manganese on Fe/V ratio of particles containing iron in two vanadium steels V1 (K3006 – nil Mn) and V2 (K3007 – 1 wt-%Mn) aged at 955 K for 72 h**

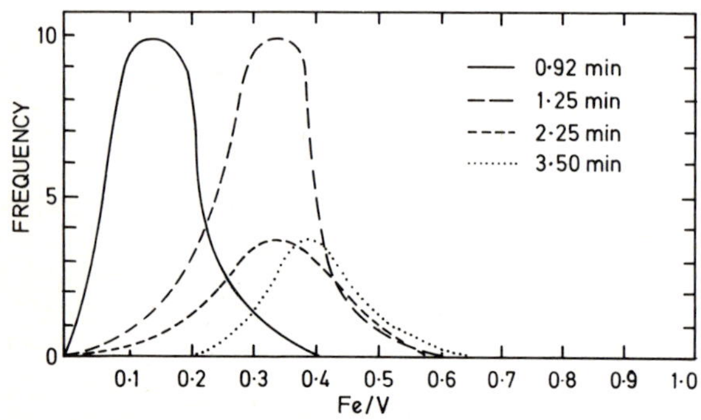

2 **Schematic: Fe/V ratio vs. number of particles for steel V1 aged at 873 K for various times**

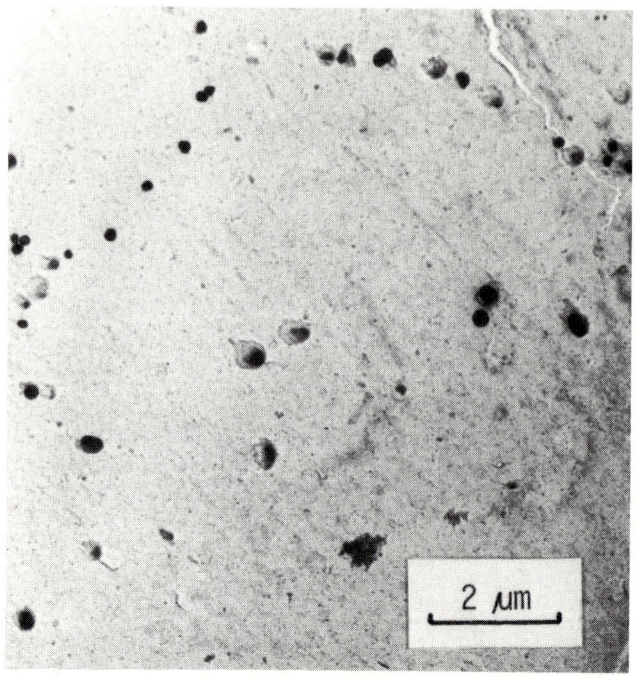

4 **Carbon extraction replica taken from steel V3 solution-treated 1473 K, aged 923 K for 60 min; particles outlining grain boundary**

Foils and carbon extraction replicas were prepared by standard techniques. Great care was taken in the washing of the replicas and in the preliminary examination to ensure reproducibility in the efficiency of particle extraction.

Electron microscopy and analysis of the carbon replicas was carried out on a Philips EM400T instrument fitted with a STEM unit and with an EDAX 9100 energy-dispersive analyser. A single-tilt low atomic number (Be) insert specimen holder was used for the entire analysis, which was carried out at 100 kV using a 20° take-off angle. With the exception of the smallest particles (<4 nm), the probe diameter was chosen to be approximately half the particle diameter. Preset counting times were 50, 60, or 100 s live time. An EDAX semi-quantitative analysis programme which included ZAF corrections was used. The programme incorporated escape peak removal and background subtraction.

RESULTS

Steels V1 and V2

The chemical analysis was obtained for a series of specimens which had been solution-treated at 1473 K, directly quenched to 873 K and isothermally aged for times between 55 s and 35 min followed by water quenching. Figure 1 is a histogram of a typical frequency vs. Fe/V ratio. Between 20 and 50 analyses were recorded for each specimen, and the results are summarized schematically in Fig. 2. The Fe/V ratio (*R*) increased from 0.1 after 55 s aging to 0.3 after 230 s. Both nanoprobe and STEM modes were used and the reproducibility in *R* for successive 60 s counts on the same particle was ±5%.

A second experiment studied the influence of manganese on *R* at long aging times for steels V1 and V2 which had been solution-treated at 1473 K, aged at 955 K for 72 h and quenched. Twenty or more particles were analysed for each steel. The particles in the manganese steel were found to contain 5–10% iron compared with 10–20% iron in the steel V1 particles (Fig. 3) but no manganese peaks were detected in the spectra from particles in V2 steel.

The trend of compositional variation with particle size was not evident in either of the above experiments.

Steel V3

The composition of this steel is such that the calculated volume fraction of vanadium carbide or carbonitride (based on stoichiometry) is about a quarter of that expected in steels V1 and V2.

The samples were again studied after solution treating (1373 K) for 1 h and directly isothermally aged for 1 h at temperatures in the range 973–873 K. The preponderance of grain boundary nucleation is seen in Fig. 4 for the 973 K aging. Large particles up to 0.5 μm were present both in the matrix and on the grain boundaries. Several groups of particles from both regions were analysed and showed the same trend; a remarkable degree of consistency within the group but a wide variation from group to group (Table 2). However, groups of particles having *R* values in excess of 2, the highest recorded during analysis of any particles in the present series of experiments, were found in this steel. Typical spectra are given in Figs. 5 and 6.

The size range of the particles which were analysed was 100–200 nm, with an average diameter of 170 nm (Fig. 7). Analysis of the matrix for a counting time of 50 s gave, as expected, very much lower counts than particle analysis. Typical values were

	Total counts	Total counts of vanadium	Total counts of iron
Matrix	3 750	23	470
Particle	20 000	5800	4300

Steel V4

This steel contained a very low carbon content of 0.007% and about the same level of nitrogen (0.011%) with a 1.0% manganese addition. The results are given in Table 3 for the analysis of ten particles distributed randomly in the matrix. A significant iron content was found and half the particles contained manganese with an Fe/Mn ratio between 3 and 4. Although the Fe/V ratio was wide, 50% of the particles had *R* values within ±0.08 of the average value.

Steel T1

The Ti/C ratio in steel T1 was arranged to give the equilibrium stoichiometric ratio for titanium carbide. However, due to the large [Ti] [C] solubility product, the solution temperature of 1573 K took only about 70% of the titanium carbide into solution. The undissolved precipitates were present as large spherical particles of 0.5–2.0 μm diameter (Fig. 8). These particles were studied in foils using a million-volt electron microscope, AEI EM7, coupled to a LINK EDA. Figure 9 shows a typical spectrum. Despite the iron matrix, this element is present in only small quantities, from the particle analysis on specimens solution-treated, water-quenched and then tempered at 823 K.

Steel T2

Steel T2 has the same bulk chemical analysis as T1 but was hot-rolled to 1123 K following the 1573 K solution treatment. A very fine particle dispersion (4–8 nm) was

TABLE 2 Variation of Fe/V ratio for steel V3

Tempering temperature, K	Group	No. of particles	Nucleation site	Highest *R* value	Lowest *R* value	Average *R* value	Total average counts (approx.)
973	1	6	Matrix	0.92	0.79	0.85	90 000
	2	8	Matrix	2.42	1.81	2.21	–
	3	7	Grain boundary	1.5	1.25	1.32	150 000
923	1	15	Grain boundary	0.82	0.62	0.71	40 000
	2	7	Grain boundary	0.85	0.61	0.65	30 000
873	1	6	Grain boundary	1.05	0.92	0.99	20 000
	2	6	Grain boundary	1.16	0.92	1.00	20 000

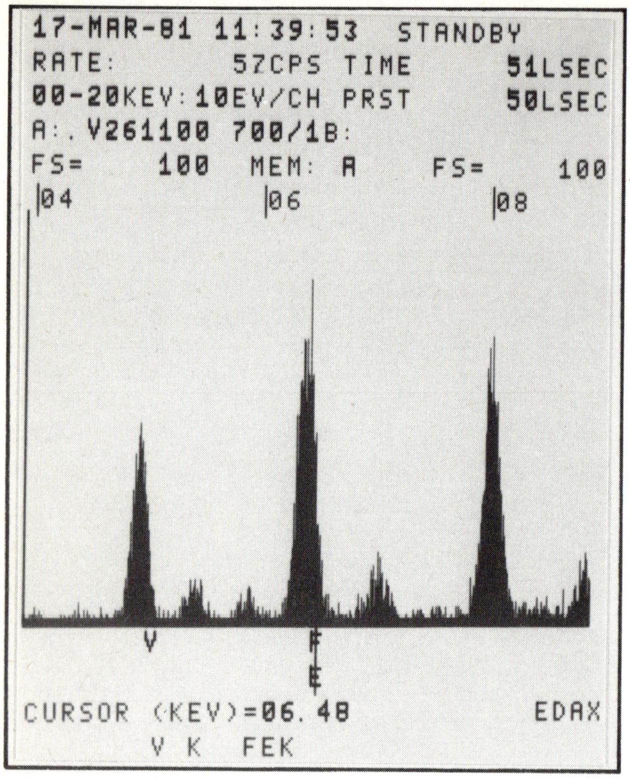

5 Spectrum from particle in steel V3 showing R value of >1

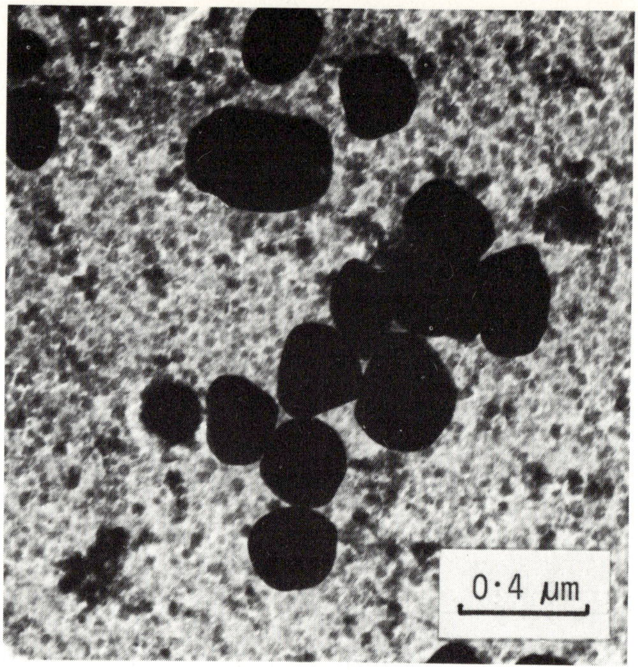

7 Carbon extraction replica of steel V3 as for Fig. 4; matrix particles

TABLE 3 Variation of Fe/V ratio for steel V4

No. of particles	Highest R value	Lowest R value	Average R value	Total average counts (approx.)
10	1.34	0.20	0.71	7500

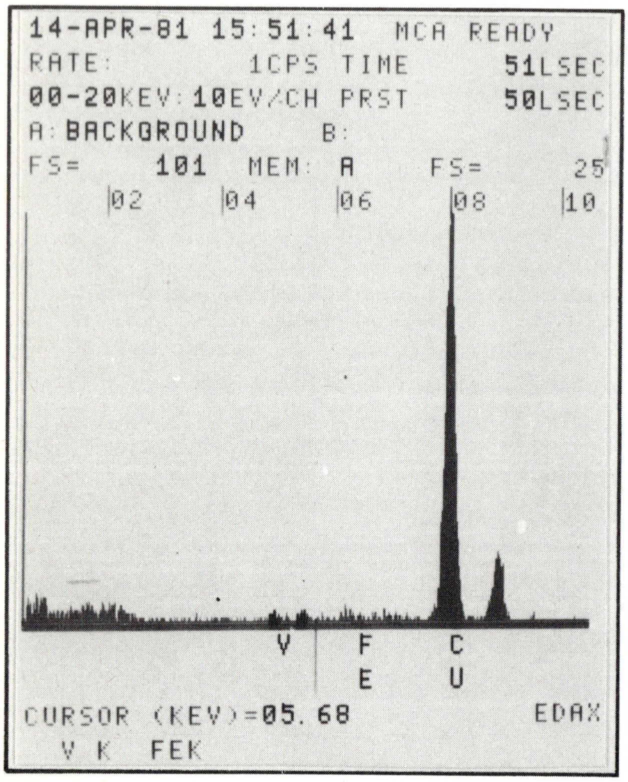

6 Spectrum giving background count for steel V3

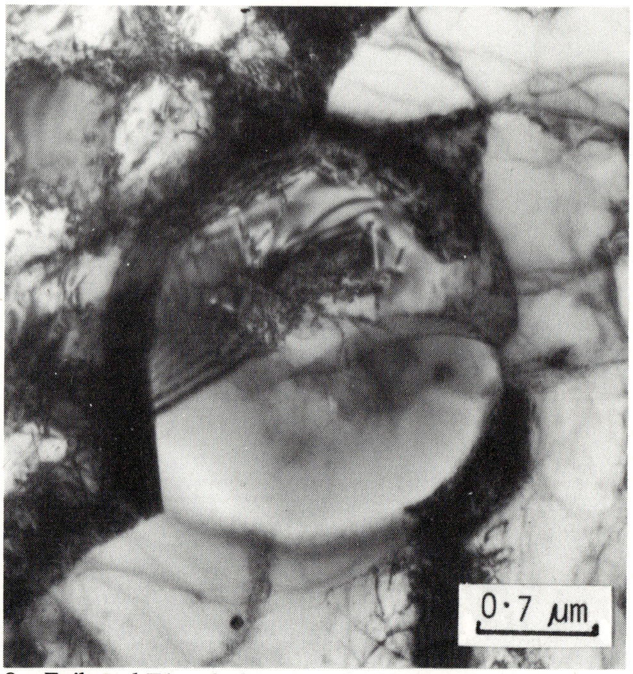

8 Foil steel T1, solution-treated 1573 K, quenched and tempered at 823 K for 100 h; undissolved titanium carbide particle

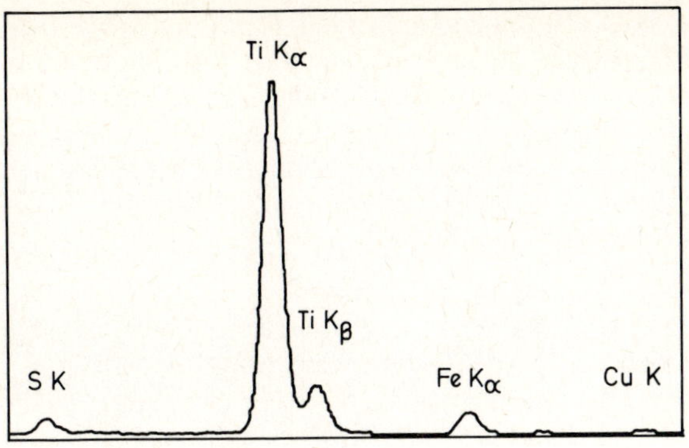

9 Spectrum from undissolved TiC particle showing small FeK_α peak

produced and a probe diameter of 4 nm used in the analysis. Fe/Ti ratios varied over a wide range of 0.3 to 1.3 (Figs. 10 and 11).

Steel T3

A Ti/V ratio of 3 was aimed at in steel T3. As with other steels, a wide Fe/Ti variation was found.

Steels N1 and N2

Particles in the range 2–15 nm were obtained in both these steels and found to contain iron. The fine particle size is shown in Fig. 12.

DISCUSSION

The aim of the present series of experiments was to examine a wide compositional variation of low-alloy steels by analysing particles present in material which had been quenched and tempered, solution-treated and directly isothermally aged, and controlled hot-rolled following solution treatment.

Previous work[9] had suggested that deviations from stoichiometry in the chemical composition of transition metal carbide and carbonitride particles could occur. This usually took the form of iron being present in the particles in addition to the expected transition metal. Further, a relationship between the iron/transition-metal ratio and the particle size was noted in low-carbon niobium steels,[10] and this also was monitored in the present study.

In the vanadium steels iron was present in almost every particle analysed, regardless of heat treatment. On the other hand, in steels containing manganese additions particles containing this element were only occasionally found; silicon was more frequently recorded.

The ideas of Hobstetter[1] on deviations from the equilibrium concentrations which were postulated for the initial stages of precipitation, particularly when the particles were coherent, also considered that with time the equilibrium stoichiometric composition would be achieved. The results for steel V1, which showed an increase in the fraction of iron present with increasing aging time, suggest that a (V, Fe) carbide may be the equilibrium composition, not vanadium carbide as previously considered.

The addition of manganese to steel is known to lower the transformation temperature and there is some evidence that the volume fraction of transition metal carbides is reduced compared with vanadium-free steels. Recent work explains the observation in terms of the increased solubility of the carbides in manganese-containing steels.[11] Further, this work has recorded a reduced tendency for iron in the carbides of the 1.5% manganese steel V2.

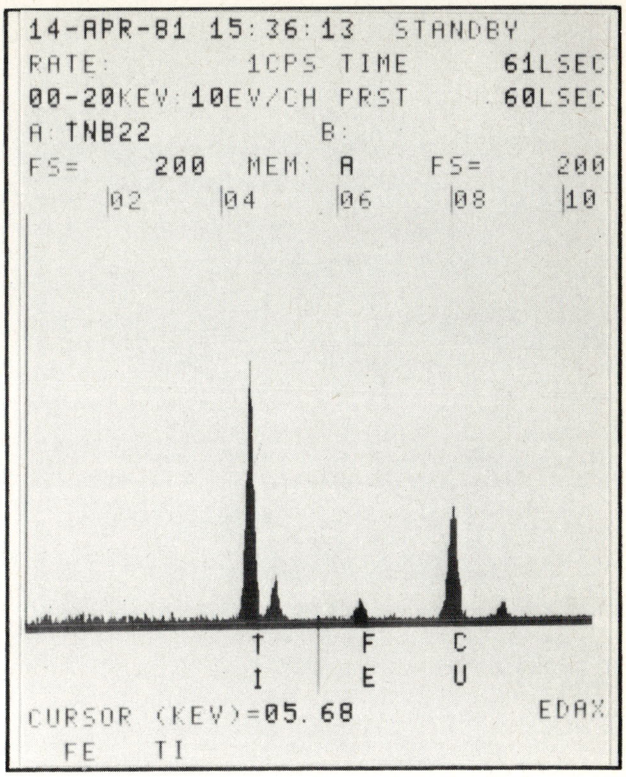

10 Spectrum from particle in steel T2 showing low Fe/Ti ratio

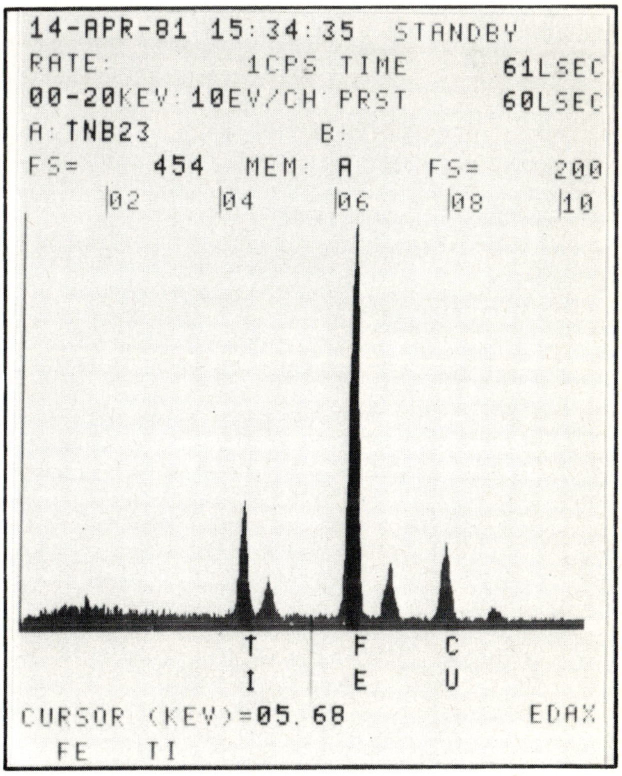

11 Spectrum from particle in steel T2 with Fe/Ti ratio >2

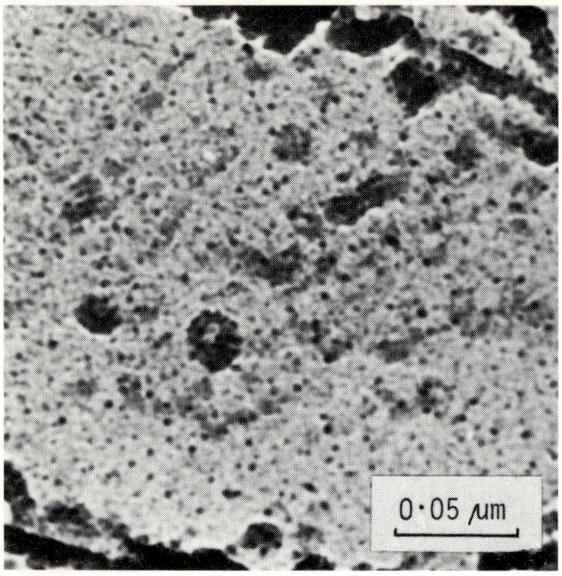

12 **Carbon extraction replica of steel N1 showing very small particle size from which spectra were obtained**

With the titanium steels, apart from the large undissolved particles examined in steel T1, large concentrations of iron were present in many particles. The titanium steels studied to date therefore appear to differ from the vanadium steels where the presence of manganese was accompanied by a lower iron level than that in manganese-free vanadium steels.

Until recently, transition metal carbides and carbonitrides have been identified in steels by comparing lattice parameters obtained from selected area electron diffraction patterns with ASTM index standards usually based on X-ray powder diffraction patterns. The limitation of ~1% on the accuracy of lattice parameter measurements from SAED together with variations arising from solubility ranges, particularly when both carbon and nitrogen were present, has led to an acceptance that the measured lattice parameter could only identify particles as MC or MCN. Many transition metal carbides and nitrides are mutually soluble and the lattice parameters of pure carbide and pure nitride are too close to allow a unique identification by electron diffraction.

The addition of iron to the NaCl lattice to give an $(Fe, M)_xC_yN_z$ compound could result in a significantly larger particle volume fraction than that calculated from bulk chemical composition assuming an M:C stoichiometric ratio. Further, where iron is present in considerable quantities, calculations of precipitation hardening contributions to the yield stress could result in erroneous conclusions.

CONCLUSIONS

1 Iron was found to be an important element in almost all the particles analysed in the steels containing the transition elements niobium, titanium and vanadium, which were studied in the solution-treated and directly isothermally aged condition after controlled hot-rolling and in the quenched and tempered state.

2 The concentration of iron was found to increase with increasing isothermal aging times in a 0.10%C-0.46%V steel.

3 The addition of 1.0% manganese to the 0.10%C-0.46%V steel was associated with a decrease in the level of iron recorded in the carbide particles. No significant levels of manganese were recorded in these particles.

4 Manganese had less influence on the Fe/Ti ratio in 0.11%C-0.53%Ti-1.26%Mn steels. A significant level of iron was found in most particles.

5 A similar trend was noted in the niobium steels studied in this work.

6 The relationship between particle size and Fe/M ratio was not substantiated in the experiments completed to date.

REFERENCES

1 J. N. HOBSTETTER: *Trans. AIME,* 1949, **180**, 121–130
2 S. R. GOODMAN, S. S. BRENNER, and J. R. LOW Jr: *Metall. Trans.*, 1973, **4**, 2371–2378
3 G. L. DUNLOP and P. J. TURNER: *Met. Sci.*, 1975, **9**, 370–374
4 R. J. TUNNEY, G. W. LORIMER, and N. RIDLEY: *Metal Sci.*, 1978, **12**, 271–276
5 J. M. TITCHMARSH: AERE Report R9661, Dec. 1979
6 P. DOIG, D. LONSDALE, and P. E. J. FLEWITT: *Micron,* 1979, **10**, 277–281
7 R. G. MENZIES, R. H. BRICKNELL, and A. J. CRAVEN: *Philos. Mag. A*, 1980, **41**, 493–508
8 S. LONGAVER, A. NORDGREN, B. BENGTSSON, and B. LOBERG: *Electron Microscopy*, 1980, **3**, 42–43
9 T. N. BAKER: *Electron Microscopy,* 1980, **3**, 44–45
10 Y. C. HIRSCH and B. A. PARKER: *Micron,* 1980, **11**, 40
11 M. G. AKBEN, I. WEISS, and J. J. JONAS: *Acta Metall.*, 1981, **29**, 111–121
12 A. J. LAPOINTE: 1975, PhD thesis, University of Strathclyde

DISCUSSION of paper by Baker and Duckworth

R. C. COCHRANE:
With reference to the figure showing the change in iron content as a function of aging time, did you observe any changes in crystal structure as a function of aging?

T. N. BAKER:
As we have only recently commenced this study, the work has been concentrated on the microanalysis aspect, and as yet we have not carried out microdiffraction in a systematic manner.

D. B. WILLIAMS:
The magnitude of the Cu peaks in the X-ray spectrum obtained with the primary beam on the carbon support film is strong evidence for a high flux of *bremsstrahlung*

radiation from the EM400 illumination system. This is indeed characteristic of an 'as-received' EM400, and is probably unimportant in the study of small extracted particles on a thin C film. However, I would suggest that if you ever wish to perform quantitative microanalysis from self-supporting disc specimens you ensure that the manufacturers clean up your spurious X-ray problem. For example, does your instrument have extra-thick (~1 mm) Pt apertures in the condenser system?

T. N. BAKER:
Thank you for your comments. This aperture problem will certainly be taken care of in the immediate future.

P. R. HOWELL:
How well do your results correlate with those obtained using the atom probe field-ion microscope?

T. N. BAKER:
As far as I am aware atom probe work on particles similar to those in our work has been confined to a single investigation reported by Dunlop and Turner.* They detected iron after very short aging times, but this element was not found in specimens aged for several hours. However, their compositions and heat treatment were different from those in the present work, and the number of particles analysed was very restricted. We have carried out over a thousand particle analyses to date.

A. J. GARRATT-REED:
The work at MIT has been on strained specimens aged at much higher temperatures than those Dr Baker has reported. Nevertheless we agree that iron is always present in the precipitates we observe (both intergranular and on the prior austenite boundaries) but at low concentrations. We are looking forward to obtaining atom probe data from these precipitates to help clarify exactly what is happening, for it has been suggested that the iron can be a surface layer left after the replication.

T. N. BAKER:
We are very conscious of the possibility of the iron being associated with surface contamination and have taken precautions to alleviate this problem. Further, the very low iron and vanadium counts recorded from the matrix adjacent to the particles does give us more confidence in that the presence of iron in significant quantities is a genuine observation. Thirdly, the reproducibility of the cps from particles within a given group which we observe would be very unlikely if the iron was entirely associated with a surface contaminated layer. Finally, it is worthwhile mentioning that iron has already been detected in particles similar to these we have studied – in the atom probe by Dunlop and Turner, and in Fe-Cu alloys by Goodman, Brenner, and Low.†

D. A. PORTER:
A comment regarding the high iron concentrations you have measured: there is a danger that very thin surface (oxide) layers with a composition different from that of the extracted precipitate can drastically affect the apparent composition. You could therefore have an iron-rich layer around the precipitates.

T. N. BAKER:
This is certainly an aspect which should be considered. However, the regular variation in Fe/V ratio in the steels V1 and V2 would not be expected if the iron is present only in an oxide layer around the particle. One would expect the iron content in an oxide layer to be independent of the aging treatment and dependent only on the specimen preparation technique. Further, it is very difficult to accept that, in the cases where the iron peak was twice that of the titanium peak (a result which was obtained from more than twenty particles in steel T3), the iron counts originate only from a surface oxide film. It may well be the case that some iron could be associated with an oxide film, if indeed this is present, and this source may account for the very low iron levels recorded at very long aging times.

* G. L. DUNLOP and P. J. TURNER: *Met. Sci.*, 1975, **9**, 370–374

† S. R. GOODMAN, S. S. BRENNER, and J. R. LOW Jr: *Metall. Trans.*, 1973, **4**, 2371–2378

Analytical electron microscopy of partitioning in pearlite

N Ridley and G W Lorimer

Some of the practical problems which are encountered during the study of the partitioning of alloying elements such as manganese, chromium, and silicon during pearlite growth using analytical electron microscopy are considered. The problems which must be overcome include limitations in spatial resolution associated with the fineness of lamellar pearlite, absorption and fluorescence effects within the specimen and the presence of solute-rich surface films. Once identified all of these problems can be overcome and quantitative partitioning data obtained.

The authors are in the Department of Metallurgy, Joint University of Manchester/UMIST, Manchester, UK.

The technique of analytical electron microscopy has been used by the authors to study the partitioning of various alloying additions during the austenite-pearlite transformation in eutectoid steels, and the partitioning data obtained to date have been discussed in relation to pearlite growth kinetics.[1–3] In the present paper some of the practical problems which are encountered during the analysis of the pearlite microstructures will be discussed.

PEARLITE REACTION IN STEELS

To provide a background some of the more important features of the pearlite reaction in steels are outlined.

Pearlite is a lamellar transformation product, consisting of ferrite (α) and cementite (Fe_3C), which forms when austenite (γ) undergoes eutectoidal decomposition (Fig. 1). Pearlite nucleates heterogeneously, primarily at austenite grain boundaries, and growth occurs on a spherical front (Fig. 2). The important growth parameters are V, the radial growth velocity, and λ, the interlamellar spacing (Fig. 3*a*). Phase relationships for binary Fe-C alloys are shown in the part of the Fe–Fe_3C phase diagram relating to pearlite formation (Fig. 3*b*). It can be seen from this diagram that carbon has a low solubility in the α phase (~0.02 wt-% maximum), whereas cementite contains a substantial amount of carbon (~6.67 wt-%). Hence, during the transformation of austenite to pearlite there must be a considerable redistribution, or partitioning, of carbon at or near the advancing pearlite/austenite transformation front. If it is assumed that local equilibrium exists during transformation to pearlite, then the important carbon concentration terms are those shown in Figs. 3*a* and *b*. In practice the terms $C^{\gamma/\alpha}$ and $C^{\gamma/c}$ will be modified by interface local curvature.

The pearlite interlamellar spacing is dependent on transformation temperature, although for a given temperature the spacing does show a distribution of values about a mean true value. It has also been shown that the mean value is equal to 1.65 times the minimum spacing measured.[4] With decreasing transformation temperature the mean interlamellar spacing decreases and several studies have confirmed the earlier prediction of Zener[5] that there is a linear relationship between the reciprocal of mean interlamellar spacing and the transformation temperature (or the undercooling) – see Fig. 4. For a steel of eutectoid composition (0.8 wt-% C) it can be readily shown that the cementite lamellae have a thickness which is approximately equal to one-eighth of the interlamellar spacing.

In Fe–C alloys pearlite growth proceeds relatively rapidly. However, the growth rate can be markedly retarded by the presence of relatively small concentrations (up to 4%) of alloying additions such as manganese, nickel, chromium and molybdenum, and this effect is of great commercial significance in relation to the heat treatment of steels. Investigations by the authors are concerned with examining the ways in which the alloying additions partition at the austenite/pearlite transformation front, as a first step to understanding the mechanisms whereby the additions retard pearlite growth.

Work to date has shown that there are two different types of partitioning behaviour:

(1) The alloying addition partitions preferentially to either pearlitic cementite (Cr, Mo) or pearlitic ferrite (Si) at the transformation front over the full temperature range of pearlite formation, although the extent of partitioning decreases progressively as the temperature falls.
(2) The alloying addition, e.g. Mn, partitions preferentially at the transformation front at high reaction temperatures but the extent of partitioning decreases until a temperature can be identified below which there is no partitioning of the alloying additions at the front.

Clearly, different rate-controlling processes are likely to be involved for circumstances where substitutional elements, which diffuse relatively slowly, are partitioning at the reaction front, compared with the case where only carbon is partitioning. Nevertheless, in the latter case the alloying additions influence the rate of pearlite growth. All alloying additions partition, or continue to partition, behind the transformation front, although this will not influence the rate at which the interface moves.

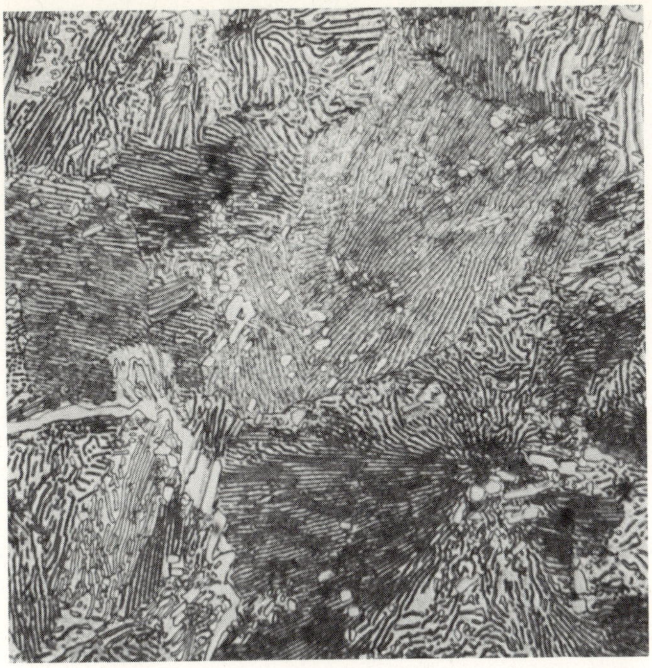

1 Lamellar pearlite in plain carbon eutectoid steel (× 345)

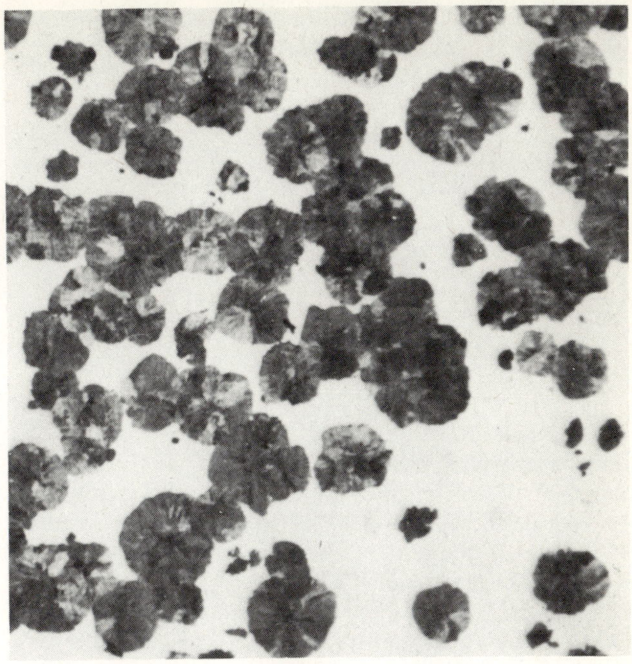

2 Heterogeneous nucleation of pearlite nodules in 0.34% Mo eutectoid steel transformed at 590°C for 8 s (× 140)

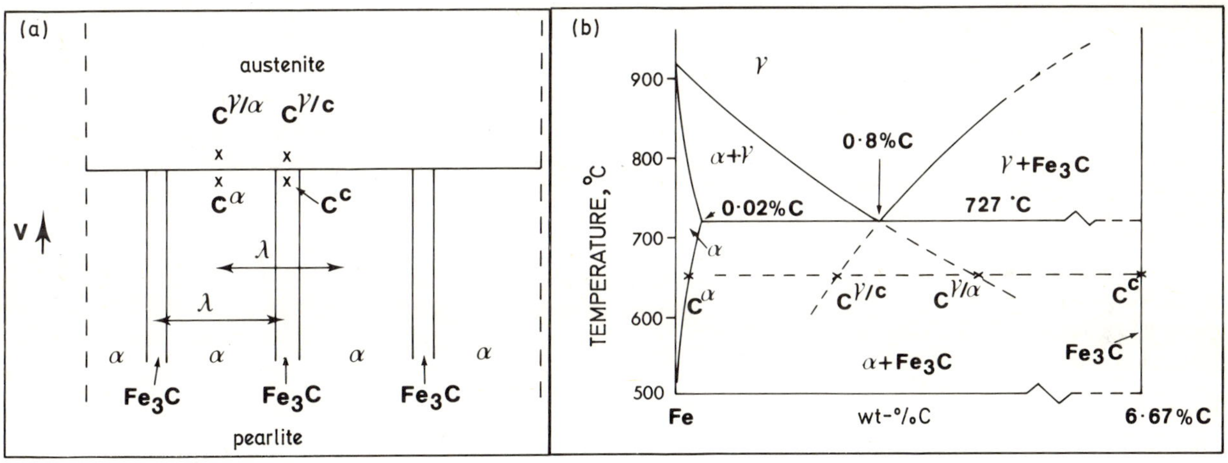

3 (a) Schematic: pearlite growth front showing important growth parameters and carbon concentration terms; (b) part of Fe–Fe_3C phase diagram relating to pearlite formation

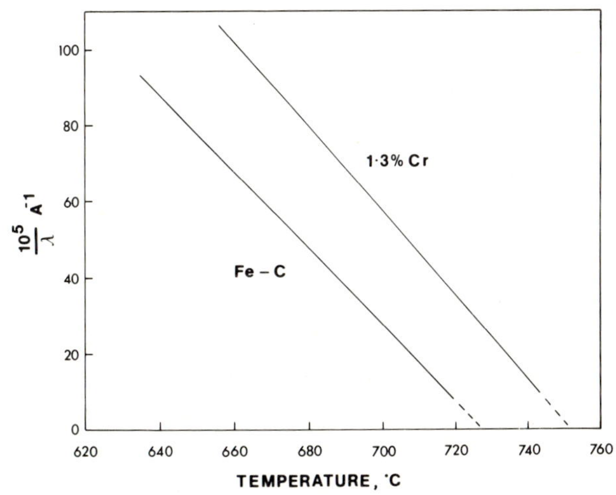

4 Reciprocal of mean interlamellar spacing vs. transformation temperature

TABLE 1 Beam broadening b as function of thickness for cementite ($\rho = 7.7$ g cm^{-3}) and ferrite ($\rho = 7.8$ g cm^{-3})

t, Å	$b_{cementite}$, Å	$b_{ferrite}$, Å
100	5	6
500	58	69
1000	163	194
2000	460	549
5000	1819	2168

EXPERIMENTAL

The partitioning studies were carried out using the analytical electron microscope EMMA-4. The diameter of the analysis probe from which sufficient X-ray counts could be obtained from a thin specimen to give a quantitative analysis was approximately 1200 Å.

Partitioning studies have been made using both thin foils and two-stage extraction replicas. Thin foil specimens were prepared from 3 mm diameter discs which were electropolished using an electrolyte of 5% volume perchloric acid in acetic acid. For the preparation of two-stage extraction replicas a thin film of cellulose acetate, which had been softened in acetone, was pressed onto the surface of a polished and etched specimen. When the film had dried it was pulled from the surface and some of the cementite plates were mechanically extracted from pearlite colonies. A carbon film was then deposited on the extraction replica and the plastic was removed by dissolving in acetone.

ANALYTICAL TECHNIQUE

During analysis the intensities of the characteristic X-ray lines from Fe K_α and the partitioning element being investigated were recorded simultaneously. The ratios of the partitioning element and iron characteristic X-ray intensities I_X/I_{Fe} were converted into weight concentrations of the partitioning element and iron (C_X and C_{Fe}, respectively) using the relationship

$$\frac{C_X}{C_{Fe}} = \frac{k_X}{k_{Fe}} \cdot \frac{I_X}{I_{Fe}} \qquad (1)$$

where k_X and k_{Fe} are determined from alloy and mineral thin standards.[6] The results of each analysis were expressed as a partition coefficient which was defined as the ratio of the weight-% of X in cementite, $X_{(cem)}$, to the weight-% of X in ferrite, $X_{(\alpha)}$:

$$K_\alpha^{cem}(X) = \frac{X_{(cem)}}{X_{(\alpha)}} \qquad (2)$$

The partition coefficients could be readily computed when measurements were made on the two phases contained in a thin foil. However, for measurements made on extracted cementite it was necessary to calculate the concentration of the alloying element in ferrite by using a mass balance procedure.[3]

LIMITS TO SPATIAL RESOLUTION

There are distinct advantages in carrying out analyses of both the ferrite and cementite phases in a thin foil specimen. In particular if a reaction front (Figs. 2 and 3) can be located then solute distributions can be determined at the pearlite/austenite interface and also at various positions behind the interface in the ferrite and cementite lamellae.

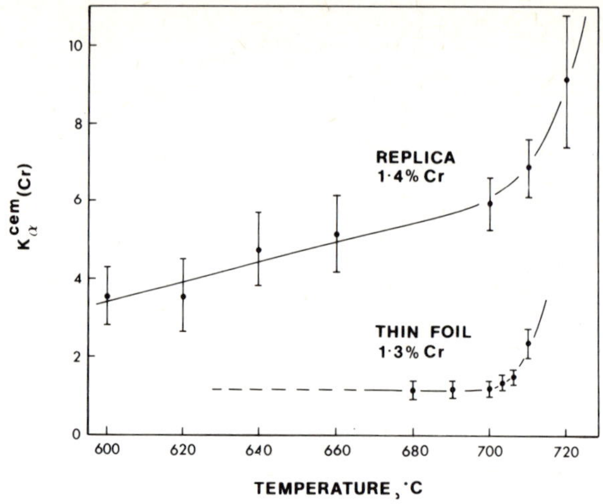

5 **Comparison of partition coefficient at transformation front vs. transformation temperature for two chromium eutectoid steels; 1.4% Cr – extraction replica, 1.3% Cr – thin foil**

When the partitioning reaction proceeds rapidly, as for example in a steel containing silicon, it is essential to carry out the analysis at the reaction front, otherwise it is impossible to determine whether partitioning has occurred at the front or within the pearlite colony behind the front. In practice when the foil is prepared by electropolishing the probability of obtaining a thin specimen which contains a pearlite/austenite interface is low.

The other limitation of the thin foil technique is that of resolution. When the mean interlamellar spacing approaches the probe diameter, 1200 Å in EMMA-4, it is difficult to obtain an analysis from exclusively the ferrite or cementite phases. This is a particular problem for the cementite phase which, as was mentioned earlier, has a thickness which is only one-eighth of that of the mean interlamellar spacing. For example when the mean interlamellar spacing is 1200 Å the cementite width is only 150 Å.

Even if the probe diameter was less than 100 Å, beam spreading within the cementite could be in excess of its thickness. A reasonable estimate of the broadening b as a function of cementite lamellae thickness can be obtained using the expression given by Goldstein *et al.*:[7]

$$b = 6.25 \times 10^5 \frac{Z}{E_0} \left|\frac{\rho}{A}\right|^{1/2} t^{3/2} \text{ cm} \qquad (3)$$

where Z is the atomic number, A the atomic weight, E_0 the incident electron energy, and ρ the density.

Table 1 shows the predicted value of broadening as a function of thickness t in cementite and ferrite for an infinitely small incident probe diameter.

Partitioning data measured using thin foils are compared in Fig. 5 with the corresponding values obtained from two-stage extraction replicas for two chromium steels of similar compositions and similar transformation and pearlite growth characteristics. The higher resolution of the carbide extraction technique shows that chromium partitions preferentially to cementite over a wide range of temperatures, whereas the thin foil study showed less partitioning at all temperatures and the existence of a 'no-partitioning' temperature. Since the mean true interlamellar spacing at the no-partitioning temperature (~703°C) was ~1700 Å (Fig. 4) the limitations of incident probe diameter and beam broadening within the specimen have resulted in partitioning not being detected below this temperature.

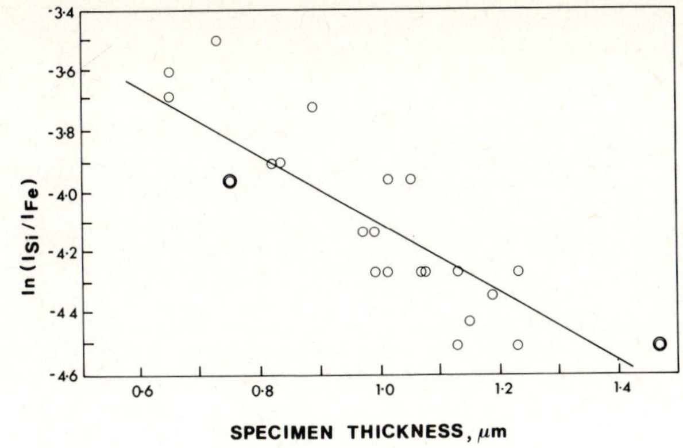

6 Logarithm of ratio of intensities of Si K to Fe K_α radiation vs. thickness for 2 wt-% Si steel

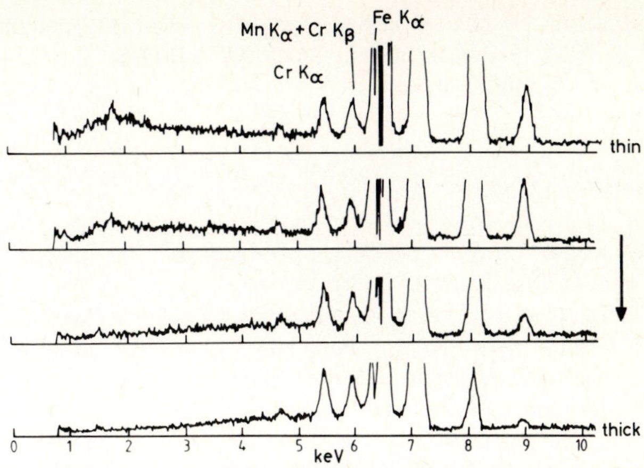

7 Series of X-ray spectra from martensitic foils of 1 wt-%Cr–1 wt-%Mn eutectoid steel showing effect of progressively increasing specimen thickness (top spectrum from thinnest region) on relative heights of Cr K_α and Mn K_α peaks

A carbide extraction replica does not suffer the limitations due to resolution, since there is only one phase present. Atom probe studies have confirmed that partitioning of chromium in pearlitic steels can occur at temperatures of 597°C.[8]

ABSORPTION EFFECTS

If the partitioning element is of a low atomic number then it is necessary to correct for the preferential absorption within the specimen of X-rays from this element. Effects due to preferential absorption of Si K X-rays have been observed by the authors during an investigation of the partitioning of silicon.

Figure 6 shows the observed variation in ln (I_{Si}/I_{Fe}) as a function of specimen thickness for a 2 wt-% Si eutectoid steel which had been quenched to produce a martensitic microstructure of uniform composition. The line has a slope, as determined from a least-squares analysis, equal to 2090 cm^2 g^{-1}. This is the approximate difference between the X-ray mass absorption coefficients of iron (2502 cm^2 g^{-1}) and silicon (710 cm^2 g^{-1}) X-rays in the specimen. This value is in agreement with that predicted by the expression:

$$\frac{I_{Si}}{I_{Fe}} = \frac{I_{0(Si)}}{I_{0(Fe)}} \cdot \frac{\mu_{Fe}}{\mu_{Si}} \left[\frac{1-\exp(-\mu_{Si}\rho t \operatorname{cosec}\alpha)}{1-\exp(-\mu_{Fe}\rho t \operatorname{cosec}\alpha)}\right] \qquad (4)$$

where I_{Si}/I_{Fe} is the observed X-ray intensity ratio, $I_{0(Si)}/I_{0(Fe)}$ the X-ray intensity ratio from an infinitely thin specimen, ρ the density, t the thickness, α the angle between the centre line of the detector and the specimen surface, and μ_{Fe} and μ_{Si} are X-ray absorption coefficients.

FLUORESCENCE EFFECTS

During the investigation of partitioning in a 1 wt-%Cr–1 wt-%Mn eutectoid steel it was observed that the height of the Cr K_α peak was almost always greater than that of Mn K_α. In addition, there was clearly an effect due to thickness; the thinner the specimen the closer the ratio of the Mn K_α to Cr K_α peak was to unity. Figure 7 shows a series of energy-dispersive X-ray spectra from a martensitic specimen of the Cr–Mn alloy. The analyses were obtained from progressively thicker regions of the foil.

The model of Nockolds *et al.*[9] for calculating the fluorescence corrections predicts that in a thin binary specimen AB the intensity of A radiation fluoresced by B radiation is given by

$$I_F^A = I_B \frac{T}{2}\left|0.923 - \ln(\mu_B T)\right| C_A \mu_B^A \omega_{K(A)} \frac{r_A - 1}{r_A} \qquad (5)$$

where μ_B^A and μ_B are the mass absorption coefficients of X-radiation from element B in element A and in the specimen, respectively; C_A is the weight fraction of element A; T is the mass thickness; $\omega_{K(A)}$ is the fluorescence yield of element A; and r_A is the absorption-edge jump-ratio. The ratio of characteristic X-ray intensities I_A and I_B can be calculated using the formula for K-shell ionization cross-sections proposed by Green and Cosslett[10] and used to modify the above expression to give the fluorescence enhancement ratio for A radiation I_F^A/I_A in a thin foil:

$$\frac{I_F^A}{I_A} = C_B \cdot \omega_{K(B)} \cdot \frac{r_A - 1}{r_A} \cdot \frac{A_A}{A_B} \cdot \mu_B^A \cdot \frac{U_B \ln U_B}{U_A \ln U_A} \cdot \frac{T}{2}\left|0.923 - \ln(\mu_B T)\right| \qquad (6)$$

where U_A and U_B are the overvoltage ratios for elements A and B respectively, and A_A and A_B are the relevant atomic weights. (Note that if the sample is tilted through an angle α then this expression must be multiplied by sec α to allow for the resultant increase in I_B.)

It is important to point out that the model of Nockolds *et al.*[9] predicts a $T \log T$ enhancement due to fluorescence while a model due to Tixier[11] predicts a T^2 variation. The

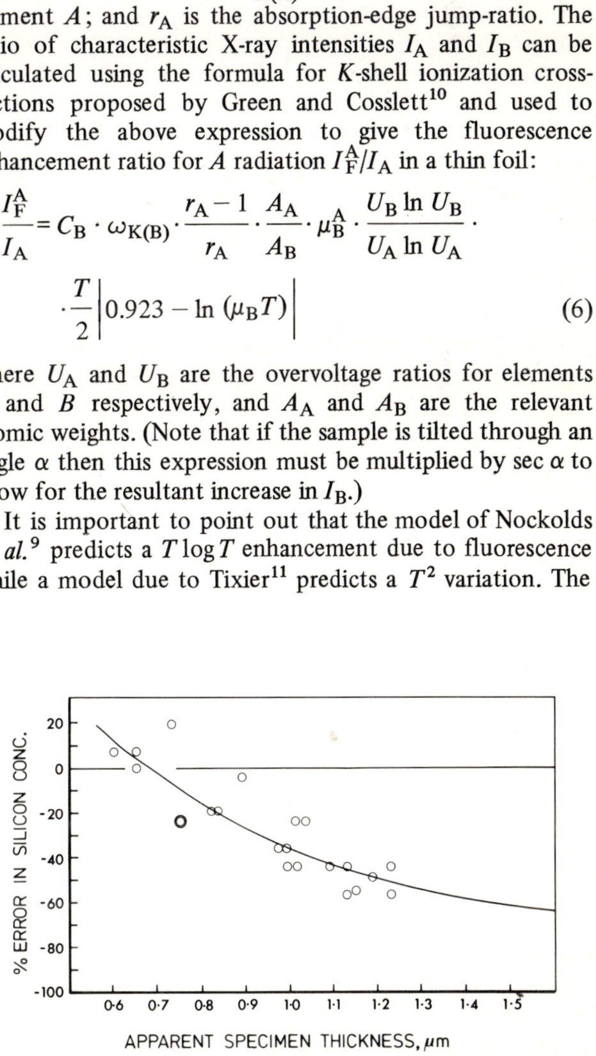

8 Error in measured concentration of silicon as function of specimen thickness for martensitic foil of 2 wt-% Si eutectoid steel; apparent increase in Si concentration of edge of foil is due to presence of Si-rich surface film

behaviour of the two models is very different for thin (<2000 Å thick) specimens. The model of Tixier does not predict a fluorescence effect as large as that shown in Fig. 7.

SURFACE FILMS

The work of Morris *et al.*[12] and others on aluminium alloys has clearly shown that during electropolishing solute-rich layers can be formed on the surface of the specimen, and that these will produce a marked change in the I_X/I_{Al} ratio near the edge of the specimen. Similar effects have been observed also on electropolished samples of eutectoid steels. Figure 8 shows the effect in a thin martensitic specimen of 2 wt-% Si eutectoid steel. The apparent silicon concentration is less than 2% in the thicker regions of the foil, due to the preferential absorption of Si *K* radiation (discussed above) but near the edge of the sample the effect is reversed: i.e. the apparent silicon content is greater than 2 wt-%.

In this and other steel specimens it has been found that the solute-rich layer can be removed by exposing the electropolished specimen to a 10–15 min treatment in an ion beam thinning apparatus (4 kV, 20 to 25° inclination).

SUMMARY

Analytical electron microscopy is shown to be an effective technique for obtaining alloy partitioning data during pearlitic growth. The practical problems which must be overcome include limitations in spatial resolution associated with the fineness of lamellar pearlite, absorption and fluorescence effects within the specimen, and the presence of solute-rich surface films. Once identified all of these problems can be overcome and quantitative partitioning data obtained.

ACKNOWLEDGMENTS

The authors are grateful to Dr S. A. Al-Salman for permission to use Figs. 5, 6, and 7 which are from his PhD thesis. EMMA-4 was purchased with a grant from SRC.

REFERENCES

1 N. A. RAZIK, G. W. LORIMER, and N. RIDLEY: *Acta Metall.*, 1974, **22**, 1249

2 S. A. AL-SALMAN, G. W. LORIMER, and N. RIDLEY: *Acta Metall.*, 1979, **27**, 1391

3 S. A. AL-SALMAN, G. W. LORIMER, and N. RIDLEY: *Metall. Trans.*, 1979, **10A**, 1703

4 G. E. PELLISSIER, M. F. HAWKES, W. A. JOHNSON, and R. F. MEHL: *Trans. ASM*, 1942, **30**, 1049

5 C. ZENER: *Trans. AIME*, 1946, **167**, 550

6 G. CLIFF and G. W. LORIMER: *J. Microsc.*, 1975, **103**, 203

7 J. I. GOLDSTEIN, J. L. COSTLEY, G. W. LORIMER, and S. J. B. REED: SEM 1977, **1**, Proc. Workshop on Analytical Electron Microscopy (ed. O. Johari), 315; 1977, Chicago, IITRI

8 P. R. WILLIAMS, M. K. MILLER, P. A. BEAVAN, and G. D. W. SMITH: 'Phase transformations', **2**, 98; 1979, London, Inst. Metallurgists

9 C. NOCKOLDS, M. J. NASIR, G. CLIFF, and G. W. LORIMER: 'Developments in electron microscopy and analysis' (ed. T. Mulvey), 417; 1979, London, Inst. Physics

10 M. GREEN and V. E. COSSLETT: *Proc. Phys. Soc.*, 1961, **78**, 1206

11 R. TIXIER: 1975, PhD thesis, Université de Paris

12 P. L. MORRIS, N. C. DAVIES, and J. A. TREVERTON: 'Electron microscopy and analysis' (ed. D. Misell), 374; 1977, London, Inst. Physics

DISCUSSION of paper by Ridley and Lorimer

R. TIXIER:

Is it possible to obtain details of the way your formula for fluorescence correction has been derived?

G. W. LORIMER:

The fluorescence correction is published in the EMAG 1979 Proceedings (Ref. 9 of our paper).

Redistribution of substitutional alloying elements during formation of pearlite

R A Ricks

The effect of substitutional alloying elements on the kinetics and resultant microstructure of pearlite formed in eutectic steels has been studied for a considerable number of years. Only recently have techniques been developed to obtained information concerning local chemical compositions with sufficient spatial resolution to allow a study to be made of the behaviour of these alloying additions during the austenite-pearlite transformation. The aim of this investigation is to determine the diffusional path taken by substitutional alloying elements (e.g. Ni, Cr, Mn) during the formation of pearlite in a number of eutectic steels using STEM/EDS. The results of this study will be correlated with kinetic information with a view to understanding the growth mechanism of pearlite and related microstructures in greater detail.

The author is in the Department of Metallurgy and Materials Science, University of Cambridge.

Since the original observation of pearlite by optical microscopy, the apparent simplicity of the lamella microstructure has allowed the development of numerous models to account for the growth mechanism. The early models, concerned with plain carbon steel pearlites, all involved the interfacial diffusion of carbon, since the measured growth rates were some fifty times faster than that allowed by volume diffusion of carbon.[1,2]

The effects of additions of such substitutional alloying elements as Cr and Mn have been incorporated into these growth models by assuming that a degree of solute partitioning occurs at the interface.[3,4] Recent investigations[5,6] have shown that the observed growth kinetics of pearlites in Cr-containing steels may be adequately explained in terms of a solute no-partition temperature above which the interfacial diffusion rate of the substitutional element controls the reaction, while below this temperature it was suggested that the kinetics were best described by carbon diffusion.

However, the application of microanalytical techniques to the formation of pearlite in Cr-Mn steels failed to reveal any no-partition temperature.[7] It was therefore concluded that at low transformation temperatures the observed partitioning of Cr and Mn merely accompanied the transformation and was not the rate-controlling process, which was again assumed to involve carbon diffusion.

Some preliminary results of a combined TEM/STEM/EDS investigation into pearlite formation in two alloy steels containing Cr and Mn will be presented. Specifically it will be shown that under certain transformation conditions the formation of alternative microstructures to lamella pearlite may be more favourable.

EXPERIMENTAL

The alloys used in this investigation had the following nominal compositions (wt-%): Fe–13%Mn–0.8%C and Fe–10%Cr–0.2%C. They were prepared from high-purity base elements melted in an argon arc furnace and homogenized under a partial pressure of argon for 48 h at 1200°C. The material was then swaged to 3 mm rod and austenitized at 1150°C for 30 min prior to heat treatment in a deoxidized tin bath preset at the required isothermal transformation temperature. Specimens for electron microscopy were slit from the heat-treated rod under conditions of flood lubrication and electropolished in a twin-jet polisher using an electrolyte consisting of 70% ethanol, 20% glycerol, and 10% perchloric acid. The specimens were thoroughly washed in cool methanol prior to examination. No influence of a solute-rich polishing layer on the specimen surface was detected using this method (and see Ref. 8).

Carbon extraction replicas of the Fe–10%Cr–0.2%C alloy were prepared from polished specimens lightly etched with Vilellas reagent prior to the evaporation of a thin carbon film onto the etched surface. The film was electrolytically removed in 20% nital at a voltage of 5–10 V and finally washed in distilled water and collected on copper grids. All electron microscopy was performed on a Philips EM400 TEM/STEM using a Link 860 analyser system incorporating standard thin foil correction programmes to compensate for the effects of X-ray detection efficiency and absorption within the specimen. In all cases a probe size of 10 nm was used.

RESULTS

Fe–13%Mn–0.8%C

The time-temperature-transformation (*TTT*) diagram for this alloy is shown in Fig. 1. Since the pearlite transformation temperature range exists only within a three-phase austenite + ferrite + carbide field[9] only ~15% transformation is possible. This austenite retention to ambient temperatures has allowed the crystallography of the pearlite transformation to be accurately determined.[10] Partially transformed specimens were used to investigate the behaviour of Mn during the formation of pearlite in this alloy; Fig. 2 shows a STEM image of a small pearlite nodule which has nucleated on an austenite grain boundary (the

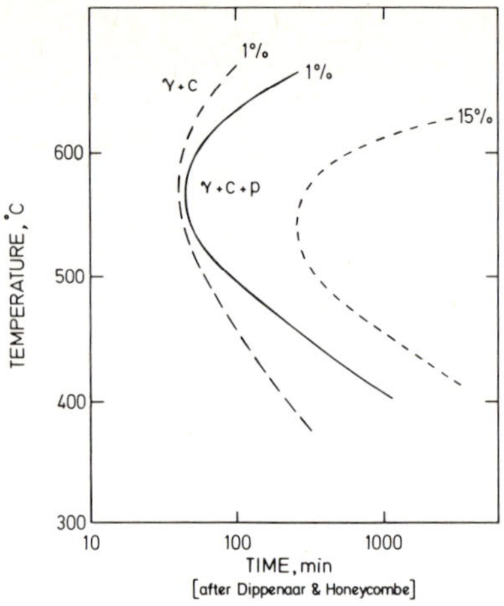

1 *TTT* diagram for Fe–13%Mn–0.8%C alloy used in this investigation

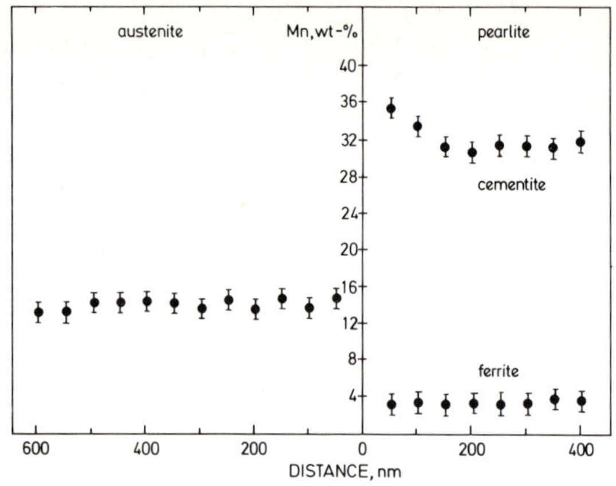

3 Mn concentration across growth interface of pearlite nodule shown in Fig. 2; note increase in Mn concentration in cementite near boundary (discussed in text)

figure captions give transformation times and temperatures). Growth of the pearlite colony has occurred primarily into the upper austenite grain and considerable branching and irregularities of the cementite lamella may be observed. This was found to be a typical morphology at this early stage of growth. The compositional profile across the growth interface of the pearlite colony shown in Fig. 2 is presented in Fig. 3. Two points are immediately obvious:

(a) there is considerable partitioning of Mn to the cementite lamellae;
(b) there is no change in the Mn concentration in the austenite up to the interface, which indicates that the partitioning of Mn is confined to the vicinity of the interface.

Furthermore the results indicate that there may well be a further enrichment of the cementite in Mn near the interface.

Figure 4 shows a well-formed pearlite colony growing into the parent austenite. Although some branching of the cementite lamellae is still apparent, in general the morphology of the pearlite was found to be much more regular than that shown in Fig. 2.

Figure 5 shows the Mn profile across the interface shown in Fig. 4. The Mn concentration ahead of the interface is again constant and partitioning of the Mn to the cementite is observed. The enrichment of the cementite lamella near the pearlite/austenite interface has now increased the carbide composition from ~32%Mn to ~38%Mn, which will be referred to under 'Discussion' below.

Fe–10%Cr–0.2%C

The decomposition of austenite to ferrite/carbide aggregates in this alloy has been extensively studied as a function of transformation temperature.[11,12] Although not of eutectoid composition, at high transformation temperatures an 'alloy

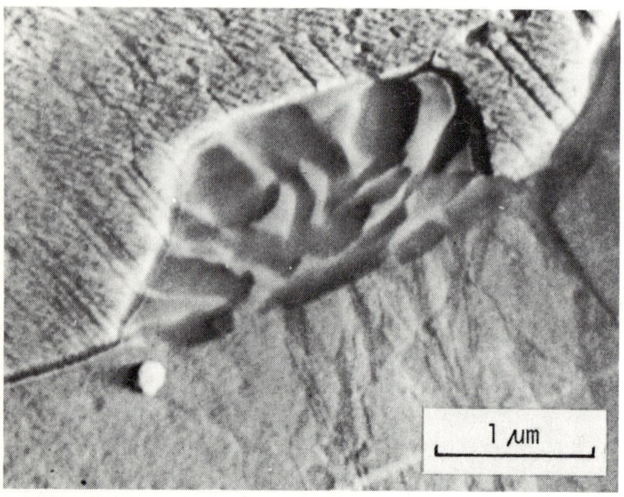

2 STEM image of small pearlite nodule, formed in Fe–13%Mn–0.8%C alloy, which has nucleated on austenite grain boundary and grown into upper austenite grain. Isothermally transformed at 600°C for 4 h

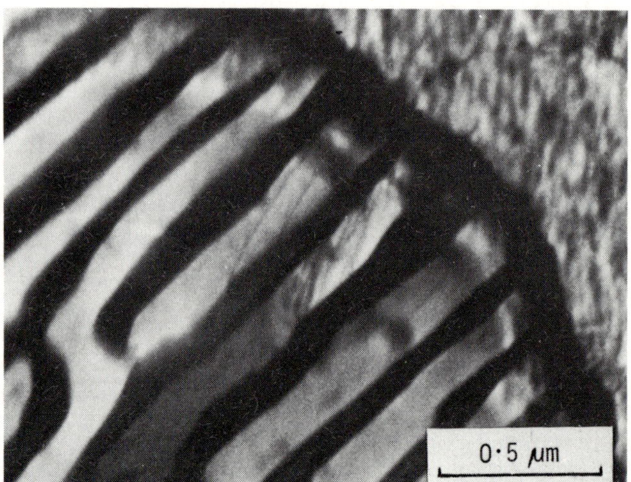

4 STEM image of austenite/pearlite interface in Fe–13%Mn–0.8%C alloy formed after isothermal transformation at 600°C for 8 h

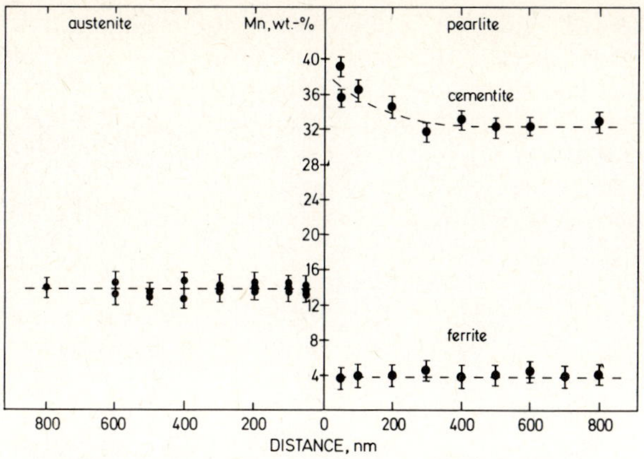

5 **Mn concentration across austenite/pearlite interface shown in Fig. 4. Again a high Mn level adjacent to interface in cementite is observed**

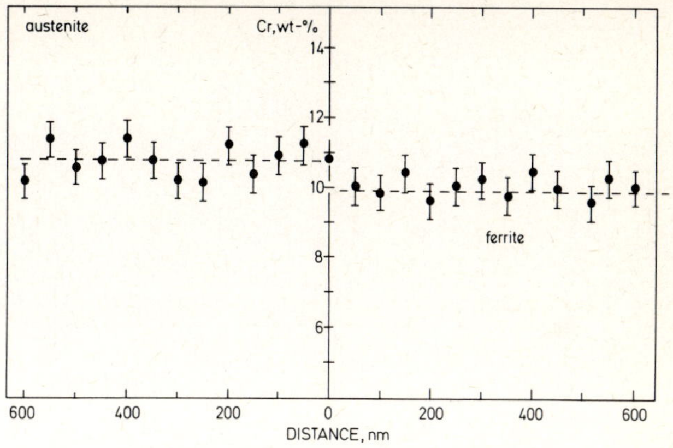

7 **Cr concentration across austenite/alloy pearlite interface shown in Fig. 6**

pearlite' is known to form[11] which is replaced by a particulate or fibrous precipitate dispersion of $M_{23}C_6$ in ferrite at lower temperatures. The results outlined in this subsection will refer predominantly to the high temperature alloy pearlite, which is composed of ferrite and $M_{23}C_6$ plates or lamellae.[11]

Figure 6 shows a typical pearlite transformation interface in this alloy after isothermal transformation at 730°C for 15 min. Although the majority of the carbides exhibit a lamella morphology, in certain regions of the pearlite colony (e.g. at *A*) the carbides become more particulate in appearance, and adopt a morphology often known as 'degenerate pearlite'.[13] The Cr profile across the pearlite/austenite interface is shown in Fig. 7. Again two points are apparent:

(a) the Cr level ahead of the interphase boundary indicates that no volume diffusion of Cr in the austenite occurs during the transformation; and
(b) the Cr level in the pearlitic ferrite is seen to drop immediately behind the interphase boundary.

Attempts to obtain Cr concentration profiles in the carbides, away from the phase boundary, were found to be affected considerably by dilution effects due to the overlapping ferrite. It was therefore necessary to use carbon extraction replicas to obtain the true Cr concentration of the carbides in the pearlite colonies (Fig. 8). The results of an EDS investigation into the composition of both the particulate and lamella carbides using such extraction techniques are shown in Fig. 9. These results indicate that the particulate carbides (or degenerate pearlite) have appreciably higher Cr concentrations than the adjacent lamellae carbides (Fig. 9*b*).

Previous studies[11,12] have shown that as the transformation temperature in this alloy is lowered the pearlitic growth mode is almost entirely replaced by the formation of fine alloy carbide fibres and eventually by particulate interphase precipitation. This change in reaction product as a function of transformation temperature will be discussed in the next section.

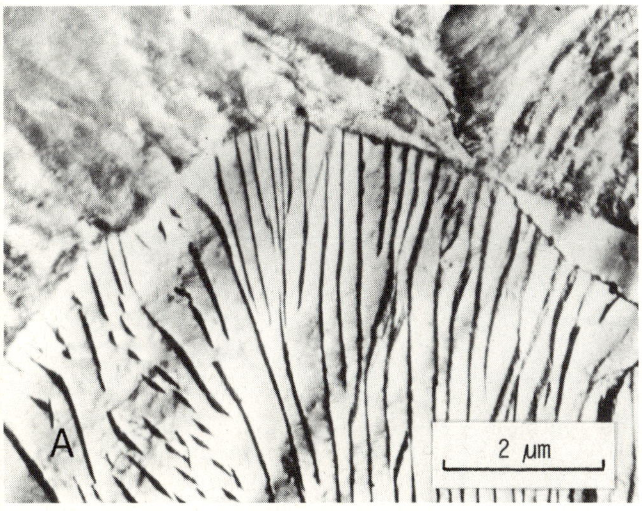

6 **STEM image of austenite/alloy pearlite interface in Fe–10%Cr–0.2%C alloy formed after isothermal transformation at 730°C for 15 min; there is considerable degeneracy of pearlite (e.g. at *A*)**

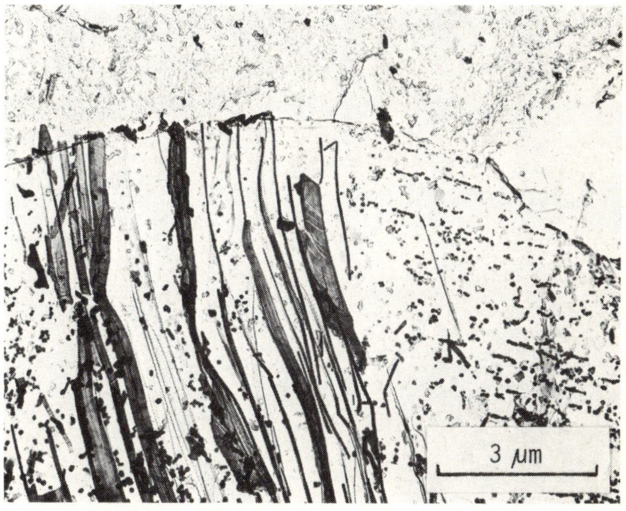

8 **Carbon extraction replica showing austenite/pearlite growth interface producing lamella and particulate carbides**

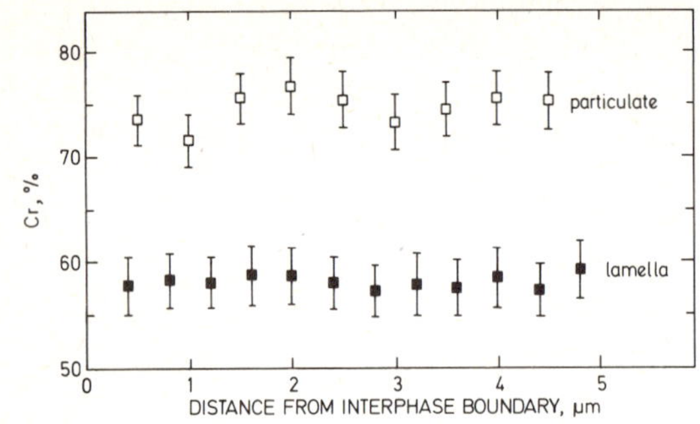

9 Cr concentration of lamella and particulate carbides shown in Fig. 8

DISCUSSION

The results obtained from the application of STEM/EDS techniques to the formation of pearlitic microstructures in a high-purity Fe-13%Mn-0.8%C have shown that the partitioning of Mn occurs in close proximity to the interface, since no change in composition ahead of the pearlite/austenite boundary was observed. The limited spatial resolution of EDS does not permit the unambiguous determination of the Mn diffusion path, but it would seem most likely that the partitioning occurs within the interphase boundary itself.

Since this particular alloy system only transforms to ~15% pearlite before the reaction stops, the remaining austenite must be enriched in solute to stabilize this latter phase. Thus the slow growth rate of the pearlite colonies observed in this alloy is most probably controlled by this volume diffusional process into the austenite. Because the growth rate is thus very low, the pearlite may form under almost equilibrium conditions with the austenite and full partitioning of the Mn to the carbides may take place at the interface. One explanation of the Mn enrichment of the cementite lamellae close to the austenite/pearlite interphase boundary may be that the cementite attempts to maintain dissimilar equilibria across the cementite/austenite and cementite/ferrite interphase boundaries which are present.

The results obtained from the Fe-10%Cr-0.2%C alloy substantiate the aforementioned interfacial diffusion mechanism since, again, no change in austenite composition was detected ahead of the interface. The pearlite transformation in this alloy is most probably controlled by the redistribution reactions occurring during the transformation, in the same manner as that proposed in low-alloy eutectoid steel pearlites.[3,4] Thus the rate-controlling mechanism of austenite decomposition in this alloy closely resembles that observed in low-alloy transformable steels. The application of microanalytical techniques to the carbides formed during the high temperature decomposition of austenite in this alloy clearly showed that substantial partitioning of the Cr to the carbides occurred. Furthermore the degenerate pearlite (or particulate precipitation) had appreciably higher Cr concentrations than the lamella pearlitic colonies. One particular interpretation of these results is as follows.

The growth rate of a lamella pearlite colony in a transformable alloy steel will be ultimately controlled by the interfacial diffusion rate of the slowest moving atomic species. Since the spacing of the lamella is dictated by the diffusion distance of the solute atom, it is clear that as the driving force for the transformation is increased the spacing of the lamellae must decrease if partitioning is to continue. However, there is a thermodynamic minimum to the allowed lamella spacings[1] and once this dimension is reached the partitioning reaction must be reduced. At this stage, if the driving force for the transformation is high and the transformation temperature is low enough, it may well become energetically more favourable to *nucleate* carbides at positions on the interphase boundary which already have high solute concentrations, instead of attempting to partition the solute to existing carbide lamellae. Thus by forming a degenerate pearlite colony instead of a lamella colony a high partitioning ratio of Cr between the ferrite and carbide can be maintained.

Furthermore, the results of Ref. 7, which indicate that a reduction in the partitioning of Cr to the carbides at the interface occurs as the transformation temperature is reduced, imply that a full partitioning reaction cannot be maintained by reducing the lamella spacing. Thus the partitioning of solute becomes less efficient as the driving force for the transformation is increased.

The above conclusions indicate why particulate precipitate dispersions may become more favourable at low transformation temperatures (e.g. Refs. 11 and 12). This phenomenon is most readily observed in alloy steels containing

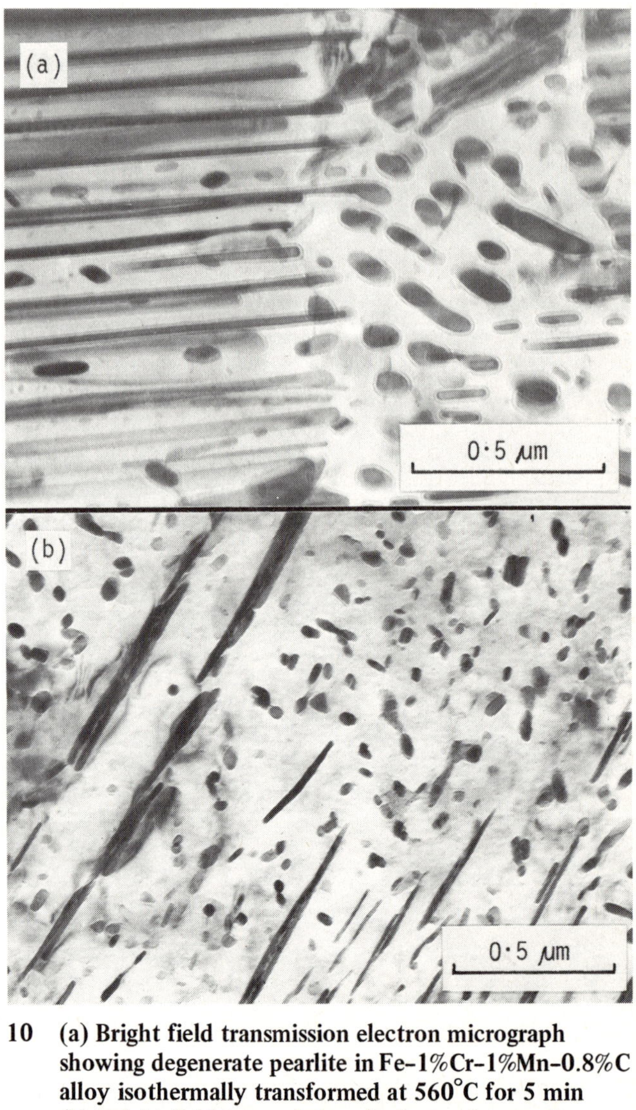

10 (a) Bright field transmission electron micrograph showing degenerate pearlite in Fe–1%Cr–1%Mn–0.8%C alloy isothermally transformed at 560°C for 5 min (b) Bright field transmission electron micrograph showing degenerate pearlite in Fe–1%Ni–0.4%C alloy isothermally transformed at 550°C for 5 s

strong carbide formers, but the formation of degenerate (i.e. particulate) pearlites is a commonly observed reaction product morphology, as shown in Fig. 10, which illustrates this microstructure in an Fe-1%Cr-1%Mn-0.8%C steel (10*a*) and an Fe-1%Ni-0.4%C steel (10*b*) (see also Refs. 13 and 14).

SUMMARY

The results of the previous sections have shown the following:

1 In an Fe-13%Mn-0.8%C alloy the observed slow pearlite growth rates allow full partitioning of alloying elements to the cementite lamellae in the pearlite colonies. Furthermore, it is suggested that cementite is able to maintain dissimilar equilibrium concentrations across cementite/ferrite and cementite/austenite boundaries.

2 In a high-purity Fe-10%Cr-0.2%C alloy the decomposition of austenite to ferrite/carbide aggregates at high reaction temperatures can take place with a greater degree of Cr partitioning if the carbide morphology is particulate instead of lamella.

3 In both the alloys investigated no volume diffusion of solute in austenite was detected adjacent to the austenite/pearlite interface, which implies that the partitioning of solute to the carbides occurs by an interfacial diffusion mechanism.

ACKNOWLEDGMENTS

The author would like to thank Professor R. W. K. Honeycombe for the provision of laboratory facilities and to acknowledge the provision of a maintenance award by the SRC. Discussions with Dr P. R. Howell are gratefully acknowledged.

REFERENCES

1 J. W. CANN and W. C. HAGEL: 'Decomposition of austenite by diffusional processes', 131, 1962, Interscience, New York.
2 D. CHEETHAM and N. RIDLEY: *J. Iron Steel Inst.*, 1973, 648
3 C. R. BROOKS and E. E. STANSBURY: *J. Iron Steel Inst.*, 1965, 514
4 D. BROWN and N. RIDLEY: *J. Iron Steel Inst.*, 1966, 811
5 N. A. RAZIK, G. W. LORIMER, and N. RIDLEY: *Metall. Trans. A*, 1976, **7A**, 209
6 R. C. SHARMA, G. R. PURDY, and J. S. KIRKCALDY: *Metall. Trans. A*, 1979, **10A**, 1129
7 S. A. AL-SALMAN, G. W. LORIMER, and N. RIDLEY: *Metall. Trans. A*, 1979, **10A**, 1703
8 R. A. RICKS, P. D. SOUTHWICK, and P. R. HOWELL: *J. Microsc.*, 1981, in press
9 R. J. DIPPENAAR: PhD thesis, University of Cambridge, 1970
10 R. J. DIPPENAAR and R. W. K. HONEYCOMBE: *Proc. R. Soc. A. (London)*, 1973, **333**, 455
11 K. CAMPBELL and R. W. K. HONEYCOMBE: *Met. Sci.*, 1974, **8**, 197
12 J. V. BEE, P. R. HOWELL, and R. W. K. HONEYCOMBE: *Metall. Trans. A.*, 1979, **10A**, 1207
13 Y. OHMORI, A. T. DAVENPORT, and R. W. K. HONEYCOMBE: *Trans. Iron Steel Inst. Jpn*, 1972, **12**, 129
14 Y. OHMORI and R. W. K. HONEYCOMBE: Proc. I.C.S.T.I.S., *Trans. Iron Steel Inst. Jpn*, 1971, **11**, 1160

DISCUSSION of paper by R. A. Ricks ('Substitutional alloying elements in pearlite')

N. RIDLEY:
My question relates to the increase in Mn concentration in the cementite at the Fe_3C/γ interface. What was the interlamellar spacing of the pearlite; and whereabouts, with respect to the Fe_3C lamellae, were the concentrations measured? Atom probe work by Smith at Oxford University has shown that there is a substantial Mn concentration profile across the Fe_3C, the Mn levels adjacent to the ferrite being significantly higher than in the centres of the lamellae.

R. A. RICKS:
At the transformation temperature used in this study the interlamellar spacing is approximately 0.2 μm. All EDS measurements were taken from the centre of each lamella but some results have been obtained which suggest that no change in Mn concentration exists across any cementite lamella at a constant distance behind the interface. Obviously such results are limited by the spatial resolution of EDS since at the edge of the lamellae severe dilution effects from the adjacent ferrite will predominate.

Fine-scale microanalysis of carbonitride inclusions in superalloys

W E Voice and R G Faulkner

Fine-scale microanalysis of the central regions of large carbonitride inclusions (>1 μm diameter) in Nimonic 80A has revealed the core regions to be rich in magnesium. Using thin foil specimens with a scanning transmission electron microscope with energy-dispersive X-ray analyser facilities (STEMMA), the resolution of the microanalysis in this work is estimated at about 1000 angstrom units. The magnesium-rich regions are between 0.3 and 0.6 μm diameter. Thus, concentration profiles are produced with sufficient spatial resolution for details of the concentration gradient between the magnesium-rich regions and the surrounding carbonitride to be elucidated. The composition of the magnesium-rich regions can also be unequivocally estimated. Hence the origins of the magnesium-rich particles can be more clearly defined. Effects of magnesium particle depth in the foil on spatial resolution is discussed with respect to theoretical predictions of beam spreading in thin films. Comparisons are made between the capabilities of the STEMMA technique and that of the electron probe microanalyser (EPMA) for this kind of work.

The authors are in the Department of Materials Engineering and Design at Loughborough University of Technology, Leicestershire.

Carbonitrides are large inclusions measuring up to 20 μm diameter which are present in many ferrous and nickel-base alloys. They are insoluble at temperatures where conventional solution treatments are carried out. They contain variable ratios of carbon to nitrogen.[1] Increasing solution treatment temperatures from 1140 to 1180°C causes an increased precipitation of secondary carbonitrides with a high carbon content.[2] Nucleation of the primary carbonitrides in the melt requires some form of nucleation site. Commonly these sites can be small inclusions and particles from the furnace lining or from deoxidants used in casting. Nucleation has been observed on alumina, silica and titanium oxide inclusions.[3]

In this work, magnesium oxide has been observed at the centres of carbonitrides in Nimonic 80A. Since the lattice parameter and crystal structure of MgO is almost identical to that of Ti(C,N) (MgO, cubic-NaCl type, $a = 4.22$ Å; TiN, $a = 4.2$ Å[4,5]), it is not inconceivable that MgO could be the nucleating agent. The nucleation process has therefore been studied more closely by fine-scale microanalysis of the MgO nuclei (only 0.5 μm diameter), the surrounding carbonitrides and the matrix. A JEOL 100CX STEM with a Link Systems 860 energy-dispersive analyser (STEMMA) has been used.

ELECTRON METALLOGRAPHY OF CARBONITRIDES

A collage of SEM, STEM and X-ray maps of carbonitride with a small MgO inclusion at the centre is shown in Fig. 1. This shows a titanium-rich particle in the nickel-chromium matrix containing small magnesium-rich regions <0.5 μm diameter.

Figure 2 shows transmission electron micrographs of these small magnesium-rich regions. A rim surrounds the carbonitride. The magnesium-rich particles have 'spokes' extending into the surrounding carbonitride. These spokes were avoided during the fine-scale spot microanalysis (STEMMA) work.

OPTIMIZATION OF CONDITIONS FOR FINE-SCALE MICROANALYSIS

The largest spot size with smallest condenser aperture at 100 kV was used. This was scanned over the foil edge to estimate the overall spot size as the beam entered the foil. This analysis indicated that the beam was about 250 Å diameter. Monte Carlo data based on Rutherford scattering for beam spreading in nickel[6] were used to estimate the X-ray source size. Attempts to quantify the beam spreading process using single scattering models have led to underestimates of the beam spreading in very thin (<1000 Å) foils.[7] Doig and Flewitt[8] have overcome this problem to some extent by considering an exponential relationship between beam spreading and foil thickness. However, over the whole range of important foil thicknesses in TEM (1000-3000 Å), Monte Carlo methods[6,9] still appear to give reliable predictions to within 200-300 Å. Using the definition for X-ray spatial resolution as the 95% beam trajectory spread at a depth equal to half the foil thickness,[6] the resolution of the analysis in this work was estimated at about 3000 Å. The foil thickness (7000-9000 Å) used to predict this value was measured by the burn mark technique.

The X-ray spatial resolution was relatively poor in this case and so a beam spot size of 200-300 Å has very little effect on final resolution.

The interfaces between various phases were aligned parallel to a line connecting the detector to the specimen. Hence, overlap of signals from material on either side of the interface, other than that arising from beam spreading, was overcome.

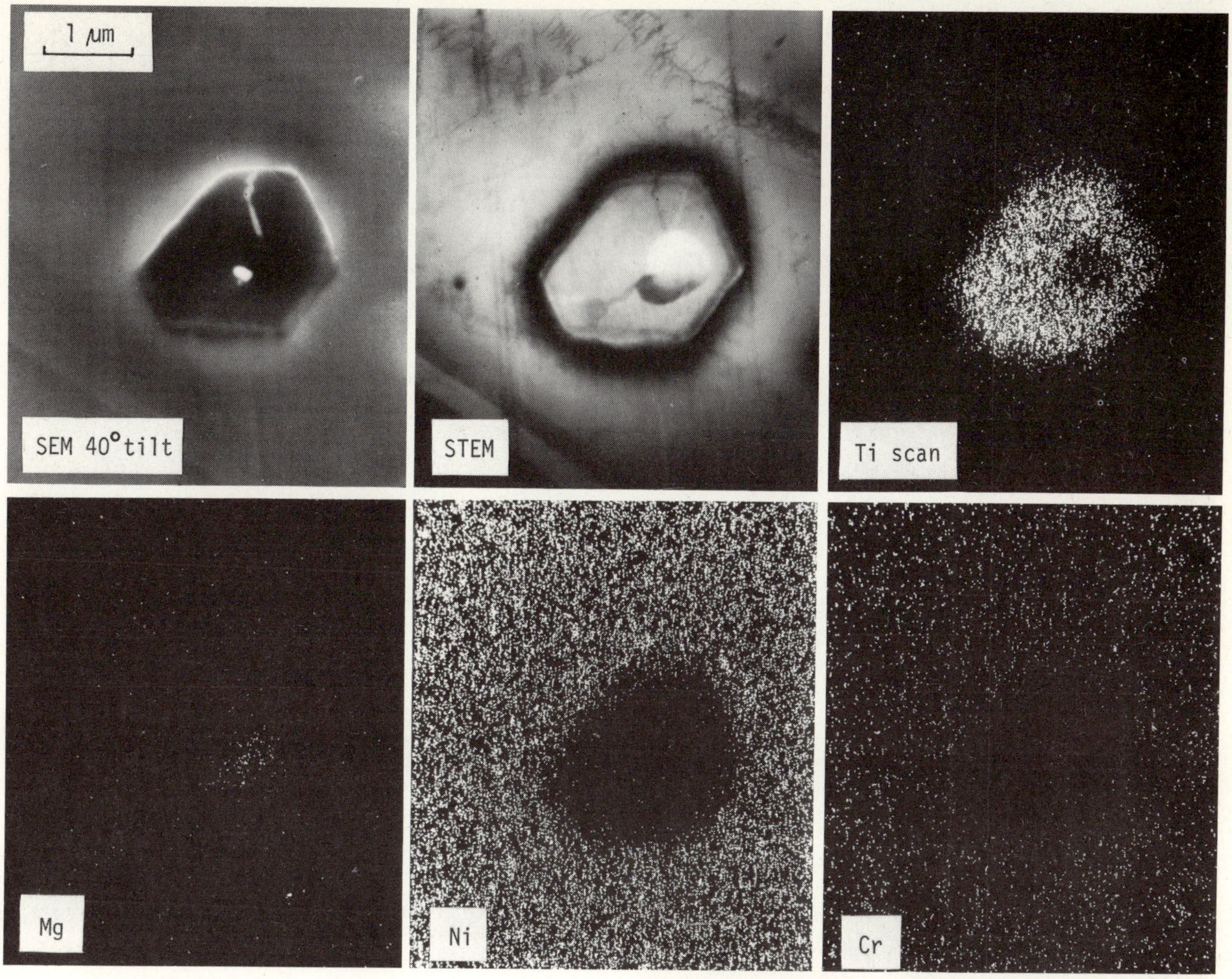

1 **X-ray and STEM pictures of carbonitrides**

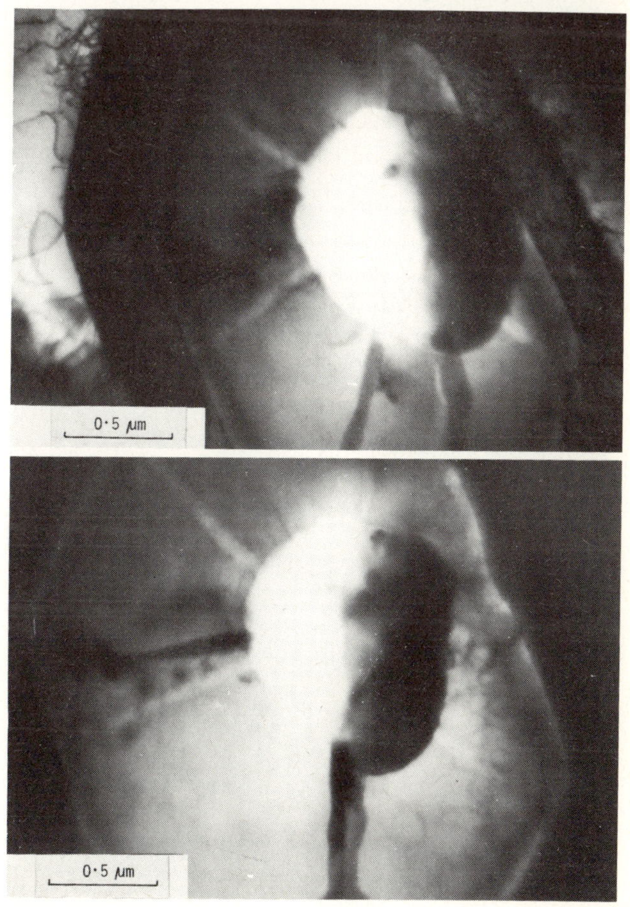

2 **(*right*) TEM micrographs of magnesium oxide core of carbonitride**

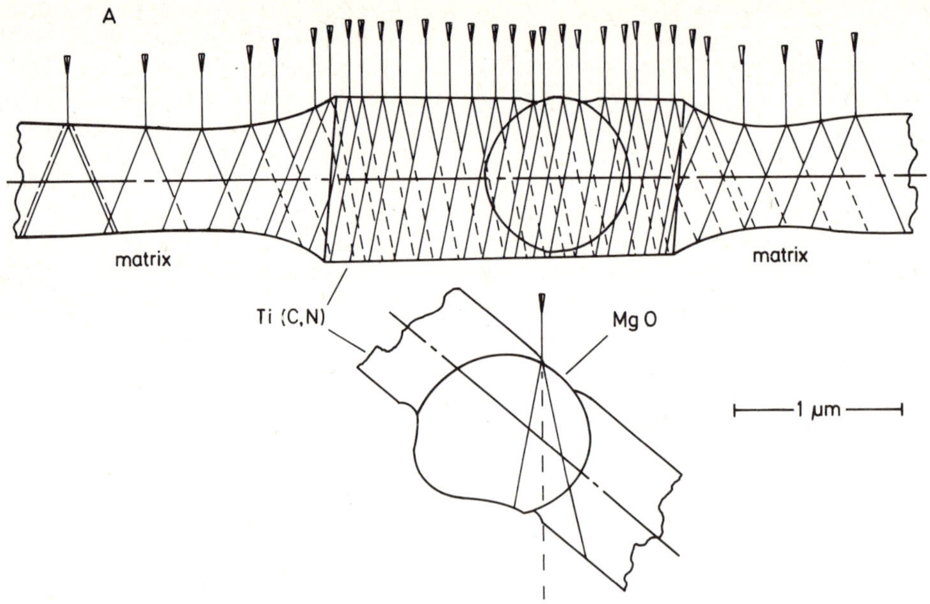

Specimen tilt 40°; traverse 90° to beam direction.
Section A contains line of traverse and beam direction

3 **Sections through Ti(C,N) particle showing 95% beam trajectories**

FINE-SCALE MICROANALYSIS OF CARBONITRIDES

By analysing the arrangement of MgO particles in the upper surface with scanning electron microscopy (SEM) and the interception of the particle with the bottom surface with transmission electron microscopy (TEM) imaging, an impression of the position of the particle in the foil was created. This is shown in Fig. 3. The foil thicknesses are assessed from burn marks. Beam spreading from each of the analysis points is indicated and the additional spreading arising from using a finite beam size is shown for the analysis point on the far left. The reduced atomic number of the material in the magnesium-rich regions is taken into account as shown by the reduced spreading in these areas. The detector was arranged perpendicularly to the plane of section A. Figure 4 shows the actual analysis spots on the specimen and the corresponding relative proportions of the elements analysed at each spot. A similar analysis was performed on another particle. The following deductions can be made from both sets of data.

1 The observed titanium enrichment around the carbonitride occurs in areas that could not have originated from the carbonitride particle (see Fig. 3).
2 There is an iron enrichment at the magnesium-oxide/carbonitride interface.
3 The titanium is stoichiometric within the carbonitride, i.e. there is no substitution within the carbide with other elements.
4 The intensity ratio of titanium in the carbonitride to magnesium in the magnesium oxide (8:1) is very high when it is simply considered that the Ti K_α and Mg K_α X-rays originate from similar sized volumes (see Fig. 4).

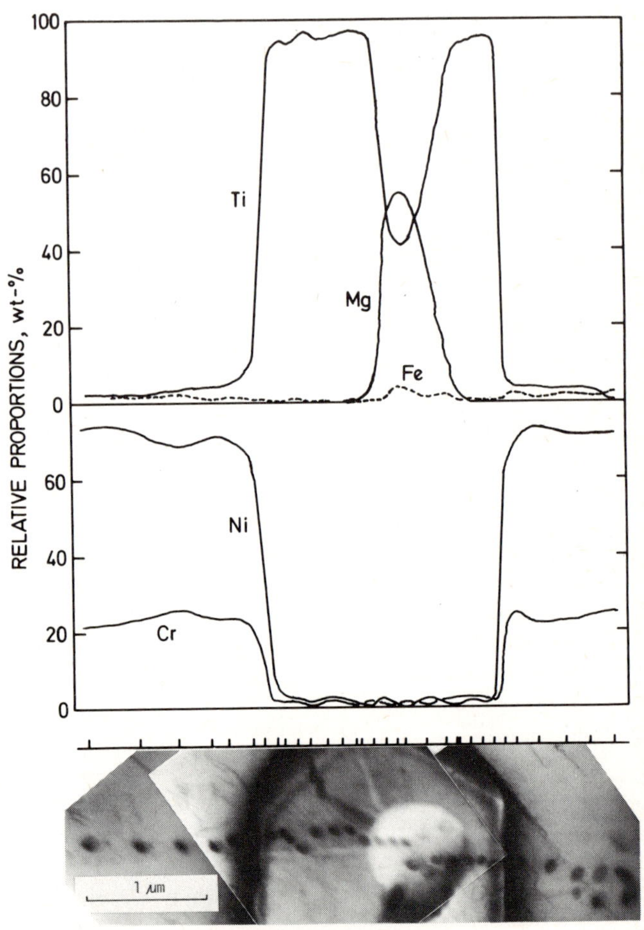

4 **Relative proportions of elements across Ti(C,N) particle in Nimonic 80A**

DISCUSSION

The results presented show an ideal example of where high spatial resolution microanalysis with scanning transmission electron microscopes (STEMMA) is useful. With lower-resolution bulk sample microanalytical instruments it is unlikely that the MgO particles would have shown up on X-ray spectra, mainly because the nickel, chromium and titanium signals would have swamped the Mg signal. Indeed, this may explain why no reports of MgO in this alloy have been previously reported. With the smaller X-ray source in STEMMA, the Mg K_α component becomes larger and precise location of the magnesium-rich regions is possible.

The 3000 Å spatial resolution used in this work allows explanation of several features concerning the formation of the carbonitrides. First, there appears to be no magnesium diffusion into the carbide because the magnesium profile has decreased to zero within 1800 Å of the magnesia/titanium-carbonitride interface. Hence, the oxide is stable and has at no stage in the previous thermal history tended to dissolve. Secondly, the reasonably constant levels of titanium observed in the carbonitrides, where the X-ray signals are originating solely from this phase, indicates that the carbonitride is stoichiometric in titanium. Hence, any lattice parameter changes observed during aging[2] are purely due to changes in the carbon/nitrogen ratio. Thirdly, because of the presence of titanium to levels of 2–3% at 0.5–1 μm away from the carbonitride interface into the matrix, it appears that either dissolution of titanium from,[10] or segregation of titanium to,[11] the interface has occurred.

Finally, the titanium/magnesium counts ratio of 8:1 observed from the pure titanium and pure magnesium parts of the sample can be explained in terms of ionization efficiency and absorption. The number of photons per incident electron in a thin foil of thickness Δz is given by:[12]

$$I = wQ \cdot \Delta z \cdot \delta$$

where w is the fluorescent yield (0.028 Mg, 0.202 Ti; Q is the ionization cross-section; and δ is the number of atoms per unit volume.

Since δ is the same for the two regions because of the similar densities of magnesia and titanium carbonitride

$$\frac{I_{\text{Ti}}}{I_{\text{Mg}}} = \frac{w_{\text{Ti}}Q_{\text{Ti}}}{w_{\text{Mg}}Q_{\text{Mg}}}$$

Q can be calculated from the Bethe law:[13]

$$QE_c^2 = \frac{a \ln (bU)}{U}$$

where w is the fluorescent yield (0.028 Mg, 0.202 Ti; Q is (Ref. 14), and where U is the over-voltage, i.e. the ratio of the working accelerating voltage to that of the critical excitation potential for the X-ray energy concerned. Hence

$$\frac{w_{\text{Ti}}Q_{\text{Ti}}}{w_{\text{Mg}}Q_{\text{Mg}}} = 1.31$$

To fully predict the observed ratio of titanium and magnesium intensities, the absorption of magnesium and titanium X-rays needs to be considered. Using the Philibert absorption correction[15] and assuming 50% titanium and 50% magnesium, the absorption correction for magnesium is 8.3 times larger than for titanium at 100 kV. Therefore the magnesium signal can be expected to be 8.3 times smaller than the titanium signal for the same amount of excited volume and the ionization efficiency-based ratio (1.31) should be multiplied by 8.3. This leads to a predicted ratio of 10.8, which is close to the observed value of 8. Clearly, absorption is occurring in the thin film. However, quantification of absorption and fluorescence in thin films is still imprecise.[16,17] Work is proceeding at Loughborough in an attempt to quantify absorption more accurately on the basis of square-shaped $\phi(\rho z)$ curves.

CONCLUSIONS

Concentration profiles around small magnesia inclusions (0.5 μm diameter) contained within titanium carbonitride particles in Nimonic 80A have been measured by STEMMA. Using established Monte Carlo data on beam spreading the X-ray spatial resolution used in this study was estimated at ~3000 Å. The carbonitrides are found to consist of stoichiometric amounts of titanium. No diffusion of magnesium into titanium carbonitride has occurred. Some segregation of titanium to within 1 μm of the titanium carbonitride particles is present. Slight iron enrichment around the magnesia particles is observed.

The ratio of measured intensities of the Ti K_α and Mg K_α X-rays from the thin film sample can be explained in terms of the relative ionization efficiencies of the two elements and heavy absorption of Mg K_α by titanium. Quantification of this latter effect in thin films is still imprecise and more work is required to resolve the problem.

ACKNOWLEDGMENTS

The assistance of George Adie with the STEMMA work and the provision of laboratory facilities by Professor I. A. Menzies of the University of Loughborough are gratefully appreciated.

REFERENCES

1 E. A. FELL: *Metallurgia*, 1961, (4), 157
2 W. E. VOICE: unpublished work, 1980
3 A. OSTROWSKI and E. W. LANGER: *Scand. J. Metall.*, 1979, **8**, 153
4 ASTM Powder Diffraction File 1980, 6, A. Swanson and B. Tatge (1949), J. C. Fel. Reports, NBS
5 H. J. BEATTIE and F. L. VERSNYDER: *Trans. ASM*, 1953, **45**, 397
6 R. G. FAULKNER and K. NORRGÅRD: *X-ray Spectrometry*, 1978, **7**, 184
7 J. I. GOLDSTEIN *et al.*: SEM 1977, **1**, ed. O. Johari, 315–324, IITRI, Chicago
8 P. DOIG and P. J. FLEWITT: this volume
9 D. F. KYSER and R. H. GEISS: 8th Int. Conf. X-ray Optics and Microanalysis (1977), **1**, 110A
10 M. J. WHELAN: *Met. Sci. J.*, 1969, **3**, 95
11 R. G. FAULKNER: *J. Mater. Sci.*, 1981, **16**, 373
12 S. J. B. REED: 'Electron microprobe analysis', 329; 1975, Cambridge Univ. Press, Cambridge
13 H. A. BETHE: *Ann. Phys. (Leipz.)*, 1930, **5**, 325
14 C. R. WORTHINGTON and S. G. TOMLIN: *Proc. Phys. Soc.*, 1956, **A69**, 401
15 J. PHILIBERT and H. H. PATTEE *et al.*: 'X-ray optics and microanalysis', 379; 1963, New York, Academic Press
16 G. LOVE *et al.*: *J. Phys. D. (Appl. Phys.)*, 1977, **10**, 7
17 H. E. BISHOP: *J. Physics D.*, 1974, **7**, 2009

STEM microanalysis of intermetallic phases in an Al – Fe – Si alloy

D A Porter and H Westengen

The binary phases Al_3Fe, Al_6Fe, Al_xFe and Al_mFe found in direct-chill cast commercial aluminium have been analysed using STEM/EDS thin foil microanalysis techniques. Besides aluminium and iron small quantities of silicon and nickel were present in the phases. The aluminium and iron contents have been established using annealed Al_3Fe as a standard. The small silicon and nickel concentrations have been determined using a standardless approach. X-ray absorption has been corrected for using a method based on characteristic X-ray intensity and beam current measurements. Systematic errors arising from 'surface layers' have been reduced by analysing a wide range of particle thicknesses.

Dr Porter is in the Department of Engineering Materials at the University of Luleå, Sweden; Dr Westengen is with Årdal og Sunndal Verk a/s, Sunndalsøra, Norway.

Commercially pure aluminium is typically 99.5 wt-% aluminium with about 0.5 wt-% of impurities, mainly iron and silicon. During solidification these impurities can segregate into a variety of different intermetallic compounds: α-AlFeSi,[2] α′-AlFeSi,[3] α″-AlFeSi,[1] β′-AlFeSi,[1] Al_3Fe,[2] Al_6Fe,[2] Al_mFe,[4] Al_xFe[1] and Si.[2] Which phases form depends on the alloy composition and cooling rate. The most common method of producing aluminium ingots for rolling or extrusion is by semicontinuous direct-chill (DC) casting in which cooling is achieved by a water spray acting directly on the surface of the partially solidified ingot. Therefore the rate of cooling through the solidification range increases drastically from the middle of the outer layers of the ingot and this results in the formation of different intermetallic phases at different depths from the surface.[1] Of particular interest is the distribution of the binary phases based on aluminium and iron. These are the equilibrium phase Al_3Fe and the metastable phases Al_6Fe, Al_xFe and Al_mFe. Al_3Fe predominates in the central part of the ingot where the cooling rate is low, while Al_mFe predominates in the outer layers where the cooling rate is greatest. Al_6Fe and Al_xFe are formed at intermediate cooling rates. The object of this work has been to determine the chemical composition of the metastable phases using STEM/EDS microanalysis. This has not previously been possible due to the small dimensions of the particles involved.

It is now recognized that accurate quantitative thin foil microanalysis can usually be achieved only if absorption, fluorescence and surface layer effects are either negligible or correctable. In the case of light elements such as aluminium, absorption effects can be significant even in relatively thin specimens; whereas fluorescence effects can often be ignored. There have also been several reports of the presence of surface layers on thin foils that have a different composition to that of the bulk of the foil.[5–10] Accurate analysis of light elements also requires careful background subtraction since the continuous X-ray background changes rapidly below ~3 kV due to the combination of increasing generation efficiency, decreasing detector efficiency, and absorption in the specimen. In the present work all of the above effects have been taken into account.

METHODS OF MICROANALYSIS

Absorption correction and foil thickness determination

Following Ref. 11, the observed characteristic X-ray intensity from element A, $I(A)$, is related to the intensity which would have been observed in the absence of absorption, $I_0(A)$, by the equation

$$\frac{I_0(A)}{I(A)} = \frac{\mu_A \rho t \operatorname{cosec} \Psi}{1 - \exp(-\mu_A \rho t \operatorname{cosec} \Psi)} \tag{1}$$

where μ_A is the mass absorption coefficient for characteristic X-rays from A in the foil ($m^2\,kg^{-1}$), ρ the density of the foil, t its thickness, and Ψ the angle between the foil surface and the line connecting the detector and the point analysed, i.e. the take-off angle. It is important to note that this equation is valid only for a flat, parallel-sided foil and under the condition that the X-ray intensity generated per unit depth is the same throughout the thickness of the foil. This latter condition is strictly true only if there is no beam spreading and no energy loss through the foil thickness. Theoretical calculations indicate that this should be a good approximation[12] although recent experimental measurements indicate that those calculations may not be valid.[13] However, provided the magnitude of the absorption correction is kept small enough, the errors involved in adopting equation (1) rather than an improved equation will probably be negligible in comparison to other sources of error.

Application of equation (1) requires a knowledge of the mass thickness of the foil (ρt). Various methods are available for the measurement of foil thickness t.[14] The most accurate, the convergent beam technique, is very time-consuming, while the simpler contamination spot method considerably overestimates the foil thickness.[15,16] Even if care is taken to obtain an accurate value of t the density of the specimen is usually an unknown variable dependent on the chemical composition and crystal structure of the area analysed.

The difficulties and uncertainties in determining ρ and t separately can be overcome by obtaining the mass thickness of the specimen (ρt) directly from the recorded X-ray spectrum. Two ways of achieving this have been suggested by Morris *et al.*[17] The first, known as the $K:L$ Intensity Ratio Method, has the advantage of being independent of specimen geometry, but it is limited in application to alloys with at least one element giving detectable K and L lines ($Z > 27$). The second approach, known as the Multigeometry Method, requires several spectra to be recorded from a single analysis point at different tilt angles. This method is slow and prone to errors caused by contamination.

Nockolds *et al.*[18] have evaluated two more techniques. One, based on the transmitted electron intensity, requires frequent beam current measurement and is likely to suffer from errors due to contamination build-up. The other is based on the measurement of the characteristic X-ray intensity from the specimen relative to that from a bulk standard. This latter method has the advantage of allowing the mass thickness to be calculated 'on-line' directly from a recorded spectrum. However, the use of a bulk standard for intensity calibration is questionable in light of the large correction factors that are required at high accelerating voltages. A better approach is to use a thin foil of known mass thickness and composition (preferably a pure element) to calibrate intensity in terms of thickness. This approach has been adopted for the purposes of the present analyses. Since aluminium predominates in all of the phases analysed, thicknesses were calculated using the Al K intensity.

The mass thickness of any specimen whatsoever containing element A is given by

$$\rho t = \frac{\cos\theta}{e_A m_A} \cdot I_0(A) \qquad (2)$$

where θ is the tilt angle, m_A is the mass fraction of element A in the specimen, e_A is the number of A counts recorded per incident electron per unit mass thickness of A in the absence of absorption, and $I_0(A)$ is the absorption-corrected intensity of A from the specimen in counts s^{-1} nA^{-1}. The calibration constant e_A is simply obtained from a thin foil of, for example, pure A with a known mass thickness (see below).

Initially $I_0(A)$ is unknown, but combining equations (1) and (2) gives

$$I_0(A) = -u_A^{-1} \ln [1 - u_A I(A)] \qquad (3)$$

where

$$u_A = \mu_A \operatorname{cosec} \Psi \cos\theta / e_A m_A \qquad (4)$$

Similarly for another element, B, the corrected intensity $I_0(B)$ can be obtained from the measured intensity $I(B)$ using

$$I_0(B) = u_B I(B)\, I_0(A)/\{1 - \exp[-u_B I_0(A)]\} \qquad (5)$$

where

$$u_B = \mu_B \operatorname{cosec} \Psi \cos\theta / e_A m_A \qquad (6)$$

and μ_B is the mass absorption coefficient for B radiation in the specimen.

Of course μ_A, μ_B and m_A depend on the foil composition and an iterative procedure is required to calculate the corrected intensities $I_0(A)$ and $I_0(B)$.

In order to achieve as good an absorption correction as possible using this method, the following experimental precautions must be taken:

(i) *No strong diffraction*. Equation (2) does not hold close to low-index Bragg reflections.[19] Therefore spectra should always be recorded under kinematical diffraction conditions.

(ii) *No bent foils*. Thinned discs should be examined in a light microscope to check that the thin areas are not bent out of the plane of the disc.

(iii) *No rapid thickness changes*. Inclusions are often wedge-shaped, e.g. thin in the middle and thick around the edges. Since equation (1) only applies to a parallel-sided foil, areas with large wedge angles must be avoided.

(iv) *X-rays must travel through foil of known composition*. In deriving equations (3)–(6) it is assumed that the X-rays are absorbed in a homogeneous single-phase foil. Therefore, when an inclusion is analysed the analysis point must be located at a sufficiently large distance from the edge of the inclusion (measured towards the detector) to allow all absorption to occur within the inclusion. Thus for accurate quantitative analysis where an absorption correction must be applied the X-ray exit path, for most practical foil thicknesses, will be resolution-limiting.

(v) *Keep absorption correction small*. Systematic errors in the absorption correction are best mitigated by analysing sufficiently thin areas. In this work the absorption correction was kept below about 15%.

Fluorescence effects

Thin foil fluorescence correction models have been proposed by Philibert and Tixier[20,21] and by Nockolds *et al.*[18] In the case of Al_3Fe, Al K radiation is fluoresced by the Fe K X-rays. For a foil of 1 g m^{-2} (~257 nm) thick these models predict values of the fluorescence enhancement ratio of 2×10^{-6} according to the model of Tixier and 3×10^{-4} according to that of Nockolds *et al.* Even when using a bulk specimen, characteristic fluorescence at 200 kV should only amount to about 2×10^{-3} of the total intensity.[22] Fluorescence corrections have therefore been neglected.

Procedure for reducing errors due to surface layers

If the surface of the specimen is covered with a thin layer that has a different composition to the bulk, the apparent composition of the foil becomes a function of foil thickness. The error induced by ignoring the effect will depend on the difference in composition between the surface and bulk, the thickness of the layer and the thickness of the foil. In the case of dilute solutions of iron in aluminium, for example, it has been shown that less than a monolayer of iron has an overwhelming effect on the apparent composition of the foil.[10]

The influence of a *uniformly thick* surface layer on recorded intensity ratios is illustrated in Fig. 1. The slope of the line OP is given by the concentration ratio in the surface layer, and the length OP by the thickness. The absorption (and fluorescence)-corrected measured intensities will lie on a line PQ, that extrapolates to a point R, displaced from the origin. The true composition of the bulk is given by the slope of PQ, whereas if measurements

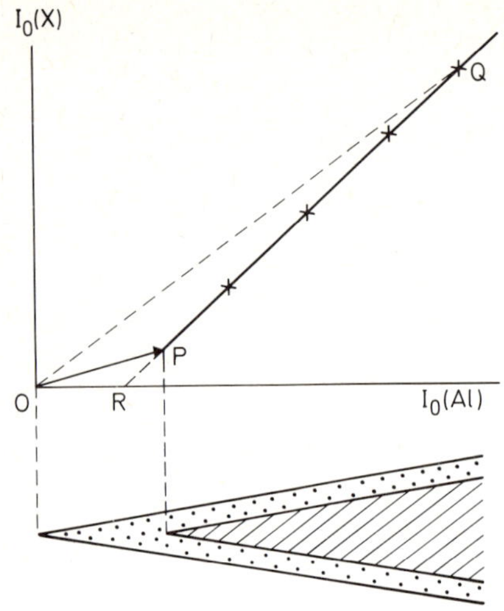

1 **Effect of surface layer with different composition to bulk on X-ray intensity as function of foil thickness**

are limited to a single foil thickness the apparent composition will correspond to a slope such as *OQ*.

Surface layers originate from specimen preparation, for example during electropolishing or ion milling, and as yet there is no simple and convenient way of preparing foils that can be guaranteed free from surface inhomogeneities. At present, therefore, it is best (where possible) to make measurements over a wide range of foil thicknesses and fit the best straight line through the data as shown in Fig. 1.

If this technique is used, however, it is important that the microscope used has an in-hole count that is (say) less than 1% of the signal that originates from the thinnest regions, which may be only ~10 nm thick. It is reported[23] that a large hole count cannot be simply subtracted from the total spectrum when the hole count itself accounts for a large fraction of the total spectrum. This was no limitation on the microscope used, however, where even at 200 kV the total in-hole count rate using an aluminium disc specimen is only ~4 cps in the whole spectrum, 0–10 keV. Furthermore this background is free of any statistically significant characteristic peaks.

Background subtraction

The two relevant parts of a typical spectrum from one of the phases of interest is shown in Fig. 2. Note that besides the major elements aluminium and iron, small quantities of silicon and sometimes nickel are also present. These elements have also been included in the analysis.

In the case of the small peaks from silicon and nickel the accuracy of the background subtraction is important. At the low-energy end of the spectrum the background has been measured on the high-energy side of the aluminium and silicon peaks and extrapolated under the peaks, taking into account:

(i) the variation of the continuum with energy which goes as $(E_0 - E)/E$;
(ii) absorption in the beryllium window, silicon dead layer and gold layer using the manufacturer's data; and
(iii) absorption in the specimen using equations (5) and (6).

In the high-energy region of the spectrum absorption in the specimen is small and the background variation was found to agree well with the theoretical function [*detector efficiency*] × $(E_0 - E)/E$. This function was therefore used to interpolate the background under the iron and nickel peaks.

EXPERIMENTAL TECHNIQUES

The particular alloy investigated was a commercially pure aluminium containing 0.47 wt-% Fe, 0.08 Si, 0.03 Ti and impurities totally less than 0.07 wt-%. The alloy was melted and poured into a vertical mould (60 × 25 × 250 mm) with a removable bottom. After a thin layer of aluminium had solidified, the bottom of the mould was removed and a water spray was directed onto the exposed solid aluminium to simulate the solidification conditions found with DC casting. Specimens for electron microscopy were then cut from the solidified ingot at distances of 18 (specimen I), 40 (specimen II), and 200 mm (specimen III) from the chilled surface. The cooling rates through the solidification range, as recorded on submerged thermocouples, were close to 40, 10 and 1 K s^{-1} for specimens I, II, and III respectively. Some material cut 40 mm from the chill surface was given an annealing treatment of 24 h at 873 K (600°C). This material will be referred to as specimen IIA.

The microstructure of the specimens examined is reported elsewhere.[1] Al_mFe is predominant in specimen I, II contains Al_6Fe and Al_xFe, while III contains Al_3Fe. In the annealed specimen (IIA) Al_6Fe and Al_xFe have been transformed into equilibrium Al_3Fe.

Thin foils for electron microscopy and microanalysis were prepared in a Tenupol jet polisher using a solution of

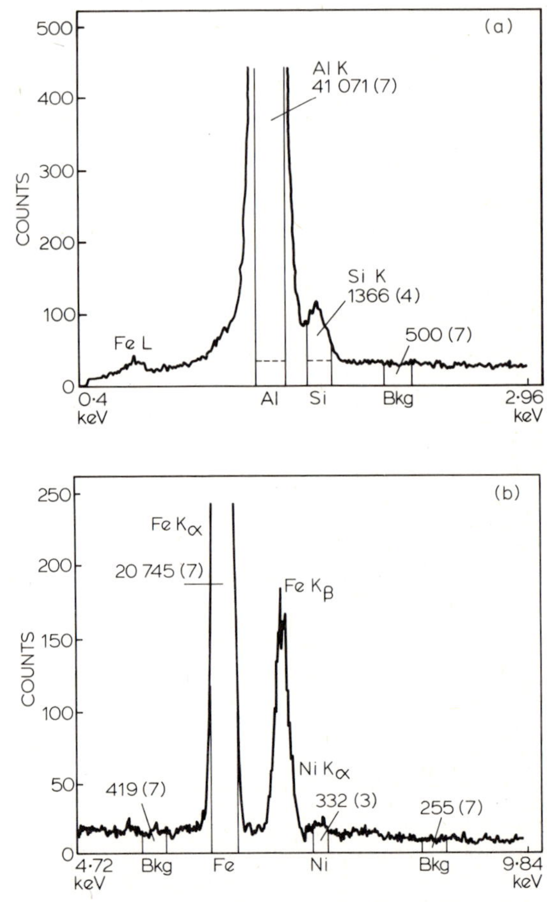

2 **Parts of X-ray spectrum from Al_mFe using 9.3 nA beam current and 50 s recording time. Numerals give counts in each window; those in parentheses show number of hole counts in same windows (a) 0.40–2.96 keV; (b) 4.72–9.84 keV**

TABLE 1 Data used to determine e_{Al}

Analysis point	Thickness t, nm	Mass thickness ρt, g m^{-2}	$I(Al)$,* s^{-1} nA^{-1}	$I_0(Al)$,† s^{-1} nA^{-1}	e_{Al},‡ s^{-1} nA^{-1} g^{-1} m^2
1	570	1.543	612	658	350
2	572	1.538	627	674	358
3	552	1.489	602	645	355
				mean	354 (± 1%)

* Number of counts in energy window of 1.41 × FWHM of Al peak; beam current 1.2 nA, count time 50 s, tilt angle θ = 35°
† Corrected for absorption using equation (1) and absorption data of Heinrich.[24] For detection geometry of Jeol 200B cosec $\Psi = \sqrt{2}/\sin\theta$
‡ From equation (2), m_{Al} = 1

33% concentrated nitric acid in methanol at a temperature of −20°C and a potential difference of 13 V. All specimens were carefully washed, dried and examined by light microscopy to ensure that the thin areas around the central hole were not bent or buckled. Electron microscopy and microanalysis was then carried out at 200 kV using a Jeol 200B TEM/STEM equipped with a Kevex 7000 EDS system.

RESULTS AND DISCUSSION

The calibration constant e_{Al} in equation (2) was determined from a thin foil of pure aluminium. The thickness of the foil was determined from a grain boundary aligned parallel to the main tilt of the specimen stage. The projected width of the boundary was measured at several tilt angles on both sides of 'vertical'. The orientation of the boundary in the foil and the foil thickness were then obtained by a linear regression analysis. The beam current was recorded using a Faraday cage located at the level of the fluorescent screen. The current in the probe was stable to ±1% over several hours and ±2% from day to day. Errors due to beam current drift were therefore insignificant.

The results of measurements at three positions along the same grain boundary are summarized in Table 1. For comparison the foil thicknesses at each analysis point were also measured using the contamination spot method. This gave values consistently greater than given by the grain boundary method by about 35 nm. From this experiment it can be concluded that, for the detection system used, 1 g m^{-2} of aluminium along the electron path gives a signal of 354 ± 4 counts s^{-1} nA^{-1} in an energy window of 1.41 × FWHM of the aluminium peak.

Following the procedures and precautions outlined earlier many analyses were made from many different particles of each of the phases of interest. The results are summarized in Figs. 3–9 and in Tables 2 and 3.

Figure 3 shows the recorded intensity ratios $I(Fe)/I(Al)$ for annealed Al_3Fe (specimen IIA) as a function of specimen thickness. There is a clear tendency for the relative concentration of aluminium to increase as the foil becomes thinner, consistent with the presence of an aluminium-rich surface layer. The same data, both before and after correction for absorption, are shown in Fig. 4. The linear regression line through the corrected data is drawn and this gives a positive intercept on the aluminium axis as expected. Similar aluminium-rich surface layers are found on the other phases (Figs. 5 and 6). The data for silicon and nickel can also be plotted in this way. Two examples are shown in Figs. 7 and 8.

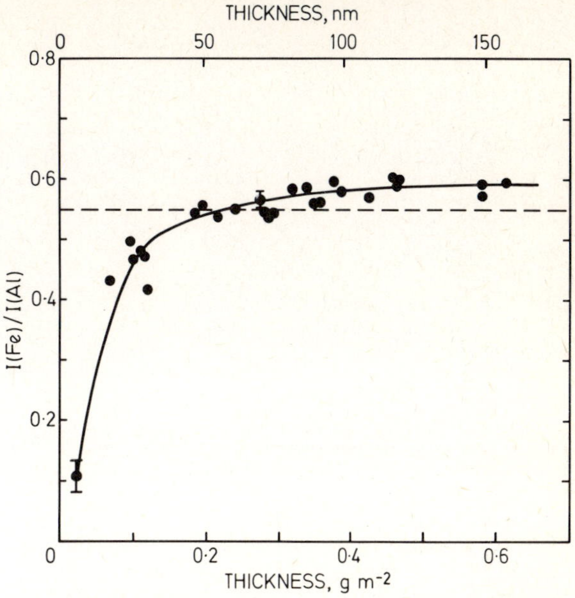

3 **Recorded intensity ratios as function of foil thickness for annealed Al_3Fe; dashed line shows 'true' ratio**

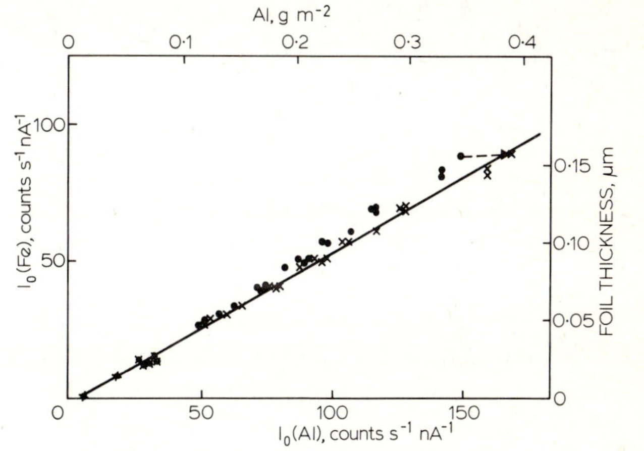

4 **Intensities of iron and aluminium X-rays from annealed Al_3Fe both before (•) and after (×) correction for absorption**

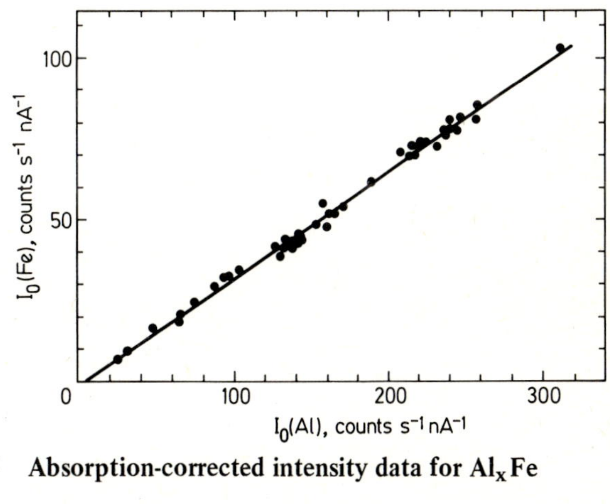

5 **Absorption-corrected intensity data for Al_xFe**

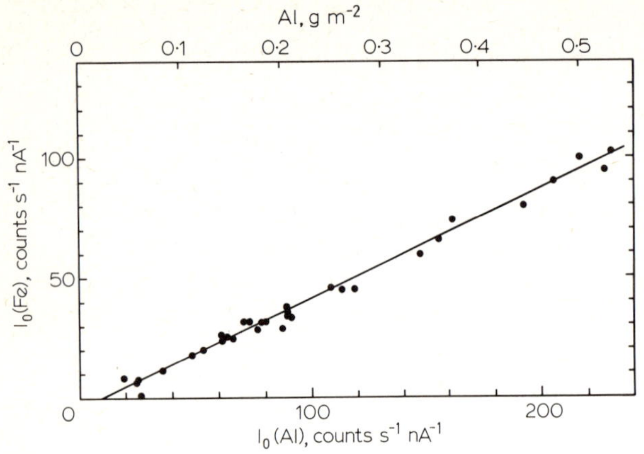

6 Absorption-corrected intensity data for Al_mFe

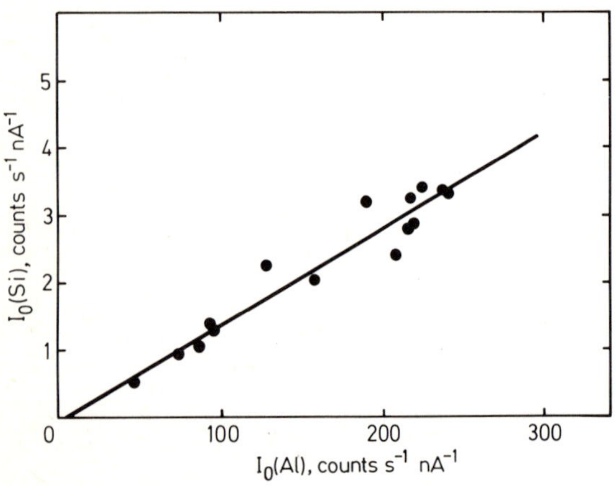

7 Intensities of silicon and aluminium X-rays from Al_xFe

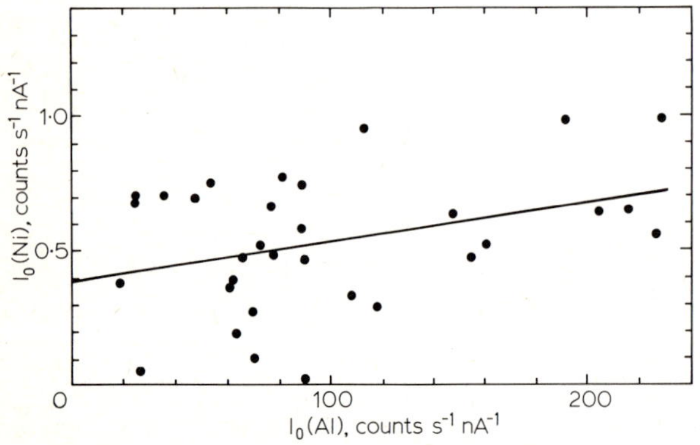

8 Intensities of nickel and aluminium X-rays from Al_mFe

In all of the above cases the best line through the data can be written as

$$I_0(X) = a_x + b_x I_0(Al) \tag{7}$$

where X stands for Fe, Si or Ni, and a_x and b_x are the regression coefficients as determined by the least-squares method. The magnitudes of a_x and b_x for each element in each phase analysed are listed in Table 2, together with the correlation coefficient r^2. The surface excess of aluminium is indicated by the negative values of a_{Fe}. If it is assumed that the surface layer contains no iron then the mass thickness of aluminium on the surface S_{Al} is given by

$$S_{Al} = \frac{-a_{Fe}}{b_{Fe}} \cdot \frac{\cos\vartheta}{e_{Al}} \tag{8}$$

Values of S_{Al} are listed in Table 2. For example, in the case of annealed Al_3Fe there appears to be ~9 mg m^{-2} of aluminium on the surface. (This would correspond to a total thickness of ~4 nm of pure Al_2O_3, i.e. ~2 nm on each surface.)

The values of a_{Si} are of interest since silicon can accumulate in the contamination spot. This has the same effect as silicon enrichment in the surface and results in positive values of a_{Si}. In the case of the Al_3Fe inclusions analysed there may therefore have been silicon in the contamination or in the surface oxide (see Table 2). However, with the Al_xFe and Al_mFe inclusions there was no indication of any silicon other than in the bulk. (Insufficient data points were available for Al_6Fe for which it was assumed $a_{Si} = 0$.)

The positive values of a_{Ni} imply that nickel was probably enriched on the surface. This is also illustrated in Fig. 8 where it can be seen that although the nickel signal is quite significant it appears to be almost independent of the specimen thickness.

In the presence of surface layers the usual ratio equation, i.e.

$$C_x/C_{Al} = k_{x/Al} \cdot I_x/I_{Al} \tag{9}$$

is replaced by the equation

$$C_x/C_{Al} = k_{x/Al} \cdot b_x \tag{10}$$

The proportionality constant $k_{Fe/Al}$ has been obtained by taking the annealed Al_3Fe inclusions (specimen IIA) as a standard. Unfortunately Al_3Fe is not stoichiometric, but according to the equilibrium phase diagram[25] the composition in equilibrium with aluminium at 873 K and below is ~36.5 Fe–63.5 wt-%Al. Substitution of these values into equation (10) gives the values for $k_{Fe/Al}$ shown in Table 2. It has been assumed that the small concentration of silicon and nickel in the Al_3Fe does not significantly affect the C_{Fe}/C_{Al} ratio.

The values for $k_{Si/Al}$ and $k_{Ni/Al}$ shown in Table 2 are calculated values using the 'standardless' program of Bengtsson.[26] The accuracy of the analysis of silicon and nickel is limited by the statistical fluctuations in b_{Si} and b_{Ni}. The error introduced by using theoretical values for k is therefore not significant (the standardless program predicts values for k to within about 10% accuracy).

The final analyses are summarized in Table 3. Conversion to atomic percentages give the Al:Fe atom ratios shown. The error bars are 95% confidence limits based on the regression analyses and the Student t distribution. Note that there is no significant difference in composition between the Al_3Fe that solidifies in the middle of the ingot and that formed by transformation of Al_6Fe and Al_xFe.

TABLE 2 Experimental data (95% confidence limits from regression analysis)

Phase	X	Pts	i_B, nA	τ, s	a_X, counts s^{-1} nA^{-1}	b_X	r^2	S_{Al}, mg m^{-2}	$k_{X/Al}$*	C_X/C_{Al}
Al_3Fe (annealed) specimen IIA	Fe	28	1.2	50	−2.04 ±1.22	0.5489 ±0.0130	0.997	9 ±5	1.047 ±0.024	0.5748† ±0.0131
	Si	28	1.2	50	0.20 ±0.23	0.0067 ±0.0025	0.543	–	0.92	0.0062 ±0.0023
	Ni	29	1.2	50	0.20 ±0.18	0.0012 ±0.0017	0.069	–	1.33	0.0016 ±0.0023
Al_3Fe (as-solidified) specimen III	Fe	10	8.8	50	−1.22 ±2.83	0.5404 ±0.0360	0.994	5 ±12	1.047 ±0.024	0.5658 ±0.0400
	Si	10	8.8	50	0.14 ±0.11	0.0074 ±0.0012	0.962	–	0.92	0.0068 ±0.0011
	Ni	10	8.8	50	0.09 ±0.07	0.0011 ±0.0009	0.501	–	1.33	0.0015 ±0.0012
Al_6Fe specimen II	Fe	4 3	8.2 1.2	50 50	−0.24 ±0.54	0.3128 ±0.0193	0.997	2 ±4	1.047 ±0.024	0.3275 ±0.0216
	Si	4	8.2	50	(0)	0.0012 ±0.0008	–	–	0.92	0.0011 ±0.0008
	Ni	4	8.2	50	(0)	0.0017 ±0.0031	–	–	1.33	0.0022 ±0.0042
Al_xFe specimen II	Fe	34 10	0.78 9.2	50 50	−3.45 ±2.09	0.3404 ±0.0108	0.991	24 ±14	1.041 ±0.024	0.3544 ±0.0139
	Si	15	9.2	50	−0.06 ±0.45	0.0143 ±0.0028	0.909	–	0.92	0.0132 ±0.0026
	Ni	28 10	0.78 9.2	50 50	0.47 ±0.80	0.0037 ±0.0025	0.220	–	1.31	0.0048 ±0.0033
Al_mFe specimen I	Fe	30 2	0.77 9.3	50 50	−4.77 ±1.98	0.4553 ±0.0171	0.990	24 ±10	1.041 ±0.024	0.4740 ±0.0209
	Si	30 2	0.77 9.3	50 50	−0.05 ±1.3	0.0189 ±0.0041	0.756	–	0.92	0.0174 ±0.0038
	Ni	30 2	0.77 9.3	50 50	0.39 ±0.45	0.0014 ±0.0014	0.130	–	1.31	0.0018 ±0.0018

Col. 3 'Pts' = number of points analysed
Col. 4 'i_B' = beam current used
Col. 5 'τ' = analysis time per point
* For Al_3Fe and Al_6Fe analyses window widths were 1.41 FWHM for Al, 1.80 FWHM for Fe, 1.22 FWHM for Si, and 1.01 FWHM for Ni.

For Al_xFe and Al_mFe analyses window widths were 1.38 FWHM for Al, 1.66 FWHM for Fe, 1.20 FWHM for Si, and 1.00 FWHM for Ni.

For Fe k values are from the Al_3Fe standard, for Si and Ni k values were calculated using the standardless program of Bengtsson.[26]
† This value is that for pure binary Al_3Fe (36.5Fe–63.5 wt-%Al).

TABLE 3 Summary of analyses (weight percent with 95% confidence intervals)

	Al_3Fe (annealed)	Al_3Fe (as solidified)	Al_6Fe	Al_xFe	Al_mFe
Al	63.19±0.54	63.53±1.6	75.14±1.2	72.86±0.77	66.97±0.96
Fe	36.32±0.53	35.95±1.6	24.61±1.2	25.82±0.77	31.74±0.96
Si	0.39±0.14	0.43±0.07	0.08±0.06	0.96±0.19	1.17±0.26
Ni	0.10±0.14	0.10±0.08	0.16±0.31	0.36±0.25	0.12±0.12
Al/Fe	3.60±0.08	3.66±0.27	6.32±0.43	5.84±0.24	4.37±0.20

Al_6Fe is, within experimental error, stoichiometric. This is not unexpected as it is isomorphous with the stoichiometric equilibrium phase Al_6Mn. The new type of inclusion Al_xFe is also within experimental error of the formula Al_6Fe, although it has a higher silicon content than Al_6Fe and a significant nickel content.

The metastable Al_mFe phase has a composition implying it may best be described as Al_4Fe. (However, the most appropriate formula can be obtained only from a determination of the atomic structure of the unit cell.) Al_mFe also has a high silicon but a low nickel content.

REFERENCES

1 H. WESTENGEN: to be published
2 I. MIKI, H. KOSUGE, and K. NAGAHAMA: *J. Jpn Inst. Light Met.*, 1975, **25**, 1
3 K. ROBINSON and P. J. BLACK: *Philos. Mag.*, 1953, **44**, 1392
4 E. H. HOLLINGSWORTH, G. R. FRANK Jr, and R. E. WILLETT: *Trans. Met. Soc. AIME*, 1962, **224**, 188
5 M. N. THOMPSON, P. DOIG, J. W. EDINGTON, and P. E. J. FLEWITT: *Philos. Mag.*, 1977, **35**, 1537
6 P. L. MORRIS, N. C. DAVIES, and I. A. TREVERTON: 'Developments in electron microscopy and analysis, 1977', 377-380, ed. D. L. Misell, Proc. EMAG, 1977, Bristol, Inst. Physics
7 P. L. MORRIS and H. J. LAMB: *J. Phys. D. (Appl. Phys.)*, 1978, **11**, L73
8 K. J. SAWLEY, G. CLIFF, and C. HAWORTH: *ibid.*, L76
9 J. M. POUTNEY and M. H. LORETTO: Proc. 7th Eur. Congr. Electron Micr., The Hague, Netherlands, Aug. 1980, **3**, 180
10 D. A. PORTER, J. STRID, and K. E. EASTERLING: to be published
11 R. KÖNIG: 'Quantitative X-ray microanalysis of thin films', in 'Electron microscopy and mineralogy', ed. H. R. Wenk, 526; 1976, Berlin, Springer-Verlag
12 J. I. GOLDSTEIN, J. L. COSTLEY, G. W. LORIMER, and S. J. B. REED: 'Scanning Electron Microscopy, 1977', vol. 1, ed. O. Johari, 315-324, 1977, Chicago, SEM/IITRI
13 N. STENTON, M. R. NOTIS, J. I. GOLDSTEIN and D. B. WILLIAMS: this conference
14 G. LOVE, M. G. C. COX, and V. D. SCOTT: 'Developments in electron microscopy and analysis', 347-350, ed. D. L. Misell, Proc. EMAG, 1977, Bristol, Inst. Physics
15 P. L. MORRIS: 'Quantitative thin-foil X-ray analysis of inorganic materials', Alcan Technical File, Banbury, Oct. 1979
16 D. A. RAE, V. D. SCOTT, and G. LOVE: this conference
17 P. L. MORRIS, M. D. BALL, and P. J. STATHAM: 'Electron microscopy and analysis', 413-416, ed. T. Mulvey, Inst. Phys. Conf. Ser. No. 52, 1980, Bristol
18 C. NOCKOLDS, M. J. NASIR, G. CLIFF, and G. W. LORIMER: *ibid.*, 417-420
19 D. CHERNS, A. HOWIE, and M. H. JACOBS: *Z. Naturforsch.*, 1973, **28a**, 565
20 R. TIXIER: PhD thesis, Université de Paris, 1975
21 J. PHILIBERT and R. TIXIER: 'Physical aspects of electron microscopy and microbeam analysis', eds. B. M. Siegel and D. R. Beaman, 333-354; 1975, New York, J. Wiley
22 S. J. B. REED: 'Electron microprobe analysis', 288-292; 1975, Cambridge, Cambridge Univ. Press
23 N. J. ZALUZEC: 'Introduction to analytical electron microscopy', eds. J. J. Hren, J. I. Goldstein, D. Joy, 121-167; 1979, Plenum Press
24 K. F. J. HEINRICH: 'The electron microprobe', eds. T. D. McKinley, K. F. J. Heinrich, D. B. Wittry, 296-378; 1966, New York, Wiley
25 L. F. MONDOLFO: 'Aluminium alloys: structure and ;properties', 282-289; 1976, London, Butterworths
26 B. BENGTSSON and K. E. EASTERLING: 'Scanning electron microscopy 1978', vol. 1, 655-662, ed. Om Johari, 1978, AMF O'Hare, Scanning Electron Microscopy Inc.

DISCUSSION of paper by Porter and Westengen

L. M. BROWN:
Did you take any electron diffraction patterns from the particles?

D. A. PORTER:
Yes. The particles were first identified on the basis of their electron diffraction patterns. The new microstructural constituent Al_xFe gave very characteristic patterns but they could not be indexed consistently. All the other phases have well-documented crystal lattices.

Characterization of supported metal catalysts using STEM

H Dexpert, E Freund, and J Lynch

A V.G. HB5 STEM equipped with an ED X-ray spectrometer has been used to characterize heterogeneous supported metallic catalysts. The possibilities in the field of high resolution microanalysis are exemplified. In the simple case of a monometallic catalyst supported on gamma alumina, the STEM allows localization of the metallic phase, even when highly dispersed, by X-ray emission microanalysis, and a qualitative description of the structure of metal crystallites and their orientation relative to the carrier by microdiffraction. The formation of alloys in multimetallic catalysts can be checked in a wide range of crystallite sizes, down to small clusters of less than 100 atoms. Catalyst aging can be followed in detail, including for instance local chemical and structural modifications, localization of poisons (sulphur, lead and the like) or carbon deposits. Improvements of the microscope which can lead to even better microanalytical capability are discussed.

The authors are with the Institut Français du Pétrole, Rueil-Malmaison, France.

The physico-chemical characterization of heterogeneous catalysts has two main purposes:

(1) to give a qualitative picture which may be used as a guide to improvement in the preparation procedure (for instance to ensure alloying in supported multimetallic catalysts); and
(2) to understand catalyst activity, or at least to relate the observed activity to fundamental studies carried out with idealized systems.

Two kinds of method may be considered:[1] global (macroscopic) methods and local (microscopic or nanometric scale) techniques involving high spatial resolution. The latter are particularly relevant in the case of industrial heterogeneous catalysts. These highly divided solids (specific areas generally larger than $10^2\ m^2\ g^{-1}$) may be considered as composed of small particles of typical dimension 10 nm, comprising light oxide carrier and supported heavy metal(s). These are complex systems, the exploration of which requires a spatial resolution of at least 1 nm (and preferably 0.2-0.3 nm) as well as microanalytical capabilities such as the determination of (i) local elemental composition, (ii) microstructure, (iii) nature of exposed surfaces, and (iv) chemical state of the elements, for both carrier and metal.

It is the purpose of this paper to present some applications of STEM to the characterization of certain industrial supported metallic catalysts.

EXPERIMENTAL

A V.G. HB5 STEM[2] equipped with a Kevex energy-dispersive X-ray spectrometer was used. The analytical pole pieces compatible with this spectrometer have a point-to-point resolution of about 0.45 nm. The electron spectrometer used for EELS studies has a 1.5 eV resolution. The microdiffraction facilities are currently under modification to allow simultaneous acquisition of diffraction pattern and bright field image.

The catalysts are laboratory-scale preparations.[3] Most have been studied in the 'fresh' state, i.e. after the proper activation procedure (a reduction in hydrogen). Some have been characterized after aging under reaction conditions in bench scale testing units. They are listed in Table 1, with relevant information on preparation procedure, composition and testing procedure.

The samples were studied either as such, after grinding and collection onto holey carbon films from suspension in ethanol, or as carbon replicas after elimination of the oxide carrier in dilute hydrofluoric acid. In all cases where contamination might have been a problem (spot mode analyses) the grids were baked overnight in the microscope (423 K, 10^{-6} Pa).

RESULTS

Three kinds of samples will be considered:

1 fresh monometallic catalysts,
2 fresh bimetallic catalysts, and
3 used (aged and polluted) catalysts.

1 Monometallic systems

A simple use of the STEM's microanalytical capabilities is in localizing the metallic phase of a supported monometallic catalyst in situations where conventional imaging cannot provide unambiguous information. This is the case if the metal is well dispersed (as very small crystallites or uniformly) and the carrier strongly diffracting so that the contrast in the image is insufficient to distinguish the metal crystallites. The distribution may be studied at high resolu-

TABLE 1

Fig. No.	Catalyst composition	Preparation	Treatment
1	0.3wt-%Pd/γ_t Al_2O_3	Dry impregnation ($Pd(NO_3)_2$); air calcination 400°C	
2	as Fig. 1	as Fig. 1 + H_2 reduction 100°C	
3	[Pt/Pd$^{(at.)}$ = 1]/γ_t Al_2O_3	Co-impregnation; air calcination; H_2 reduction	
4	Rh (0.2wt-%)-Pt (0.58wt-%)/γ_c Al_2O_3	Ion exchange (H_2PtCl_6 + $RhCl_3$); air calcination; H_2 reduction	
5	0.6wt-%Pt/γ_c Al_2O_3 + 27wt-%C	Ion exchange (H_2PtCl_6); air calcination; H_2 reduction	nC_7 reforming
6	as Fig. 3	as Fig. 3	Oxidation of car exhaust gases (leaded petrol) 400–800°C

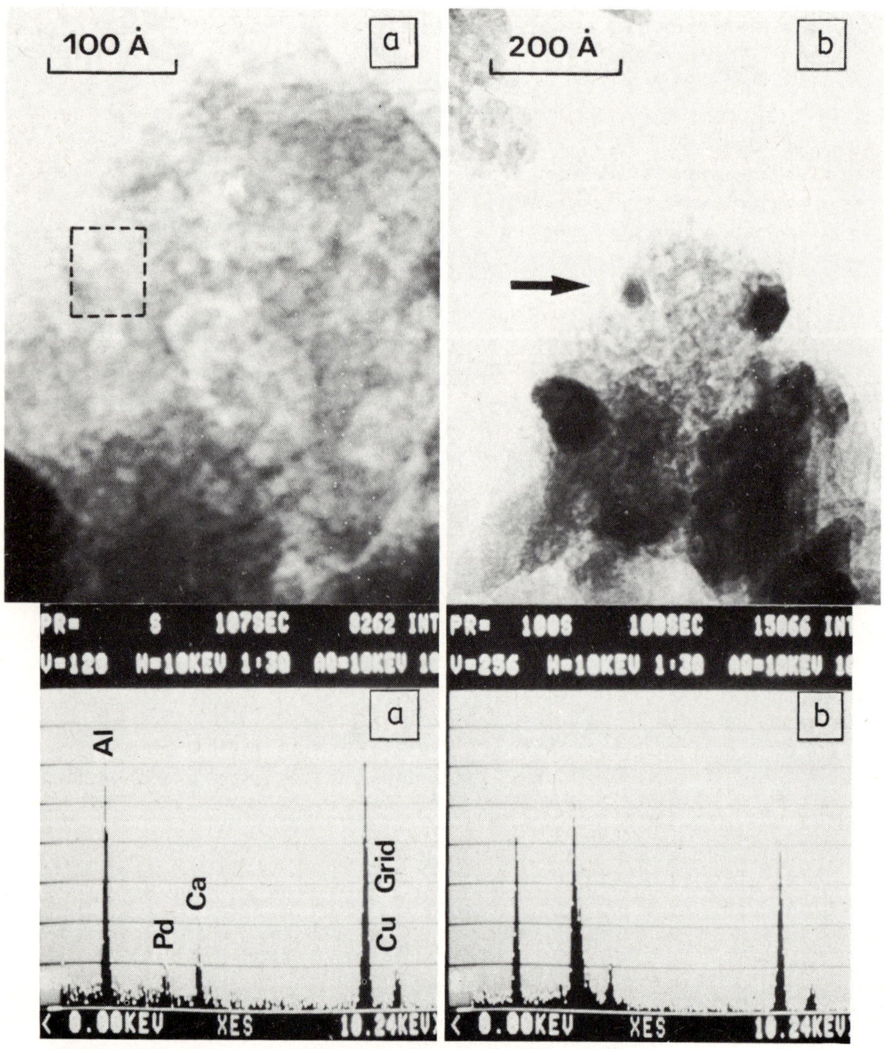

1 **Pd/alumina: (a) area of support showing no distinguishable particles and analysis of boxed area; (b) same area at lower magnification after exposure to beam, showing segregation of metal**

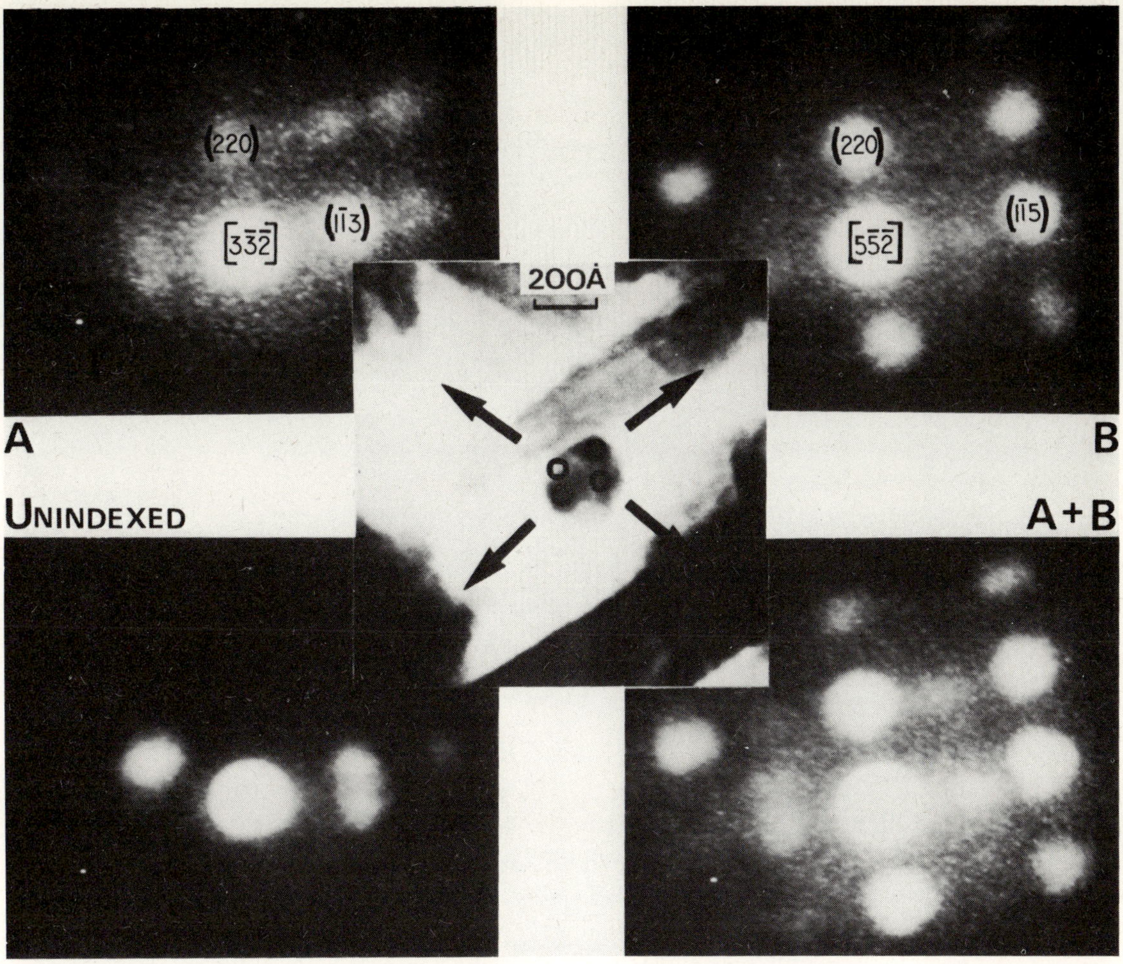

2 **Pd/alumina: large particle with four microdiffraction patterns taken from different regions**

tion using either X-ray emission or EELS for sufficiently high metal loadings.

An example is shown in Fig. 1 of an alumina-supported palladium precursor (unreduced catalyst). Here the palladium is found in the form of

(1) a highly dispersed phase not visible in the image (the analysis of 1*a*);
(2) small (1-1.5 nm) particles, occasionally distinguishable from the support, most probably the oxide; and
(3) large (10-20 nm) particles showing a defect in the preparation technique.

The co-detection of calcium is typical of the information supplied by X-ray spectroscopy analysis in this sort of system. Metal crystallites are often associated with cationic impurities in the carrier.

Figure 1*b* shows the type of very strong beam-specimen interaction often encountered in such specimens. When the probe is stopped for some minutes on the area analysed in Fig. 1*a*, a dark patch is seen (arrowed) which gives a much larger palladium response. A metal particle has formed under the heating effect of the beam.

Microdiffraction may be used to characterize both the structure of the metallic crystallites and their orientation relative to the carrier (if this is not amorphous). Figure 2 gives an example of the microstructural characterization of a palladium particle. Several microdiffraction patterns have been acquired from the same crystal of which two are indexable with zone axes a few degrees apart. A third shows both of these orientations and a fourth a different orientation. This series suggests that the particle is either under strain or polycrystalline as a result of the coalescence of smaller crystallites.

2 Multimetallic systems

In such catalysts the main problem is to check whether an alloy has been obtained or not. The answer is readily obtained by X-ray emission microanalysis provided the crystallites contain sufficient numbers of atoms. Figure 3 presents results obtained for a bimetallic Pt/Pd catalyst on alumina (atomic ratio Pt : Pd = 1). X-ray emission from the support alone shows no trace of either metal. The larger particle analysed shows an atomic ratio Pt : Pd = 0.7 : 1 when fluorescent yield and cross-section variations are corrected.[4] The smaller (diameter 2 nm) consists solely of platinum. The very low background count (0.07 counts per channel per second compared with the peak count rate 0.23 c/ch/s is due to the use of a 'virtual' objective aperture and leads to easy identification in 100 seconds. Significant peaks should be obtainable from crystallites of one-third of the volume, so that the resolution limit may be estimated at about 1.25 nm, or 100-atom particles, for this counting time.

About two-thirds of the particles analysed are alloy particles (showing the preparation to be largely successful) although wide variations in composition are observed. These variations must be considered in any attempt to

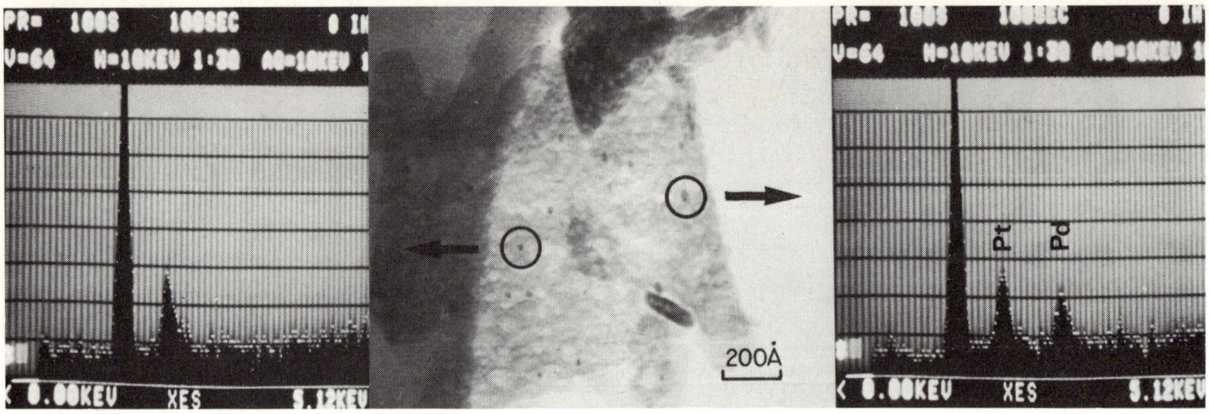

3 **Pt-Pd/alumina: analysis of two particles showing smaller to be pure Pt, larger to be an alloy**

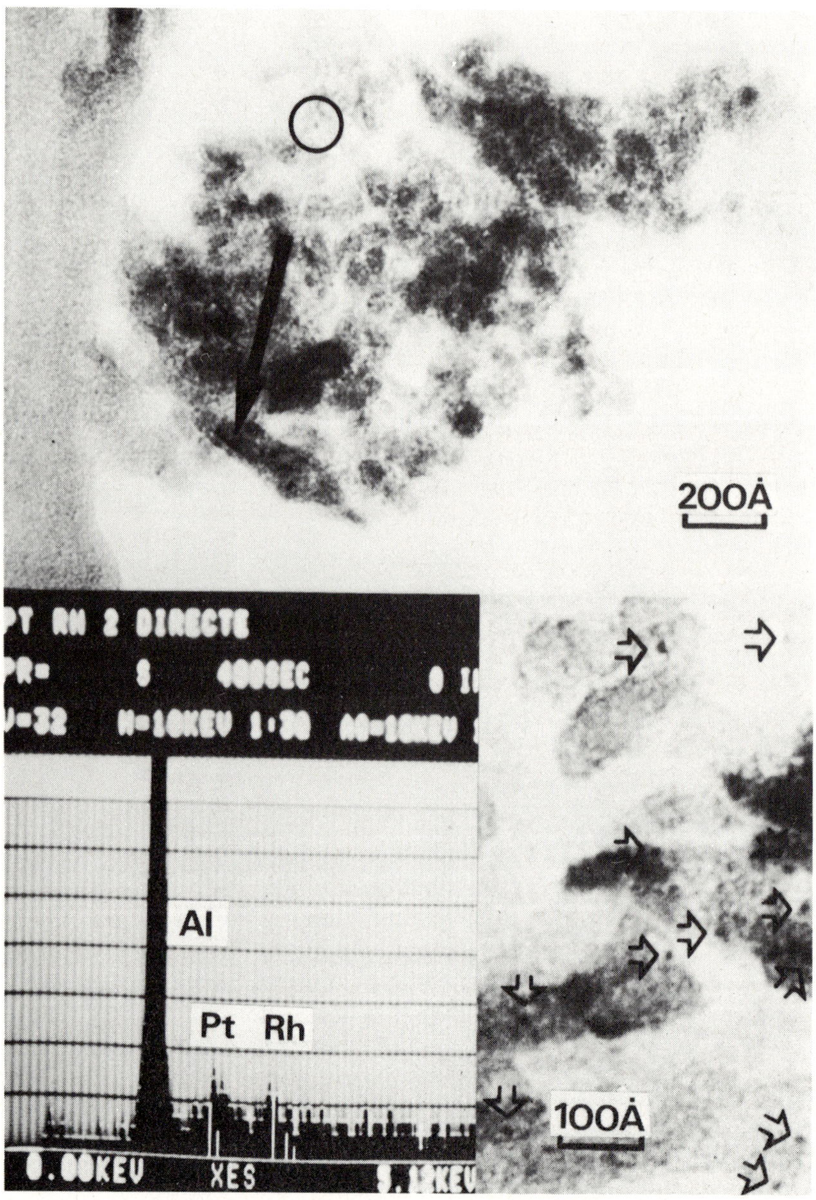

4 **Pt-Rh/alumina: images showing typical particle size to be 1.25 nm and spectrum obtained from single particle**

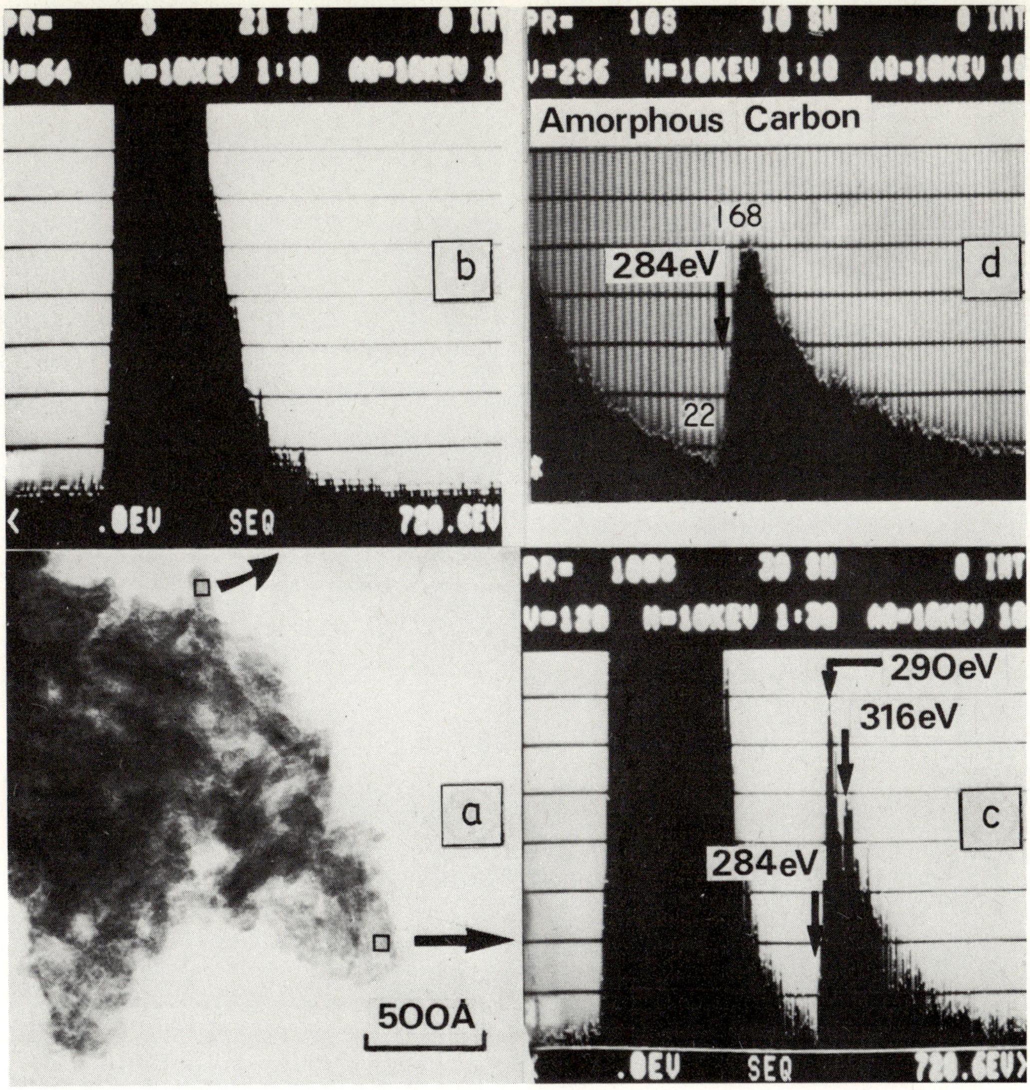

5 Pt/alumina used catalyst: (a) bright field (BF) image; (b, c) energy loss spectra of two areas of support showing heterogeneity of coke deposit; (d) energy loss spectrum of carbon support film for comparison with (c)

correlate the catalytic properties with the global composition of the catalyst.

The ultimate sensitivity for specimens of this kind has been ascertained with another bimetallic system: Pt–Rh/γ_c alumina. This catalyst shows metallic crystallites of about 1–2 nm diameter. An analysis is given in Fig. 4. The Pt and Rh lines are clearly visible on the X-ray emission spectrum, but the standard criterion:[5]

$$R = \frac{\text{signal}}{(2\ \text{background})^{1/2}} \geqslant 3$$

is not met. In fact R is 2.7 for the Pt line and 1.6 for the Rh. By increasing the counting time (up to about 15 minutes) this criterion could very likely be met. The sensitivity would then be set to about 30 atoms.

A third platinum-based bimetallic system to be considered is Pt–Ru/γ_c alumina (a reforming catalyst). This system has been previously characterized by hydrogen and oxygen chemisorption and related methods by Blanchard *et al.*[6] Depending on the preparation procedure, platinum either was or was not alloyed with ruthenium. These results were confirmed in the STEM with additional information:

(i) in the apparently unalloyed catalysts, platinum is found in the larger crystallites (1.5–2.5 nm) which are in fact dilute alloys (up to 10 at.-%), ruthenium being present as non-alloyed, very small crystallites (less than 1 nm);
(ii) in the alloyed catalysts most particles are alloyed, but again large variations of composition are found.

The preceding examples correspond to low content catalysts (typically less than 0.5 at.-%). Another limitation of X-ray emission microanalysis is apparent for heavily loaded catalysts (more than 1 at.-%). In the case of a carbon-supported Pt–Pd catalyst (used as a hydrogen oxidation catalyst in hydrogen–air alkaline fuel cells) it was not possible to analyse particles singly due to lack of spatial resolution. Thus the Pt:Pd ratio observed in the emission spectrum (1:1) cannot be interpreted as proof of alloying.

General limitations and possible improvements of X-ray emission microanalysis are discussed in the next section.

3 Poisoned metallic catalysts

Two kinds of poisoning can be schematically distinguished:[7]

(1) *Intrinsic poisoning*, which results from lack of selec-

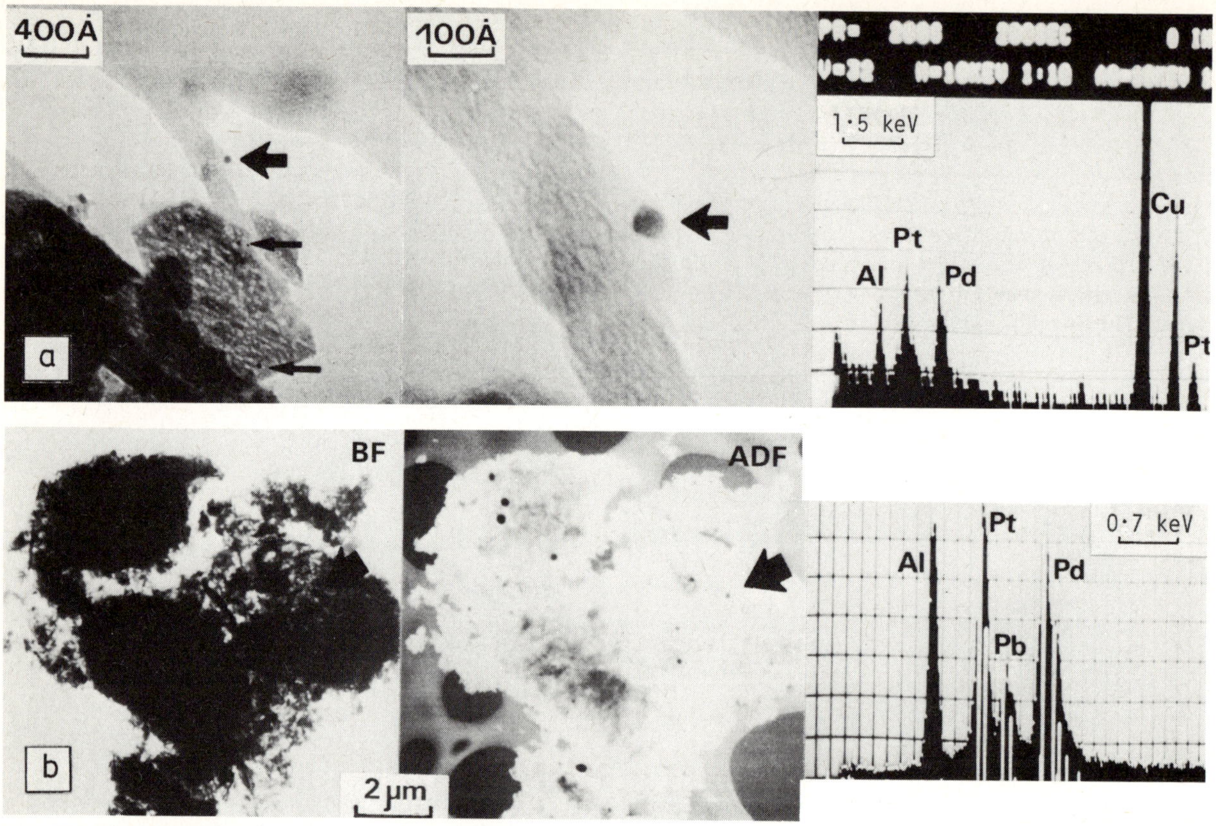

6 Pt–Pd/alumina: (a) new catalyst with analysis of typical particle; (b) bright field (BF) and annular dark field (ADF) images of used catalyst with analysis of arrowed particle

tivity; some by-products of the reaction may poison the catalyst. In the case of hydrocarbon conversion catalysts (used in petroleum refining and petrochemical processes) the main intrinsic poison is 'coke' (a general description of carbonaceous deposits).

(2) *Extrinsic poisoning*, due to impurities contained in the reactants, sulphur, arsenic, lead, etc.

The STEM has proven effective for the study of coke in reforming catalysts (generally platinum based bi- or tri-metallic alumina supported systems) using EELS. When appropriate illumination and collection conditions are adopted (large objective and collector apertures) adequate spatial resolution (better than 2 nm) and high sensitivity can be attained. Figure 5 shows an example of localization of carbon using the *K* edge.

Results obtained clearly demonstrate the heterogeneity of carbon deposition (and this is confirmed by other methods). Furthermore the EXELFS variations observed in these deposits differ from those seen on spectra taken from the amorphous carbon support film. They resemble rather those obtained by Batson and Craven from graphite,[8] suggesting a pre-graphitic structure for the coke. This has also been confirmed by X-ray diffraction for high coke contents and Raman spectroscopy. Work is now in progress on more detailed studies of relationships between coke and metal, coke and acid centres, etc., which can be carried out in the STEM.

Very little can be done to characterize lead-poisoned exhaust post-combustion catalysts such as the Pt–Pd/γ alumina system considered above, by macroscopic methods. From work in real conditions with leaded petrol it can be observed that a large amount of lead is fixed by the catalyst (up to 8 wt-%). These deposits are not, however, detectable in X-ray diffraction studies. The STEM gives a very clear picture of the used catalyst.

Figure 6 shows some of the analysis carried out on (a) fresh and (b) used catalysts. Morphological changes of the support are seen, indicating partial recrystallization to α alumina. Lead is found fixed on the γ alumina in varying degrees and on a morphologically different phase, up to concentrations consistent with a $PbAl_2O_4$ composition (the stable phase).

During use, sintering of the metals occurs, leading to a much smaller number of large (100–200 nm) particles, difficult to detect in the bright field (BF) image against the aggregates of support crystallites. Due to their large scattering at very high angles, they remain dark in the annular dark field (ADF) image whilst the alumina appears bright. Analysis showed all the particles to contain small amounts of lead (2 wt-% on average).

DISCUSSION

From the above results it is clear that the STEM is well suited to the study of a large range of problems in industrial catalysis. The advantages of STEM are based on the nature of the specimens, typically heavy metal clusters on light supports. This means that novel imaging techniques (such as annular dark field and *Z* contrast imaging) can be valuable and leads to fairly easy elemental identification and distribution 'mapping'. In addition local structural information is available and will be of interest, particularly in the case of crystalline supports.

There are, however, many problems and areas of study in this field for which the currently available commercial STEMs are limited. Improvements would be welcomed in all of the analytical facilities which follow listed as (a)-(d).

(a) Image

It suffices to say that a 2 nm crystallite is considered as very large in the field of supported noble metal catalysts. Ultra-high resolution (of the order available with a 200 keV CTEM[10]) is clearly needed just to visualize what will be analysed. This would require the development of high voltage STEM.

(b) X-ray emission analysis

At present the practical sensitivity of the HB5 is about 100 atoms, although with stable specimens and longer counting times a 30 atoms limit seems attainable. Again this corresponds to sizeable particles or high metal loadings of the carrier. This limit may be brought down by improved lens design (the current for a given probe size may be increased for lenses of lower C_s) and improved detector geometry (to increase the solid angle of collection). Many of the pollutants of interest are of low atomic number and cannot be studied with the present detector system. Windowless detectors now being developed should greatly increase the range of study in the field of poisoning.

(c) Microdiffraction

The Grigson (scanning) detection system has a low quantum efficiency (10^{-4}-10^{-6} depending on the angular resolution and therefore on the collector aperture size chosen). Parallel recording can be made much more efficient with careful choice of the optical system used.[11] Trials of parallel recording methods are currently being carried out at the IFP.

(d) EELS

Many of the noble metals studied present characteristic losses in the low energy loss (plasmon-dominated) region of the spectrum, due to M and N shell ionizations.[12] The edges are therefore not sharp and are superimposed on a high background. Again, improved detection capability – by the use of post-specimen lenses to overcome spectrometer limitations, and parallel recording[13] – would lead to increased utility in catalysis problems.

ACKNOWLEDGMENT

J. Lynch would like to thank the Royal Society and the Centre National de la Recherche Scientifique for an exchange fellowship.

REFERENCES

1 'Characterization of catalysts', eds. J. M. Thomas and R. M. Lambert, Wiley, 1981
2 L. M. BROWN: Inst. Phys. Conf. Ser., 1977, **36**, 141
3 R. L. MOSS: in 'Experimental methods in catalytic research', eds. R. B. Anderson and P. T. Dawson, Academic Press, 1978
4 S. J. B. REED: 'Electron microprobe analysis', Cambridge Univ. Press, 1975
5 J. I. GOLDSTEIN and H. YAKOWITZ: 'Practical S.E.M.', Plenum Press, New York, 1975, 457
6 G. BLANCHARD *et al.*: accepted for publication *J. Cat.*
7 R. E. MONTARNAL: Proc. 2nd Int. Symp. on Prep. of Cat., 1978, **2**, 519
8 P. BATSON and A. J. CRAVEN: *Phys. Rev. Lett.*, 1979, **42**, 893
9 J. P. BRUNELLE *et al.*: in 'Catalyst deactivation', eds. B. Delmon and G. F. Froment, Elsevier, 1980
10 E. D. BOYES *et al.*: Inst. Phys. Conf. Ser., 1980, **52**, 445
11 J. M. COWLEY: Inst. Phys. Conf. Ser., 1980, **52**, 239
12 W. LOTZ: *J. Opt. Soc. Am.*, 1970, **60**, 206
13 D. G. JENKINS *et al.*: Inst. Phys. Conf. Ser., 1980, **52**, 81

Use of scanning transmission electron microscopy in estimation of remanent life of pressure parts

R B Carruthers and M J Collins

Low-alloy ferritic steels find widespread application in operational conditions where creep is the main mode of deformation and it is of great economic value to the operators of power plant to be able to estimate the lives of components operating in such a regime. Whilst the stresses imposed on large-section components can often be estimated quite accurately, the detailed thermal history is rarely available. Hence, much effort has been devoted to attempting to estimate the effective temperature of these components. Optical metallographic examination of carbide spheroidization has been widely used to estimate temperatures but suffers from the limitations imposed by the resolution of light microscopes and the necessity for large extrapolation beyond experimentally determined data. The authors have attempted to extend the technique by the application of scanning transmission electron microscopy with energy-dispersive X-ray analysis to monitor changes in carbide composition as a function of exposure conditions. Results indicate that the change in substitutional solute concentration in cementite may be a viable technique for the estimation of effective temperature.

The authors are in the Scientific Services Department of the CEGB North-Eastern Region, Harrogate, Yorkshire

Low-alloy ferritic steels are extensively used in the power generation industry in heavy-section pressure vessels and steam pipes. These operate under conditions of high pressure and temperature where creep rupture is the probable ultimate failure mechanism. Such components have in the past been designed for lives of 100 000 h or longer and current philosophy seeks to increase safe operational lifetime beyond the original conservative limits. Frequently it has been possible to extend service life in this manner, although occasionally changes in operation from design conditions have necessitated early replacement.

Considerable savings can be achieved by obtaining maximum possible life from these components, but at the same time it is essential to preserve the safety of personnel and integrity of plant.

In order to estimate the life of a component, a knowledge of the stresses acting on it and the effective temperature it is subject to during its operational life is required. Whilst the stresses and operational life are usually well known detailed retrospective data on temperatures are seldom available. Because of the large number of plant items which need to be assessed, a simple and quick way of estimating temperatures experienced is an attractive proposition.

Some years ago, a simple optical metallographic technique was developed, based upon observations of the rate of spheroidization of pearlitic cementite.[1] This has been adopted as a standard technique by the CEGB and finds extensive application both in the UK and abroad. However, besides being a subjective technique it also suffers from the limitations imposed by the resolution of light microscopes and the necessity for large extrapolation beyond the experimentally determined data. Recently some effort has been devoted towards extending or refining the technique by the application of high resolution electron microscopy. This paper presents details of one investigation which suggests that the changes in carbide composition during aging may provide a means of estimating the temperature to which a sample has been exposed.

EXPERIMENTAL DETAILS

In order to examine changes in carbide composition systematically it is necessary to have access to a number of samples from several casts of the relevant steels which have been exposed for precisely known times at well-documented temperatures. Such samples have been provided in the form of broken stress-rupture samples from testing programmes conducted by E.R.A. Technology Ltd at Leatherhead and British Steel Corporation at Corby. Details of stress and thermal history were provided so that an examination of both gauge-length and threaded portions of the specimens would allow any influence of applied stress on the aging process to be investigated. To date, specimens have been examined from casts of two commercial alloys, a $\frac{1}{2}$Cr–$\frac{1}{2}$Mo–$\frac{1}{4}$V and a 1Cr–$\frac{1}{2}$Mo steel, the analyses of which are given in Table 1.

TABLE 1 Chemical analysis of material (wt-%)

Cast	C	Cr	Mo	V	Mn	Si	S	P
M	0.11	0.35	0.55	0.25	0.42	0.28	0.026	0.010
N	0.13	0.36	0.56	0.32	0.52	0.19	0.019	0.011
P	0.11	0.92	0.52	–	0.52	0.31	0.017	–
Q	0.10	0.94	0.47	–	0.47	0.25	0.036	0.035

balance Fe

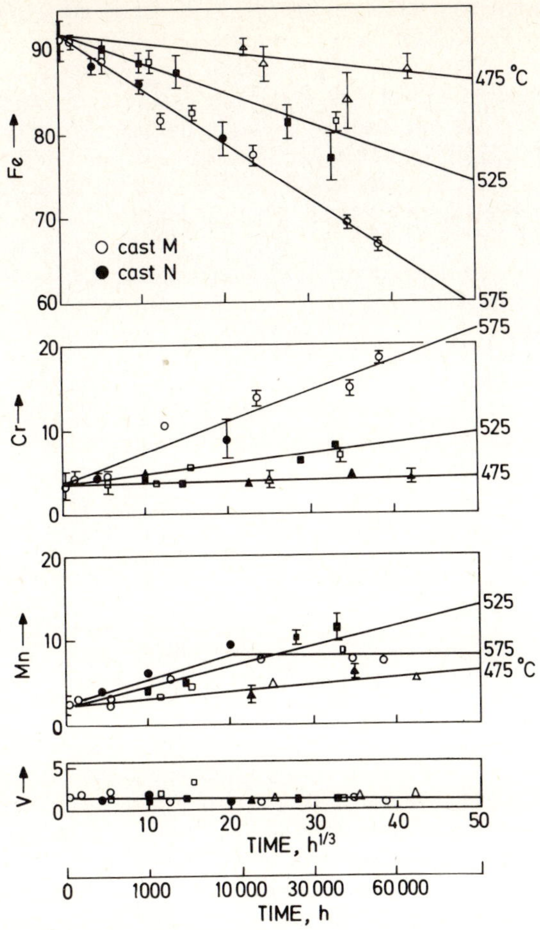

1 **Changes in cementite composition for $\frac{1}{2}$Cr-$\frac{1}{2}$Mo-$\frac{1}{4}$V casts *M* and *N***

Sections from the rupture specimens were mounted in thermal setting resin and mechanically ground and polished to 1 μm finish. They were then lightly etched in a 2% solution of nitric acid in ethanol and carbon extraction replicas prepared for examination in a Hitachi 700-H scanning transmission electron microscope equipped with energy-dispersive X-ray analysis facilities. When used in the STEM mode, the spot size of about 50–100 Å diameter allowed excitation of a very small volume of material and thus analysis of individual carbides. The examination of replicas was chosen to reduce any complications due to excitation of the surrounding matrix.

Generally the heat treatment given to these alloys results in a ferrite/pearlite or ferrite/granular-bainite structure, depending on section size and hence on cooling rate from the austenitizing temperature. The cementite (nominal composition Fe_3C) in the pearlitic colonies in these steels has been studied since it is easily identifiable and also, being formed at one temperature (eutectoid temperature), should not show the range of compositions and morphologies displayed by the alloy carbides.

RESULTS

Alloys $\frac{1}{2}$Cr-$\frac{1}{2}$Mo-$\frac{1}{4}$V and 1Cr–1Mo

Both steels studied have exhibited an important feature which can be related to exposure conditions. It has been observed that as time at temperature increases chromium, manganese and molybdenum dissolve in the pearlitic cementite. This variation in substitutional element composition is progressive and is shown in Figs. 1 and 2 for the casts of each material. As can be seen there is close agreement within each alloy group and similar characteristics are displayed by each group; Fe concentration falling as Cr rises. The situation is more complex in the case of Mn where the evidence suggests that a limiting composition may be reached; but at present there are insufficient data to confirm this. It is also clear that in the relevant material the solubility of V in cementite is not great and the composition does not change very much with exposure.

Attempts to measure the Mo content resulted in much greater scatter than was observed in the other analyses, especially at prolonged exposures. The cause of this variation is shown in Fig. 3; Mo_2C particles nucleate on the pearlitic cementite and if the volume of material excited by the electron beam includes such particles an anomalously high Mo count results.

An attempt has been made to determine the activation energy of the mechanism controlling the process responsible for the change in cementite composition. A simple Arrhenius relationship was assumed and the time to reach 85%Fe was plotted against reciprocal temperature. The apparent activation energy for both steels was approximately 60 kcal/mol, i.e. similar to substitutional element diffusion in α-Fe.

In addition to the common features just described, $\frac{1}{2}$Cr-$\frac{1}{2}$Mo-$\frac{1}{4}$V exhibits an additional effect because of the presence of vanadium. It has been found that during service at temperatures below 600°C the V_4C_3 particles initially present in the ferrite as a result of the tempering treatment are progressively replaced by Mo_2C which frequently

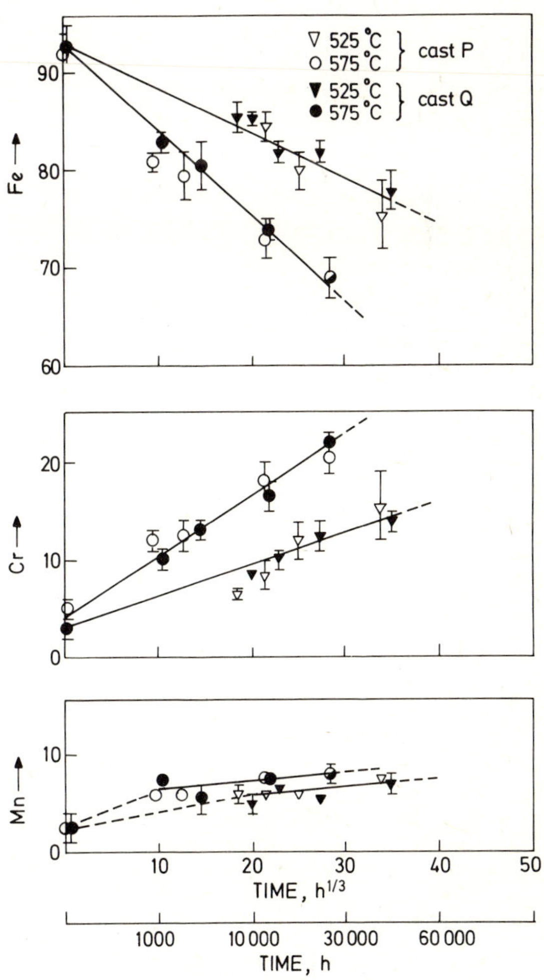

2 **Changes in cementite composition for 1Cr-$\frac{1}{2}$Mo casts *P* and *Q***

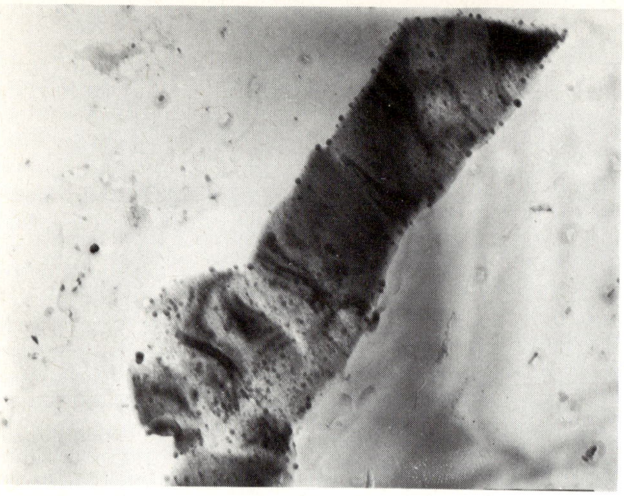

3 **Secondary precipitation on cementite particle (carbon extraction replica, × 39 000)**

nucleates at the ends of V_4C_3 particles, leading to the characteristic 'H' carbide described by a number of authors.[2-5]

General comment on analyses
It should be noted that the data presented in Figs. 1 and 2 are integrated counts corrected for background and peak overlap, rather than actual percentages of alloying elements. This is because no fully quantitative analytical software was available and it is hoped that modifications currently being made will improve the analysis. Preliminary work, analysing some of our figures with a quantitative software package, suggests that the results presented are accurate to within 1 or 2%.

Also, to achieve consistent results it has been necessary to select particles for analysis with some care. Generally carbides of constant contrast (and hence uniform thickness) about 1000–2000 Å wide (Fig. 4), not overlapped by other particles, and free from secondary precipitation were chosen. The 200 kV accelerating voltage proved most useful in this matter as it permitted easy identification of secondary particles overlaid by cementite which could have been opaque at 100 kV.

DISCUSSION
It is clear from the results presented in the previous section that carbide analysis using scanning transmission electron metallography and energy-dispersive X-ray analysis may provide a powerful tool for estimating effective temperature. Specific element analysis of this type has not been possible until recently. Earlier techniques of carbide identification relied on lattice parameter measurements of individual carbides or bulk extractions. Difficulties in precise identification of composition then arose because some elements dilate the lattice while others contract it.

Although the present technique has been demonstrated on cementite in low-alloy ferritic steels, it is probably applicable to other alloys in which the carbide composition changes as a consequence of service exposure. In one sample from another cast currently under examination, the Fe_3C-base carbide appeared to have been replaced by one based on $M_{23}C_6$. In the former carbide the V level was always about $1 \pm 0.5\%$ whilst in the latter it was about 14%. Such changes in solubility may be used to generate time- and temperature-dependent points to allow the extent of aging to be quantified as a function of the microstructure.

The STEM used in the current investigation is heavily committed and limited access has restricted the investigation of the influence of stress on the aging process. One sample from each alloy has been examined and the analyses of carbides in gauge-length and grips are the same, within the experimental scatter. Clearly, this is a field of great potential interest, as processes producing a lattice dilation are likely to be accelerated by an applied stress and vice-versa. Further work is planned.

CONCLUSIONS
The technique described provides a rapid means of generating semi-quantitative analyses of individual carbides which are not readily obtained by electron diffraction. The results obtained to date suggest that it may prove a powerful tool for estimating effective temperature of plant, allowing more accurate determination of remanent life. This would allow more cost-effective plant operation with improved faith in integrity and more accurate planning of replacement schedules. The latter could result in savings of hundreds of thousands of pounds per annum yet still maintain adequate safety standards.

ACKNOWLEDGMENTS
This paper is published by permission of the Central Electricity Generating Board. The authors wish to thank E.R.A. Technology Ltd and British Steel Corporation (Corby) who generously donated material for this investigation. Colleagues at the Electricity Council Research Centre (Capenhurst) are also thanked for provision of facilities and helpful discussions.

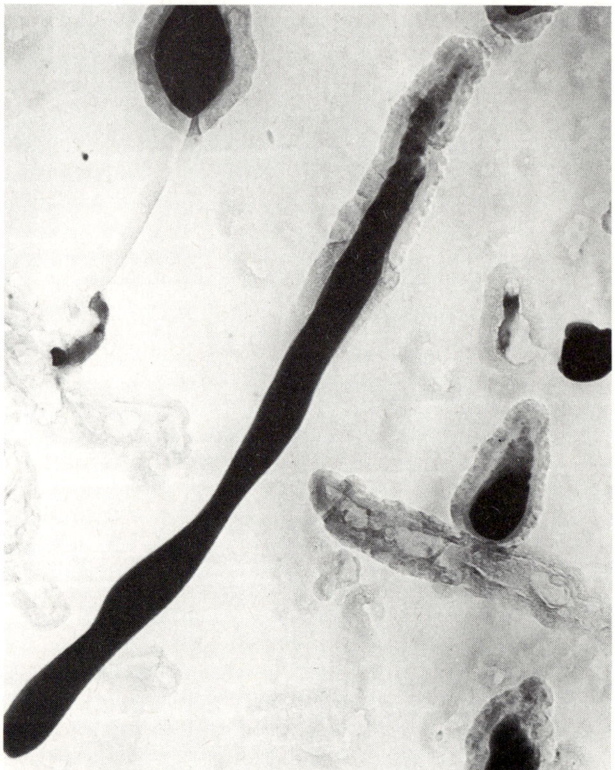

4 **Carbon extraction replica typical of carbon particles analysed (× 56 000)**

REFERENCES

1 R. B. CARRUTHERS and R. V. DAY: CEGB Report SSD/NE/R138, 1968
2 R. B. CARRUTHERS and M. J. COLLINS: CEGB Note NER/SSD/R423, 1980
3 M. J. COLLINS: *Met. Technol.*, 1978, 5, 325
4 C. CASTANI and M. LIVRAGHI: Proc. 18th Nat. Conf. (AIM), 1978, Verona; CEGB Trans. 7435
5 M. C. MURPHY and G. D. BRANCH: *J. Iron Steel Inst.*, 1969, **207**, 1347

DISCUSSION of paper by Carruthers and Collins

G. D. W. SMITH:
This is clearly an experimental approach of some considerable importance, but are you not limited by the need to section the component for examination? And should you not be looking at particles in grain boundaries rather than in grain interiors? Surely changes at grain boundaries are largely responsible for eventual failures.

R. B. CARRUTHERS:
There is no limitation imposed by a need to section components under investigation because techniques are available which allow samples to be taken with minimal damage to a component.

It is true that grain boundaries play a major role in creep but the purpose of this work is to develop a means of estimating an operating temperature and not to study creep processes. Pearlitic carbides are examined because they are readily identified and because it was considered that they would have more uniform composition and change in a more uniform manner than grain boundary particles whose initial compositions could be variable and which would be subject to changes influenced by relatively rapid grain boundary diffusion processes. Work is planned to verify these suppositions.

P. DOIG:
Adequate sampling of large plant components can be and is achieved by use of dry plastic extraction replicas. In addition very fine machining processes are being applied to the removal of thin, shallow saucer-shaped samples which do not result in significant stress concentrations.

Application of STEM/EDS to study of microstructural development in HSLA steel weld metals

G S Barritte, R A Ricks, and P R Howell

The results of a considerable number of investigations into the mechanical properties of HSLA steel weld metals have shown that the strength and toughness of such materials are critically dependent on the microstructure of the weld. In particular the role of the so-called 'acicular ferrite' phase in promoting improved properties has been associated with the inclusion content of the weld metal. However, some doubt still exists concerning those factors which control the formation of this microstructure and it has been suggested that either the inclusion type and/or the exact matrix composition is the important factor. The authors have analysed the composition of both the inclusions and the adjacent austenite using STEM/EDS techniques. Specifically, it is shown that the inclusion chemistry is controlled, to a large extent, by the nature of the welding flux and that these inclusions do not appreciably alter the local composition of the austenite. Further, it is suggested that the most important factor controlling the formation of acicular ferrite is the austenite grain size.

The authors are in the Department of Metallurgy and Materials Science, University of Cambridge.

The development of ferritic microstructures in HSLA steel weld metal has received considerable attention in recent years.[1–6] Of particular importance is the formation of the fine, so-called 'acicular ferrite' since this microstructural constituent has been shown to be associated with improved weld metal toughness.[6] However, there is still some doubt concerning the nature of this microstructural constituent although it has been suggested[3] that it forms above the B_s and M_s temperatures. It therefore seems likely that acicular ferrite consists, to a large extent, of intragranularly nucleated Widmanstätten ferrite.

Similarly, there is some disagreement concerning the factors which promote the formation of this reaction product. Abson and Dolby[3] have proposed that inclusions are wholly responsible for the nucleation of intragranular ferrite. However, it has been found that increasing the weld metal oxygen content (and hence the inclusion volume fraction) from 0.03% to 0.06% actually suppresses the formation of acicular ferrite. To account for this observation, it has been suggested that inclusion type as well as distribution can influence the effectiveness of such sites as ferrite nucleants.

In contrast, Cochrane and Kirkwood[2] claim that the intragranular nucleation of acicular ferrite can be adequately explained by classical theories of homogeneous nucleation. They concluded that the lack of this microstructural constituent in high oxygen welds ($>\sim 0.06\%$) could be explained in terms of enhanced nucleation of proeutectoid ferrite at inclusions located at prior austenite grain boundaries.

Finally, Farrar and Watson[5] have considered that the role of inclusions is to determine the matrix composition. In particular, they suggest that acicular ferrite will develop if a threshold value of $\sim$1.1%Mn is exceeded in the weld metal matrix.

In this paper, weld metal (from both manual metal arc and submerged metal arc welds) has been investigated to establish:

(i) the nature of the acicular ferrite constituent and the role of inclusions in the nucleation of intragranular ferrite;
(ii) the composition of weld metal inclusions as a function of oxygen content and flux type; and
(iii) the effect of inclusion type and distribution on the resultant composition of the matrix.

EXPERIMENTAL

Three dissimilar weld metals have been used in this investigation: a manual metal arc weld (MMA) and two submerged arc welds (S.A.80 and S.A.121). The bulk compositions of these weld metals are given in Table 1. Two different oxygen levels in the submerged arc welds were produced by employing an acidic fused flux (S.A.80) and a basic agglomerated flux (S.A.121).

TABLE 1 Bulk compactions of weld metals (wt-%) (measured by X-ray fluorescence)

	S.A.80	S.A.121	MMA
Mn	1.31	1.32	1.42
Si	0.33	0.29	0.51
O	0.06	0.02	0.02
C	0.13	0.14	0.09

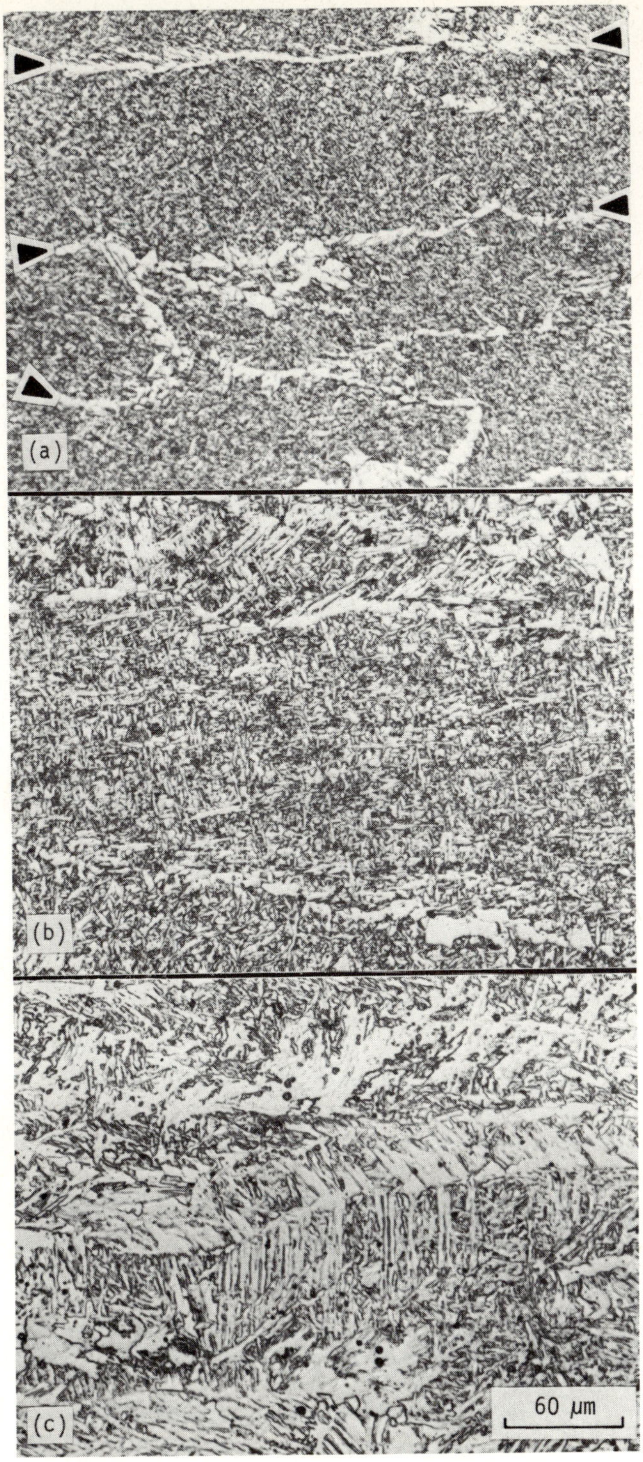

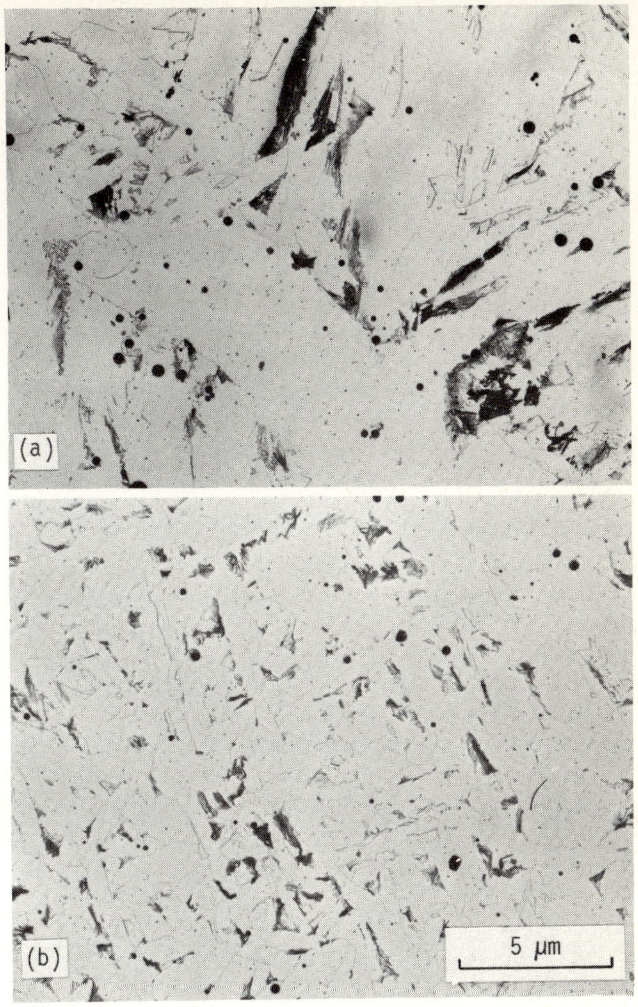

2 **Carbon extraction replicas of S.A.80 and S.A.121 illustrating distribution of inclusions: (a) S.A.80 showing relatively high number density; (b) S.A.121 showing relatively low number density of inclusions**

1 **Light optical micrographs of three weld metals investigated:**
(a) Manual metal arc weld
Body of prior austenite grains comprise acicular ferrite constituent whilst limited growth of austenite boundary nucleated ferrite is observed (*arrowed*). Average austenite grain size (measured normal to columnar grain structure) ~100 μm
(b) Submerged arc weld S.A.121, illustrating development of acicular ferrite and grain boundary ferrite
Relative coarseness of this structure (compared to Fig. 1*a*) is reflection of slower cooling rate experienced by submerged arc welds. Average austenite grain size ~150 μm
(c) Submerged arc weld S.A.80
Structure consists mainly of austenite grain boundary nucleated ferrite side-plates; average austenite grain size ~60 μm

TABLE 2 Inclusion parameters in submerged arc welds

	Number density	Volume fraction
S.A.80	1.6×10^8 mm^{-3}	0.4 vol.-%
S.A.121	1.1×10^8 mm^{-3}	0.2 vol.-%

For microstructural examination, optical specimens were etched in 2% nital, whilst direct carbon extraction replicas were produced from lightly etched sections of the welds. Thin foil specimens were prepared for transmission electron microscopy from 0.25 mm thick discs slit from the welds under conditions of flood lubrication. The discs were ground to 0.1 mm thickness and subsequently thinned in a twin jet electropolishing unit using a 5% perchloric/95% acetic acid electrolyte at room temperature and a potential of 90 V.

Both thin foils and carbon extraction replicas were examined in a Philips EM400 TEM/STEM using a Link 860 microanalysis system. Standard thin foil correction programmes were used to compensate for X-ray detection efficiency and absorption. For quantitative analyses the specimen was tilted 45° towards the detector. For comparative purposes, however, a tilt angle of 35° was often employed since this reduced astigmatism in the image and improved the working resolution.

RESULTS

Weld metal microstructures

Figures 1*a–c* show typical examples of the microstructures produced in the three welds. The medium oxygen welds (MMA and S.A.121) produce typical acicular ferrite microstructures (Figs. 1*a* and *b*) and only limited growth of the austenite boundary nucleated ferrite is observed. In contrast the high oxygen weld (S.A.80) consists mainly of austenite grain boundary nucleated ferrite side-plates and the acicular ferrite content is low. These figures also serve to show the smaller austenite grain size associated with the high oxygen weld metal (S.A.80).

Figures 2*a* and *b* illustrate the differences in inclusion content between the high oxygen weld metal and the medium oxygen weld metal. As can be seen, the inclusions are larger and more numerous in the high oxygen S.A.80 than in the medium oxygen S.A.121. Table 2 gives the results of a quantitative analysis* concerning the number density and volume fraction of inclusions in the two welds.

These results show that the volume fraction of inclusions in weld S.A.80 is approximately twice that observed in weld S.A.121, which reflects the lower oxygen content of the latter specimen.

In the next two sections the hypotheses outlined above concerning the effects of inclusion chemistry and adjacent austenite composition upon the formation of acicular ferrite will be re-examined using STEM/EDS.

Inclusion chemistry

This section presents the results of a STEM/EDS investigation into the chemistry of the inclusions formed during the welding process as a function of the oxygen content of the weld metal and the flux type used.

Figures 3*a* and *b* are typical EDS spectra obtained from single inclusions in S.A.80 and S.A.121 welds respectively. These spectra are characteristic of the vast majority of inclusions observed in these two welds, although occasionally almost pure MnS inclusions were detected in S.A.80. The spectra are very similar in terms of the elements present,** the major difference being the presence of the small Ti peak in the spectrum of S.A.121 (Fig. 3*b*). The presence of Al, Si and Mn indicates that the inclusions are mixed in character and most probably consist of $MnSiO_3$, $Al_2(SiO_3)_3$ and Al_2O_3. The sulphur peak in both spectra implies that some MnS is also present.

Although the spectra obtained from S.A.80 and S.A.121 indicate that there is little difference in the nature of the inclusions, differences in the actual concentrations of Al, Si and Mn are observed. For example, Fig. 3*a* indicates that the inclusions in S.A.80 contain more Si than Al, while the opposite is true for S.A.121 (3*b*). These differences may be explained in terms of the composition of the fluxes used in the welding processes.

Table 3 shows the average inclusion composition of welds S.A.80 and S.A.121. The high Si/Al ratio of the inclusion in weld S.A.80 is associated with the acidic flux used during the process which contained ~36% SiO_2 and ~12% Al_2O_3. Similarly the low Si/Al ratio of weld S.A.121 is due to the use of a basic flux containing ~14% SiO_2 and ~18% Al_2O_3. The presence of the small Ti peak in the spectra obtained for weld S.A.121 (3*b*) reflects the higher Ti content of the basic flux (1.0% TiO_2 in the basic flux, 0.5% TiO_2 in the acidic flux). Secondary electron imaging of extracted inclusions often revealed that they were covered by an outer coating of material which was dissimilar in appearance to the bulk of the inclusion (Fig. 4).

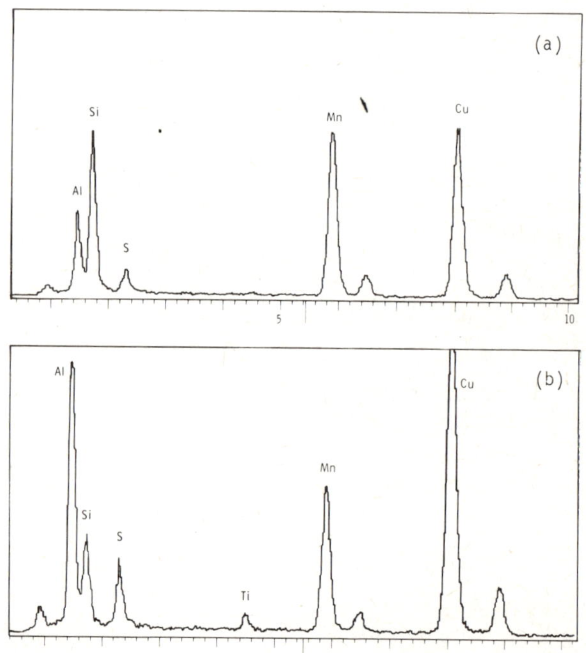

3 **Typical spectra from inclusions in submerged arc welds; in both instances Cu peaks are characteristic of support grid: (a) S.A.80 – major elements detected are Al, Si, S and Mn; (b) S.A.121 – in addition to Al, Si, S and Mn, small Ti peak is also observed**

* The data contained in Table 2 were collated using a Quantimet 720 image analysing computer equipped with full pattern recognition facilities.

** The copper peaks in both spectra originate from the support grid of the carbon extraction replica and are not characteristic of the inclusions.

TABLE 3 Inclusion analysis in submerged arc welds (wt-%)

	Mn	Si	Al	S	Ti
S.A.80	53	26	17	4	<0.3
S.A.121	33	11	47	5	3.0

Figures 5*a* and *b* show sulphur profiles across such inclusions indicating that the outer layers were richer in S than the interior. However, in both instances the S level is seen to be much higher at one side of the inclusion (especially in Fig. 5*b*). This may be explained by the orientation of the inclusion such that the high S level is associated with that side which is closer to the X-ray detector. Under these conditions the effects associated with absorption and secondary fluorescence by the bulk of the inclusion will be minimized. It therefore seems likely that the inclusions in both welds are similar in character and are composed of a central core of mixed (Al, Mn) oxides and (Al, Mn) silicates which are coated with a layer of MnS. This is also consistent with considerations of the melting points of the constituents of the inclusion. Thus the mixed silicates which tend to have high solidification temperatures would form relatively early in the thermal history of the weld and subsequently act as nuclei for the formation of MnS.

Similar results were obtained from inclusions extracted from weld MMA. In addition to the Al, Si, S and Mn peaks present (Fig. 6), a large Ti peak is also observed and this is due to the high Ti content of the flux used during this welding process (~5% TiO_2).

Matrix chemistry

Farrar and Watson[5] have suggested that the exact *matrix* Mn content (rather than the *bulk* Mn content) is the important factor in controlling the formation of acicular ferrite. Specifically they calculate that for an acidic manganese silicate-based flux, one-eighth of the weld metal

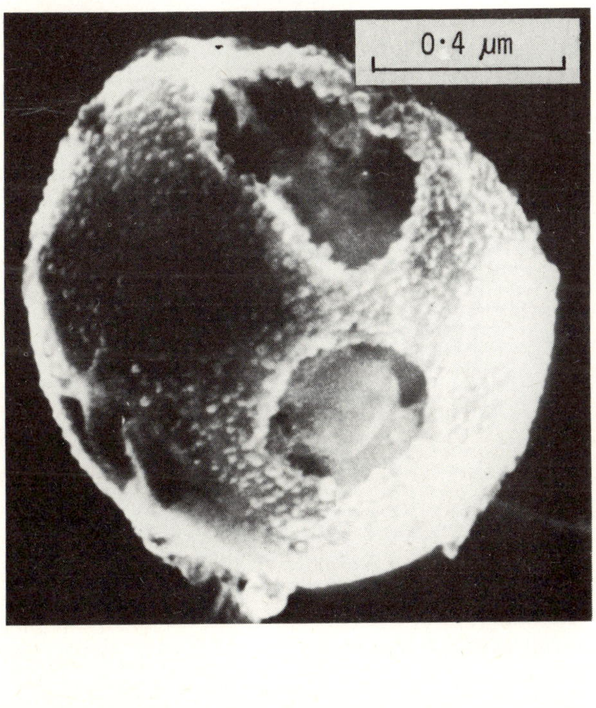

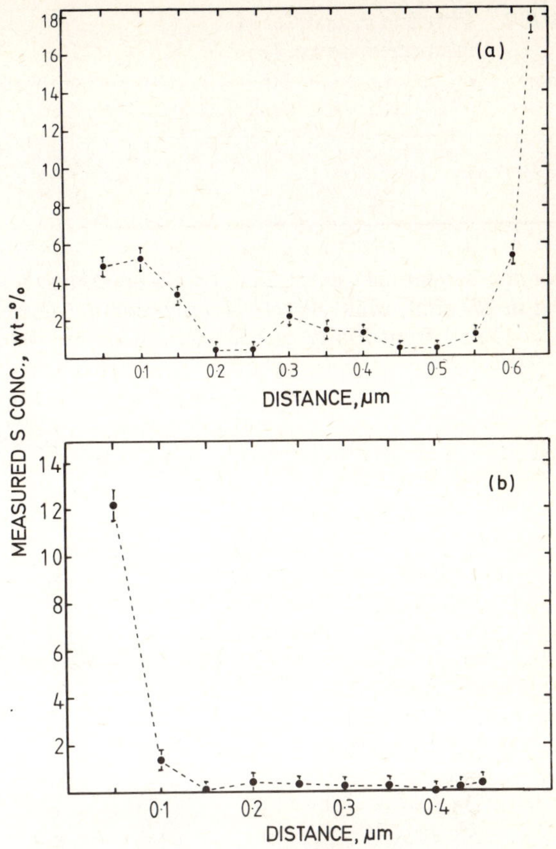

5 **Sulphur profiles across inclusions in S.A.121; in both cases, S level tends to increase towards periphery of inclusion (*see text*)**

manganese is in the inclusions at 0.1 wt-% oxygen. Farrar and Watson[5] also suggest that for a basic flux the bulk analysis will correspond closely to the matrix composition.

In order to test this hypothesis, the average matrix compositions have been determined for all three weld types. Similarly, the existence of any change in matrix composition in the vicinity of the inclusions has been investigated.

The results of the STEM/EDS analyses of the weld metal matrix compositions are given in Table 4. The figures quoted are the means of a large number of individual analyses and the errors quoted are the standard deviations in the mean. It is clear from Table 4 that there is no detectable difference in Mn levels between S.A.80 and S.A.121 and that these levels (and that characteristic

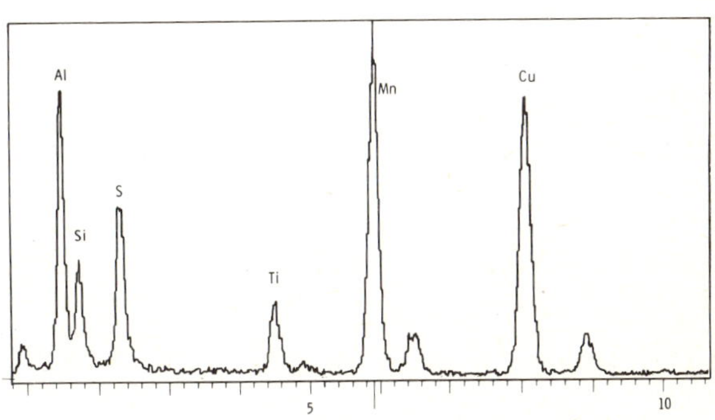

6 **Spectrum from inclusion in manual metal arc weld; increased Ti level is seen compared to spectra shown in Fig. 3**

TABLE 4 STEM/EDS analyses of weld metal matrix (wt-%)

	S.A.80	S.A.121	MMA
Mn	1.3±0.1	1.3±0.1	1.48±0.04
Si	0.5±0.2	0.3±0.1	0.50±0.10

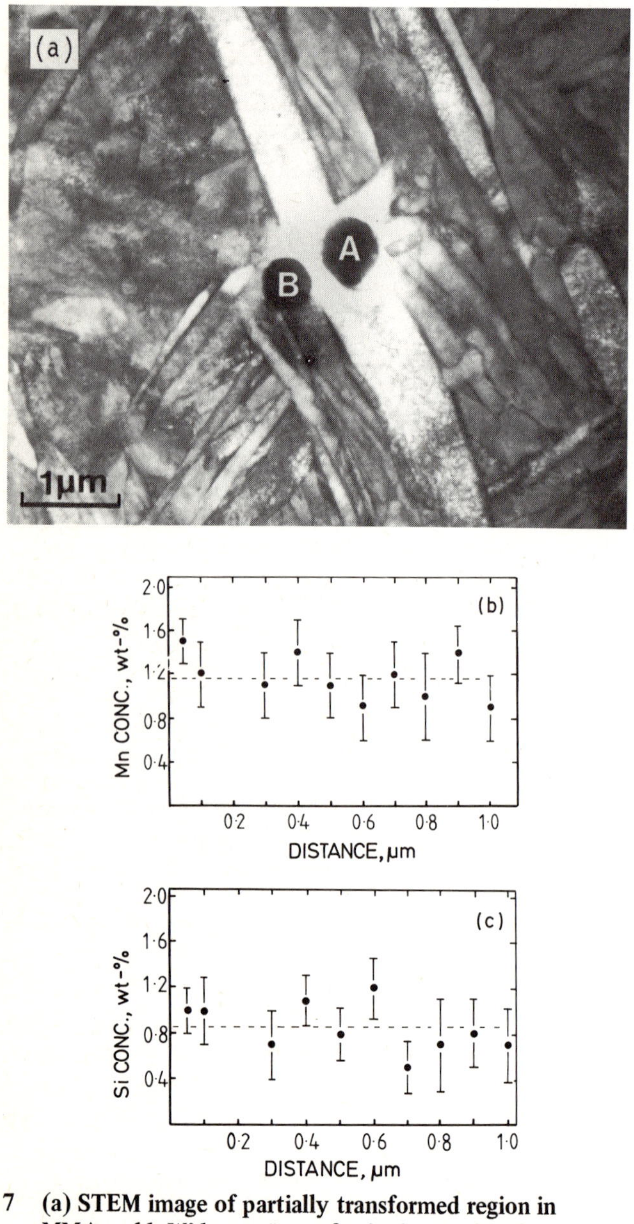

7 **(a) STEM image of partially transformed region in MMA weld; Widmanstätten ferrite is associated with inclusions *A* and *B*; matrix phase is lath martensite (b) Mn profile away from inclusion *A* (located at origin) along Widmanstätten plate; no systematic variation in Mn level is observed (c) Si profile corresponding to Fig. 7*b*; again, no systematic variation in Si level is observed**

of the manual metal arc) correspond very closely to the bulk compositions quoted in Table 1. Hence it would appear that the nature of the inclusion distribution has little observable effect on the matrix composition of the weld metal. That little difference between S.A.80 and S.A.121 is apparent may be explained in terms of the similarity of the inclusion types in the two welds. Similarly, calculations based on the work of Bailey and Pargeter[7] with the known inclusion chemistry have shown that at the oxygen levels of interest (i.e. between 0.03% and 0.06%) little detectable change (with respect to either Mn or Si) should be found.

However, the possibility still exists that the inclusions alter the composition of the adjacent matrix. Figure 7*a* is a STEM image of a partially transformed region in the MMA weld. As can be seen, there is a close association between the intragranular Widmanstätten ferrite and the inclusions at *A* and *B*. Figures 7*b* and *c* are Mn and Si profiles† respectively, away from inclusion *A* into the Widmanstätten ferrite. No decrease in either Mn or Si concentration is observed in the immediate vicinity of the inclusions. A further example of the distribution of Mn in the vicinity of an inclusion is given in Fig. 8. Again, no depletion in Mn is observed.

DISCUSSION

The results detailed in the previous section have shown that there is little difference in either the inclusion type or matrix composition between S.A.121 (which does produce an acicular ferrite microstructure) and S.A.80 (which does not). Hence, these results would seem to suggest that the inclusions have little or no effect on the formation of acicular ferrite. However, the examination of partially transformed specimens has supplied direct evidence for the nucleation of Widmanstätten ferrite on inclusions.

Figure 9 is an example of a partially transformed structure in MMA which was produced by iced-brine quenching immediately welding was completed. The development of grain boundary ferrite can be seen on the

† These profiles were conducted with the specimen tilted at 35° towards the detector rather than 45°. This resulted in a systematic decrease in the measured Mn concentration.

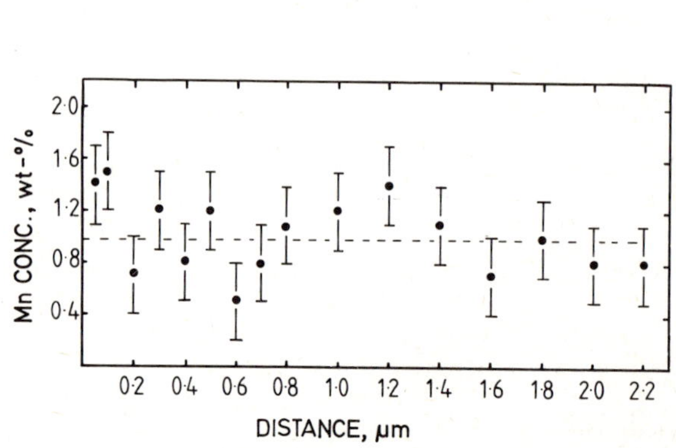

8 **Mn concentration profile away from inclusion (located at origin) in S.A.121. Increased Mn levels immediately adjacent to inclusion most probably result from either electron beam spreading or from secondary fluorescence**

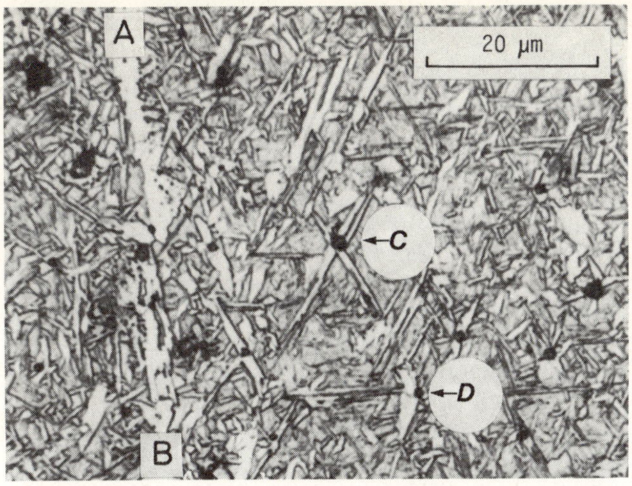

9 **Light optical micrograph of partially transformed MMA weld; grain boundary ferrite has developed (along *A–B*) together with formation of intragranular Widmanstätten ferrite (e.g. at *C*, *D*); there is also an apparent association between Widmanstätten ferrite at *C* and an inclusion**

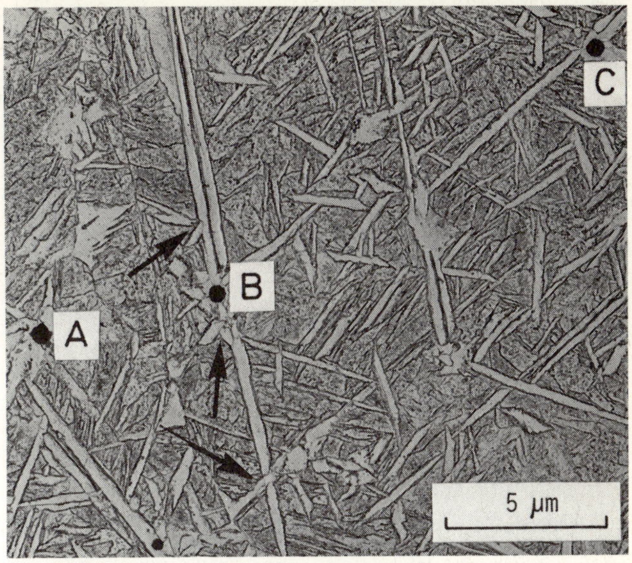

11 **TEM image of S.A.121 subsequently tungsten-inert-gas remelted and quenched; inclusions at *A*, *B* and *C* are seen to be associated with intragranular Widmanstätten ferrite; evidence for sympathetic nucleation of ferrite *arrowed***

prior austenite grain boundary (*A–B*) together with the formation of intragranular Widmanstätten ferrite (e.g. at *C*, *D*).‡

Figure 10 is a transmission electron micrograph of the quenched MMA weld. Here, a clear association between the inclusion (at *D*) and the intragranular Widmanstätten ferrite can be seen. It is also of interest to note that multiple nucleation has occurred (e.g. at *A*, *B* and *C*). Further examples of this nucleation of ferrite at inclusions are given in Fig. 11, which shows a TEM image of a carbon extraction replica taken from a partially transformed weld, and also in the STEM image shown as Fig. 7*b*. In the former case inclusions at *A*, *B* and *C* are seen to be associated with intragranular Widmanstätten ferrite. Again, multiple nucleation may be observed (e.g. at *A*). Evidence for the sympathetic nucleation of ferrite is also observed (*arrowed*) as discussed by Aaronson and Wells.[8]

These microstructural results support the hypothesis that acicular ferrite is composed of intragranularly nucleated Widmanstätten ferrite and that this reaction product can nucleate both on inclusions and by a sympathetic nucleation mechanism. The question still remains, however, why acicular ferrite forms in MMA and S.A.121 and not in S.A.80.

One fact which has received little attention is the effect that the inclusions have on the austenite grain size and the subsequent effect this has on the transformation of austenite to ferrite. One of the major differences between MMA, S.A.121 and S.A.80 is the much finer austenite grain size associated with S.A.80 (e.g. compare, for instance, Fig. 1*c* with 1*b*). This finer grain size in S.A.80 is associated with the increased number density and volume fraction of inclusions since they provide more effective barriers to austenite grain growth. The fine grain size will exert a profound influence on the transformation of austenite to ferrite, since

(i) the total number of available intergranular nucleation sites will be increased as the grain boundary area increases;
(ii) the growth of the ferrite will be completed in a shorter time interval (and hence a smaller temperature range during continuous cooling) since each nucleus will have less austenite to consume before impingement occurs;
(iii) although the Widmanstätten ferrite start temperature will be decreased,[9] the formation of grain boundary nucleated ferrite will occur at higher temperatures, thus reducing the likelihood of intragranular nucleation; and
(iv) the carbon content of the untransformed austenite will be enriched more rapidly in the fine grained material. This enrichment will tend to suppress any intragranular nucleation of ferrite.

‡ It is interesting to compare this image with the fully transformed microstructure (Fig. 1*a*). From this it can be suggested that the impingement of the intragranular Widmanstätten ferrite (Fig. 9) can lead to the development of the acicular ferrite constituent shown in 1*a*.

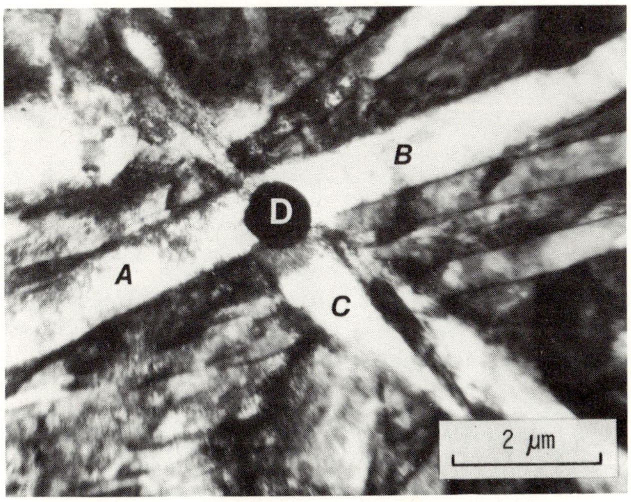

10 **Transmission electron micrograph of partially transformed MMA weld metal; multiple nucleation of Widmanstätten ferrite (at *A*, *B*, *C*) has occurred at inclusion (*D*)**

The overall effect of such a reduction in grain size by increasing the inclusion content will be to promote the development of austenite grain boundary nucleation reactions. Furthermore, the result of the inclusion pinning during austenite grain growth will cause many of the inclusions to be associated with grain boundaries, thus providing even more effective sites for grain boundary nucleation (as suggested by Cochrane and Kirkwood[2]). Consequently in S.A.80 the predominant microstructural constituent will be coarse grain boundary nucleated ferrite.

In contrast, the larger austenite grain size associated with MMA and S.A.121 allows the development of intragranularly nucleated Widmanstätten ferrite, primarily on inclusions and then by a sympathetic nucleation process. This development of a fine 'acicular' constituent in the austenite is therefore due to both a large austenite grain size and sufficient intragranular inclusions to initiate the production of acicular ferrite. However, it should be realised that the presence of intragranular inclusions is essential to initiate the formation of acicular ferrite. Thus if the inclusion content of the weld metal is very low, although a large austenite grain size will be formed no intragranular nucleation sites will exist and the production of acicular ferrite will be suppressed.

SUMMARY

The results detailed in this paper have shown that

1 The inclusions in the two submerged arc welds comprise mixed aluminium/manganese silicates/oxides and are coated by a thin layer of manganese sulphide.
2 The inclusions in the manual metal arc weld contain, in addition to the above, a significant TiO_2 content.
3 The inclusions have little effect on the composition of the matrix.
4 Inclusions can nucleate ferrite. However, increasing the volume fraction of such inclusions can suppress the intragranular nucleation of ferrite by promoting grain boundary reactions.
5 The microstructure of acicular ferrite has been shown to be composed of inclusion-nucleated Widmanstätten ferrite together with sympathetically nucleated ferrite sideplates.

ACKNOWLEDGMENTS

The authors are grateful to Professor R. W. K. Honeycombe for the provision of laboratory facilities and for his continuing advice and encouragement. Financial support was received from the Welding Institute and from the SRC. The authors would also like to express their gratitude to The Welding Institute for preparing the weld deposits.

REFERENCES

1 R. E. DOLBY: Weld. Inst. Res. Report No. 14/1976/m, 1976
2 R. C. COCHRANE and P. R. KIRKWOOD: in 'Trends in steels and consumables for welding', Weld. Inst. Conf. 1978, 103
3 D. J. ABSON and R. E. DOLBY: Weld. Inst. Res. Report No. 19/1978, 1978
4 Y. ITO and M. NAKANISHI: The Sumitomo Search No. 15, 1976
5 R. A. FARRAR and M. N. WATSON: *Met. Constr.*, July 1979, 285
6 J. G. GARLAND and P. R. KIRKWOOD: *Met. Constr.*, May 1975, 320
7 N. BAILEY and R. J. PARGETER: Weld. Inst. Res. Report No. 70/1978/m, 1978
8 H. I. AARONSON and C. WELLS: *Trans. AIME*, 1956, 8 1216
9 P. R. KRAHE, K. R. KINSMAN, and H. I. AARONSON: *Acta Metall.*, 1972, **20**, 1109

DISCUSSION of paper by Barritte, Ricks, and Howell

R. C. COCHRANE:

A comment on the microstructure of weld metals. I would not like to have participants left with the impression that weld metal microstructures are simply explained in terms of prior austenite grain size. Both the inclusion distribution and composition are a function of the flux formulation and welding conditions. If you make a judicious choice of these variables you can arrive at a series of weld deposits with more or less constant austenite grain size.* In this case you will still see differences in the microstructure of the weld metal deposits along the lines given in the preceding presentation. Whilst austenite grain size is clearly a significant variable there are still some overriding effects of inclusions in the development of the weld metal microstructure. The essential details of this process given in the presentation agree with earlier studies by Kirkwood and myself.†

P. R. HOWELL:

In reply to Dr Cochrane's comments, it should be emphasized that, as stated in the presentation, although the prior austenite grain size is an important parameter in determining the final ferrite microstructure, the inclusion population is paramount in the initial development of the acicular ferrite constituent. The results of the study presented here refer to equivalent welding conditions and, apart from oxygen content, equivalent weld metal compositions. Thus the only two variables which have been investigated are the oxygen content of weld metal and the austenite grain size – both of which, it should be appreciated, are intimately connected.

In the results presented in the two papers cited by Dr Cochrane, the following inconsistencies with the present investigation should be noted:

(i) The high energy barrier associated with homogeneous nucleation of ferrite would effectively prohibit the formation of acicular ferrite by this means as suggested by Dr Cochrane's second reference.†
(ii) Direct microstructural evidence for inclusion nucleation of ferrite has been presented here, and it is suggested that the fine interlocking microstructure associated with acicular ferrite is a result of subsequent sympathetic nucleation events.
(iii) The lack of acicular ferrite in very low oxygen weld metals supports the above hypothesis that intragranular inclusions are necessary to *initiate* the formation of this microstructural constituent.

*P. R. KIRKWOOD and R. C. COCHRANE: *Met. Technol.*, to be published
†P. R. KIRKWOOD and R. C. COCHRANE: 'Trends in welding consumables', Welding Institute Conference, Nov. 1978

Optimization of instrumental conditions for TEM analysis of hydrated cement

D Chescoe and P J Goodhew

Analysis of the hydration products of cements is particularly difficult because of chemical decomposition and elemental redistribution under the electron beam. The three major controllable variables are electron energy, beam current density, and specimen temperature. The authors report the results of their extensive study of a calcium sulphoaluminate hydrate similar to ettringite and typical of a common hydration product of Ordinary Portland Cement (OPC). Elemental analyses have been carried out at 40, 100 and 200 keV on a JEOL 200 CX Temscan at room temperature and at 100 keV on a Philips EM400T at both room temperature and low temperature. The optimum conditions for analysis depend on a compromise between degradation of the specimen and poor counting statistics. It is therefore not generally possible to realise the potential advantages of operating at high kV where ionization damage should theoretically be minimized. Cooling the specimen to near liquid nitrogen temperature does not reduce the damage rate significantly. The effect of beam current density can be assessed most easily using the TEM and STEM modes of the EM400T. The TEM mode gives wider control over the beam current density, which must be kept below $10^5\ A/m^2$ for this type of specimen; it also permits the operator to detect unacceptable levels of beam damage or specimen drift during the analysis.

The authors are in the Microstructural Studies Unit, Department of Metallurgy and Materials Technology, University of Surrey, Guildford, UK.

Cements are potentially cheap constructional materials predominantly used in compression. Although their mechanical properties in tension could well be described as 'pathetic', there is a great deal of interest in the possibility of increasing their tensile strength. Unfortunately it is not clear what is the underlying microstructural reason for the weakness in tension, a topic of some controversy at the present time.[1,2] It is in order to investigate the mechanical and chemical interlocking of the many microstructural phases that we have been studying the hydration of Portland cements.[3]

During the setting of cement the major type of reaction involves the hydration of a mixture of lime, silica and alumina to which may be added gypsum and oxides such as magnesia or Fe_2O_3. Cement chemists have developed a shorthand terminology for the initial compounds and their common product phases. However, we can avoid the jargon and represent a typical (unbalanced) hydration reaction as:

$$3CaO \cdot Al_2O_3 + CaSO_4 \cdot 2H_2O + nH_2O$$
$$\rightarrow 3CaO \cdot Al_2O_3 \cdot Ca(SO_4, (OH)_2) \cdot xH_2O$$

The hydration products have been extensively studied by relatively large-scale techniques such as chemical analysis, X-ray diffraction and DTA and a variety of crystalline phases have been identified.[4] It is generally agreed that in the early stages of hydration fibrous growths emerge from the original cement clinker particles (Fig. 1) but the mechanism by which this occurs is still a matter for speculation.[5,6,7] It would be possible for such growths to form as a result of the bursting of an osmotic membrane or by a through-solution precipitation mechanism. In order to attempt to distinguish between these mechanisms it would be helpful to know how the composition of the fibres changes along their length. As can be seen from Fig. 1 the diameter of the fibres is typically less than 100 nm and hence a TEM-based analysis appears to be the most suitable approach.

ANALYSIS OF CEMENTS IN ELECTRON MICROSCOPE

Several workers[8,9,10] have studied the composition of hydrated tricalcium silicate by electron microprobe analysis. In every case, however, the hydration reaction was well advanced (mainly in order to give a solid specimen) and no single microstructural feature could be analysed. The specimen appeared to be stable under the electron beam and no reports of compositional variation at the time of analysis were made. The only published report of the analysis of cement products in the TEM was by Cliff *et al.*,[11] who used an EMMA instrument. These workers only examined ground samples of well-hardened cement and again reported no evidence of instability. However, during our analytical TEM studies of individual cement fibres during the early stages of hydration[3,12] we have frequently observed that fibres can be damaged by the electron beam and that this may lead to variation of the apparent composition with analytical condition. It is the object of this paper to document these variations and to map out the usable analytical regions on two commonly used analytical electron microscopes.

1 **Typical fibrillar growth of hydrated cement**

Electron beam damage is most frequently reported on polymer and biological specimens.[13] Although the mean atomic weight of cement products is three times that of most biological tissue, and hence the fraction of electrons which suffer inelastic scattering is reduced, we might expect that techniques which enhance the stability of biological materials would also be appropriate for cement hydration products. The most obvious technique is to reduce the irradiating beam current but the limit to this is set by the efficiency of generation and detection of X-rays. In general many analyses are already performed with such poor X-ray counting statistics that a further reduction in X-ray signal could not be tolerated. The two other possible variables are then electron energy and specimen temperature. In principle, an increase in electron energy should lead to a reduced ionization damage rate without reducing the X-ray peak-to-background ratio significantly.[14] Indeed, there is experimental evidence that damage rate is proportional to (electron velocity)$^{-2}$, at least in some polymers.[15] Many workers have also investigated the effect of lowering specimen temperature and the results, well summarized by Isaacson,[13] indicate that at least with regard to imaging most materials suffer a lower damage rate at low temperatures although the effect is not dramatic. We have therefore investigated the analysis of one type of cement product as a

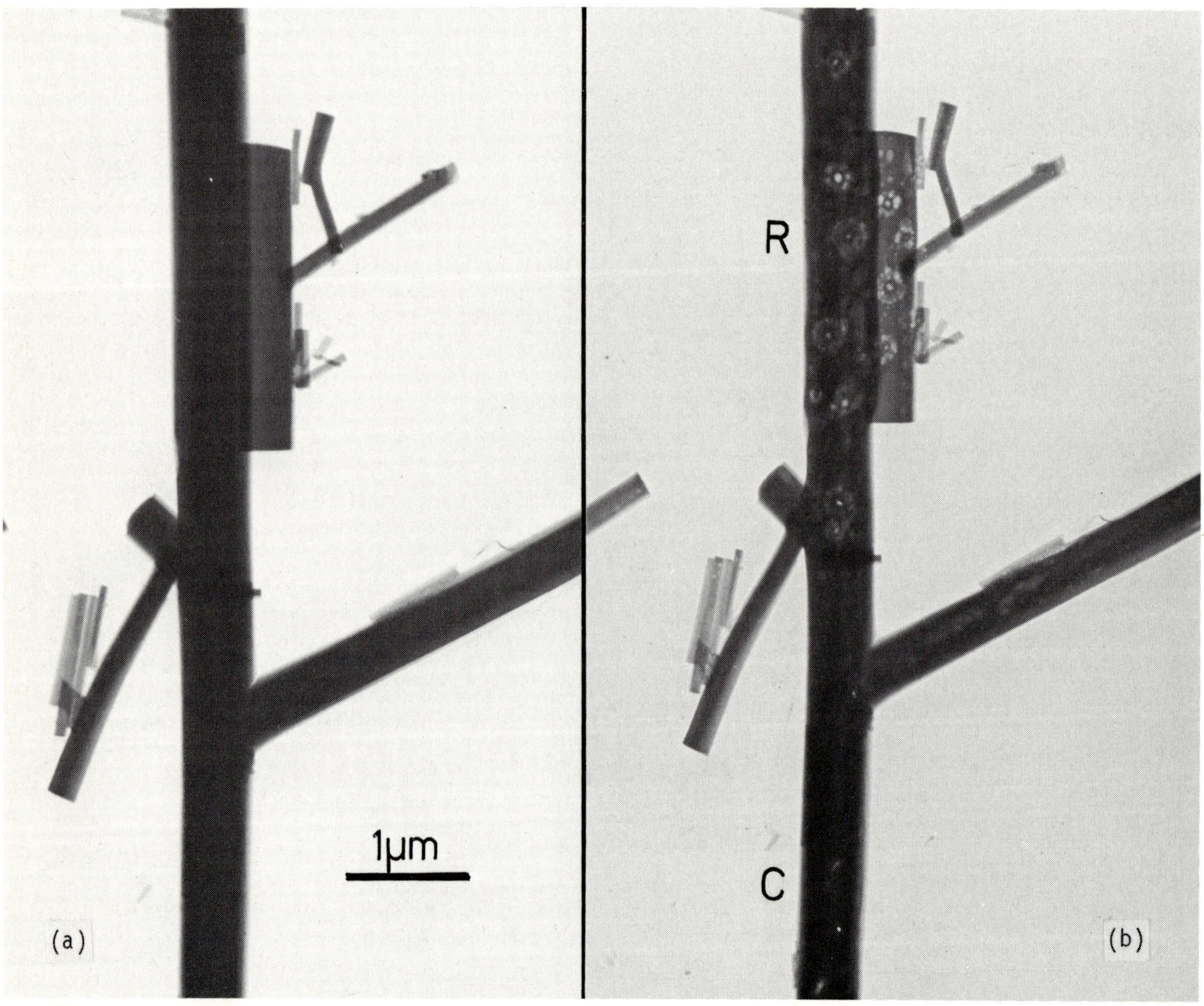

2 **(a) Fibrous growths of ettringite, supported on formvar film; (b) same area after 100 s spot analyses with specimen at RT (lower region) and 85 K (upper region)**

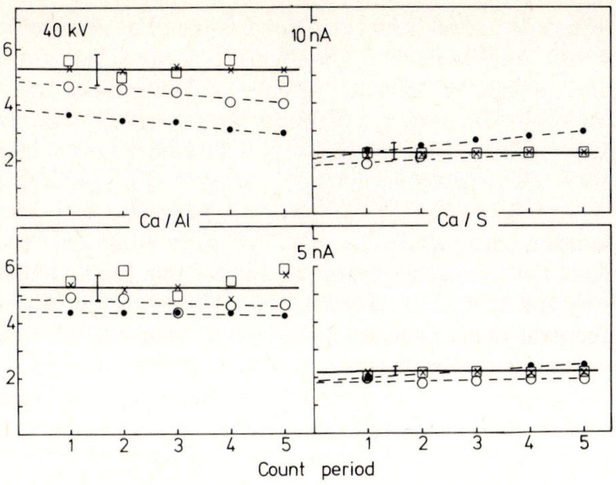

Counting periods:
× 100 s raster
□ 30 s raster
● 100 s stationary spot
○ 30 s stationary spot
Nominal beam dia. 10 nm; solid line drawn through 100 s raster results

3 Measured Ca/Al and Ca/S peak ratios in a variety of analytical conditions at 40 kV; upper plots collected at beam current of 10 nA, lower plots at 5 nA

function of electron energy, beam current, current density and specimen temperature.

EXPERIMENTAL TECHNIQUE

All experiments were conducted on a single batch of a single type of fibre, chosen because it was possible to identify it by electron diffraction. The hydrated product phase is known as ettringite and has a nominal composition $Ca_6(Al(OH)_6)_2(SO_4)_3 26H_2O$. Ettringite has frequently been reported in hydrated Ordinary Portland cement,[4] and grows in a fibrous form (Fig. 2*a*).

The specimen for analysis was produced by adding water to a ground mixture of tricalcium silicate and gypsum. This water/cement mixture was then enclosed in a glass cell with a formvar-coated electron microscope grid.[3] After a short (2.5 h) period of hydration the cell was broken open and the grid removed, immediately coated with a thin layer of carbon, and then placed in the electron microscope.

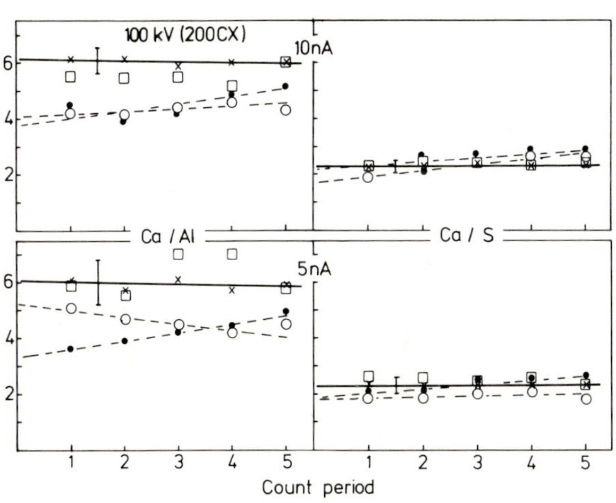

Symbols and conditions as for Fig. 3

4 Peak ratios at 100 kV using JEM 200 CX

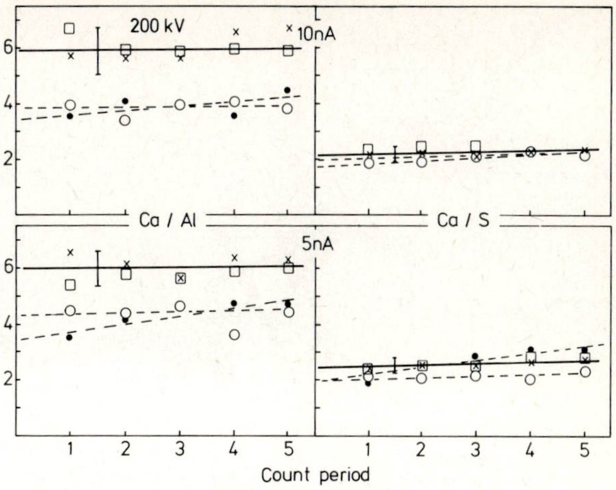

Symbols and conditions as for Figs. 3 and 4

5 Peak ratios at 200 kV

Analyses were performed using a JEOL 200 CX Temscan at 40, 100 and 200 keV and a Philips EM400T at 100 kV. Beam currents were rendered comparable by calibration of the photographic exposure meter and spot sizes quoted below are nominal according to manufacturers' data. X-ray spectra were collected using a Link detector/analyser chain with integral windows established on Ca, Al and S K_α peaks and at four background positions. The majority of data is presented as elemental ratios (after subtraction of background) in order to eliminate variations due to fibre diameter.

RESULTS AND DISCUSSION

Figures 3–6 show the effect of accelerating potential and beam current on the measured elemental peak ratios. For each set of parameters four analytical conditions were used:

(i) focused spot, nominal diameter 10 nm, stationary

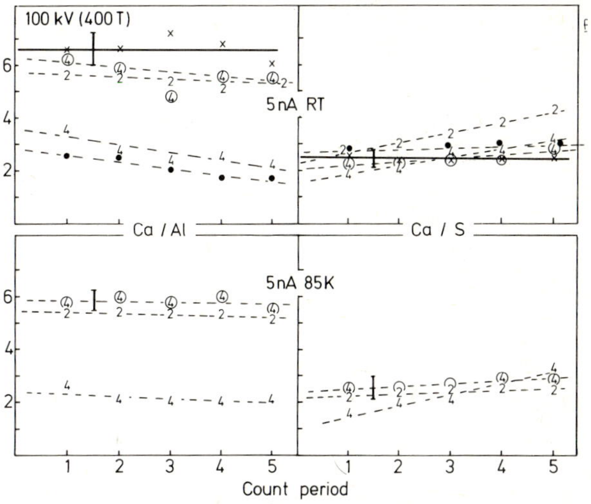

Analytical conditions: 10 nm spot rastered (×) and stationary (●); 20 nm stationary spot focused (4) and spread to 100 nm dia. (circled numeral 4); 100 nm stationary focused spot (2); all count periods 100 s

6 Peak ratios at 100 kV using EM400T; all data collected using beam current at 5 nA; upper plots show results at room temperature; lower data collected with specimen cooled to 85 K

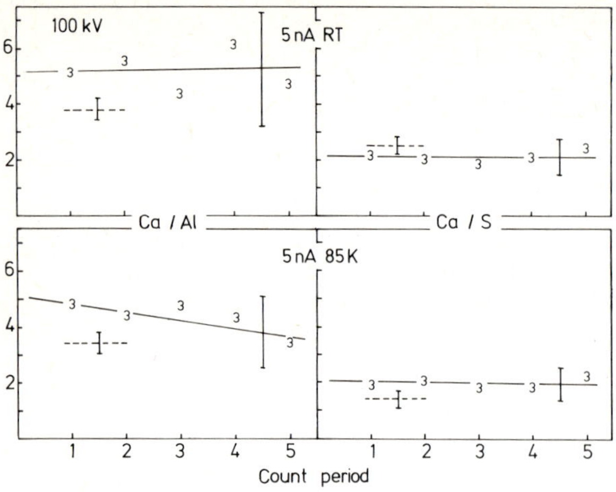

7 **Peak ratios at 100 kV for very short counting times. Nominal beam dia. 40 nm, counting conditions 10 s followed by 20 s with beam off fibre (3); dashed line shows results of same total dose in same conditions collected in one 50 s period; upper plots at room temperature, lower plots with specimen at 85 K**

within a single fibre; five consecutive 100 s count periods (*symbol* ●)

(ii) as (i) with 5 x 30 s count periods (*symbol* ○)

(iii) same spot rastered over an area of 1 x 1 μm, spanning several fibres; five consecutive 100 s count periods (*symbol* x)

(iv) as (iii) with 5 x 30 s count periods (*symbol* □)

For each condition five regions were analysed. After taking the (peak-background) elemental ratios for Ca/Al and Ca/S (all K_α peaks) the five readings were averaged. Each plotted point in Figs. 3-6 is thus the mean of five measurements and scatter bands have been indicated. This procedure was repeated for beam currents of 5 nA and 10 nA at 40 kV, 100 kV and 200 kV on the JEM 200 CX and for conditions (i) and (iii) at a beam current of 5 nA at 100 kV on the EM400T. The peak-to-background ratio for Ca K_α was found to be approximately 25 for all conditions on both microscopes except at 200 kV on the JEM 200 CX, where values of 12 at 10 nA and 7 at 5 nA were recorded. These low ratios probably arise from a high 'background' count at this voltage, since we have so far been unable to eliminate the 'hole count' at 200 kV. The effect on the present experimental results can be seen in the large scatter of the results at 200 kV (Fig. 5).

The raster scans gave, not surprisingly, the most stable analyses. The results from 30 s and 100 s count periods are similar and we shall concentrate on the 100 s raster results (*symbol* x) since the counting statistics were best in this condition. It can be seen from Figs. 3-6 that there was no consistent change in either the Ca/Al or Ca/S ratios in raster analysis conditions. Extrapolation of the results to zero count time (which is equivalent to averaging the results in these stable conditions) indicates an uncorrect Ca/S ratio of 2.3±0.2 for all the analyses and a Ca/Al ratio of 6.0±0.2 for the 100 kV and 200 kV conditions. Only at 40 kV was a significantly different Ca/Al ratio detected (5.3), although this ratio remained stable over the whole analysis time (Fig. 3).

As soon as conditions (i) and (ii), using a 10 nm spot, were attempted it was evident that considerable beam damage was occurring (Fig. 2*b*). The next problem was to establish whether any set of conditions could be found in which extrapolation to zero count time would reproduce the elemental ratios found using a rastered beam. It is clear from Figs. 3-6 that it is possible to deduce a value of the Ca/S ratio within 20% of the rastered value (2.3). However, the Ca/Al ratio is always low when measured with a focused spot, on occasion dropping to one-half of the raster value (e.g. Fig. 6). Extrapolation to zero

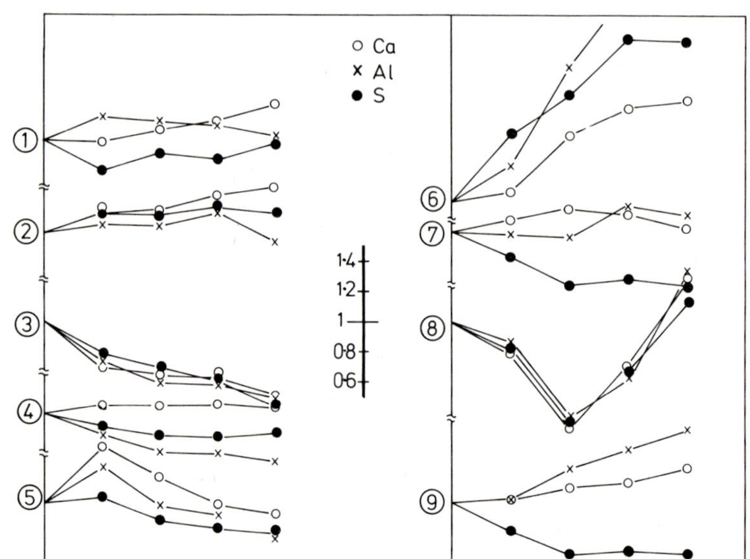

8 **Changes in peak counts for elements Ca, Al and S for nine individual fibres at 100 kV. Beam current 5 nA, spot size 10 nm, stationary for 100 s; counts have been normalized to first count period so that variations are shown**

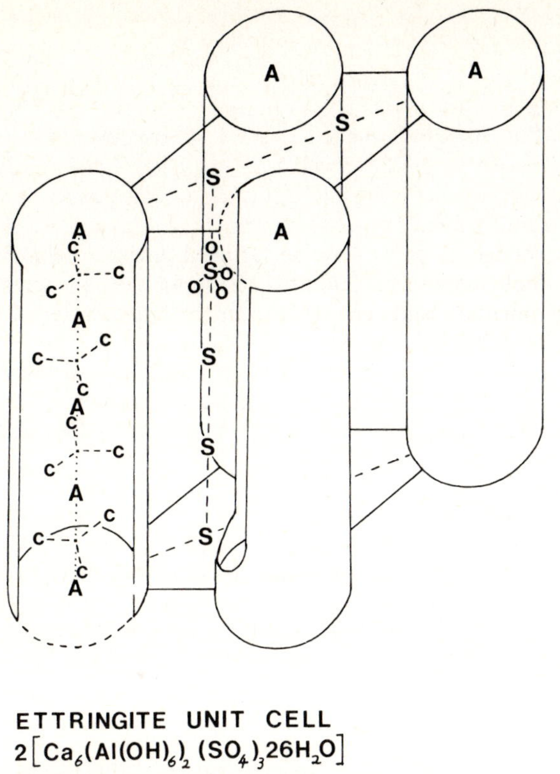

9 Ettringite unit cell, showing rod-like calcium aluminohydroxide rods and less tightly bound sulphate sites, only 75% of which are occupied; most of oxygen and hydrogen atoms omitted for clarity

count time does not eliminate the effect of the beam damage and indeed the Ca/Al ratio was observed both to increase and decrease with time in quite similar analysis conditions (Fig. 4). The difference between the Ca/Al ratios with stationary and rastered beams appears to be unaffected by a factor-of-two change in beam current (upper and lower plots in Figs. 3–5) and clearly does not improve with increase in accelerating potential. Indeed the 40 kV results are the 'best' in the sense that the stationary spot results differ least from, and extrapolate most nearly to, the raster results.

We studied the effect of local dose rate by comparing the spot and rastered spot results at 100 kV on the EM400T with those obtained with a focused 20 nm spot, the same beam defocused to span the whole fibre (~100 nm diameter beam) and a focused 100 nm spot. The results, shown in Fig. 6 (*symbols* 4, circled-4, and 2) are effectively identical to the spot and raster analyses. Clearly a spread beam and a rastered focused beam deliver an equivalently small amount of damage and, as is shown in Fig. 7, a recovery period after intense irradiation is useful in minimizing damage effects.

We also investigated the possibility of minimizing the analysis time. Figure 7 (upper) shows the results of stationary spot analyses using a nominal 40 nm diameter beam at 100 kV in the EM400T. Each point is derived from the average of five analyses as before; however, in this case each analysis period was 10 s, followed by a period of 20 s in which the beam was removed from the fibre. For comparison the results obtained using a single 50 s count period are superimposed. Although the scatter band is large for such poor counting statistics it is interesting that both the Ca/Al and Ca/S ratios extrapolate to almost the same values as were found at 40 kV (5.3 and 2.3). The short analysis periods appear to give an appreciably 'better' analysis than the 50 s count.

Finally we examined the effect of cooling the specimen to 85 K. The lower halves of Figs. 6 and 7 show that no improvement in the analysis with a stationary spot was obtained, although it did appear from the micrograph that the beam damage extended over a smaller volume within the fibre than at room temperature (Fig. 2*b*). With a specimen of such low thermal mass connected via a poor conducting path to the specimen holder it is perhaps not surprising that cooling the grid has such a negligible effect.

Let us now consider what the beam damage is actually doing to the fibres when a focused spot is used. Figure 8 shows the change in peak count rates for the three elements of interest from *individual* fibres during five consecutive 100 s count periods at 100 kV with 5 nA in a 10 nm stationary spot. In order to eliminate the effect of fibre diameter the counts have been normalized to unity at the first count period. It can be seen that in two cases (Nos. *3* and *8*) all elements changed together, consistent with a total mass loss or gain from the region of interest without significant change in composition. However, in all the other cases the elemental ratios changed during analysis and it is difficult to detect a very consistent pattern. We can see from Figs. 3–6 that the Ca/Al ratio is always lowered by damage and it seems from Fig. 8 that quite often Ca is the most stable (or perhaps the least unstable!) element while Al increases during irradiation. However, from a consideration of the structure of ettringite (schematically shown in Fig. 9) we would expect that S would be the most mobile species, while Ca and Al were more likely to move in association with one another. Since this is not what is observed we must speculate that the rod-like Ca/Al/OH structures are destroyed by electron irradiation at the doses used here ($\sim 10^7$ A/m^2).

Although the low doses needed to make a meaningful analysis possible ($\sim 10^5$ A/m^2) can be obtained either by rastering a fine spot or by using a larger stationary spot there are considerable practical advantages to analysing in the TEM mode. The most important of these are that the extent of beam damage or specimen drift can be assessed, in most cases, during the analysis.

CONCLUSIONS

1 Ettringite fibres, a typical cement constituent, cannot be analysed successfully at accelerating voltages between 40 and 200 kV at electron beam doses as high as 10^7 A/m^2. This precludes analyses using a stationary spot of diameter much below 100 nm.

2 Reproducible analyses for Ca, S and Al can be carried out by rastering a 10 nm spot or by spreading the beam to a diameter of 100 nm or greater. For practical reasons the latter is generally to be preferred.

3 Cooling the specimen to 85 K is of no assistance in stabilizing the specimen for analysis.

4 More work is needed to establish which of the Ca/Al ratios found at 40 kV and higher voltages is more accurately representative of the actual composition.

REFERENCES

1 J. D. BIRCHALL, A. J. HOWARD, and K. KENDALL: *Nature*, 1981, **289**, 388
2 J. E. BAILEY and D. D. HIGGINS: *Nature*, to be published
3 J. E. BAILEY and D. CHESCOE: *Proc. Brit. Ceram. Soc.*, 1979, **28**, 165
4 H. F. W. TAYLOR: *Proc. Brit. Ceram. Soc.*, 1979, **28**, 147
5 D. D. DOUBLE, A. HELLAWELL, and S. J. PERRY: *Proc. R. Soc. London*, 1978, **A359**, 435

6 J. D. BIRCHALL, A. J. HOWARD, and J. E. BAILEY: *Proc. R. Soc. London,* 1978, **A360**, 445

7 P. BARNES, A. GHOSE, and A. L. MACKAY: *Cem. Concrete Res.,* 1980, **10**, 639

8 M. W. GRUTZEK and D. M. ROY: *Nature,* 1969, **223**, 492

9 S. GOTO, M. DAIMON, G. HOSAKA, and R. KONDO: *J. Am. Ceram. Soc.,* 1976, **59**, 281

10 M. S. STUCKE and A. J. MAJUNDER: *Cem. Concrete Res.,* 1977, **7**, 711

11 G. CLIFF, J. A. GARD, K. MOHAN, and H. F. W. TAYLOR: *J. Am. Ceram. Soc.,* 1978

12 J. E. BAILEY and D. CHESCOE: Proc. 7th Int. Cong. Chemistry of Cement, Paris, 1980, in press

13 M. S. ISAACSON: in 'Principles and techniques of electron microscopy', ed. M. A. Hayat, Van Nostrand Reinhold, 1977, **7**, 1

14 P. J. GOODHEW and D. CHESCOE: *Micron,* 1980, **11**, 153

15 M. W. GRUBB and G. W. GROVES: *Philos. Mag.,* 1971, **24**, 815

DISCUSSION of paper by Chescoe and Goodhew

E. E. LACHOWSKI:
We have tried Dr Loretto's suggestion of using a distorted beam to analyse whole ettringite particles and have obtained results which are similar to those presented in this paper. Consistent analyses were obtained, but the compositions determined differed somewhat from the expected values.

P. J. GOODHEW:
It is encouraging that your experience confirms ours. However, the elliptical beam technique is of little use for our purposes since we are trying to measure any change in composition along a single fibre. The average composition of the fibre is easily obtained by spreading or scanning the beam, as we show in the paper.

Part 3
Electron energy loss spectrometry (EELS)

Electron energy loss spectroscopy

D C Joy

Electron energy loss spectroscopy (ELS) when carried out in a transmission or scanning transmission microscope provides a sensitive microanalytical technique of high spatial resolution. The shape and energy loss of the characteristic ionization 'edges' permit an unambiguous identification of elements from Li upwards in the periodic table, with a detection sensitivity of the order of 10^4 atoms in an analysed area of a few hundred angstroms diameter. Standardless, computer-based quantitation routines give a relative accuracy of 20% for the common K *and* L *edges. As with energy-dispersive X-ray spectrometry (EDS) in the electron microscope there are a variety of instrumental and spectral artifacts that can occur and reduce the sensitivity and accuracy of the technique and more work is required to assess their significance.*

The author is with Bell Laboratories, Murray Hill, New Jersey, USA.

In a period of a few years electron energy loss spectroscopy (ELS) has been transformed from a laboratory curiosity to a widely utilized microanalytical technique. This upsurge of interest has been the result of a widespread need for a high sensitivity, high spatial resolution, microanalytical method applicable to the low Z elements. However, as with any technique there are both advantages and drawbacks and as the use of the method increases it is necessary to consider both in order to correctly establish its relationship and potential relative to other approaches. The purpose of this paper therefore is to outline the basic physics of ELS for microanalysis, to look at its practical application, and finally to identify some of the significant problems that establish limits on its usefulness and accuracy.

ELECTRON INTERACTIONS AND ELECTRON SPECTROSCOPY

The impact of an energetic electron beam (30–200 keV) on an electron-transparent solid produces a spectrum of events ranging from those in which the incident electron is totally unscattered in its passage through the sample to those in which it is either backscattered or brought to rest as the result of multiple inelastic scattering events. In addition the de-excitation of atoms and molecules, energized as the result of an inelastic scattering event with an incident electron, results in the production of other radiations (photons and electrons) at energies characteristic of the atom or molecule. Many familiar analytical techniques (Auger, fluorescent X-ray spectroscopy, cathodo-luminescence) rely on these de-excitations because the radiations are readily collected and analysed. However, this approach has several drawbacks. Firstly most de-excitations are accomplished by competitive processes (for example Auger electrons and fluorescent X-rays) and consequently the yield of the species collected depends not only on the number of excitations but on the efficiency of the particular decay event of interest. Since this is usually a sensitive function of the parent atom, there can be wide differences in the detection sensitivity achieved. Secondly these de-excitations may follow excitations that were not caused by an incident electron but by a previously scattered electron or even by another X-ray. As a result the spatial limits of the analysed volume are not determined by the irradiated volume but by factors such as the nature and density of the material and the type of emission being studied.

These problems can be circumvented by studying the primary event, the scattering of the incident electron. All the available information about the sample can be obtained by an analysis of the observed angular and energy distribution of the transmitted electrons in terms of some suitable model. The number of events observed will now be directly related to the number of interactions, and because the characteristic scattering angles for electrons in the kilovolt range are only a few milliradians, the analysed volume is essentially defined by the irradiated volume. The most general description of the scattering is in the momentum change suffered by the incident electron,[1] which requires simultaneous measurement of the energy loss E and the scattering angle θ, but for most microanalytical purposes it is adequate to collect all the transmitted electrons scattered up to some maximum angle β and to obtain their energy loss spectrum (Fig. 1). This loses the information about momentum transfer, but has the practical advantages of requiring much simpler instrumentation and making more efficient use of the electron signal. All the subsequent discussion will refer to this mode of operation.

CORE-LEVEL EXCITATIONS

While all the elastic and inelastic interactions between the incident electron and the solid are potentially microanalytical tools,[2] the most specific and useful is that related to the ionization of an electron from an inner shell of an atom.

The signal $I(E)$ detected at some energy loss E through a spectrometer collecting electrons scattered through angles up to β can be written as

$$I(E) = I \cdot N \cdot \eta \cdot \sigma(\beta, E, E_0) \qquad (1)$$

where I is the intensity of the incident beam (electrons/s), N is the number of atoms in the irradiated area, η is an

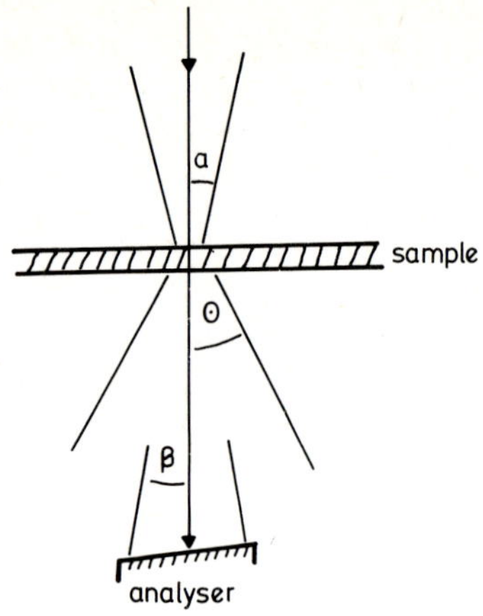

1 **Definition of geometrical parameters for electron energy loss analysis; α_0 is incident beam convergence, β spectrometer acceptance angle, θ scattering angle**

TABLE 1

Element	Z	E_k, eV	Element	Z	$E_{L_{23}}$, eV
Li	3	55	Al	13	73
Be	4	110	Si	14	100
B	5	188	P	15	136
C	6	284	S	16	165
N	7	400	Cl	17	202
O	8	532	Ar	18	250
F	9	684	K	19	296
Ne	10	865	Ca	20	348
Na	11	1075	Sc	21	405
Mg	12	1305	Ti	22	459
Al	13	1510	V	23	517
Si	14	1832	Cr	24	580
			Mn	25	645
			Fe	26	714
			Co	27	786
			Ni	28	863
			Cu	29	941
			Zn	30	1031

L_{23} energy is unweighted mean of $L2$ and $L3$ edges

Figs. 2–5

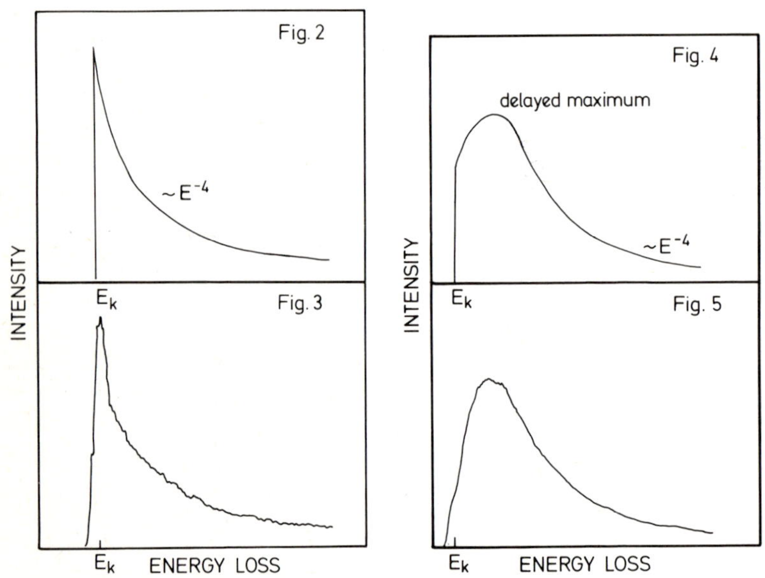

2 **Intensity profile expected on passing through K ionization edge of energy E_k in transmission energy loss spectrum. After the edge, intensity falls away as about E^{-4} where E is energy loss**

3 **Experimentally observed K edge from carbon. Taken at 100 keV in STEM mode with $\beta = 7.5$ mrad and resolution 5 eV. Recording time 200 ms/ch, channel increment 0.9 eV/ch**

4 **Intensity profile predicted for L edge of onset energy E_k. 'Centrifugal potential' arising from angular momentum of L shells causes delayed maximum in intensity followed by fall-off with energy loss E as E^{-4}**

5 **Experimentally observed L_{23} edge from silicon, recorded at 100 keV in STEM mode with $\beta = 7.5$ mrad and resolution of 5 eV. Recording time 200 ms/ch, channel increment 0.9 eV/ch**

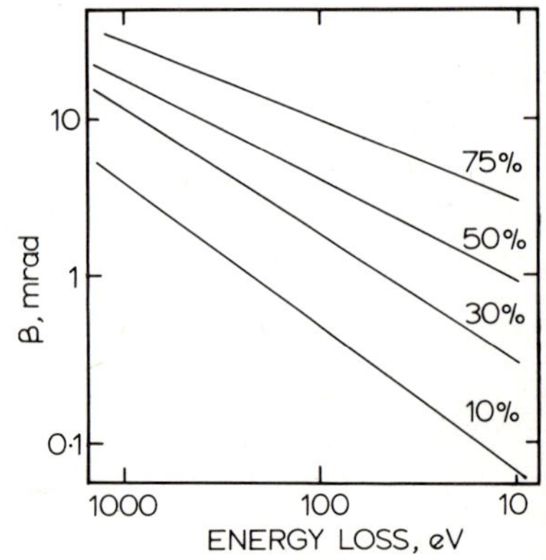

6 **Relative percentage detection efficiency for inelastically scattered electrons of energy loss E (eV) for spectrometer acceptance angle β (mrad)**

efficiency factor (which may depend on α, β and other parameters as discussed later but which is usually close to unity), and σ is the interaction cross-section representing the probability of the incident electron being scattered through an angle less than β while suffering an energy loss E. Some minimum energy E_k is needed to ionize a particular inner shell of an atom. No electron losing energy $E < E_k$ can ionize the atom so the cross-section for this event will be zero, but for energy losses $E > E_k$ a finite cross-section for the ionization exists and consequently there will be an increase in the signal $I(E)$ when $E \geqslant E_k$. This will produce a discontinuity or 'edge' in the spectrum at E_k. Since the energy E_k closely approximates the binding energy for the particular shell ionized it is uniquely characteristic of the atom and a single measurement of the energy (loss) E_k is in principle sufficient for an elemental identification.

Any energy loss by an electron of $E > E_k$ will still be able to produce the ionization, so the 'edge' could extend from E_k to E_0, the incident beam energy. However, the cross-section falls as about E^{-4} for $E > E_k$ so to a first approximation the inner-shell ionization edge would appear in the energy loss spectrum as a 'triangle' (Fig. 2). This simple model is followed very closely by K-shell (i.e. the $1-s$ orbitals) ionizations observed experimentally (Fig. 3). For L shells, however, this situation is made more complex by additional energy terms arising from the angular momentum of the $2p$ orbitals. This has the effect of increasing the apparent energy needed to ionize the atom giving a 'delayed maximum' in the edge (Fig. 4), an effect seen in the L_{23}-shell edge from silicon (Fig. 5) and other first-row elements. The transition elements, however, give L edges which look much more like those from K shells.

There are two pieces of information that can thus be used for elemental analysis from the energy loss spectrum, the *energy* and the *shape of the edge*. Table 1 lists the energies E_k and the types of edge most commonly met in the loss range 50-2 keV. Only edges from K and L_{23} shells are shown. The $L_1(2s)$ shell edges are comparatively weak (1/8 of the L_{23} strength) and rarely visible, and the $M_{23}(3p)$ and $N_{23}(4p)$ edges lying in this range from high-Z elements have poorly defined onsets at E_k and so are of limited use for analysis.

Electrons which have generated an inner-shell excitation have an angular distribution $I(\theta)/I(O)$ which can be written, for a collimated incident beam, as

$$\frac{I(\theta)}{I(O)} = \frac{F(E,\theta)}{\theta^2 + \theta_E^2} \tag{2}$$

where θ_E is the characteristic inelastic scattering angle given for non-relativistic accelerating voltages as $E/2E_0$ and $F(E,\theta)$ is the 'oscillator strength' representing the number of electrons per atom participating in the energy loss process. The term $F(E,\theta)$ is approximately constant for small values of θ but falls to zero for $\theta_{max} \simeq \sqrt{(2\theta_E)}$. The scattering is thus seen to be forward peaked with a width varying as θ_E. Taking as an example the carbon K edge ($E_k = 284$ eV) at $E_0 = 100$ keV, $\theta_E \simeq 1.5$ mrad while $\theta_{max} \simeq 55$ mrad, suggesting that it should be possible to collect a high fraction of all the scattered electrons of interest. Making the oversimplification that $F(E,\theta)$ is constant for $\theta < \theta_{max}$, and zero for $\theta \geqslant \theta_{max}$, equation (2) can be integrated to give a collection efficiency η_β

$$\eta_\beta = \frac{\ln(1 + \beta^2/\theta_E^2)}{\ln(2/\theta_E)} \tag{3}$$

This quantity is plotted in Fig. 6 as a function of the collection angle β and the energy loss E and shows that for relatively modest collection angles ($\beta \sim 10$ mrad) the collection efficiency is typically between 30 and 75% which compares very well with the 1% figure which is usual for X-ray detectors.

It is not necessarily true, however, that the largest possible collection angle into the spectrometer is the best. This is because the discussion so far has made no mention of any 'background' on which the edges might be riding. The cross-section σ in equation (1) contains contributions from other interactions besides the wanted ionization edges. These other interactions, which include noncharacteristic scattering events of many kinds, combine to produce a signal[3]

$$I(E) = I \cdot A \cdot E^{-r} \tag{4}$$

where A and r are constants (for limited ranges of the spectrum) whose values depend on both E and β as well as on the material.

This background plays a dominant role in energy loss spectroscopy for a variety of reasons. Firstly, the inelastic electrons comprising the background have an angular distribution which although similar to that of the characteristic edge electrons rises more rapidly with β and saturates for higher values of θ_{max}. As the angular range β for collecting inelastic electrons is varied therefore the visibility or signal-to-background ratio of the edge will change. For small values of β ($\beta \ll \theta_E$) both the edge and the background will be small signals and thus noisy and of little use. For large values of $\beta (\gg \theta_E)$ the edge signal has reached its maximum value but the background is still increasing so the signal/noise ratio of the edge will fall. At some intermediate value of β there is thus an optimum condition for detecting the edge, and experimentally this is found to be when $\beta \simeq \theta_E$.[4] (It is worth noting that other problems also restrict the value of β which is useful; for example on crystalline samples it is desirable to keep $\beta < \theta_{BRAGG}$ to avoid problems from diffracted beams.)

Secondly, the fact that the edges ride on a background whose magnitude is both considerable and changing rapidly with energy loss means that the true shape of the edge is not always obvious, and in some cases even that an edge may be practically invisible because of the background shape. Figure 7 shows the silicon L_{23} edge from Fig. 5 as it actually appears on the spectrum. Although it is still easily visible it is certainly less well defined and it is obvious that any operations requiring a knowledge of the true shape of the edge or of the integrated intensity contained in the edge will necessitate the removal of the background. This point is discussed in a later section.

Finally, the dominant contribution to the background will often be the 'tail' of some previous major edge. This means that when performing microanalysis for elements in a matrix there may be a considerable difference in the sensitivity depending on whether the element of interest has an edge which is above or below the energy loss of a major edge from the matrix. Edges at high energy losses (>1.5 keV) will typically be riding on a background comprised of tails from many preceding edges and this, coupled with the fact that the ionization cross-sections fall as about E^{-2}, means that the detection sensitivity in this regime is very limited and it is then preferable to switch to EDS for the elements affected.

The magnitude of the cross-section in equation (1) is typically only of the order of 10^{-21} cm^2 per atom. Since a carbon foil a few hundred angstroms thick contains about

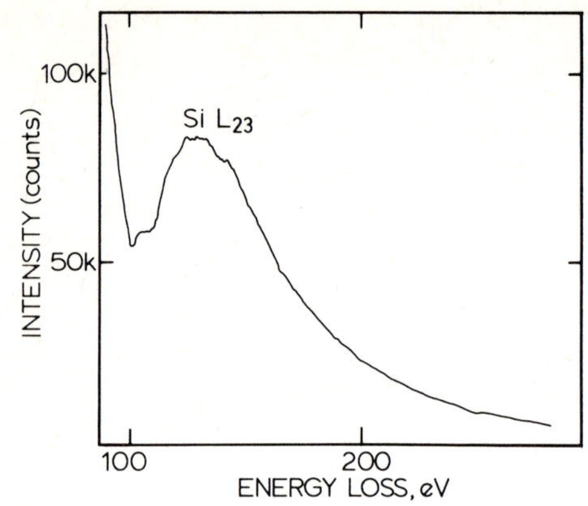

7 Appearance of Si L_{23} edge on background intensity. Background falling as about E^{-3} reduces visibility of edge giving edge jump ratio (peak/minimum) of only two to one. Conditions as Fig. 5

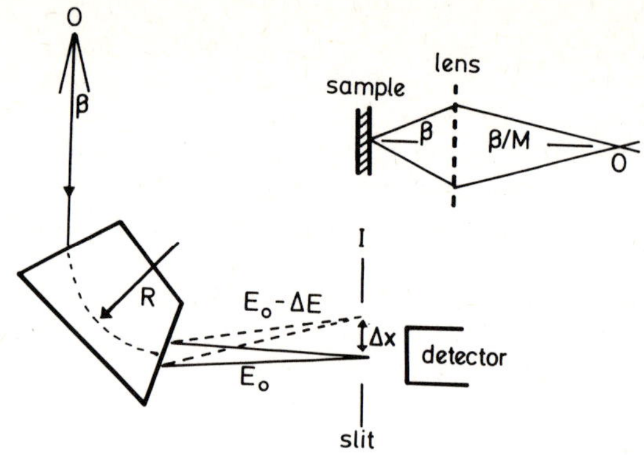

9 Basic layout of analyser for ELS. Electrons leaving object point O are deflected through 90° in radius R by magnetic field and form dispersed line image at I. Dispersion $\Delta E/\Delta x$ is typically a few μm/eV. Lens of magnification M placed between specimen and object point reduces pencil of rays through spectrometer from β to β/M

10^{18} atom/cm^2 this indicates that the characteristic edge signal is only about $10^{18} \times 10^{-21}$, that is 10^{-3}, of the intensity of the incident beam. Since the incident beam intensity used for microanalytical purposes is itself relatively weak (typically 10^{-9} A) this means that it will be necessary to have efficient low noise detection systems. It is also evident that if it is desired to record the whole energy loss range, including the zero-loss portion of the spectrum containing the unscattered and elastically scattered components, the recording and display system must have a very dynamic range. Figure 8 shows such a complete spectrum from the zero-loss peak to the carbon K edge including a gain change of $\times 250$. This shows clearly that 90% of the total incident intensity is either unscattered or elastically scattered and that, of the remainder formed by the inelastically scattered electrons, it is only the smallest component – that in the edges – which contains significant microanalytical information.

ELECTRON SPECTROMETERS AND THEIR INTERFACE TO THE MICROSCOPE

The energy loss spectrum is obtained by the addition of an electron spectrometer to the microscope. Although a wide variety of designs are described in the literature the great majority of all current systems use magnetic section prisms because of their simplicity and high performance. In operation the magnetic prism is analogous to the glass prism and its effect on white light (Fig. 9). Electrons at E_0 diverging from the object point O are deflected through an angle of typically 90° and brought to a focus at the image point I. Electrons of a different energy $E_0 - \Delta E$ are again brought to a point focus but at a position Δx away from I, so forming a line dispersion of the spectrum. Three parameters define the performance of such a prism spectrometer, the dispersion $\Delta E/\Delta x$ (eV/cm), the solid angle of acceptance $\pi\beta^2$, and the energy resolution δE. The dispersion is directly related to radius of curvature R of the analyser and

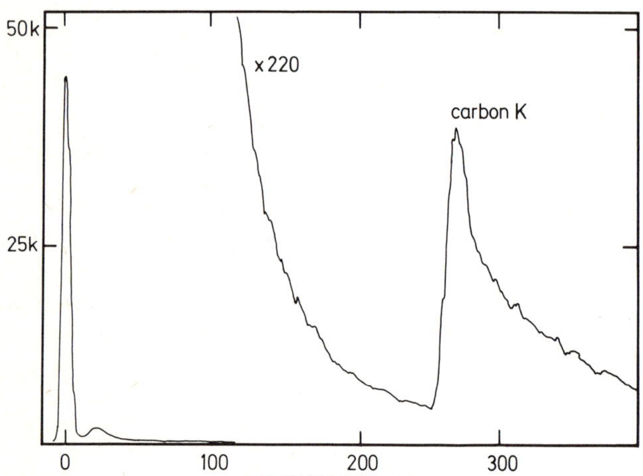

8 Energy loss spectrum from 200 Å thick carbon film including zero-loss and plasmon peaks. Gain change of $\times 200$ is required to make K edge visible. Conditions as in Fig. 3

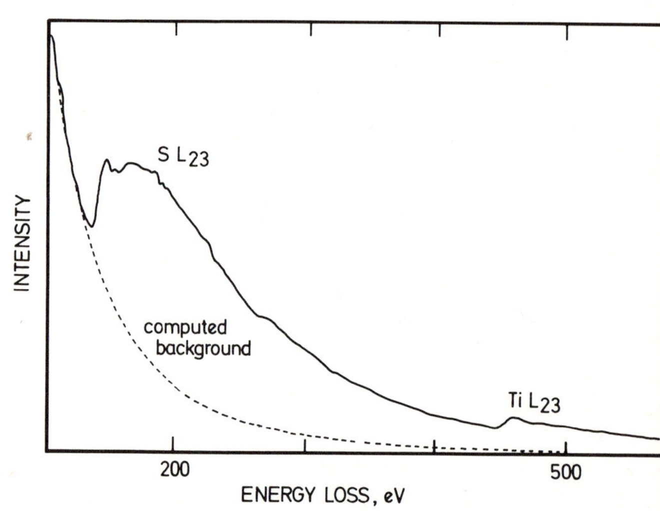

10 Spectrum from TiS_2 together with computer modelled background fit and extrapolation. Fit to model $A \cdot E^{-r}$ is optimized using least-squares model. Here $r = 3.65$ (for $\beta = 12.2$ mrad, 100 keV). Fit made by LSI-11 on ORTEC EDS-II multichannel analyser

inversely to the incident beam energy, and is typically 4 μm/eV for $R = 15$ cm and $E_0 = 100$ keV. The resolution δE is the smallest difference in energy that can be distinguished at the image point. This in turn will depend on the physical size of points in the image plane and their spacing. Since the spectrometer is in essence a lens the image point size will be a function of the object point size and of aberrations in the analyser which increase the image diameter. While substantial reductions in the aberrations of magnetic prisms can be made by shaping the pole pieces[5] it can be shown quite generally[6] that

$$\pi\beta^2 \cdot \frac{E_0}{\delta E} \simeq 1 \qquad (5)$$

It was shown earlier that for effective microanalysis β needs to be about 10 mrad, which indicates that at $E_0 = 100$ keV, $\delta E \simeq 30$ eV. This order-of-magnitude estimate is too high for most purposes, so in order to preserve both the efficiency and the resolution of the spectrometer one or more lenses are placed between the specimen and the object point of the analyser. If the magnification of the lenses is M then for a collection angle of β measured at the specimen the angle into the spectrometer is only β/M so in equation (5) the best energy resolution is improved by a factor M^2. The intermediate and projector lenses on a TEM/STEM are conveniently available for this purpose and by an optimum choice of conditions even simple spectrometers can combine high efficiency and good resolution.

There are, however, obvious limits to what can be achieved in this way. As the magnification M is increased the size of the object point increases, so that eventually the dominant factor limiting resolution will be the diameter of the image, which will be approximately $M \times$ the analysed area of the sample, divided by the dispersion. Consequently either the analysed field of view or the magnification must be limited. It must also be noted that chromatic aberrations in these coupling optics can cause problems, some of which are noted later. The optimum design criteria for these post-specimen lenses are still a matter of discussion.[7]

In practice the spectrometer is not operated at its limiting resolution but at some larger value S/D where S is the spacing of the selector slits ahead of the detector and D is the dispersion. By contrast with the EDS case the ability to set the resolution is an important experimental variable. Ideally the smallest value would always be best because of the better definition of detail, such as the 'white-line' structure in edges, which can aid identification. However, the lower the energy window accepted by the slit, the smaller the signal collected, and this factor can be crucial in microanalysis at high spatial resolution. A resolution of between 3 and 6 eV is a reasonable compromise although at higher energy losses (>1 keV) a figure of 10–20 eV may be even better. For some special applications (plasmon loss studies, fine-structure effects, etc.) resolutions of 1 eV or less may be required but at this level other factors such as the energy spread of the source, the stability of the high voltage and lens supplies, and AC ground loops or electromagnetic inference are usually very important and may mandate special procedures.

Most spectrometers will function in either TEM or STEM mode. STEM operation allows a high spatial resolution and by virtue of the small illuminated area a high coupling magnification into the spectrometer and hence good collection efficiency. These advantages are somewhat offset by the lower current available in STEM mode and the increased problem of contamination, and TEM operation is preferable where signal statistics rather than spatial resolution is the main concern.

DATA COLLECTION AND REDUCTION

The final component in an ELS system is the data collection, display and reduction system. It is now standard practice to record the spectra into a multichannel analyser (MCA) operated in multiscaling mode so that successive channels contain all the counts accumulated in some dwell period τ. The spectrum is obtained by scanning the dispersed signal over the analyser slits, by a separate set of coils or ramping the prism magnet, using a scan generated in synchronism by the MCA.

In order to get the signal into the MCA it is necessary to count the electrons directly, or to suitably digitize the signal. While electron counting, using a suitable scintillator and photomultiplier, is undoubtedly the most attractive approach, substitution of typical experimental parameters in equation (1) will show that count rates in excess of 100 MHz are encountered over much of the usual spectral range. Since this is well beyond the capability of the best current counting chains, the alternative of analog-to-digital conversion using voltage-to-frequency convertors is commonly employed. The convertor produces a pulse train whose instantaneous repetition rate is linearly proportional to the input voltage. As these units are both cheap and offer a very wide dynamic range (>10^5 to 1) they are ideally suited to this application.

With the spectrum stored in the MCA data processing is possible using a computer connected to the core memory. This is important because in many cases some operations on the spectrum are necessary in order to obtain the required information. The most basic of these is background modelling and stripping, which relies on the analytic model of equation (4). Usually the background region ahead of an edge is read into the computer over some energy range, e.g. 50–100 eV, and a least-squares minimization procedure used to obtain the parameters A and r. These are then used to plot the fitted background and to extrapolate this fit beyond the edge (Fig. 10). Using a small (PDP 11-03) computer of the kind often supplied with an MCA, this procedure only requires one or two seconds for each edge. Modelling, and if necessary stripping, the background not only aids in the identification (and quantitation)

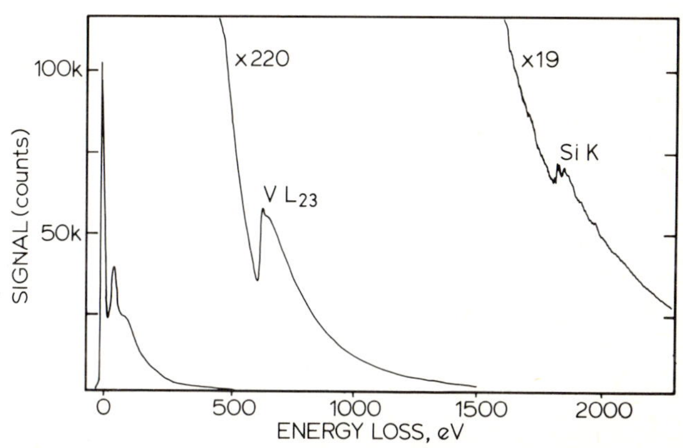

11 Complete spectrum from V_3Si showing zero loss, plasmon peak, V L_{23} edge at 510 eV and Si K edge at 1830 eV loss. No Si L_{23} edge observed under these conditions even though it should be strongest edge (due to interferences from both plasmon-tails and V M edges)

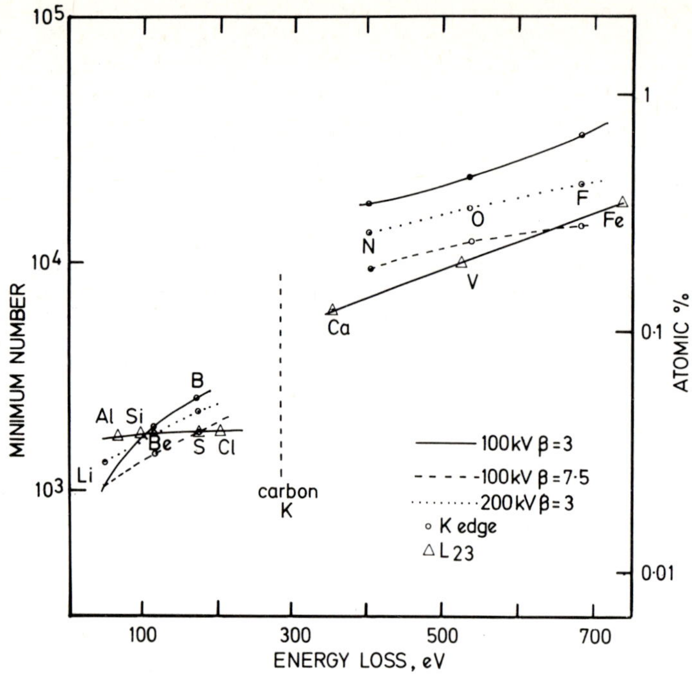

12 **Calculated detection limits for elements in 800 Å thick carbon matrix, plotted as minimum number of atoms, or as equivalent atomic fraction, for two operating voltages and two acceptance angles. For all cases beam current of 10^{-10} A into 300 Å diameter probe is assumed, with recording time of 200 ms/ch**

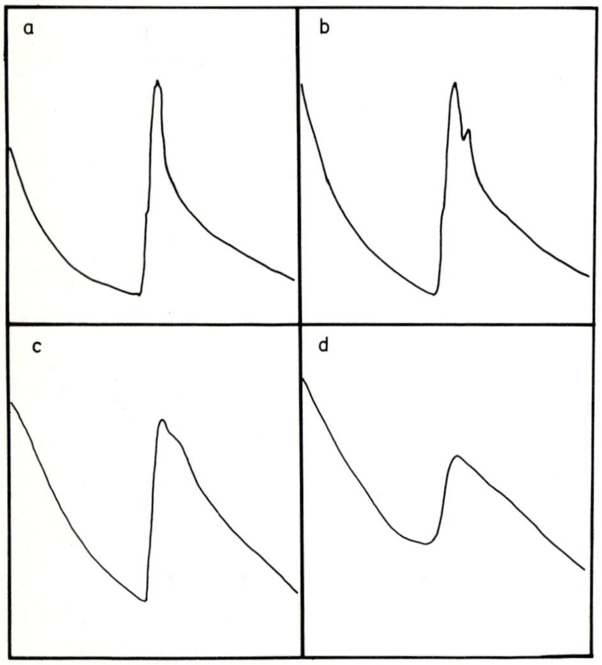

13 **Change in appearance of *K* edge as sample is increased in thickness. True edge shape in (a) suffers first plasmon replication (b) and then multiple scattering (c) and increase in background intensity with a fall in edge visibility (d). In extreme cases edge will merge into background leaving only change in slope**

of edges, but is also a valuable diagnostic tool for the performance of the spectrometer.[8] The other commonly required functions, such as calibration and scaling of the display, integrals of the spectrum, etc., are invariably provided by current MCAs as either firmware or software options.

PRACTICAL MICROANALYSIS

Microanalysis using ELS is as straightforward as the equivalent EDS procedure although there are more experimental variables. Once the spectrometer has been aligned and the calibration (eV/channel) found, analysis is a matter of placing the beam (assuming STEM operation) on the area of interest and starting the spectral scan. This task is greatly aided if the microscope is set up so as to permit easy switching between the spectrometer and say a dark-field detector, or if a separate detector can be placed in the beam path for imaging purposes. Since the ELS data are acquired sequentially the analysis of elements widely separated in energy loss is not only a time-consuming process but can lead to misleading results if radiation damage or contamination occurs during the scan. In such cases, after a spectrum has been obtained at a minimum signal/noise ratio it is desirable to arrange to scan only the spectral regions of interest located from the display, either by limiting the scanned range and repeating the scan for each edge of interest or by using a two-speed scan of the type provided on some new MCAs. It is also desirable to avoid scanning the zero-loss peak region except where this is required for calibration purposes, since the magnitude of this signal can cause a long-term distortion in the characteristics of the recording system.[8] Where the zero-loss peak is required, it is preferable to scan the spectrum in reverse from high energy losses to low.

In addition to a choice of the scan conditions it is also necessary to select the acceptance β. Although a compromise value of between 3 and 10 mrad is satisfactory, the ability to maximize the signal/noise ratio by optimizing β is sometimes desirable particularly when one edge is very weak or the edges occur at widely separated energy losses. Finally it is necessary to be aware that experimental conditions may occasionally make an edge almost invisible and lead to a misleading conclusion unless either some other edge from the same element (or the EDS spectrum if available) is detectable. This is illustrated in Fig. 11 which shows spectra from V_3Si. Under the experimental conditions used the Si L_{23} edge is essentially invisible, only the V L_{23} edge and the Si K edge being apparent. It is fair to say that experience is an important asset in ELS microanalysis, and that whenever possible the simultaneous acquisition of an EDS spectrum is very desirable. Failing this the ELS data must be taken over the widest possible energy range so as to include as many edges from the heavier elements as possible.

SENSITIVITY OF ELS MICROANALYSIS

The sensitivity of a microanalytical technique is an important factor in determining its usefulness. For ELS, as for EDS, the limiting condition will be the statistical requirement for a minimum signal-to-noise level between the edge and the background. Because of the dominance of the background in the spectrum the sensitivity will depend very much on the composition and mass-thickness of the matrix as well as on the experimental parameters such as beam current, recording time and acceptance angle. Consequently detailed calculations are necessary in order to obtain realistic results. Figure 12 shows results calculated for a carbon matrix, under a variety of experimental conditions typical of STEM operation.[9] The most obvious features are the very wide variation in sensitivity as the atomic number changes, and the pronounced fall in sensitivity for edges on the high side of the carbon K edge, reflecting the fact that the backgrounds on either side of the edge are different by almost a factor of 10. Similar data can be calculated for other model systems with approximately the same level of sensitivity. Although the absolute detection limits are excellent, typically 10^{-19} g, the mass or atomic fraction this represents is only between 0.5 and 5% in most cases. This indicates that ELS is best suited for the study of precipitates, segregation effects and similar situations where the local mass fraction is high, or can be made so by choosing the probe diameter. It can also be seen that much of the potential gain in microanalytical sensitivity derived from ELS operation is sacrificed when the normal sequential mode of spectrum recording is used since the analysis time is only of the order of a few hundred milliseconds per channel compared with 100 s or more for EDS. However, recent developments in parallel detector arrays which acquire data simultaneously in many channels will ultimately remove this disability and yield a substantial gain in sensitivity.

SPATIAL RESOLUTION

Because the acceptance angle β into the electron spectrometer is so small ($\sim 10^{-2}$ rad) the volume of material analysed is defined geometrically by the cone of semi-angle β about the beam impact point. In principle therefore very high spatial resolutions are possible. In practice the need to obtain an adequately high signal/noise ratio in the spectrum sets the minimum usable probe size, a value of 300 Å being practical (for losses up to ~1.5 keV) when using a thermionic source, and perhaps 20 Å for a field emitter. Once again this limit is the result of a sequential acquisition of the ELS data. In cases where a parallel detector could be used, or where only one channel is measured as for example when imaging in a specific energy loss, the resolution is correspondingly enhanced and recent results have demonstrated values as low as 5 Å.[10] If the statistically determined limit can be circumvented the absolute limit will be set by the closeness of approach needed for an electron to ionize an atom.[11] Quantum mechanical calculations suggest this to be typically a few angstroms for most core edges, the value falling as the ionization energy increases. For all practical purposes it is adequate to assume that ELS can achieve the highest level of spatial resolution of any volume analytical technique, and in most cases rival the resolution of the corresponding micrograph.

PROBLEMS AND LIMITATIONS OF ELS

The problems and limitations of ELS fall into two classes: those that are inherent and those that are related to instrumental factors. The most persistent problem is that of sample thickness since, unlike the insidious breakdown of the 'thin-film approximation' in EDS analysis, its effect is directly visible on the spectrum. The assumption underlying ELS is that each electron is only scattered once before the sample. If this is not true then the measured energy loss is not characteristic of any one event but a convolution of multiple events. Figure 13 shows the progressive changes in the appearance of an edge as the sample thickness is increased. The initial edge profile first gives way to an edge followed by a 'replica' of the plasmon or low loss spectral detail, representing electrons that have suffered both plasmon and core-level excitations. At still greater thicknesses the edge profile becomes flatter as more losses both characteristic and non-characteristic are convoluted onto the edge, and finally the edge disappears.[12] The mean background level is also seen to steadily increase with thickness. An excessive sample thickness therefore results not only in a fall in the visibility of the edge, but in a distortion of its shape and apparent magnitude, so adversely affecting the information required for microanalysis and quantitation.

The useful upper limit in sample thickness is about one total mean free path for inelastic scattering. This is best assessed by a measurement of the plasmon ratio, and experimentally a plasmon peak height/zero loss peak ratio of less than 0.2 is required for good results. For most common metals and alloys this corresponds to thickness of between 500 and 1000 Å at 100 kV. This value is lower than that which would normally be used for imaging or EDS analysis but is not usually an impossible one to attain. Larger thicknesses can be examined at higher accelerating voltages, and 200 keV operation should offer a factor of at least 1.5. Computer deconvolution of the scattering is also now readily done in relatively short times (~1 min) but secondary effects of the deconvolution such as the increase in noise limit the value of this procedure for general microanalysis.

Diffraction effects can also seriously distort the spectra from crystals, modifying both the background and the edges. These effects are very orientation-dependent[13] and the only safe procedure is to orient the sample so as to excite no strong diffracted beams while using a divergent incident beam (to damp out diffraction effects) and a small acceptance angle.

While the effects of electron and X-ray scattering on EDS data are now well understood, it is not always appreciated that analogous effects can occur in ELS systems.[8] The most serious is that arising from the scattering of the intense zero-loss beam from the walls of the spectrometer flight tube, or from obstructions such as apertures in the beam path. The result can range from fairly minor modifications to the slope of the background to major discontinuities in the spectrum. A 'spectrum' recorded in the absence of a specimen is sufficient to reveal such scattering and this test is as basic as the 'hole count' in EDS studies.

Fluorescent X-rays generated from the beam in the spectrometer can also reach the detector, creating an additional spectral artifact which has a relatively constant intensity with energy loss (unlike electron scattering which varies rapidly with energy loss). After the elimination of gross electron scattering, the quality of a spectrometer is best judged by measurements of spectra from well characterized materials. Edge jump ratios (peak edge intensity/background at edge onset) are sensitive to X-ray scattering and values of less than 5 from K edges of pure elements (e.g. C, Si) is indication of a problem. Poor background fits, or anomalous fitting parameters (A and r in equation (4)) usually indicate electron scattering.

QUANTITATION OF ELS DATA

Absolute quantitation (atom/cm^2) of ELS data is possible by an application of equation (1). This gives

$$N = \frac{I_k}{I\sigma\eta} \tag{6}$$

Because of practical limitations the edge integral (i.e. the integral under the edge and above the background) can only be obtained for some energy window Δ, so the quantitation relation becomes

$$N = \frac{I_k(\beta, \Delta)}{I\sigma_p(\beta, \Delta)\,\eta} \tag{7}$$

where the variables β, Δ indicate that the integral and cross-section relate to scattering angles up to β, and energy losses in the window E_k to $E_k + \Delta$. When an 'absolute' quantitation is required I is conventionally replaced by $I_0(\beta, \Delta)$, the integral in the energy window Δ under the zero-loss peak for the acceptance angle β. This procedure not only gives a first-order correction for backscattering and plasmon-scattering in the sample but allows the necessary integral to be evaluated directly from the spectrum. However, in microanalysis element ratios are required, thus

$$N_A/N_B = \frac{I_{k,A}}{I_{k,B}}\frac{(\sigma_{p,B}\cdot\eta_B)}{(\sigma_{p,A}\cdot\eta_A)} \tag{8}$$

which is functionally identical to the 'k-factor' formulation for EDS quantitation. It is implicit here that all the integrals are evaluated for the same window Δ, and acceptance angle β.

Interactive computer programs are now available for quantitation from spectra stored in an MCA.[14] These rely on the hydrogenic cross-section models[15] which permit rapid first-principle calculations of K- and L-shell partial ionization cross-sections. Under all usual experimental conditions the accuracy of these models is high, probably within $\pm 25\%$.[14,16] Since, however, there are no agreed standard compounds whose homogeneity or stoichiometry can be guaranteed for spatial resolutions of a few hundred angstroms estimates of the accuracy are confined to a few model systems. It is probable that in many situations the 'error' in the cross-sections is less than that from the uncertainty in the efficiency factors η. It has been customary to assume that η is always unity, but in fact a variety of factors in the system can lead to modifications in the integrals measured.

Aside from problems in the data collection system, such as gain shifts and dynamic non-linearities, there are two important corrections which should be mentioned. Firstly, in systems with finite incident beam convergence α_0, the theoretical cross-section is reduced by a factor which can be considerable. This is especially significant in high spatial resolution microanalysis where α is typically ~10 mrad, i.e. of the order of β or larger. For $\alpha_0 \gg \beta$ the correction reduces to an analytical form[14]

$$\eta = \frac{\ln(1+\alpha_0^2/\theta_E^2)}{\ln(1+\beta^2/\theta_E^2)}\cdot\frac{\beta^2}{\alpha^2} \tag{9}$$

giving a correction of ~0.3 to 0.75 depending on θ_E, suggesting that in some cases the accuracy of quantitation may be limited by the precision with which α_0 can be determined. Secondly, in microscopes which use coupling lenses between the sample and the spectrometer, chromatic aberrations can lead to substantial errors both because they change the object point of the spectrometer, causing defocus at high energy losses, and because the chromatically shifted optics modify the apparent acceptance angles into the spectrometer. The severity of these effects will depend on the design and mode of operation of the coupling optics but preliminary experiments indicate that corrections of 25–50% may be required.

Under properly controlled conditions, however, ELS quantitation is already a rapid and precise procedure, and in many respects it is easier than EDS quantitation since no matrix corrections are required. However, many applications of the programs and developments in standards will be required before a level of confidence equivalent to that for EDS is obtained.

REFERENCES

1 J. SILCOX: 'Introduction to analytical electron microscopy', 295; 1979, New York, Plenum
2 D. B. WILLIAMS and J. W. EDINGTON: *J. Microsc.*, 1976, **108**, 113
3 D. M. MAHER, D. C. JOY, R. F. EGERTON, and P. MOCHEL: *J. Appl. Phys.*, 1979, **50**, 5105
4 D. C. JOY and D. M. MAHER: *Ultramicroscopy*, 1978, **3**, 69
5 A. V. CREWE, M. ISAACSON, and D. E. JOHNSON: *Rev. Sci. Instrum.*, 1971, **42**, 411
6 M. ISAACSON: Proc. 11th Ann. SEM Symp., **1**, 763; 1978, Chicago, SEM Inc.
7 D. E. JOHNSON: Proc. 37th Ann. Meeting EMSA, 528; 1979, Baton Rouge, La, Claitors
8 D. C. JOY and D. M. MAHER: 'Scanning electron microscopy 1980', **1**, 25; 1980, Chicago, SEM Inc.
9 D. C. JOY and D. M. MAHER: *Ultramicroscopy*, 1980, **5**, 333
10 F. P. OTTENSMEYER, D. P. BAZETH-JONES, and K. M. ADAMSON-SHARPE: 'Microbeam analysis 1980', 280; 1980, San Francisco, San Francisco Press
11 A. J. CRAVEN, M. GIBSON, A. HOWIE, and J. SPALDING: *Phil. Mag. A*, 1979, **38**, 519
12 P. REZ: 'Microbeam analysis 1979', 117–121; 1980, San Francisco, San Francisco Press
13 N. J. ZALUZEC, J. HREN, and R. W. CARPENTER: Proc. 38th Ann. Meeting EMSA, 114; 1980, Baton Rouge, La, Claitors
14 D. C. JOY and D. M. MAHER: *J. Microsc.*, **124** (in press)
15 R. F. EGERTON: *Ultramicroscopy*, 1979, **4**, 169
16 D. C. JOY, R. F. EGERTON, and D. M. MAHER: 'Scanning electron microscopy 1979', **2**, 817; 1979, Chicago, SEM, Inc.

DISCUSSION of paper by D. C. Joy

S. A. COLLETT:

May I question the role of β in the energy efficiency factor:

$$\eta_E = \frac{\ln(1 + \beta^2/\theta_E^2)}{\ln(2/\theta_E)}$$

as used by Dr Joy. If a convergent beam is used, as implied, then β, as used in η_E, is not the spectrometer acceptance angle but an angle dependent on both the spectrometer acceptance angle and the angle of convergence.

D. C. JOY:

This is quite true. When a convergent beam is used it is convoluted with the scattering distribution and leads to a broadening of the scattering profile, which in turn causes a fall in collection efficiency. This can be calculated in a straightforward way and the results agree well with experiment. In the case where $\alpha \simeq \beta$ then the correction is typically 20%; when $\alpha \gg \beta$, as may happen in a dedicated STEM, then the correction factor can be found analytically as

$$\eta_\alpha = \frac{\ln(1 + \beta^2/\theta_E^2)}{\ln(1 + \alpha^2/\theta_E^2)} \cdot \frac{\beta^2}{\alpha^2}$$

The difficulty is that it is now necessary to know both α and β accurately, and the chromatic aberrations in lenses will make it hard to achieve this, especially at large energy losses.

An advanced electron energy loss spectrometer

O L Krivanek and P R Swann

A compact electron energy loss spectrometer with corrected second-order aberrations and full electromagnetic alignment has been developed for TEM and STEM applications. Its efficient coupling to the microscope and noise-free single-electron detection system enable it to achieve ~1 eV energy resolution at energy losses up to 1 keV, and to cover energy losses up to 10 keV.

O.L.K. is with the Center for Solid State Science and the Department of Physics at Arizona State University, Tempe, Az., USA; P.R.S. is with Gatan Inc. in Warrendale, Pa, USA.

Among all possible electron energy loss spectrometer designs suitable for TEM applications,[1] the single magnetic sector spectrometer has recently become predominant.[2–6] This is mainly due to its comparative simplicity, to the fact that all of its important aberration coefficients can be corrected,[6,7] and to the ease with which it can be attached to an electron microscope column. The instrument designed by us is a compact magnetic sector spectrometer which takes advantage of the above factors and introduces two innovations: electromagnetic focusing coils which replace the usual mechanical alignment, and a hybrid electron detection system which incorporates both single-electron counting and PMT current measuring, and can cover a range of electron intensities $>10^8$ without operator intervention.

DESIGN

A cross-sectional drawing of the spectrometer, including a Philips 400 matching flange, is shown in Fig. 1. The principal components are:

1 The *spectrometer–microscope flange* which contains one exchangeable and two fixed apertures. The flange permits the spectrometer to rotate around the microscope column axis so that it can be brought out of the knee-well of the electron microscope for easy access during operation, and be pushed against the back wall of the knee-well when not in use. The flange also makes possible an *x–y* translation of the whole spectrometer so that the entrance apertures and the magnet can be aligned with respect to the centre of the microscope viewing screen. The entrance apertures are designed so as to minimize the flux of reflected electrons through the three-aperture combination. The exchangeable aperture sizes are normally in the range 1-5 mm, although smaller or larger apertures can be installed if required.

2 The *alignment coil assembly* which contains a dipole, two quadrupoles, and a sextupole. The dipole has a low inductance and is used for fast, undistorted scanning of the spectrum needed for flicker-free display of the low loss region and particularly the zero loss peak on a videoscope display. The two quadrupoles give the two degrees of first-order focus necessary in a magnetic spectrometer. One ensures that electrons of the same energy are focused into a sharp line at the energy-selecting slit, the other that the line is precisely parallel to the slit. The sextupole action is similar to changing the entrance curvature of the magnetic sector. It balances one of the two important second-order aberration coefficients (Y_4 and Y_6 in the notation of Shuman[6]) against the other. The idea here is first to minimize both Y_4 and Y_6 by proper selection of the magnet shape (see next section) and then to make the remaining $Y_4 \simeq Y_6$ with the sextupole, thereby further minimizing their combined effect ($Y_4 + Y_6 \sim$ constant on changing just one curvature, cf. Ref. 8). In practice this approach leads to a spectrometer whose resolution is limited by third- rather than second-order aberrations.

3 The *window-type magnetic sector* with curved polefaces ($R_1 = 10$ cm, $R_2 = 4.96$ cm), 10 cm bending radius, 17.5° entrance angle, 45° exit angle, 1.25 cm gap, and 90° deflection angle. The sector is designed to focus on the projector lens crossover 55 cm in front of the entrance poleface, to produce a point crossover 9.8 cm after the exit poleface, to have zero second-order aberration coefficients in both the dispersion plane ($Y_4 = 0$) and perpendicular to it ($Y_6 = 0$), and the dispersion plane at 90° to the beam ($Y_3 = 0$). Its measured dispersion is 1.8 μm/eV at 100 kV. Small remaining differences between the design optical parameters and the actual optical parameters are cancelled by the weakly excited focusing coils.

The magnet winding consists of three separate coils: one provides the average field to deflect the electrons by 90°; the second accepts the energy scan ramp; and a third is supplied with a variable phase and amplitude AC current to compensate for stray magnetic fields at mains frequency. Scanning the spectrum by changing the field in the whole magnet is slower than scanning by the auxiliary dipole, but it has the advantage that electrons passing through the slit are always precisely focused at the slit, irrespective of the magnitude of the energy loss and the inclination of the dispersion plane.

4 The *electron drift tube*, electrically isolated for future electrostatic scanning.

5 The *electron detector chamber* consisting of an energy-selecting slit, a scintillator and a photomultiplier tube. The stainless steel slit is made with <0.3 μm mechanical tolerance and can be opened from 0 to 100 μm wide. The

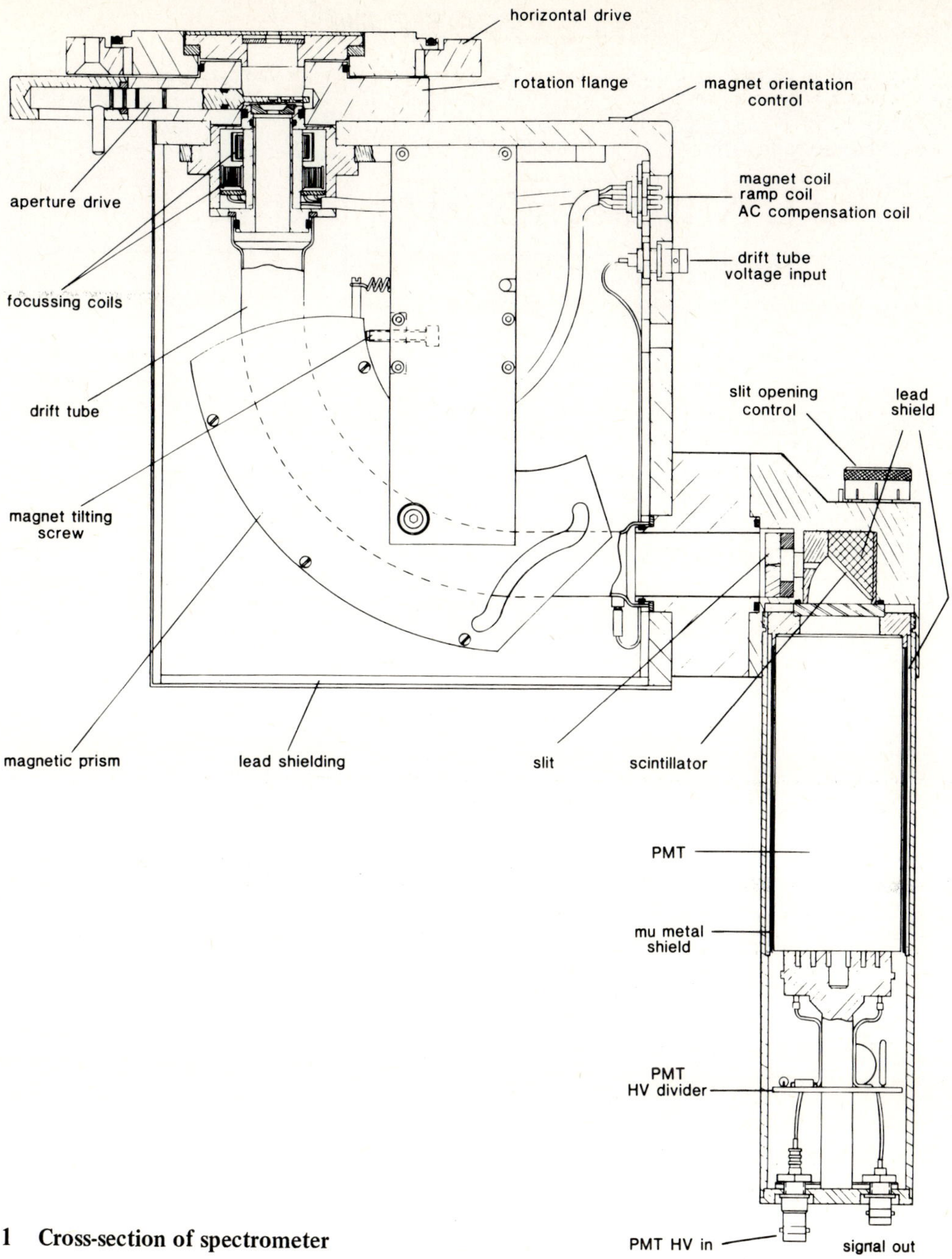

1 Cross-section of spectrometer

scintillator is the plastic NE 110 which gives 2.4 ns wide pulses,[9] ~800 photons per incident electron at 100 kV, and has a critical damage dose ~0.1C cm^{-2}. The photomultiplier tube is RCA 8575. It is mounted perpendicular to the beam to save space and to avoid direct exposure of the PMT to stray X-rays produced at slit edges.

ELECTRONICS

The basic spectrometer electronics has been divided into two NIM-compatible modules. The first one provides all the currents for the electron-optical elements. It includes a 3 ppm stability, 0 to 600 mA current supply for the main winding of the magnetic sector, the current supplies for the alignment coils, a variable phase and amplitude supply for the stray AC field compensation,[4] and a voltage-programmable, fast response current supply which sweeps the spectrum using either the pre-spectrometer dipole or the scan winding of the main coil. The currents to the focusing coils are linked to the main magnet winding current and the sweep current to maintain focus at any primary beam voltage and any energy loss.

The control voltage for the fast auxiliary dipole scan is provided by the sawtooth ramp output of a videoscope used to display the zero loss peak, while the slower but wider energy scan with the magnetic sector winding is controlled by the ramp from a multichannel scaler (MCS). A switch on the front of the NIM module accomplishes the switchover. The total width of the MCS scan can be varied from 10 eV to 10 kV by changing the gain of the programmable scan current supply. The videoscope scan is arranged so that it corresponds to the first 200 channels (1/5) of the MCS scan.

The second module is concerned with signal detection and processing. Its block diagram is shown in Fig. 2. The PMT anode is held close to ground potential so that both the PMT current and pulses due to single electrons are a valid measure of the signal strength. A simple R-C network at the input end of the module separates the pulses from

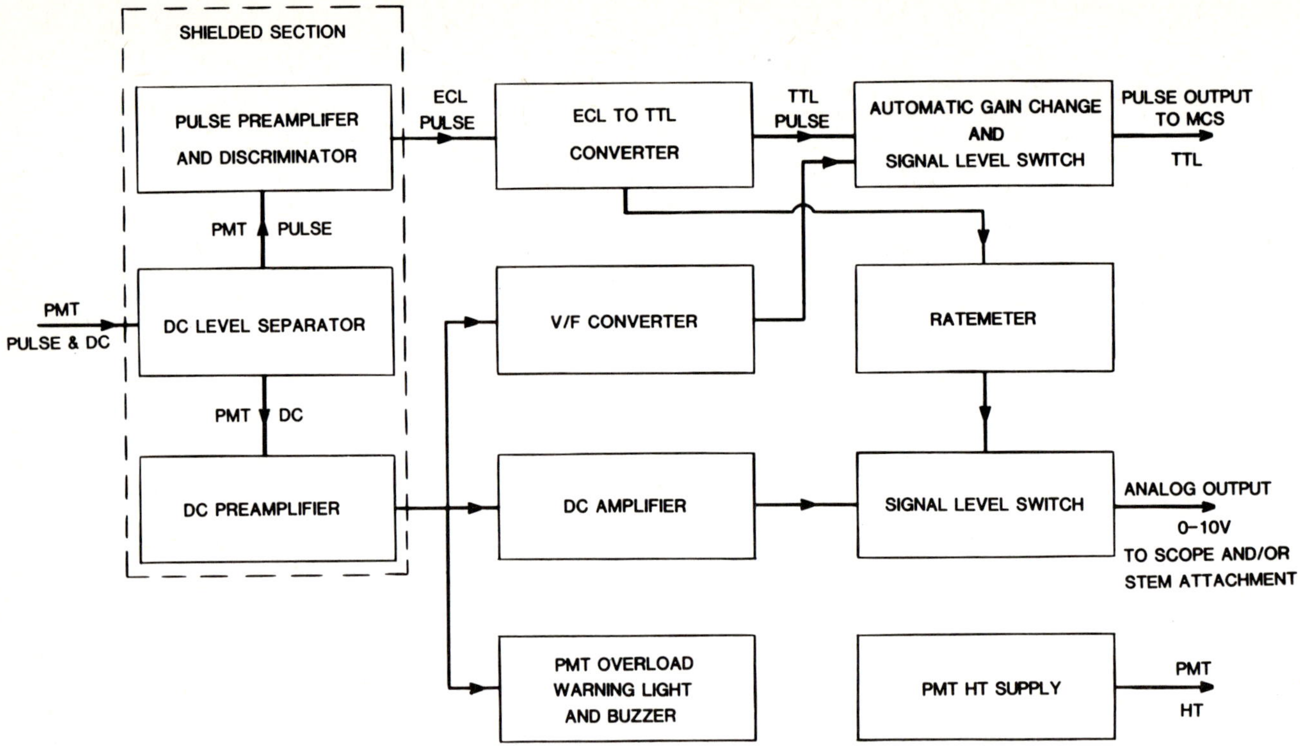

2 **Schematic of PMT signal-processing module**

the DC level. The two signals are then processed in parallel, one by pulse-counting electronics consisting of a preamplifier, a discriminator, an ECL-to-TTL converter, and a ratemeter, the other by current-measuring electronics consisting of a preamplifier, an amplifier, and a V–F converter. Each channel produces a train of 25 ns wide TTL pulses to be used by an MCS and a 0–10 V analog signal to be used by the videoscope and/or a STEM attachment. One or the other channel is selected as the output source for both the MCS and the videoscope by a 'signal level' switch. At low-to-medium electron intensities the pulse-counting channel is preferable since it discriminates out the weak noise pulses that make up the PMT dark current. In the pulse counting mode the count rate is limited to ~20 MHz by the width of the output TTL pulses (stretched from 5 to 25 ns in order to accommodate the majority of current MCS units), and pulse overlap problems become significant already at ~2 MHz. At count rates $>10^5$, however, the PMT dark current becomes a small fraction of the total PMT current. One can therefore switch over to current-measuring anywhere between $\sim 10^5$ and $\sim 5\times10^6$ Hz. The output then remains proportional to the input up to ~10 GHz (10^{-9} A), provided that the PMT gain is kept low ($\sim 10^5$) by setting the PMT high voltage to around 1200 V. A gain change of ~1000× occurs automatically at the switchover point. This arrangement leads to a dynamic range of $>10^8$, and since both signals are always available, the switchover is much faster ($<1\ \mu s$) than a gain change implemented by lowering the PMT voltage.[10] The 'signal level' switch can be operated either manually or automatically by a threshold trigger linked to the spectrum scan ramp such that a switchover from current-measuring to pulse-counting occurs at a preselected channel of the MCS scan, usually 50–100 eV past the zero loss.

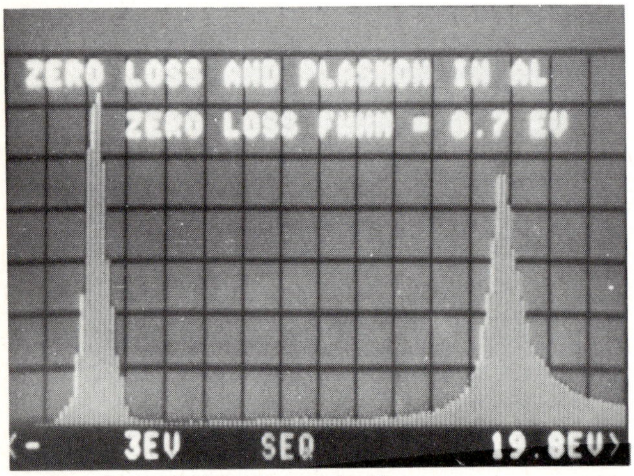

3 **Zero loss and first plasmon in Al**

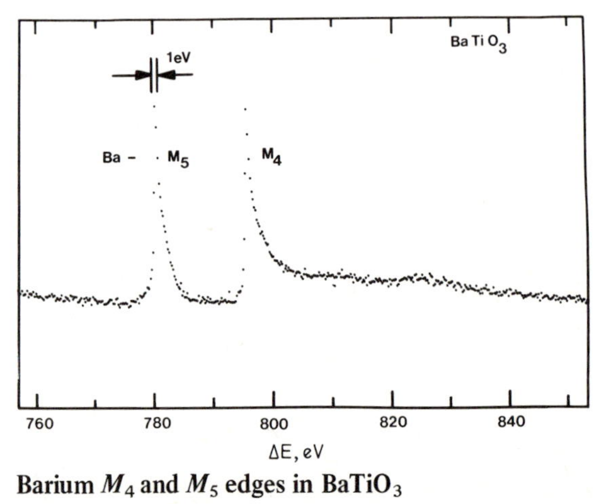

4 **Barium M_4 and M_5 edges in $BaTiO_3$**

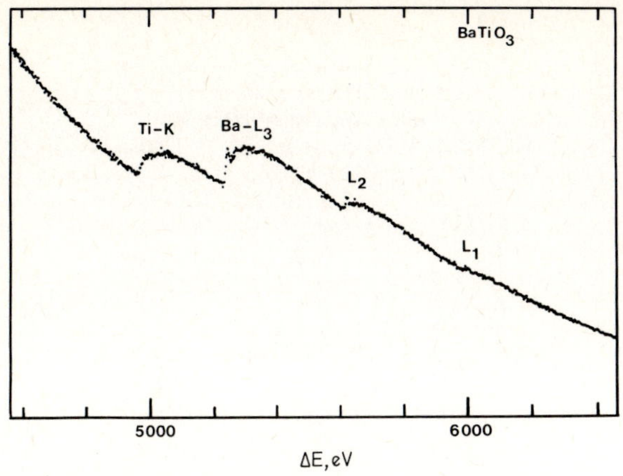

5 Ti K and Ba L_3, L_2, and L_1 edges in $BaTiO_3$

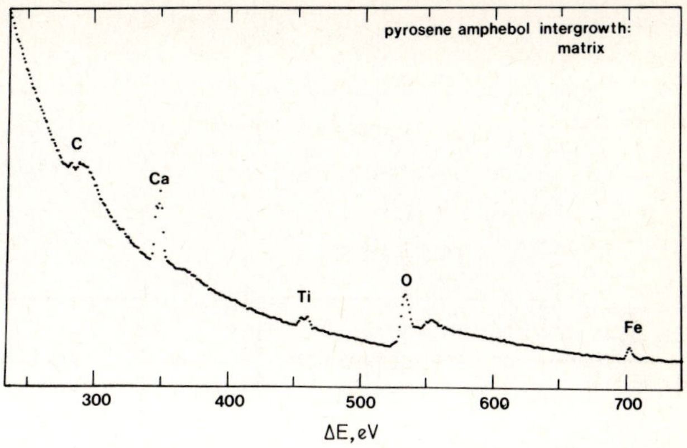

7 Spectrum of geological sample at ~2.5 eV resolution

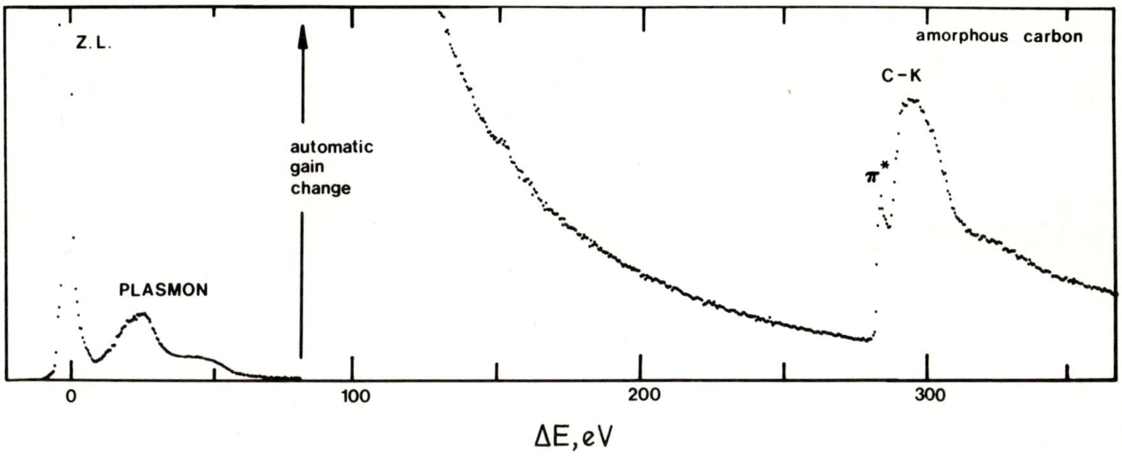

6 Spectrum of amorphous carbon illustrating automatic gain change achieved by switching from current-measuring to pulse-counting

The PMT pulses are rather weak (~10 mV) when the PMT voltage is only 1200 V, but even these pulses can be reliably discriminated from noise pulses which are weaker still (~100 μV).

Whenever the beam current incident on the scintillator exceeds 10^{-9} A, damage to the scintillator becomes quite rapid. Exposure of 3–30 h to 10^{-9} A, depending on how sharply the beam is focused on the scintillator, delivers the critical dose. The signal processor therefore incorporates a warning light and a buzzer to inform the operator whenever the safe limit (10^{-9} A) is being exceeded.

PERFORMANCE

Figures 3–7 illustrate the performance of the spectrometer.

Figure 3 shows the zero loss and the first plasmon in aluminium acquired in a single sweep at 120 kV with a spectrometer angular acceptance of 2 mrad, 0.16 eV/channel, and 1 s per 1000 channel sweep. The full-width at half-maximum of the zero loss is 0.7 eV, showing that the spectrometer energy resolution performance is ultimately limited only by the energy spread of the electron gun. LaB_6 filament was used at a total beam current <1 μA to obtain this result. During a slower scan the resolution degraded slightly to ~1 eV, probably due to 120 Hz and higher-frequency stray fields, and at 5 μA beam current it degraded further to ~2 eV due to beam energy spread.[11]

Figure 4 shows the Ba $M_{4,5}$ edge in barium titanate acquired with a 4 mrad acceptance angle into the spectrometer, 10 mrad acceptance angle at the specimen, 0.2 eV/channel and 0.1 s/channel in a single sweep. The M_5 and M_4 sharp lines are at 781 and 796 eV respectively, and are resolved as only ~1.2 eV wide.

Figure 5 shows the Ti K edge and the Ba L edges also from barium titanate. The spectrum was acquired with 5 eV and 0.2 s per channel, single-sweep, spectrometer entrance angle of 10 mrad, and specimen acceptance angle of ~100 mrad. Unlike the energy-dispersive X-ray case, Ti K at 4966 eV and Ba L_3, L_2 and L_1 at 5247, 5623 and 5989 eV respectively are clearly separated, the energy resolution of EELS even at this energy still being better than 10 eV.

Figure 6 shows an amorphous carbon K edge acquired with the automatic gain change. The π^* peak is clearly visible, and the jump ratio is ~7.5 : 1.

Figure 7 is an EELS spectrum from a geological specimen illustrating why good resolution is an advantage even in simple microanalysis: none of the sharp edge-threshold peaks which make the identification of the elements easy at 2.5 eV resolution would have been recorded at a resolution of 10 eV or more.

ACKNOWLEDGMENTS

The authors are grateful to Professor M. Isaacson and Mr J.

Schildkraut for assistance during the initial design of the magnetic sector. One of them (O.L.K.) is also grateful to the Arizona HREM Facility (NSF Grant CHE-7916098) for partial support.

REFERENCES

1 H. T. PEARCE-PERCY: SEM, 1978, AMF O'Hare, ed. O. Johari, **1**, 41
2 D. B. WITTRY, R. P. FERRIER, and V. E. COSSLETT: *Brit. J. Appl. Phys.*, 1969, **2**, 1967
3 A. V. CREWE, M. ISAACSON, and D. E. JOHNSON: *Rev. Sci. Instrum.*, 1971, **42**, 411
4 R. F. EGERTON: *Ultramicroscopy*, 1978, **3**, 39
5 D. C. JOY and D. M. MAHER: *J. Miscrosc.*, 1978, **114**, 117
6 H. SHUMAN: *Ultramicroscopy*, 1980, **5**, 45
7 N. W. PARKER, M. UMLAUT, and M. S. ISAACSON: Proc. 37th EMSA Meeting, 1979, Claitors Publ., Baton Rouge, G. W. Bailey, 578
8 N. J. ZALUZEC: Proc. Cornell AEM Workshop, ed. P. Fejes, 1978, 40
9 A. ENGEL: *Ultramicroscopy*, to be published
10 R. F. EGERTON: Proc. 38th EMSA Meeting, 1980, Claitors Publ., Baton Rouge, G. W. Bailey, 130
11 O. L. KRIVANEK: *Optik*, 1975, **43**, 361

DISCUSSION of paper by Krivanek and Swann

L. M. BROWN:
What does the aberration figure look like when fully corrected to second-order? This correction is presumably done by adjusting the sextapole?

P. R. SWANN:
When fully corrected the dispersion figure is a bright circular disc.

L. M. BROWN:
What are the machining tolerances on the magnet?

P. R. SWANN:
The machining tolerances at the entrance and exit faces of the magnet are ± 0.001 in (± 25 μm). The offset error between pole faces is less than ± 0.003 in (± 75 μm).

P. J. STATHAM:
How do you achieve your factor of 10^3 automatic gain change between V/F and pulse counting modes?

P. R. SWANN:
The gain is obtained by operational amplifiers in the signal processor. Since the pulse and DC signals are both present at the moment of switching, the switching time is very short. This method is far superior to using the PMT voltage as a gain control.

A. J. BOURDILLON:
Our spectrometer is similar to that described by Dr Swann except that the spectrometer magnet alignment is mechanical. The spectrometer performance is greatly improved when the spectrometer is perfectly aligned.

P. R. SWANN:
I agree that it is most important to properly align the spectrometer to obtain the best resolution. It is also very important to compensate for stray AC fields.

Post-specimen lenses in electron spectroscopy

A J Craven, T W Buggy, and R P Ferrier

This paper is concerned with the quantitative analysis of electron energy loss spectra when post-specimen lenses are used. Chromatic aberration effects in such lenses can cause significant errors and must be considered. Using a range of angular compressions, spectra have been recorded from an amorphous film of silicon containing nitrogen. After subtracting the stray instrumental background, the spectra are consistent with previously published work. Since the spectra were recorded using a highly convergent probe, it is necessary to take account of the convolution of the angular distribution in the incident probe with that of the inelastic scattering. An analytic solution for the convolution gives quite good agreement with the experimental data if chromatic effects are taken into account. The agreement is improved if the exact angular scattering distribution for inelastic scattering is used and a further small correction is made for the change of signal with defocus.

The authors are in the Department of Natural Philosophy, University of Glasgow.

The spatial resolution of electron energy loss spectroscopy is frequently limited by the area of specimen illuminated by the electron probe rather than by the intrinsic interaction length of the energy loss process itself; the interaction length for a 1 keV energy loss by a 100 keV electron is of the order of an atom diameter and is inversely proportional to the energy loss.[1] The practical limit to spatial resolution occurs because an adequate signal must be detected. For a given specimen and energy loss, the signal is determined by the incident probe current and the fraction of the inelastically scattered electrons collected by the electron spectrometer. The probe current can be maximized for a given probe diameter by a suitable choice of the probe half-angle α_0 (e.g. Ref. 2) provided that it does not exceed the maximum angle of scattering required in the spectrum. The latter is often chosen to maximize the ratio of the peak signal to the noise in the background (e.g. Ref. 3). The fraction of the inelastic scattering collected by the spectrometer can be maximized if the half-angle of scattering, β, accepted by the spectrometer equals the maximum angle of scattering required in the spectrum. Thus the maximum signal will be collected if both angles approach the maximum angle of scattering required. The exact choice is not well defined since there is a convolution of the probe angular distribution and the inelastic scattering distribution; however, the optimum angle of scattering to maximize the signal-to-noise ratio lies on a broad maximum so this is not critical.

The half-angle of acceptance of the electron spectrometer, β_s, is frequently smaller than the above angles and post-specimen lenses can be used to match the angular scattering distribution from the specimen to the angular acceptance of the electron spectrometer. A number of recent papers have discussed this technique.[4–7]

To analyse the resulting spectrum quantitatively, one must take into account both the convolution of the angular distribution of current in the incident probe with the angular distribution of the inelastic scattering and the chromatic effects in the post-specimen lenses; this paper considers such an analysis. Spectra have been recorded from an amorphous film 250 Å thick, obtained by sputtering silicon in an argon-nitrogen atmosphere. The spectra were recorded for a number of angular compressions by the post-specimen lenses and the results are compared with the values expected theoretically.

CHROMATIC EFFECTS IN POST-SPECIMEN LENSES

In earlier papers[7,8] we considered the design criteria for post-specimen lenses for use with an extended V.G. Microscopes HB5 scanning transmission electron microscope (STEM) with a field emission gun (FEG). In the experiments described below, the objective post-specimen field (OII), the first post-specimen lens (PI) and the third post-specimen lens (PIII) are used to vary the angular compression into the electron spectrometer. For high spatial resolution the electron probe is focused on the specimen and it is convenient to use the post-specimen lenses to provide an image of this at the object plane of the spectrometer. In this way, the influence of the source size on the energy resolution is negligible (e.g. Ref. 4). The half-angle of acceptance of the spectrometer, β_s, is defined by the collector aperture situated immediately before the spectrometer magnet. In general, the half-angle accepted at the specimen, $\beta(\Delta E)$, is equal to $C(\Delta E)\beta_s$ where ΔE is the energy loss; for elastically scattered electrons the half-angle accepted, $\beta(O)$, is $C(O)\beta_s$, where $C(O)$ is the angular compression for elastic electrons. $C(O)$ equals the magnification of the image of the specimen at the spectrometer object plane. $C(O)\beta_s$ is the value obtained by measuring the radius of the collector aperture size using the diffraction pattern which appears on it as an angular scale.

When the electrons lose energy in the specimen, the post-specimen optics give rise to three effects:

(1) the image of the specimen occurs closer to the final lens.

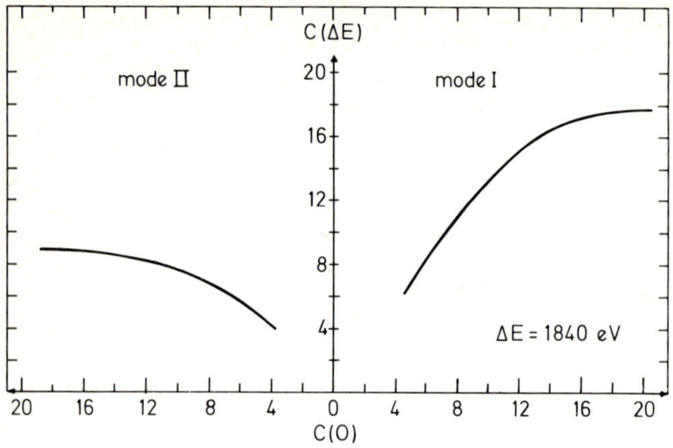

1 **Plot of factor $C(\Delta E)$ (which converts angular acceptance of spectrometer to angular acceptance at specimen for electrons which have lost energy, ΔE) against $C(O)$ (factor for elastically scattered electrons)**

(2) the magnification of that image changes; and
(3) the object plane for the spectrometer and its magnification may change.

The experiment discussed below is primarily to check the lens calculations[7] and to verify that the data can be analysed quantitatively in the presence of chromatic effects. Thus the change in the spectrometer properties with energy loss was removed by matching the strength of its magnetic field to the energy of the electrons being measured, i.e. the spectrometer object plane was a fixed plane in space. (In our case the spectrum is scanned by deflection coils after the magnet and so the particular energy loss to be measured was placed in the middle of the scan by using the zero control of the magnet current supply.)

The change in the half-angle accepted at the specimen is a combination of (1) and (2) above and both must be included in the factor $C(\Delta E)$ which is thus a function of the energy loss. Detailed calculations of this factor and the defocus expected are given in Ref. 7. Figure 1 is a plot of $C(\Delta E)$ against $C(O)$ for the silicon K edge ($\Delta E = 1840$ eV) with 100 keV incident electrons. Mode I has no crossover between PI and PIII whereas mode II does. There is an obvious difference between the two modes. In mode I the collection efficiency is higher for electrons which have lost 1840 eV than for those which have lost no energy, whereas the reverse is true in mode II. In both cases the change in collection efficiency is quite appreciable.

EXPERIMENTAL PROCEDURE

To use the calculations to interpret the data, the microscope must be set up in a reproducible manner since the properties of the post-specimen optics are very sensitive to the height of the specimen in the objective lens. The probe is first focused on the selected area diffraction (SAD) aperture using the condenser lenses. (In the HB5 the SAD aperture is between the final condenser lens and the objective lens and is thus a plane into which the crossover before the objective lens can be focused reproducibly.) The objective lens is set to the required excitation, taking care to end up at the same point on the hysteresis cycle of the lens. The specimen is then focused using the z-lift mechanism which is adequate to focus at 100 K magnification. This ensures that the objective post-field has a well defined magnification and image position. Each post-specimen lens is individually aligned to the spectrometer and, since there is a fixed spectrometer object plane, the lens excitation required to produce a focused spectrum is a check on the alignment procedure.

The angular compression $C(O)$ is measured by a rocking beam diffraction technique (e.g. Ref. 9) which shows the collector aperture diameter and the diffraction pattern of the specimen. Since the camera length is defined by the rocking before the objective lens, it need only be calibrated at a convenient angular compression. The relative sizes of the collector aperture then give the relative angular compressions. This technique also gives the angular size of the incident probe. The values obtained for the angular compression of OII, OII plus PI, and OII plus PIII give a further check on the consistency of alignment. Agreement to better than 1% has been obtained in two sessions separated by new operators learning to align the microscope and so we believe that the alignment should be highly reproducible.

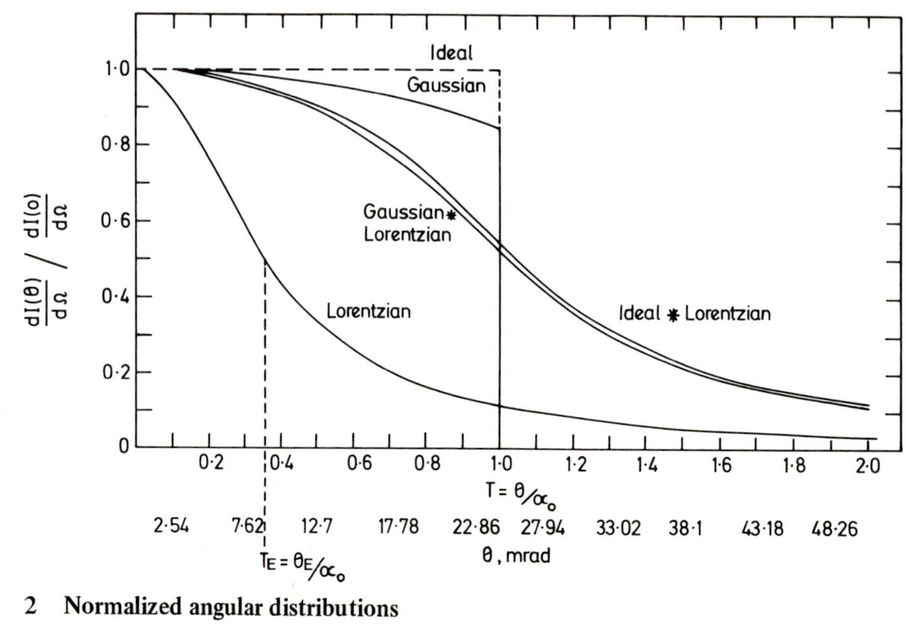

2 **Normalized angular distributions**

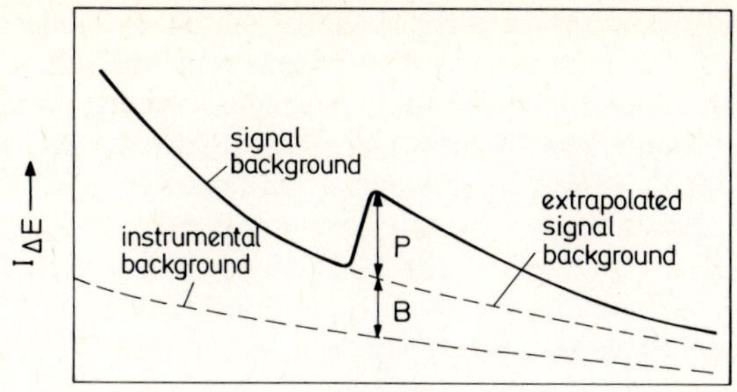

3 **Schematic energy loss spectrum**

The experiments we have performed have been limited by the absence of the data collection system which will eventually be used.[10] The results were obtained using analogue amplification of the photomultiplier output, the resulting signal being recorded on an x–y plotter with 100 s scans. This poses many problems and, to keep a large signal, the condenser lenses were adjusted to give a small source demagnification, a large probe half-angle (25.4 mrad) was used, and the slits of the spectrometer were adjusted to give a zero-loss peak 65 eV wide. The latter was several times larger than the effect of any defocus caused by chromatic effects. The low source demagnification and the large probe half-angle cause the angular current density in the probe to be non-uniform. The shape of this distribution can be recorded by scanning it over a small collector aperture using the Grigson coils.[11] Figure 2 shows the ideal (i.e. uniform) angular distribution and the actual (i.e. a truncated Gaussian*) angular distribution. Since the current from a field emission gun drops with time, the tip was flashed before every trace. In this way the probe current was the same from trace to trace to better than 5%.

Our spectrometer has a stray background and at the silicon K edge it is quite appreciable. Attempts will be made to reduce it in future but in the meantime it can be removed by subtraction. The stray scattering manifests itself as a signal when a spectrum is recorded from a hole in the specimen. It is the result of multiple scattering of the large electron intensity in the zero-loss peak. Some of this passes through the slits giving a spurious signal. Since the magnitude of the energy loss at the silicon K edge is much larger than the width of the low-loss region of any spectrum of interest, the stray background present will have the same shape as that recorded from a hole in the specimen and its intensity will be proportional to the intensity in the low-loss spectrum. Thus, to a good approximation, this stray effect can be removed by subtracting the spectrum recorded from a hole scaled by the ratio of the area of the low-loss region of the spectrum of interest to the area of the zero-loss of the spectrum from the hole. Figure 3 is a schematic spectrum showing the various contributions. The experimental procedure was to record:

(1) the spectrum in the region of the silicon K edge from a hole in the specimen;
(2) the spectrum in the same region from a point on the specimen;
(3) the low-loss region of the spectrum from the same point on the specimen;
(4) the low-loss region from a hole in the specimen; and
(5) the SAD pattern showing the size of the collector aperture.

This provides a consistent set of spectra allowing instrumental background subtraction and a value of $C(O)$.

EXPERIMENTAL RESULTS AND DISCUSSION

The area under the zero-loss peak of the spectrum from a hole in the specimen is proportional to the current within a cone of a half-angle β, i.e.

$$I_p(\beta) = 2\pi \int_0^{\beta} \frac{dI_p(\theta)}{d\Omega}\,\theta\, d\theta \qquad (1)$$

Figure 4 shows a plot of $I_p(\beta)/I_p(\alpha_0)$ against β for the probe with the Gaussian angular distribution shown in Fig. 2. The experimental points normalized to the probe current are plotted against β(O). The agreement is very good, showing that the probe current is reproducible after flashing and that our measurements of $C(O)$ are accurate.

After subtracting the instrumental background, the signal background prior to the silicon K edge was fitted to the power law $(\Delta E)^{-r}$.[12] After extrapolating and subtracting the signal background, the edge signal was fitted to $(\Delta E)^{-s}$.[12] The average value of r was 3.9 ± 0.3 and the average value of s was 3.5 ± 0.6. The errors represent the maximum and minimum values found and are caused mainly by the error in the subtraction of the instrumental background when the signal is very small. These values are consistent with those found for carbon by Maher *et al.*[12] assuming that a scaling law as proposed by Isaacson[13] applies. With the smallest angular compression the maximum angle of scattering in our spectra is 33 mrad or $3.6\theta_E$ and with the largest angular compression the maximum angle of scattering is 53 mrad or $5.7\theta_E$. Maher *et al.* found values of r from 3.8 ± 0.2 to 3.7 ± 0.2 and values of s from 4.7 ± 0.3 to 4.2 ± 0.3 for such angles of scattering. At present, any systematic variation in our values is obscured by the errors. Measuring the peak-to-background ratio P/B as defined in Fig. 3 gives a value of 4.3 ± 0.4 for the smallest angular compression, falling monotonically to 2.9 ± 0.3 for the largest. These figures, for a silicon specimen containing a large atomic fraction of nitrogen, are very similar to those obtained by Johnson[5] from a somewhat thicker aluminium on formvar specimen. Thus we believe the technique for subtracting the instrumental background is effective.

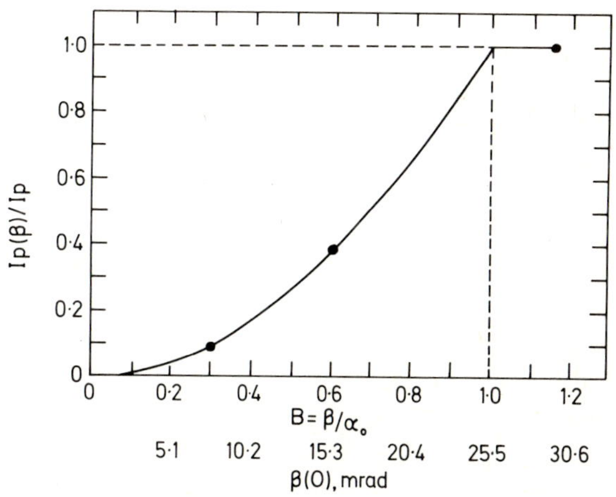

4 **Fraction of probe current detected with angular acceptance β at specimen plane assuming Gaussian probe distribution. Experimental points normalized to probe current**

* *Note in proof:* throughout this paper and in the relevant figures 'Gaussian' is to be understood as '*truncated* Gaussian'.

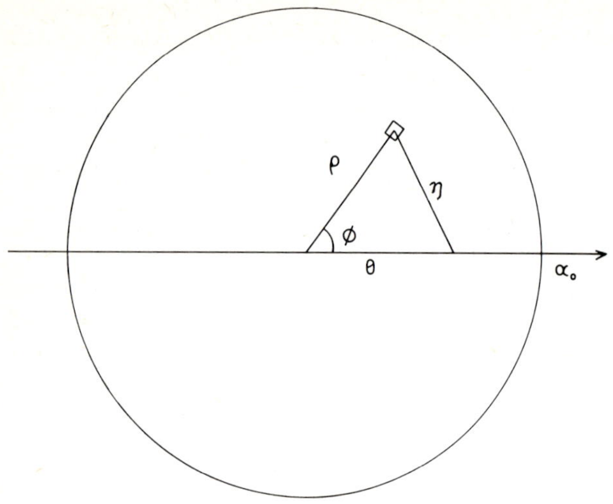

5 **Angular coordinates for convolution of incident probe with inelastic scattering distribution**

The angular distribution of the inelastic scattering leaving the specimen is the convolution of the angular distribution of the inelastic scattering with the angular distribution in the incident probe as noted by Joy, Maher and Farrow.[14] However, we disagree with the angular distribution calculated by these authors when a uniform angular distribution in the probe and a Lorentzian angular distribution for the inelastic scattering are assumed; our analysis is given below.

Figure 5 represents the angular position in the incident probe. For the inelastic scattering from an element of current at (ρ, ϕ) in the incident probe to appear at an angle θ from the axis, the inelastic scattering angle must be η. Thus the contribution to the inelastic scattering distribution, $dI_s/d\Omega$, for a small element of solid angle at (ρ, ϕ) is

$$d^2 \frac{dI_s}{d\Omega} \propto \frac{dI_p(\rho, \phi)}{d\Omega} \rho \, d\rho \, d\phi \frac{1}{\eta^2 + \theta_E^2} \qquad (2)$$

where $dI_p(\rho, \phi)/d\Omega$ is the angular current density in the probe at (ρ, ϕ) and $\theta_E = \Delta E/2E_0$ is the characteristic angle for an energy loss of ΔE and an initial energy E_0. Using the cosine formula to express η in terms of ρ and ϕ, the double integral to give $dI_s/d\Omega$ is analytic if $dI_p/d\Omega$ is constant so that:

$$\frac{dI_s}{d\Omega} \propto \ln \left| \frac{\psi^2 + \sqrt{(\psi^4 + 4\theta_E^2\theta^2)}}{2\theta_E^2} \right| \qquad (3)$$

where $\psi^2 = \alpha_0^2 + \theta_E^2 - \theta^2$. If, as in our case, $dI_p/d\Omega$ is Gaussian and the ratio x of $dI_p(\alpha_0)/d\Omega$ to $dI_p(O)/d\Omega$ is large, the exponential can be expanded to give a quadratic term in the probe angular current density and the resulting double integral is also analytic giving:

$$\frac{dI_s}{d\Omega} \propto \{\alpha_0^2 + (1-x)(\theta_E^2 - \theta^2)\} \ln \left| \frac{\psi^2 + \sqrt{(\psi^4 + 4\theta_E^2\theta^2)}}{2\theta_E^2} \right| + (1-x)(\theta_E^2 + \theta^2) - (1-x)\sqrt{(\psi^4 + 4\theta_E^2\theta^2)} \qquad (4)$$

The Lorentzian distribution and its convolutions with the ideal and Gaussian probes are shown in Fig. 2 for the silicon K edge. The probe half-angle is 25.4 mrad and the curves are normalized to the same value at $\theta = 0$. The difference between the two convolutions is small but it is accentuated when integrating over β to find the detected signal, $I_s(\beta)$. The integral has the same form as equation (1). For no values of α_0 or θ_E do such convolutions peak away from the axis.

Figure 6 is a plot of the height of the silicon K edge (P), which is proportional to $I_s(\beta)$, against the square of the half-angle of acceptance at the specimen. Two sets of points are plotted. The points marked * are plotted with $\beta = C(O)\beta_s$, i.e. no correction for chromatic effects. If the chromatic effects are included the points move to positions marked by •. Clearly the correction is important as it results in the expected monotonic variation of $I_s(\beta)$. While the amorphous film is uniform and has no diffraction contrast, there is no guarantee of an absolutely uniform thickness. Comparison of the area under the low-loss spectra ($\Delta E < 100$ eV) from the specimen and from a hole indicated thickness variations less than 5%. Error bars of 7% have been put on the experimental points to account for this and for measurement errors. When the data collection system is installed the improvement should be significant. Curve A plotted on Fig. 6 is $I_s(\beta)$ for the convolution of the Gaussian probe and the Lorentzian distribution. The initial slope is matched to the experimental points. The point at $\beta^2(\Delta E) = 740$ lies below this curve. Part of the explanation for this is that the angular dependence of the inelastic scattering is not Lorentzian. The Lorentzian distribution occurs only if the generalized oscillator strength is independent of the scattering vector, **q** (see for example Ref. 15). Figure 7 is a plot of the generalized oscillator strength close to the silicon K edge.[16] For the angles of scattering which are collected, there is clearly a significant drop in the generalized oscillator strength. Fitting a quadratic to this curve allows the real angular distribution leaving the specimen to be calculated. The integral of this distribution out to β is plotted as Curve B on Fig. 6. Again the initial slope is matched to the experimental points. The point at $\beta^2 = 740$ is now somewhat closer to the theoretical curve.

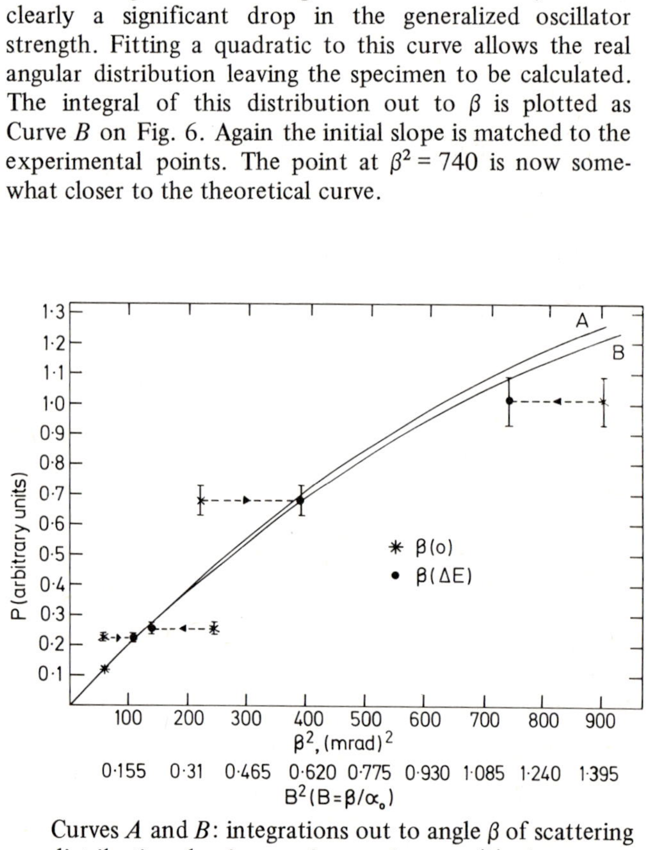

Curves A and B: integrations out to angle β of scattering distributions leaving specimen using eqn (1); A uses scattering distribution of eqn (4) i.e. convolution of Gaussian probe and Lorentzian inelastic scattering distribution; B uses convolution of Gaussian probe and actual angular scattering distribution of inelastic events. In both cases initial slope is fitted to experimental points

6 **Height P of silicon K edge plotted against half-angle of acceptance at specimen. Points marked * have not been corrected for chromatic effects and points marked • have**

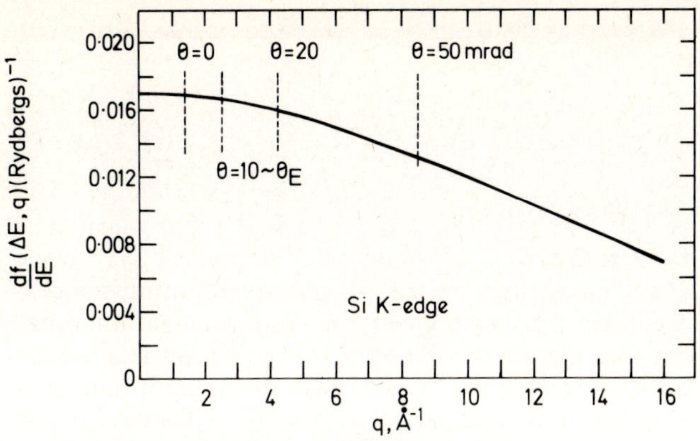

7 **Generalized oscillator strength for energy 2 eV above silicon *K* edge. Angles of scattering relevant to experimental conditions are shown**

A second reason for the deviation of this point is that the effect of defocus on the spectrum is greatest under these conditions. The difference in energy between the start of the edge and its maximum value is appreciably greater for this point than for the other points. Thus the measured height of the peak will be lowered by the convolution of the spectrum with the defocus term. An estimate using the measured increase in width of the edge (approximately 20 eV) indicates that the peak height measured for this point is 2–3% too small. There is an effect on the other points but it is much smaller. If a correction is made for this effect, then the agreement between the experiment and theory is very good. Using the proper data collection system, it is hoped that the experimental error can be reduced significantly.

CONCLUSIONS

Chromatic aberrations in post-specimen lenses can cause large errors in the quantitative analysis of electron energy loss spectra. Either the effect must be removed by adjusting all the excitation parameters after the specimen so that the ray paths for electrons which have lost energy are identical to those for electrons which have lost no energy, i.e. a sequential process; or suitable corrections must be made from a detailed knowledge of the electron optics. If a better energy resolution is required, then the defocus of the spectrum is more important and dynamic focusing is worthwhile in a sequential recording system. Without attempting a full compensation for the energy change, this can be achieved by changing the excitation of:

(1) a single lens, resulting in a modified $C(\Delta E)$;
(2) a quadrupole before the collector aperture, resulting in a modified $C(\Delta E)$ and a non-cylindrically symmetric angle of acceptance at the specimen; or
(3) a quadrupole after the collector aperture, resulting in a modified contribution to the energy resolution from the spectrometer aberrations.

Such a focus correction would remove the small change in signal mentioned at the end of the previous section.

The change in collection angle with energy loss will also have an effect on the first-order compensation for multiple scattering proposed by Egerton[17] and in common use. The size of this effect remains to be investigated.

The data presented in this paper behave in the expected fashion and agree with the angular distributions occurring in the experiments. With more sophisticated data collection, a good theoretical model for the angular distribution of the inelastic scattering and a correction for the defocus, the agreement should allow quantitative analysis even in the presence of these effects.

ACKNOWLEDGMENTS

The authors would like to thank the Science Research Council for providing the equipment used in this work and for a CASE studentship in collaboration with V. G. Microscopes for T.W.B.

REFERENCES

1 A. J. CRAVEN, J. M. GIBSON, A. HOWIE, and D. R. SPALDING: *Phil. Mag.*, 1978, **A38**, 519
2 A. J. CRAVEN: 'Electron microscopy and analysis 1979', p. 47, ed. T. Mulvey (Inst. Physics (London) Conference Series 52)
3 D. C. JOY and D. M. MAHER: *Ultramicroscopy*, 1978, **3**, 69
4 A. V. CREWE: *Optik*, 1977, **47**, 299
5 D. E. JOHNSON: *Ultramicroscopy*, 1980, **5**, 163
6 R. F. EGERTON: 'Scanning electron microscopy 1980', vol. I, p. 41, ed. O. Johari (SEM Inc., Chicago, 1980)
7 A. J. CRAVEN and T. W. BUGGY: submitted to *Ultramicroscopy*
8 A. J. CRAVEN, R. P. FERRIER, J. N. CHAPMAN, and W. A. P. NICHOLSON: 'Electron microscopy 1980', Proc. 7th European Congr. Electron Microscopy, vol. 3, p. 163, ed. P. Brederoo and V. E. Cosslett
9 A. J. CRAVEN: 'Developments in electron microscopy and analysis 1977', p. 311, ed. D. L. Misell (Inst. Physics (London) Conference Series 36)
10 A. J. CRAVEN, A. M. MacLEOD, and R. P. FERRIER: 'Electron microscopy 1980', Proc. 7th European Congr. Electron Microscopy, vol. 3, p. 170, ed. P. Brederoo and V. E. Cosslett
11 L. M. BROWN, A. J. CRAVEN, L. G. P. JONES, P. R. WARD, and C. J. WILSON: 'Developments in electron microscopy and analysis', p. 3, ed. J. A. Venables (Academic Press 1976)
12 D. M. MAHER, D. C. JOY, R. F. EGERTON, and P. MOCHEL: *J. Appl. Phys.*, 1979, **50**, 5105
13 M. ISAACSON: Proc. 38th Annual EMSA Meeting 1980, San Francisco, p. 110, ed. G. W. Bailey (Claitor Publ. Div., Baton Rouge, 1980)
14 D. C. JOY, D. M. MAHER, and R. G. FARROW: 'Microbeam analysis 1980', p. 154, ed. D. B. Wittry (San Francisco Press Inc., 1980)
15 R. D. LEAPMAN, P. REZ, and D. F. MAYERS: *J. Chem. Phys.*, 1980, **72**, 1232
16 T. W. BUGGY: unpublished work
17 R. F. EGERTON: 'Scanning electron microscopy 1978', vol. I, p. 133, ed. O. Johari (SEM Inc., Chicago, 1978)

Discussion overleaf

DISCUSSION of paper by Craven, Buggy, and Ferrier

D. C. JOY:
Similar effects to the ones you have described can also be observed on other instruments. The data on this viewgraph were obtained on a JEOL JEM100B modified so as to allow the accelerating voltage to be incremented by 780 and 1560 eV downwards without any change of lens excitations. The viewgraph shows a carbon edge recorded at its normal loss energy (284 eV) and then shifted down by 780 and 1560 eV. Two effects can be seen: firstly the spectrometer is seen to be defocused at the high losses, the edge being significantly broadened; and secondly the intensity on the K edge is seen to fall when shifted downwards in energy because chromatic errors in the coupling lenses have altered the effective collection angle β. The error is about 30% maximum.

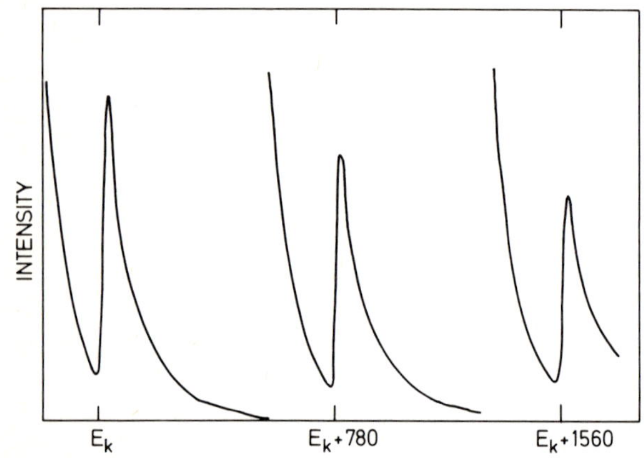

Variation observed in carbon *K* edge when recorded at nominal acceptance angle of 3 mrad but with energy shifts of 780 and 1560 eV relative to standard accelerating voltage of 100 keV. Note loss of resolution due to defocusing of spectrometer and loss of intensity due to change in effective collection angle; both effects are related to chromatic aberrations in coupling lenses [*see comment above by D. C. Joy*]

P. HAGEMANN:
If chromatic aberration causes problems in changing the acceptance angles when the electrons have left the specimen, why not scan the accelerating voltage instead of the spectrometer magnetic field?

A. J. CRAVEN:
If the accelerating voltage is scanned, the post-specimen lenses and the spectrometer remain focused for the appropriate energy loss but the pre-specimen lenses are defocused. Thus a correction of a pre-specimen lens is required to maintain the specimen in focus. If high resolution is required this is critical and, in addition, any misalignment in the column will cause translation of the probe. A correction for the change in probe current is also required. A second approach is to change the excitation of the post-specimen lenses and the spectrometer, leaving the pre-specimen lenses unchanged. However, if a high-excitation objective lens is used there is always coupling between the pre- and post-specimen optics. Changing either a lens current or the accelerating voltage is also a relatively slow process.

J. A. EADES:
For scanning an EEL spectrum, one proposed method is to vary the HT of the microscope. In many microscopes (especially the Philips EM400) the HT is so thoroughly stabilized that it can be varied only very slowly – too slowly for practical purposes in many cases.

Comparison of orientation effects in electron energy loss and energy-dispersive X-ray spectrometry

A J Bourdillon, P G Self, and W M Stobbs

Experimental results of an investigation of crystallographic orientation effects in electron energy loss spectrometry (EELS) are described and compared with the related effects already observed for energy-dispersive X-ray spectrometry (EDS). While (as is explained) such effects should only be observed in EELS at energy losses about twice the fluorescence energy of X-rays showing localization, the results obtained do not show the strong systematic trends expected from the theory. Possible reasons for this are discussed and it is concluded that background subtraction errors associated with Compton scattering are possibly responsible.

The authors are in the Department of Metallurgy and Materials Science, University of Cambridge.

Microanalyses using EDS on a TEM/STEM microscope can be up to 40% in error when the region examined is oriented close to a low-index Bragg condition. This is essentially a consequence of the electronic Borrman[1] effect: the excitation of different Block waves in the crystal is a function of orientation, and the waves can individually excite different atomic species in a compound or ordered system with probabilities dependent upon their relative excitation and form. The effect occurs in the excitation process and has been systematically observed in energy-dispersive spectroscopy.[2] The necessarily related effect in energy loss spectroscopy (EELS) has been reported by Lempfuhl and Taftø[3] and by Zanchi and Sevely[4] and Egerton.[5]

Here we discuss the reasons why smaller orientation effects might be expected in EELS than in EDS and describe the experimental results of this investigation for a number of materials. These results indicate that difficulties encountered in systematically observing the Borrman effect in EELS are probably associated with the limited accuracy of background subtraction attainable when high collection angles are employed and Compton scattering becomes important.

IMPACT PARAMETER

There is general and overwhelming evidence for the Block wave description of the electron beam as used in dynamical theory and this is used in our current interpretation of the orientation effects observed in EDS. The excellent correlations obtained thus lend support for current estimates for the impact parameter b which must be smaller than crystallographic d-spacing if localization is to occur. Such estimates suggest, for example, that the strong orientation effects found for Cu L emission (930 eV) indicate an upper limit for b of 0.028 nm at this energy for 100 kV incident electrons.

In general the uncertainty principle can be used to relate a lower limit for the impact parameter b to the energy ΔE associated with the virtual photon coupling of the passing electron to that excited. If v is the velocity of the incident electron, the approach indicates that

$$\langle b \rangle < \frac{\hbar v}{2\Delta E} f$$

where $\hbar$ is Planck's constant $\div 2\pi$, and f is close to unity (about 1.24). Thus b is a function of energy loss and is plotted for the case of 100 kV incident electrons in Fig. 1.

There are a number of reasons why the impact parameter to be expected in EELS must be larger than that associated with EDS. Firstly, in EELS the edges observed are usually, but not always, of lower energy than those associated with the fluorescence observed in EDS. Secondly, whereas in EDS excitation can occur with varying probability to any energy above a particular absorption edge before emission, thus promoting localization effects, the window of energies over which an edge integral is measured in EELS is normally defined. In fact, the mean energy of excitation in EDS is more than a factor of two higher than the energy of the absorption edge.[2] This of course implies that the impact parameter in EDS should be smaller than in EELS by the same factor. Thirdly, in EELS the collection angle may be defined. Again we can see by applying the 'uncertainty' principle[2] that if the aperture excludes electrons which have not suffered a given minimum horizontal momentum change Δp_h during the energy loss process, no localization effects can possibly be

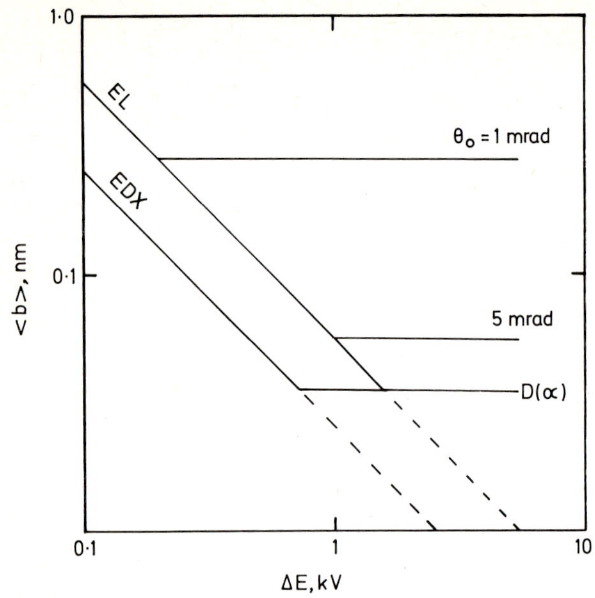

1 Plots showing calculated impact parameters applicable to EL ($\langle b \rangle = \hbar v/2\Delta E$) and to EDS ($\langle b \rangle = \hbar v/2\Delta E_c$) for 100 kV incident electrons. Cut-offs occur at Debye-Waller factor (~0.035 nm) in both cases and at restrictions imposed by collection angle effects in EL

observed. On this approach

$$\langle b \rangle^2 \langle \Delta p_h \rangle^2 > \frac{\hbar^2}{4}$$

For an aperture centred on the axis and incident beam electron

$$\Delta p_h \simeq p \cdot \theta$$

where θ is the half angle subtended by the aperture and p the momentum of the incident electrons. Thus, for example, if the collection angle is limited to 10 mrad, $\langle b \rangle$ must be greater than 0.28 nm and no localization effects will be observed unless the relevant atomic spacings are larger than this. Other limiting factors for b are shown in Fig. 1 together with the cut-off associated with lattice vibrations for room temperature observation as suggested by the appropriate Debye-Waller factor.

Using the above relations and considering the conditions for Bragg reflection it follows that if localization is observed, $\langle \theta \rangle > \theta_B/2\pi$ where θ_B is the Bragg angle for the incident electrons; and conversely if the scattering angles $\langle \theta \rangle < \theta_B/2\pi$, $b > d$ and there will be no localization.

EXPERIMENTAL MEASUREMENTS

An energy loss spectrometer has been attached to a Philips EM400 STEM microscope so as to allow large collection angles at the specimen through coupling by post-specimen lenses.[6] Energy loss spectra were observed using three configurations in our investigation of orientation effects.

Axial measurements with large acceptance angle

In the first configuration an axial probe of 4 mrad was used in conjunction with a collection semi-angle of 60 mrad. Foils of graphite, amorphous carbon, copper, titanium and titanium disulphide (materials chosen for their appropriate energy losses) were oriented with respect to the incident beam so as to excite a systematic row. Spectra were recorded either for specific orientations between the Laue condition and the Bragg orientation for 3 g for a given region or, as was the case of the two-layer materials graphite and TiS_2, for specific positions of approximately constant thickness across a bend contour. Absorption edges studied occur at 165 eV for S L, 284 eV for C K, 456 eV for Ti L, 931 eV for Cu L, and 2472 eV for S K. The accelerating voltage used was 120 kV except for the results shown in Fig. 2 when the incident electrons were given an energy of 60 kV. The intensity of the carbon edge for a window 100 eV wide is plotted as a function of orientation after background subtraction.

Our results may be generally summarized by saying that no strong orientation effects were systematically observed. In particular it was not found to be generally the case – though graphite as indicated in Fig. 2 is exceptional – that the count rate increased at the symmetric condition as might be expected. However, changes of about 10% with orientation were usually observed, though without theoretical correlation in terms of expected changes in Block wave excitation. Furthermore, changes of this magnitude were as common for different systematic rows for the S K excitation, which because of its high energy might be expected to have a small b and show strong localization effects, as for the S L with large b.

We may infer that the theory is wrong either in its application to both EDS and EELS, or only in EELS with an underestimation in this case of $\langle b \rangle$. Alternatively, given the good correlation of the theory with EDS results and the limit imposed on $\langle b \rangle$ by the uncertainty principle further experimental effects need to be considered before the EELS results can be interpreted. Evidence favours the third possibility and we have accordingly made a preliminary examination of the effects of specimen thickness acceptance angle, temperature contamination and finally Compton scattering.

The large collection angle was employed not only to increase the signal and to improve the signal-to-noise ratio,

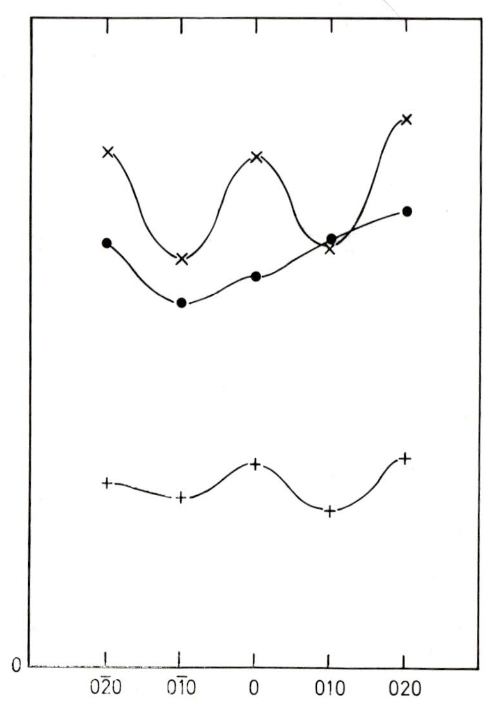

2 Plot showing C K edge intensity after background subtraction and windowed over 100 eV, against crystallographic orientation of graphite (×), and also of zero loss (•) and of their ratio, C K/0l (+)

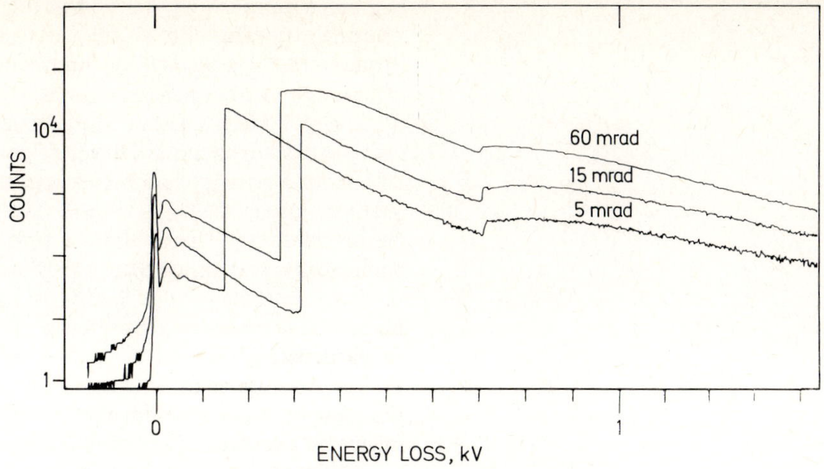

3 Energy loss spectra of iron measured using collection angles of 60, 15, and 5 mrad and 120 kV incident electrons. Gain changes between 150 and 250 eV at switch-over to pulse counting from *I* to *F* conversion

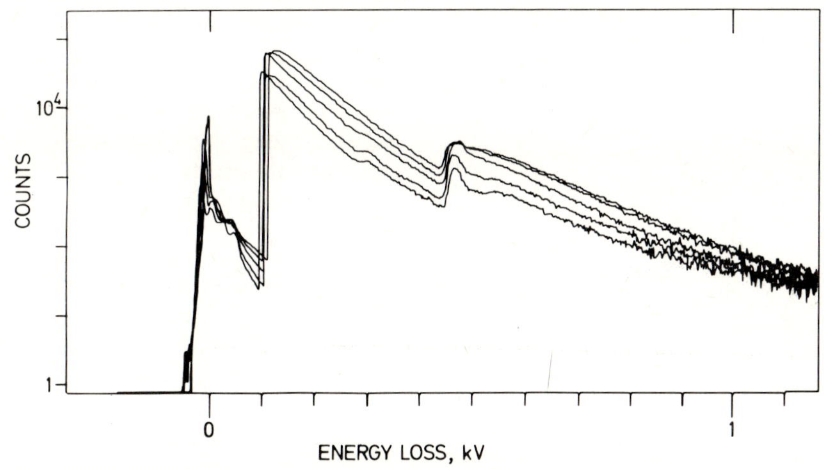

4a **Energy loss spectra of titanium at thicknesses of $\ll 1\xi_{010}$, $<1\xi_{010}$, $1\xi_{010}$, $(5/2)\xi_{010}$ and $4\xi_{010}$, in order from lowest to highest. 120 kV incident electrons were used with collection angle 60 mrad**

but also to reduce fluctuations in the signal level due to multiple elastic and inelastic/elastic scattering. Figure 3 shows the effect on the spectrum of iron produced by changing the collection angle. On changing this from 5 to 60 mrad the step height increases by a factor of about 10 while the background level increases by 20. There is an improvement in the step height signal-to-total-noise ratio of about 2. Meanwhile the slope of the background at the edge quite obviously changes, being steeper for higher collection angle. Thus the background follows a different angular distribution both to that of the iron edge and as a function of collection angle. Possible reasons for this are discussed below under 'Compton scattering measurements'. However, it is clear that if it is supposed that the electrons in the classical energy loss background have the same angular distribution as those responsible for the edge structure, another mechanism, sensitive to collection angle, must be introduced to explain the change in step height/background ratio.

Measurements of the loss spectra as a fraction of thickness also show interesting changes in the background scattering behaviour. Figure 4*a* shows spectra recorded along a wedge of Ti oriented at a low-index Bragg condition so that thickness measurements could be readily obtained using the observed fringes. It may be noted that the spectra show effects of multiple scattering, such as rounding of the Ti *L* edge, at relatively low foil thickness. Figure 4*b* shows how various signals behave as the foil thickness is increased. Predictably the zero loss intensity (and peak height) decay logarithmically. From thicknesses of only about $(3/2)\xi_{010}$, the intensity of plasmon loss (12–108 eV) and of the core loss tend to a constant. This behaviour is consistent with multiple scattering becoming important as the foil thickness approaches the mean free path for plasmon excitation.[7] Thus the first plasmon, for example, loses electrons to the second by multiple scattering as it wins others from the zero loss by single scattering. Meanwhile the background level below the Ti *L* edge continues to increase and its slope as a function of the energy loss decreases. Extrapolating this change of slope to high energies and integrating, it is calculated that with increasing thickness above $(3/2)\xi_{010}$, current is not conserved, in the acceptance angle

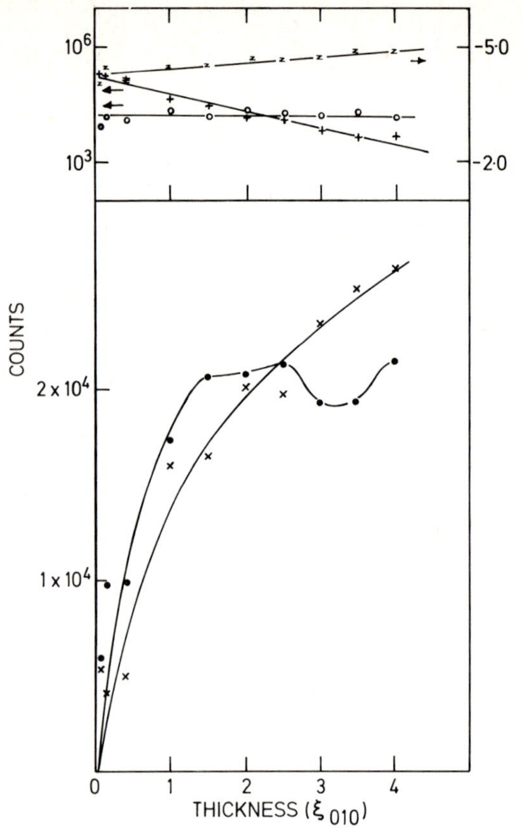

4b Analysis of spectra in Fig. 4a showing, plotted against orientation: intensities of Ti *L* with background subtraction and 100 eV window (●); height of background measured as lowest point below edge (×); zero-loss intensity (+); plasmon intensity, windowed between 12 and 108 eV (○), and negative of background slope below Ti edge (z)

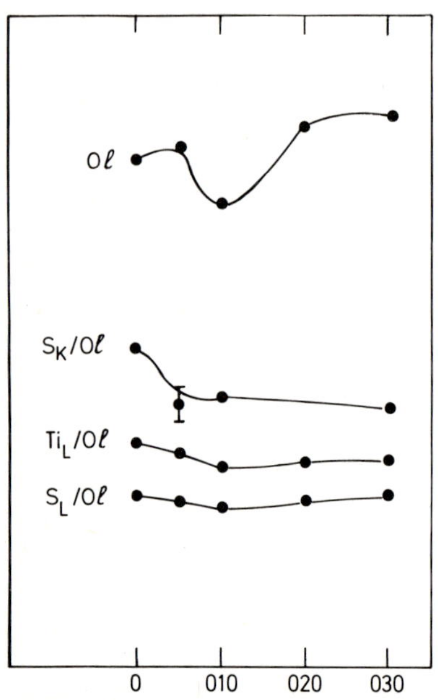

5 Plot showing measured intensities of zero-loss (0*l*) intensity, and ratios of intensities of S *K*/0*l*, Ti *L*/0*l* and S *L*/0*l* in TiS_2 plotted against crystallographic orientation. Plots, on different scales, have same zero base. Incident beam was at 120 kV. Experimental configuration described in text

employed, since only half of the increased zero loss is accounted for by the change in background slope. This suggests that the remaining half is scattered outside the collecting aperture, presumably mainly elastically.

Similar effects also result from contamination and heating. The background level of spectrum due to a foil left under a strong beam increases with time and the edges become increasingly rounded. In the measurements described earlier these effects were minimized by using comparatively weak probes.

Use of high convergence probe and small acceptance angle

Our second approach was to use a relatively convergent axial probe with a semi-angle of about 20 mrad and to accept only electrons scattered through an axial objective aperture of 3 mrad semi-angle. A bend contour in a TiS_2 foil was scanned at low thickness and the orientation dependences of the various edges are shown after background subtraction in Fig. 5. The results are more consistent with theory than those previously described. In all cases there is an increase in count rate at the Laue condition with greater changes for the S *K* absorption than for the lower energy Ti *L* and S *L* edges. Though these results appear to correlate less well with theoretical predictions than those observed by Lempfuhl and Taftø in MgO,[3] who used a similar configuration, the improvement implies that the anomalous results generally observed are associated with angular effects. It appears probable that Compton scattering is responsible.

Compton scattering measurements

The Compton scattering for carbon with an incident energy of 120 kV was measured by tilting the incident electron beam through angles of 90 mrad and 60 mrad (Fig. 6) and collecting the axially scattered electrons through an aperture of semi-angle 3 mrad. The Compton profile[8] appears as a large peak centred at 1.3 kV for a tilt of 90 mrad, and at 0.7 kV for a tilt of 60 mrad.

Consider how the Compton scattering might affect the background levels in the region of the edges of interest to us in examining localization effects, particularly when collection angles around 60 mrad are used, as was in fact the case for the large acceptance angle experiments described above. We see from Fig. 6 that, at least for relatively large scattering angles, the C *K* edge is small compared to the Compton loss profile and it is clear that the latter can contribute significantly to the background, and in a nonlinear way. Considering its integrated effect for a given solid acceptance angle of 60 mrad it appears qualitatively that the background due to the Compton scattering, which must tend to zero at both low and high energy, peaks in the region of interest, around 1 kV, where b/d is critically close to 1. Standard background subtraction routines do not account for such a peak, and it is likely that gross errors can be introduced.

CONCLUSION

Using large acceptance angles relatively low but theoretically consistent orientation effects are observed for graphite at 284 eV with an incident electron energy of 60 kV. This is not inconsistent with the theoretical cross-section analysis indicated in Fig. 1. For higher energy losses in other materials, despite the fact that stronger localization effects might then be expected no consistent effects are observed at high collection angles and only weak effects for low collection angles.

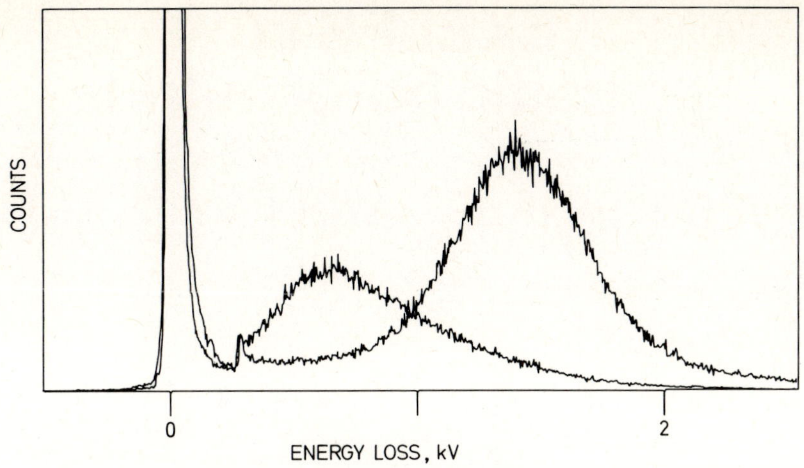

6 Plots showing Compton profiles produced from amorphous carbon by 120 kV electrons tilted to 90 mrad peaking at 1.4 kV and 60 mrad peaking at 0.7 kV. Plots normalized on C *K* edge. Collection angle 5 mrad axial

It is suggested that Compton scattering so affects the background at the energies of interest, when using a high collection angle, as to mask the orientation effects on the losses so readily observed in EDS. This would imply either that low collection angles have to be employed in examining localization effects or alternatively that new background subtraction procedures will have to be developed.

ACKNOWLEDGMENTS

A grant from the Science Research Council enabled A.J.B. to undertake this work. The authors are grateful to Dr B. Williams for comments on Compton profiles and to Professor R. W. K. Honeycombe for providing laboratory facilities.

REFERENCES

1 G. BORRMAN: *Phys. Z.*, 1941, **42**, 157
2 A. J. BOURDILLON, P. G. SELF, and W. M. STOBBS: 1981, to be published in *Philos. Mag.*
3 G. LEMPFUHL and J. TAFTØ: 'Electron microscopy', vol. 3, Proc. European Congr. Electron Microscopy, The Hague, August 24–29, 1980, 62–63
4 G. ZANCHI and J. SEVELY: *Philos. Mag.*, 1980, **41A**, 417
5 R. F. EGERTON: *Ultramicroscopy*, 1979, **4**, 169
6 A. J. BOURDILLON, N. W. JEPPS, W. M. STOBBS, and O. L. KRIVANEK: Proc. Micro 80, Brighton, August 18–22, 1980
7 P. STEPHENS: *Ultramicroscopy*, 1980, **5**, 343
8 B. G. WILLIAMS: *Phys. Scr.*, 1977, **15**, 92

DISCUSSION of paper by Bourdillon, Self, and Stobbs

G. W. LORIMER:
You have described effects of orientation on the observed ratio of Pt L to Cu K_{α} radiation in an ordered Cu-Pt alloy. Am I correct in assuming that an orientation effect would *not* be observed in a disordered structure?

A. J. BOURDILLON:
Yes, we only observe and predict such an orientation effect on X-ray intensity ratios in ordered structures.

EXELFS in graphitic boron nitride

A P Stephens and L M Brown

One use of STEM which appears attractive is its potential ability to provide radial distribution functions calculated from extended fine structure seen in energy loss spectra, particularly those for light elements for which EXAFS experiments are very difficult. To test the techniques, spectra were recorded from very thin (5 nm) micro-platelets of graphitic BN. Data from the energy loss spectrum at energies above the boron K *loss and the nitrogen* K *loss have been collected; for the former loss, the noise level is ≲0.5% after about one hour of collection time. The data from the nitrogen loss are not so good. The radial distribution functions centred on both the boron and the nitrogen have been calculated and compared with those derived from the known crystal structure. It is found that the peak due to second-nearest neighbours is missing from the experimental data. Considerable effort has been spent to show that this discrepancy is not due to a flaw in the numerical processing, nor to specimen orientation effects which alter the shape of the spectrum near the edge. It seems that much better model calculations are required for the phase shift and amplitude of backscattering of the relatively low-energy secondary electrons emitted from the excited atoms.*

The work was carried out when both authors were in the Cavendish Laboratory in Cambridge, England; A.P.S. is now with Software Sciences Ltd, Daventry, Northamptonshire.

Boron nitride is an opaque white substance, waxy to the touch, which is used in crystal growing to provide an inert lubricated mould which does not bond to reactive substances. Because both boron and nitrogen have K absorption edges in a very convenient energy range, it makes an ideal material for a preliminary study of 'extended energy loss fine structure' (EXELFS), the counterpart in electron spectroscopy of EXAFS.[1] This technique in principle permits the measurement of the radial distribution function in an amorphous or crystalline material by examination of the oscillatory structure associated with absorption edges in the electron energy loss spectrum. The spectrum is written as the product of the single-atom spectrum and a factor $[1+\chi(E)]$, where $\chi(E)$ represents modulation of the spectrum according to the formula[2]

$$\chi(E) \simeq \frac{A(K)}{K} \sum_j \frac{N_j}{R_j^2} \exp -[R_j/\lambda(K)] \exp -(\sigma_j^2 K^2/2) \times \sin[2KR_j + \phi(K)] \quad (1)$$

Here K is the wave vector of the ejected secondary electron; $A(K)$ is the atomic backscattering amplitude; N_j is the number of atoms in the jth shell at radius R_j; $\lambda(K)$ is the mean free path of the secondary electron for inelastic scattering; σ_j^2 is the mean-square amplitude of displacement due either to thermal motion or to disorder; and $\phi(K)$ is a phase shift resulting from scattering of the secondary electrons. The relationship between E and K is

$$K^2 = 2me(E - E_0)/h^2 \quad (2)$$

where E is the energy lost and E_0 is the value of the absorption edge, nominally 188 V and 402 V for boron and nitrogen respectively. Now if the radial distribution function is defined by

$$n(r) = \sum_j \frac{N_j \delta(r - R_j)}{R_j^2} \quad (3)$$

then $n(r)$ is given by

$$\chi(K) = \frac{A(K)}{K} \int_0^\infty n(r) \sin[2Kr + \phi(K)]\, dr \quad (4)$$

and $n(r)$ can be determined from the Fourier sine transform of $K\chi(K)/A(K)$. In equation (4) the attenuation term and the broadening term have both been dropped, the former because the mean free path is regarded as slowly varying over the range of interest and the latter because the width of peaks in $n(r)$ is controlled by the quality of the data, not by thermal vibrations in the crystal.

Equation (4) and its inverse apply to modulations above the boron edge, $\chi_B(K)$ and $n_B(r)$, and above the nitrogen edge, $\chi_N(K)$ and $n_N(r)$. The ability of EXELFS to provide multi-centred radial distribution functions for light elements is an important potential application of electron spectroscopy and STEM.[3]

To see how the techniques work out in practice, spectra from BN were analysed and the radial distribution functions compared with the known crystal structure.

TABLE 1 Interatomic spacings and co-ordination

	Radius R_j, Å	Co-ordination N_j	N_j/R_j^2
B–B	2.504	6	0.957
	3.631	6*	0.455
	4.338	6	0.319
	4.411	6*	0.308
	5.065	12*	0.468
	5.08	6	0.233
B–N	1.446	3	1.435
	2.892	3	0.359
	3.331	2*	0.180
	3.815	6	0.412
	4.167	12*	0.691
	5.213	6	0.221
	5.469	12*	0.401
	5.784	3	0.090

* Interplanar co-ordination

EXPERIMENTAL

A specimen of BN was prepared by scratching a hole in a 3 mm disc 250 μm thick cut from a rod of polycrystalline material obtained from the Carborundum Company, Latrobe, Pennsylvania. Small flakes of electron-transparent material were found at the edge of the scratch. Electron diffraction showed the flakes to be graphitic BN, oriented with the *c* axis perpendicular to the disc. Examples are shown in Fig. 1, in which the crystal structure according to Wyckoff[4] is also shown. Table 1 shows the interatomic spacings and co-ordination for the first few separations.

Attention must be paid to the thickness of specimen from which the energy loss spectrum is obtained. To achieve good counting statistics, the higher count rates obtainable from thicker specimens are attractive, but multiple scattering results in convolution of plasmon losses with the absorption edges, and so affects the EXELFS structure. Attempts to deconvolute the spectrum numerically, although quite successful for most purposes, do not produce a spectrum of adequate quality for the subsequent numerical processing. Energy loss spectra were therefore obtained from regions less than 10 nm thick, as judged from plasmon losses.

The energy resolution required for EXELFS is not great. The modulations have a periodicity of at least 30 V, so one can open the spectrometer slits to increase the count rate and use an energy resolution of 4.5 V. Acceptable statistics are achieved with about 10^4 counts per channel. Using the HB5 STEM with no elaborate modifications to the spectrometer, it is possible to record a useful spectrum from a flake typically 50 nm wide in about 1 hour.

1a
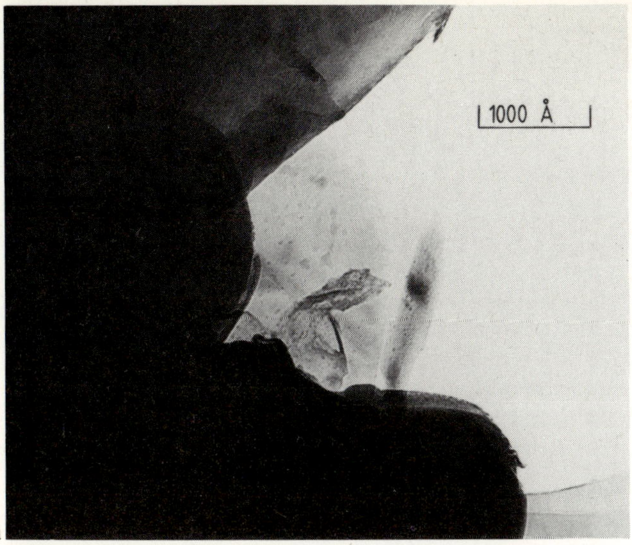

1b
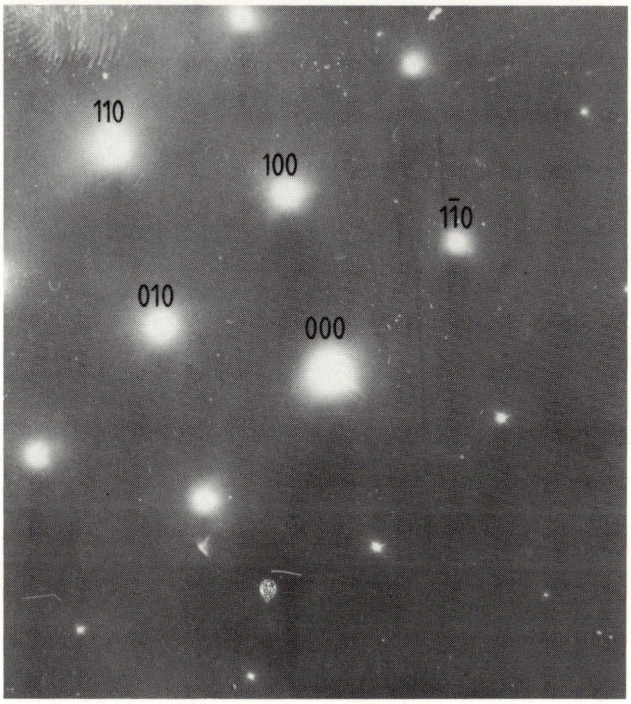

1c
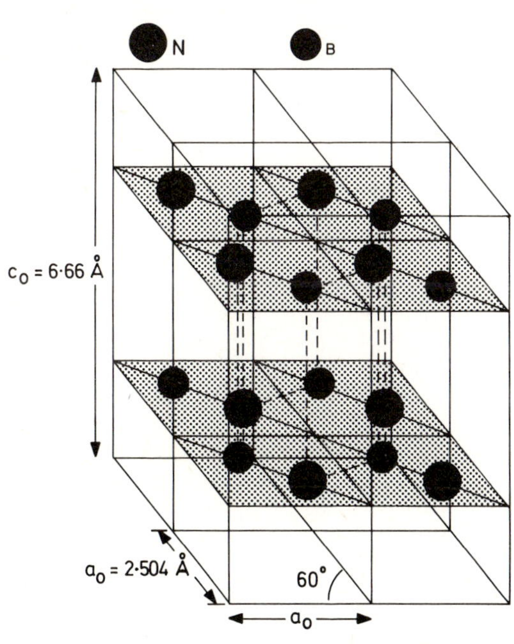

1 **(a) Bright-field micrograph of flake of BN (edge of flake may be as thin as 5 nm)**
(b) Selected area diffraction pattern from single-crystal flake oriented with *c* axis parallel to electron beam; hexagonal symmetry is obvious
(c) Structure of graphitic (hexagonal) BN [after Ref. 4]

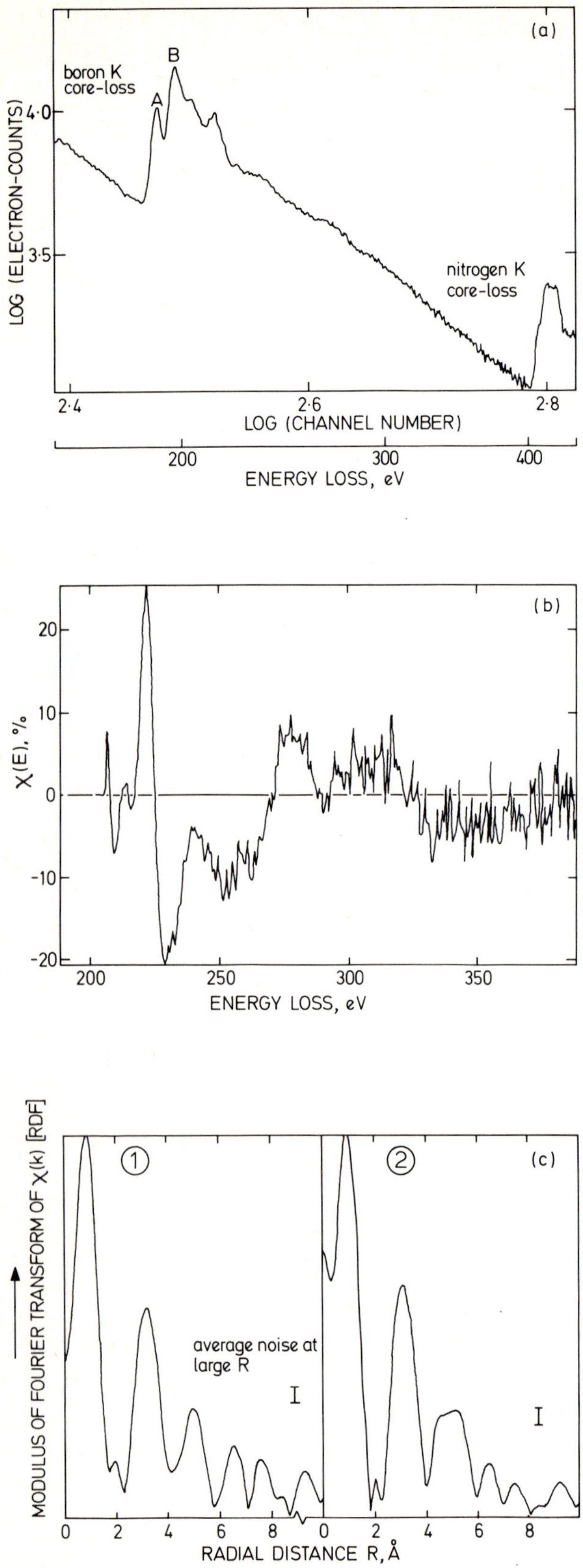

2 (a) Log-log plot of energy loss spectrum of graphitic boron nitride (specimen thickness ≃8 nm); data show range from just below boron K edge to just above nitrogen K edge; peaks A and B correspond to transitions to π^* and σ^* states [see Ref. 10]
(b) Modulations $\chi(E)$ extracted from spectrum shown in (a)
(c) 1: RDF obtained from spectrum of (a); 2: RDF from another spectrum from different specimen; note reproducibility of RDF up to third peak

NUMERICAL TECHNIQUES

Figure 2*a* shows a log-log plot of the energy loss spectrum in the energy range from 150 to 450 V. The spectrum is stored on tape and processed on the IBM 370 computer using programs developed by A. P. Stephens.[5] The processing occurs in stages:

1 correction for finite aperture size is made by dividing by the angular efficiency factor;[6]
2 the background is stripped, assuming that it follows a power-law in energy,[6] both above the edge and below it;
3 the modulations $\chi(E)$ are extracted,* and transformed to $\chi(K)$ using equation (2); and
4 $\chi(K)$ is 'windowed' (see below), and Fourier-transformed; $n(r)$ is displayed.

There are three steps in the numerical process which merit discussion. The first concerns the position of the energy loss E_0 in equation (2) which is just sufficient to promote a core electron to an unbound state. The theoretical value of E_0 depends upon the inner potential experienced by the secondary electron, as well as on the Fermi energy. Lee and Beni[7] were able to determine E_0 from a self-consistent analysis of their data on EXAFS from Ge, but our data are not good enough to follow their procedure. They observed that E_0 for the K edge of Ge should be about 5 eV below the energy corresponding to transitions to the vacuum level. We have chosen E_0 similarly 5 eV below the vacuum level. The variation of the resultant RDF with E_0 is as follows: increasing E_0 increases the position of a peak near 1.5 Å by 0.02 Å eV^{-1} and one near 4 Å by 0.5 Å eV^{-1}. Thus if E_0 is uncertain to ±5 V, the peak positions are uncertain to ±0.1 Å and ±0.25 Å respectively.

The second step of importance is the 'windowing' of $\chi(K)$ in such a manner as to exclude structure too close to the edge (band structure, peaks A and B in Fig. 2*a*). Several window functions were tried, and the one selected is shown in Fig. 3*a*. It excludes all modulations below 25 eV above the edge, and smoothly 'switches' them on over a range of 0.5 $Å^{-1}$; and the noisy large-energy modulations are reduced in intensity by a Gaussian function centred at $K = 0$ of width 18 $Å^{-1}$. The Fourier transform of the window function is shown in Fig. 3*b*; this is the 'peak-shape function' which is convoluted with the true radial distribution function as a result of the limitation in the useful data range. The resulting effective peak width is about 0.6 Å.

The final point of considerable importance in the numerical processing concerns $A(K)$ and $\phi(K)$, the scattering amplitude and phase shift of the secondary electrons. According to Lee *et al.*[8] the phase shift is given by

$$\phi(K) = a + bK + cK^2 + d/K^3 \qquad (4)$$

Values of
$a = -2r$
$b = -1.25r$, Å
$c = 0.036r$, Å
$d = 30r$, $Å^{-3}$

can be deduced by extrapolation from their values, tabulated for elements heavier than carbon. The major term in equation (4) is the term linear in K, which has the effect of shifting the position of peaks in the RDF by about 0.6 Å. A similar shift is observed in carbon EXELFS.[9] Computation shows that the other terms in $\phi(K)$ have little effect. The variation of $A(K)$ is traditionally taken to be inversely

* The modulations are found by 'adaptive' extraction, i.e. a variable power law, but with variability constrained so as not to remove structure in $\chi(K)$ finer than some chosen K.

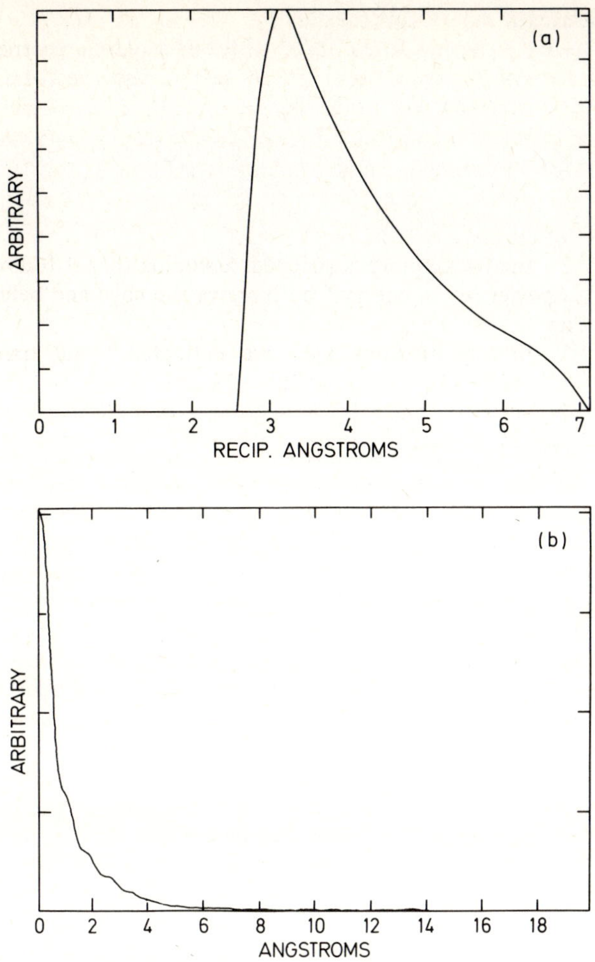

3 **(a) Window function used to exclude structure too close to edge;**
(b) Peak-shape function, which is Fourier transform of (a)

proportional to K, although one might expect from Rutherford backscattering that it should be proportional to K^{-2}. One finds little effect of this sort of variation, possibly because the high-K oscillations are noisy. More elaborate models for $A(K)$ and $\phi(K)$ are discussed below.

RESULTS

Figure 2*a* shows a log-log plot of the energy loss spectrum in the energy range from 150 to 450 V. Figure 2*b* shows $\chi(E)$ extracted from that spectrum; Fig. 2*c* shows the resulting RDF; 2*c* (1) and (2) show the RDF computed from a similar flake but the spectrum (of similar quality and appearance) recorded on a different day. Both Figs. 2*c* (1) and (2) are the result of computations in which $\phi(K)$ has been set equal to zero; 2*c* (1) and (2) are in substantial agreement, showing the reproducibility of the experiment. The RDF consists of three prominent peaks: one at 0.95 ± 0.05 Å, another at 3.2 ± 1 Å, and a third, wider peak at 5.0 ± 0.1 Å. Other peaks in the RDF are in rather poor agreement, and are probably influenced by the noise level which is shown in the diagrams. The width of the first two peaks is undoubtedly mainly controlled by the peak-shape function of Fig. 3*b*.

Figure 4*a* shows a log-log plot of the energy loss spectrum in the energy range from 400 to 650 V, namely the nitrogen K edge. Figure 4*b* shows the smoothed modulation $\chi(E)$ extracted from that spectrum, and Fig. 4*c* the

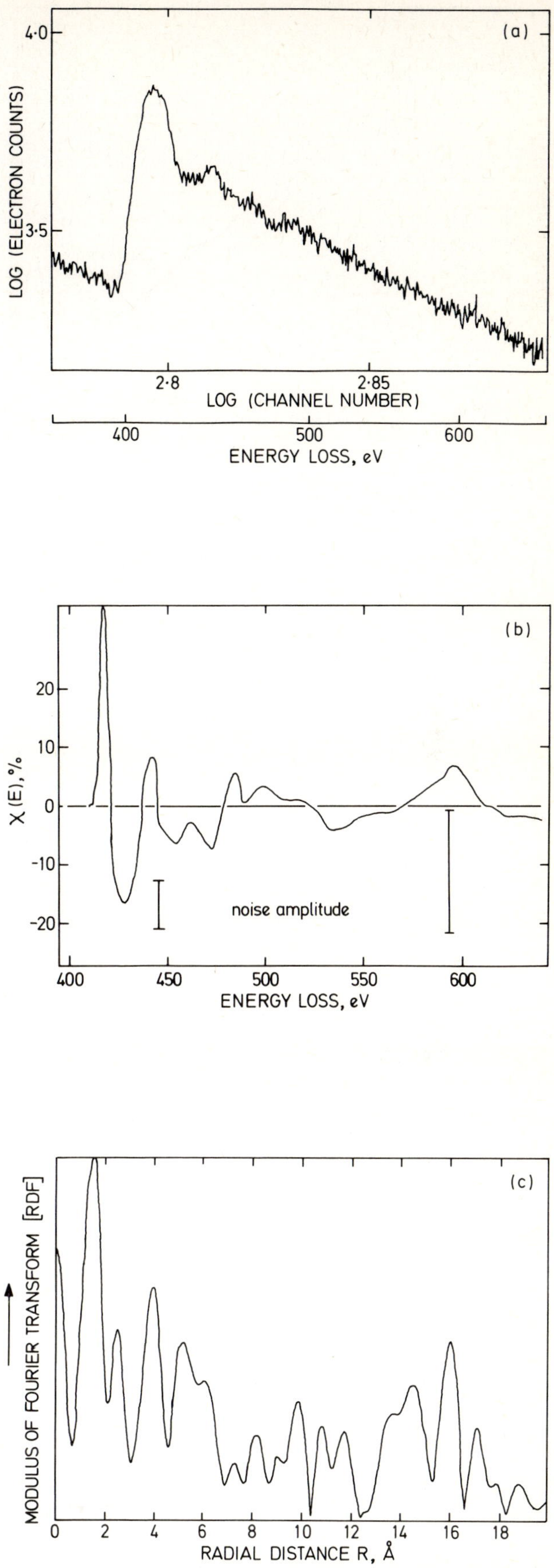

4 **(a) Log-log plot of electron energy loss spectrum showing nitrogen K edge; data have been corrected for finite aperture size**
(b) Modulations $\chi(E)$ extracted from spectrum in (a); due to very large noise content, modulations have been smoothed by eye
(c) RDF obtained from spectrum of (a)

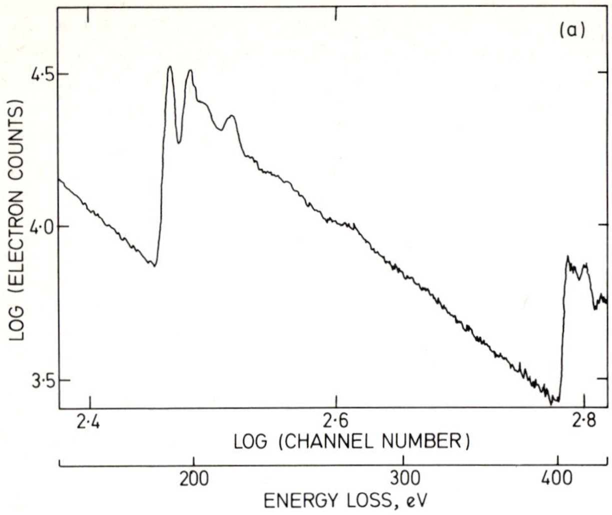

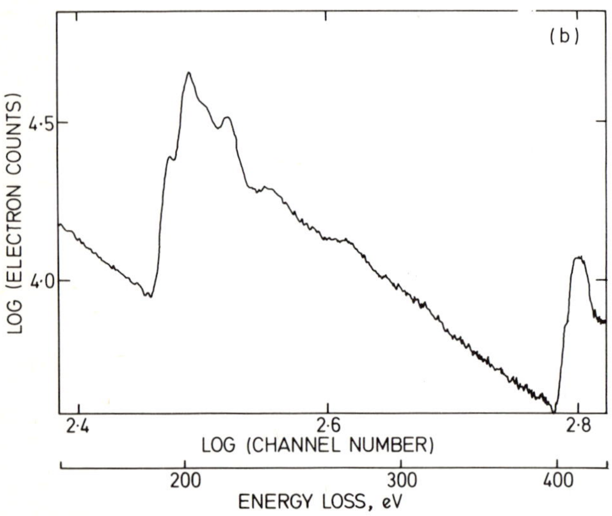

5 (a) Log–log plot of electron energy loss spectrum of graphitic BN; *c* axis of specimen is parallel to incident beam
(b) As (a) but specimen tilted by 45°

RDF derived from the unsmoothed modulations assuming $\phi(K) = 0$ as before. The noise amplitude is such that only one peak has significance: the first one at 1.6 Å. This peak is at a significantly different position from the corresponding one in Figs 2*c* (1, 2). The noise content can affect the position of this peak, but not by more than about ±0.2 Å. The noise is of course due to the difficulty in obtaining adequate statistics at higher energy losses; the limitation is on specimen stability, operator patience, and drift of spectrometer zero.

Before discussing these results, one further experiment should be mentioned: the effect of specimen orientation. Figure 5 shows two spectra from crystals whose orientations differ by approximately 45°. Although the shape of the edge is very different in the region within about 10 V of the initial rise, the spectrum further from the edge is unchanged in shape. The RDFs derived from the two spectra are indistinguishable. The shape close to the edge reflects 'band structure', in particular the π and σ character of the final states;[10] an alternative view of this structure is that it is the EXELFS resulting from the periodic spacing of distant neighbours. But the information used to derive the RDF for near neighbours is unaffected by such effects.

DISCUSSION

In order to compare the RDFs obtained from energy loss results with the known crystal structure, energy loss spectra were synthesized using the structure, and the synthesized spectra were then analysed according to the recipe outlined above (section 'Numerical techniques').

The synthesis was performed as follows. An idealized core-loss edge, given by a power law above the edge voltage, was assumed. Then $\chi(E)$ was given by equation (1), assuming the data of Table 1, and the phase shift of equation (4) or a more elaborate model to be described. The EXELFS were then generated so that the modulations had an arbitrary maximum value of 5%, in rough accord with the experimental situation (Fig. 2*b*). The modulations were convoluted to simulate the spectrometer resolution, and random noise was added to model electron arrival statistics. Then the synthesized spectrum was treated in exactly the same way as energy loss data in order to obtain the RDF. A first test spectrum was run with one 'atom' at 1.5 Å and two at 4.0 Å. The resulting RDF showed two peaks at the correct location and with roughly the correct areas. It was concluded that the procedure worked properly.

Figure 6 shows the RDF generated for the BN structure, using the phase shift $\phi(K)$ as given by equation (4). It shows a major nearest-neighbour peak at 0.9 Å, a smaller 'second-nearest' peak at 2 Å, and a wider or doubled third peak at 3.7–4.7 Å. The positions of these peaks can easily be deduced from Table 1 if one remembers that the effect of the phase shift is to reduce the atom positions R_j by about 0.6 Å. The first peak corresponds to nearest neighbour B–N spacings; the second to a mixture of B–B and B–N second and third neighbours, with the former predominant because they are more numerous; and the third peak to a mixture of several neighbours, the dominant contribution probably coming from the 12 B–N spacings at 4.167 Å.

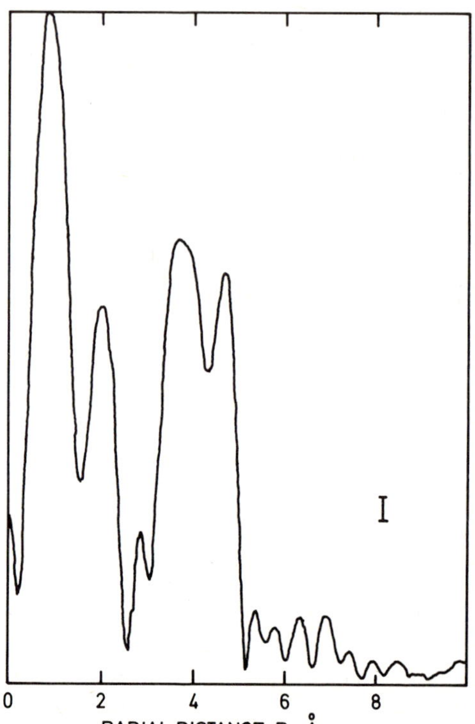

6 RDF obtained from electron energy loss data generated by using graphitic BN structure and parameterized phase shift (see text); ordinate, in arbitrary units, is modulus of Fourier transform

When the synthesized RDF is compared with the experimental ones, the following points emerge.

For boron-based RDFs (Figs. 2*c* and *d*) the first peak is correctly located, but the second peak is observed at 3.2 ± 0.1 Å rather than at 2 Å, and the third peak at 5.0 ± 0.1 Å rather thn at 3.7–4.7 Å. One might argue that in the experimental RDF the second peak is missing and the third peak rather badly represented. However, the correspondence between the EXELFS-determined structure and the known structure is very poor indeed.

For nitrogen-based RDFs (Fig. 4*c*), the experimental peaks should be shifted 0.5 Å to the left. When this is done, moderately good correspondence is achieved for all three or four major peaks. However, all but the first peak are strongly influenced by noise, and the agreement (if such it can be called) is fortuitous.

An attempt was made to use better phase shift and backscattering calculations. Dr S. J. Gurman (University of Leicester) kindly supplied us with calculations showing $A(K)$ and $\phi(K)$ for backscattering from atomic boron and nitrogen. Insertion of these values slightly altered the curves of Fig. 6, but did not improve the agreement with experiment. The calculations can be improved still further to allow for charge transfer between B and N, and it is clear that more work should be done in this area.

One possible source of the discrepancy between the EXELFS-determined RDF and the known RDF is the attenuation term in equation (1). The mean free path $\lambda(K)$ for electron excitation varies from about 2 Å at 30 V to 6.5 Å at 250 V,[11] and so it will differentially attenuate the modulations corresponding to the smaller spacings and those corresponding to the larger spacings, the latter being much more strongly attenuated. At the time of writing, this effect seems the most plausible source of the discrepancy between theory and experiment.

CONCLUSIONS

Although the distance of the nearest neighbours can be determined from the observed EXELFS and tabulated phase shifts, the position of further neighbours is not accurately given. The data seem adequate to allow these to be calculated, and the problem seems to reside in either the application of equation (1) or in effects not included in it. A prime candidate for the difficulty is the variation of $\lambda(K)$, the attenuation distance, over the energy range of useful EXELFS.

ACKNOWLEDGMENTS

The authors acknowledge comments from Drs A. Howie and P. H. Gaskell, and help with the STEM from Dr S. J. Pennycook. A.P.S. is grateful for a studentship from the Science Research Council. Thanks are due to Professor A. B. Pippard and the University of Cambridge for support for this work.

REFERENCES

1 R. D. LEAPMAN and V. E. COSSLETT: *J. Phys. D. (Appl. Phys.)*, 1976, 9, L29
2 E. A. STERN: *Phys. Rev.*, 1974, **B10**, 3027
3 M. ISAACSON and M. UTLAUT: *Optik*, 1978, **50**, 213
4 R. W. G. WYCKOFF: 'Crystal structures', vol. 1, 1960, Interscience
5 A. P. STEPHENS: PhD thesis, Univ. Cambridge, 1981
6 *See for example* S. A. COLLETT, L. M. BROWN, M. H. JACOBS: this conference; D. C. JOY: 'Analytical EM', 1979, Plenum Press, N.Y./London
7 P. A. LEE and G. BENI: *Phys. Rev.*, 1977, **15B**, 2862
8 P. A. LEE, Boon-Keng TOO, and A. L. SIMONS: *J. Am. Chem. Soc.*, 1977, **99**, 3856
9 P. E. BATSON and A. J. CRAVEN: *Phys. Rev. Lett.*, 1979, **42**, 893
10 R. D. LEAPMAN and J. SILCOX: *ibid.*, 1361
11 Calculated according to R. A. FERRELL: *Phys. Rev.*, 1956, **101**, 554, with the plasmon energy for BN taken to be 26 eV

DISCUSSION of paper by Stephens and Brown

D. C. JOY:
Can you please indicate what phase shift values you used in your calculation? Secondly, is it not true that the BN case is not a very good one because of the limited energy window between the B and N edges? The K vector range that is accessible in this case is only about 3 $Å^{-1}$ so the accuracy of the transform Δx will be limited to almost 0.5 Å.

L. M. BROWN:
The phase-shift values are taken from the paper by Lee *et al.* (our Ref. 7), extrapolated to boron. The most important term is the term in Φ which is linear in K, and which has a similar value in carbon. Another source has been Dr S. J. Gurman, who kindly provided values from his pseudopotential calculations.

I believe that the second limitation is likely to be true for most multicomponent glasses containing light elements. Many interesting boron-stabilized glasses (e.g. Fe_3B) are most unsuitable because of the overlap of the boron edge with the iron edges.

C. Y. F. COLLIEX:
Has anybody taken into account the symmetry of the outgoing wave (*p*-wave for a *K* shell hole) to explain the absence of second neighbours in the reconstructed radial distribution? The influence of inelastic cross-section is not enough reduced in length scale to be responsible for the absence of such peaks.

L. M. BROWN:
Dr Howie is attempting to do something about this. But the fact that tilting the crystal by large amounts does not affect the result seems to me to show that this cannot be important. Surely the secondary electron wave function is incoherent mixtures of *p*-waves corresponding (in classical terms) to the mixture of excitations caused by the fast electrons. If you regard the impact parameter as a vector, the angular momentum exchanged will have variable direction, so the modes of the *p*-waves will be similarly variably directed.

With regard to the inelastic scattering: the paper by Howie and Stern (*Z. Naturforsch.*, 1972, **27a**, 382) shows quite sharp variation of mean free path with energy for electrons of the relevant energies.

A. J. BOURDILLON:
The problem is to say why EXELFS should not work as well as EXAFS for which several highly successful correlations exist with *ab initio* calculations. The explanation for discrepancies in EXELFS in terms of the low energy of secondary electrons would apply equally to EXAFS.

L. M. BROWN:
I cannot answer this; it is hard to see why BN would be very different from, say, Cu. Of course the characteristic plasmon loss and atomic vibration frequencies are very different, but detailed calculation seems to be required to see if this can account for the discrepancies.

I. F. FERGUSON:
As we go to ever smaller particles we must consider: do we really know the crystal structure of the material studied? The platelets here are only sixteen layers of atoms thick and in the EXELFS calculations one should consider the following:

(i) Do we really know the interplanar spacing? It is scarcely a bulk specimen and edge relaxation effects must apply.
(ii) Is the surface covered with an adsorbed, epitaxial(?) carbon layer deposited from the vacuum?
(iii) Is there on top of this layer an adsorbed oxygen or nitrogen layer? The factors (ii) and (iii) together would represent an extension of the thickness of the platelet by ~10%.
(iv) Do we know the crystal structure of the BN? Has it a hexagonal or rhombohedral graphite structure?

L. M. BROWN:
By electron diffraction we know that the platelets have hexagonal graphitic structure. We do not know the interplanar spacing directly for the majority of the platelets because we see only the first Laue zone in the diffraction pattern which has the *c* axis normal. But the second peak in the RDF and even most of the third peak come from atoms within a single layer plane, so the EXELFS should be very insensitive to interplanar spacing.

We must have no carbon contamination even to contemplate doing the experiment. A monolayer of carbon would be easily detectable: it would contribute approximately 400 counts in channels above ~285 V. In fact, the EXELFS dip in this energy range. The boron-to-nitrogen ratio deduced from the spectra is unity to within 2%.

Microanalytical electron microscopy on type 304 steel: correlation between EDX and EELS results

S A Collett, L M Brown, and M H Jacobs

Microanalysis has been performed on a type 304 steel using EDX and EELS methods. A VG Microscopes HB5 scanning transmission electron microscope (STEM) was used to produce spectra by means of the two techniques. The Fe:Cr:Ni weight composition ratios in the steel were calculated from peak areas in the X-ray spectra and from edge areas in the energy loss spectra. The results from each technique are compared, and the composition ratios are found to agree to within 10%. Theoretical aspects of the correlation between the two techniques are discussed, with reference to the cross-sections required for the two methods.

S.A.C. and L.M.B. are in the Cavendish Laboratory, Cambridge; M.H.J. is with TI Research Laboratories, Saffron Walden, Essex.

Energy-dispersive X-ray analysis (EDX) using an electron microscope is an established technique for elemental microanalysis. It benefits from the fact that the interpretation of the spectra obtained, leading to mass concentration ratios, is performed with relative ease. A second technique, electron energy loss spectroscopy (EELS) is also able to give microanalytical information. Whereas in X-ray spectra peaks are observed, in EELS it is edges arising from excitation of inner shell electrons which are observed. The energy of the edge identifies the element; the area under the edge gives the quantity of the element present.

Since little has been done in terms of a direct comparison between the two techniques, this paper is devoted to such a comparison. Microanalysis has been performed using the two techniques on the same steel specimens. Mass concentration ratios have been deduced from each technique, and thereafter compared in terms of accuracy. An appendix is given concerning the theoretical cross-sections used in both techniques, pointing out how closely they are related. The final section of the paper is about the volume analysed in the specimens. It is shown that for most specimen thicknesses, EELS is superior to EDX in enabling a smaller volume to be analysed. As a microanalytical tool, with the emphasis on the prefix *micro*, EELS is shown to be generally preferable to EDX.

EXPERIMENTAL PROCEDURE

The electron microscope used to produce the X-ray and energy loss spectra was a VG Microscopes HB5 STEM. The material on which the work was done was type 304 steel, whose composition is indicated in Table 1. This was thinned fairly easily using electropolishing; the electrolyte was 5% perchloric acid (60% A.R.) in absolute alcohol (ethanol). The thinning was done at 273 K. This produced specimens of typical thicknesses 50–100 nm.

Spectra were obtained from regions well away from grain boundaries. This produced as uniform a material composition as was possible in the region analysed, since grain boundaries attract impurities and produce compositional changes in their immediate neighbourhood. (The study of grain boundary segregation is mentioned below.) Energy loss and X-ray spectra were collected (usually simultaneously) from the same region of the specimen, using a Link Systems 860.

A typical X-ray spectrum from the steel is shown in Fig. 1. The energy range is 4–9 keV, which is relevant for the K_α and K_β peaks of Cr, Fe and Ni. The mass concentration ratios of Cr and Ni to Fe were calculated as follows. The background was subtracted using average counts per channel from, if possible, channels to the left and right of the relevant K_α peak. The resulting areas under the peaks lead to mass concentration ratios from the ratio equation[1] for thin films:

$$C_A/C_B = k_{AB}(I_A/I_B) \tag{1}$$

where k_{AB} is an experimental parameter, although its approximate value can be calculated theoretically. In order to find the experimental k_{AB} values, a calibration curve of k_{XSi} versus energy was found from known standards containing silicon (as used by TI Research Laboratories). This curve is shown in Fig. 2. Then k_{AB} is given by

$$k_{AB} = \frac{k_{ASi}}{k_{BSi}}$$

where k_{ASi} and k_{BSi} are found from the curve using the known K_α X-ray energies.

A number of corrections to equation (1) are frequently used, mainly for thicker specimens. Each is now discussed and shown to be comparatively negligible for the specimens used. The first correction is for absorption: equation (1)

TABLE 1 Mass concentration ratios from EDX and EELS in type 304 steel

	EDX	EELS $L_{II/III}$ edge		$M_{II/III}$ edge
		Egerton	Formula*	(formula)*
(a) Cr:Fe (nominal value 0.27)§	0.28	0.28	0.30	0.28
	0.29†	0.27	0.29	0.29
	0.28	0.28	0.31	0.29
	0.30	0.27	0.29	–
	0.29	0.27	0.29	–
	0.29	–	–	0.28
	0.28	–	–	0.27
	0.30‡	–	–	0.29
(b) Ni:Fe (nominal value 0.12)§	0.11†	0.13	0.14	–
	0.11‡	–	–	0.12

ERRORS: for EDX measurements ±0.01; for EELS ±0.02
§ values determined by independent optical absorption spectroscopy measurements
†, ‡ identify equivalent sets of figures (i.e. for same region of same specimen)
* formula as given in equation (5)

becomes[2,3]

$$\frac{C_A}{C_B} = k_{AB}\frac{I_A}{I_B}\left[\frac{(\mu/\rho)^A_{spec}}{(\mu/\rho)^B_{spec}}\right]\left\{\frac{1-\exp\left[-(\mu/\rho)^B_{spec}\,\rho t\,\mathrm{cosec}\,\gamma\right]}{1-\exp\left[-(\mu/\rho)^A_{spec}\,\rho t\,\mathrm{cosec}\,\gamma\right]}\right\} \quad (2)$$

The factors involving mass absorption coefficients were evaluated, assuming mass concentrations for the steel of 0.70 for Fe, 0.19 for Cr, and 0.08 for Ni, as confirmed by STEM, SEM and EMMA measurements, and using 12° for the take-off angle γ.

On inserting thicknesses t of 50–100 nm in equation (2), the change in C_A/C_B is about one or two percent. Since the working accuracy seems to be about 5%, the absorption effect is marginal but would have the effect of lowering slightly the Cr:Fe ratios of Table 1 and of raising slightly the Ni:Fe ratios (as determined by EDX).

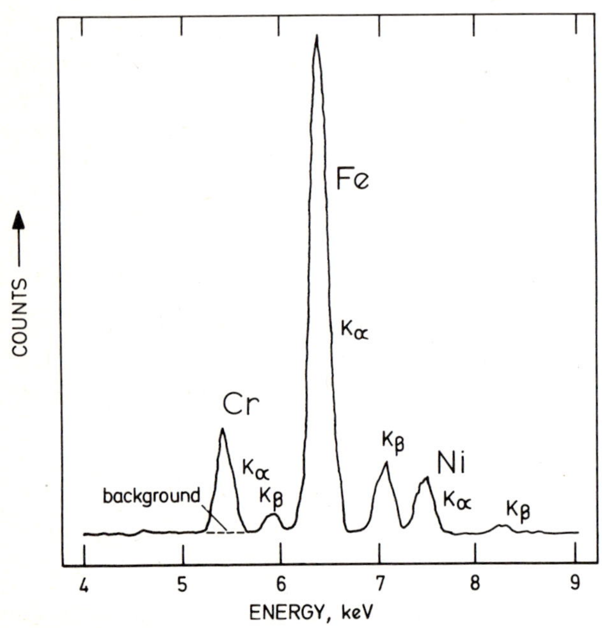

1 Typical X-ray spectrum from type 304 steel specimen; K_α peaks used for microanalysis. Background level artificially increased

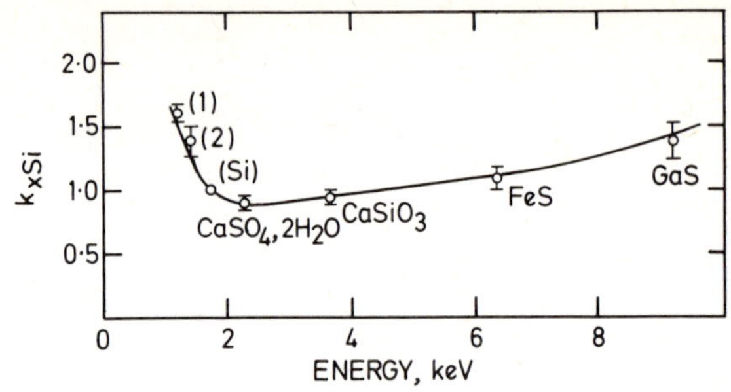

2 X-ray calibration curve for STEM; shows example of standard used for relevant element for each point: (1) refers to St. John's Island Olivine which contains Mg Si; (2) refers to Orthoclase which contains Al and Si

A correction is sometimes needed for fluorescence. This was calculated and led to a change of less than 0.1% in C_A/C_B, and so was neglected. Specimen orientation effects can also change the results of X-ray microanalysis. No evidence for such effects was observed, probably because the beam convergence was usually of the same order as a Bragg angle, and because in these foils the incoherent elastic scattering which produces beam broadening will also remove any orientation effects which depend on angular positions to an accuracy better than one Bragg angle.

We now turn to the electron energy loss spectra. A typical energy loss spectrum for the steel is shown in Fig. 3. The gain has been increased for the high-loss regions. The Cr and Fe edges are easy to see; the Ni edges are hard to see. With each of such spectra, the background under each edge was subtracted using an AE^{-r} expression as fitted to the curve before the edge, where A and r are appropriate constants. Now, the ratio of the number of atoms per unit area of one element to another is given by

$$\frac{N_A}{N_B} = \frac{\sigma_B(\alpha,\Delta)}{\sigma_A(\alpha,\Delta)}\cdot\frac{I_A(\alpha,\Delta)}{I_B(\alpha,\Delta)} \quad (3)$$

where $I_A(\alpha, \Delta)$ is the area under the edge of element A using an energy window Δ, and $\sigma_A(\alpha, \Delta)$ is the partial cross-section; α is, approximately, the maximum scattering angle of electrons, that is, the sum of the beam convergence semi-

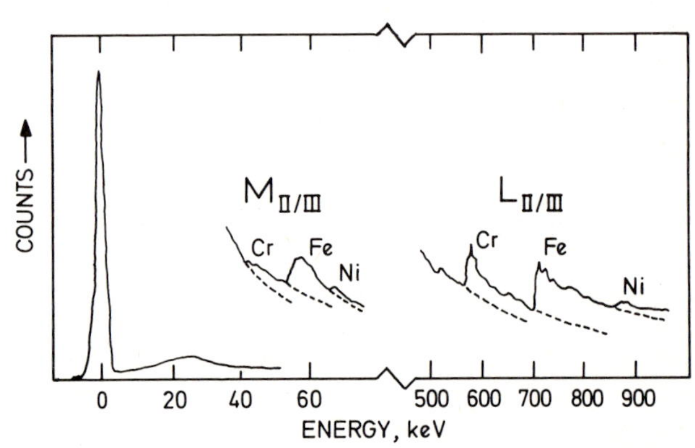

3 Typical electron energy loss spectrum from type 304 steel specimen. Gain and energy scale have been changed at higher losses; L edges have larger gain; dotted lines indicate schematic background extrapolation. Minor statistical fluctuations not shown

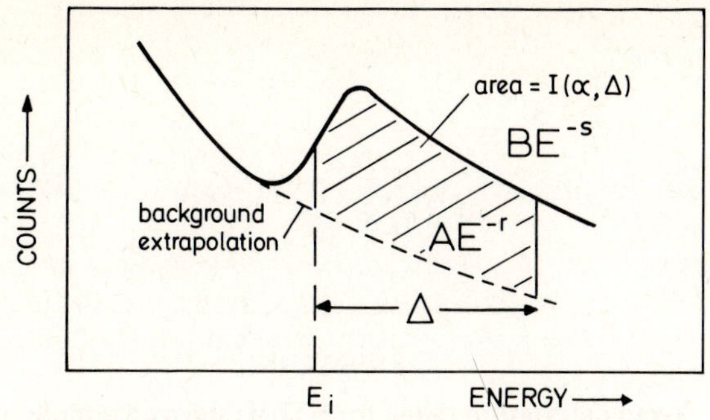

4 **Schematic to illustrate obtaining of area under edge, using energy window Δ. Edge is at energy E_i, near middle of rise in edge; AE^{-r} and BE^{-s} are fitted power laws to curves; α is approximately maximum angle of scattering of accepted single-scattered electrons**

angle and the spectrometer aperture semi-angle. Under conditions of single-scattering, α is exactly determined by a convolution of the incident and emergent angular beam distributions.

The area under each edge was found using an energy window of 100 eV for $L_{II/III}$ edges and 8 eV for $M_{II/III}$ edges. Figure 4 illustrates how the area was obtained.

The mass concentration ratios were found using the expression

$$\frac{C_A}{C_B} = \frac{A_A N_A/t\rho}{A_B N_B/t\rho} = \frac{A_A N_A}{A_B N_B} \quad (4)$$

where t is the specimen thickness, ρ is the specimen density, and A_A is the atomic weight of element A. The L.H.S. of equation (4) is the same as that of equation (1) for EDX, hence the results can be directly compared. Equation (4) requires knowledge of the relevant partial cross-sections, $\sigma(\alpha, \Delta)$. For the L edges, partial cross-sections due to Egerton[4] were used. For comparison, the following formula for the partial cross-section was also used:

$$\sigma_i(\alpha, \Delta) \propto \frac{n}{E_i E_0} \ln \left[1 + \left(\frac{\alpha}{\theta_E}\right)^2\right]\left[1 - \left(1 + \frac{\Delta}{E_i}\right)^{1-s}\right] \quad (5)$$

where n is the number of electrons per unit volume contributing to the edge, E_i is the edge energy, s is the index of the power law BE^{-s} of the spectrum above the edge, and θ_E is $E_i/2E_0$ with E_0 as the incident electron energy. Equation (5) was used for both the M and L edges. A derivation, and discussion, of equation (5) is included in the Appendix.

RANDOM ERRORS IN EDX AND EELS

The main random error for EDX with STEM occurs in the peak areas. If a peak has N counts, the standard error is $N^{1/2}$ counts. Each of the relevant peaks typically had about 10^4 counts, resulting in a 1% error. The counts are found after background subtraction, which itself produces an error. The background generally forms only a few percent of the peak area, hence the associated error is rather less than 1% (about 0.5%). When these errors are combined, one finds the total random error in the mass concentration ratio, as indicated in Table 1.

A similar process may be done for EELS, and specifically for the Cr:Fe ratio. The number of counts in an energy window is typically 10^3 after background subtraction at the edge. This leads to an error of about 30 counts, that is, 3%. The background extrapolation has an associated error of about 3%, as deduced from the range of power-law expressions that could fit the data. The location of the edge energy can be difficult, because of the slow rise of some edges, especially M edges. The consequent uncertainty in the position of the energy window gives an error of a few percent; prominent band structure effects near the edge energy aggravate this error. The total error in the area under an edge is thus about 6%; this leads to a total random error of about 8%, as indicated in Table 1.

RESULTS

The mass concentration ratios of Cr and Ni to Fe are shown in Table 1 for all techniques. It is seen that there is general agreement between most sets of figures to an accuracy of generally better than 10%. The agreements for Cr:Fe ratios are particularly good. Thus one can say that for both $L_{II/III}$ and $M_{II/III}$ edges the quantitative microanalysis with EELS is in good agreement with that from EDX, which has the best accuracy.

The results for Ni are rather less consistent. This is to be expected since the Ni edges (EELS) and peaks (EDX) sustained the largest statistical errors; in some cases, the Ni edges were invisible.

The results confirm the use in this case of Egerton's partial cross-sections for $L_{II/III}$ edges. The use of equation (5) is also confirmed, especially for $M_{II/III}$ edges where agreement with EDX is best.

One can summarize this section as follows: EELS has been shown to be just as capable of quantitative microanalysis as EDX, though with less accuracy. The exact accuracy obtained with EELS depends on which partial cross-sections are used, and the above has confirmed the use of two such partial cross-sections. The subject of cross-sections is discussed in the Appendix.

VOLUME ANALYSED: EELS AND EDX

When an electron beam passes through a specimen the beam spreads, mainly because of elastic scattering. In EDX the X-rays are collected from the whole of the spreading beam. Thus the volume analysed is basically that of the beam within the specimen. In EELS the electrons collected are those that pass through the slit of the spectrometer; these in turn are electrons accepted by the collector aperture. If these electrons are only single-scattered, thus implying a thin specimen, the volume in the specimen from which they are collected will generally be smaller than that of the beam. The situation is depicted in Fig. 5.

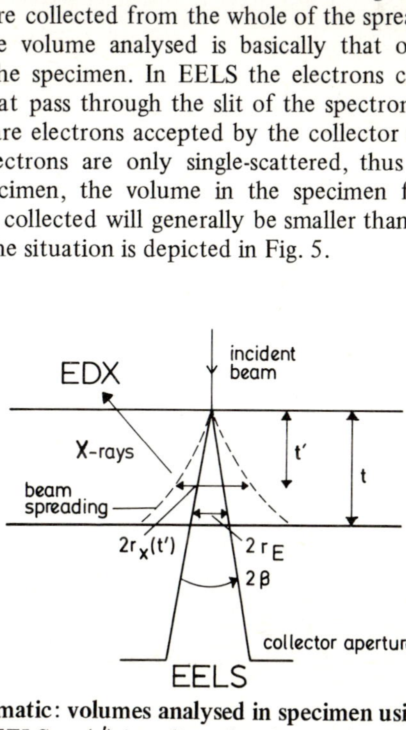

5 **Schematic: volumes analysed in specimen using EDX and EELS; $r_x(t')$ is radius of region analysed by EDX at t'; r_E refers to radius of cone of acceptance, as relevant for EELS; β is collector aperture angle**

Thus EELS generally analyses smaller volumes inside the specimen than does EDX. The exact details of the volume comparison will form the subject of a later paper; however, a short discussion is now given. The volume analysed is

$$V = \int_0^t A \, dt'$$

where t' is the vertical distance from the incident beam side of the specimen, and A is the area analysed at t'. This area is

$$A = \pi [r(t')]^2$$

where $r(t')$, the radius analysed, is given by the beam spreading for EDX and by the cone of acceptance for EELS. A zero-width, non-converging incident beam is assumed throughout. For EDX, it can be shown[3,5] that

$$2r(t') \simeq 6 \cdot 10^{-6} \frac{ZN^{1/2}}{E_0} \left(\frac{t'}{a}\right)^{3/2} \tag{6}$$

where Z is the atomic number and a^3/N is the volume per atom. Equation (18) has units of μm when E_0 is in keV. For EELS, the radius of the acceptance cone is

$$r(t') = \beta_{\text{eff}} t' \tag{7}$$

where β_{eff} is the result of the angular distribution of the (mainly) inelastically scattered electrons accepted within angle β into the spectrometer. One can see that the radius analysed depends differently on t' for the two techniques; in general, the EELS radius is smaller than the EDX radius for most specimen thicknesses, as depicted in Fig. 5. Now, the above equations are the simplest possible; further considerations such as the effect of beam convergence lead to some complicated expressions. All these, however, indicate that EELS is better at analysing small volumes than EDX.

An example illustrates the point. A number of grain boundaries have been studied with the two techniques and a depletion of chromium has been observed. The depletion observed was more prominent when using EELS than when using EDX.

When simultaneous EELS and EDX data are collected from a grain boundary, the ratio of signals from a grain boundary segregated layer to those from the whole volume analysed is found from

$$\frac{I_{\text{gb}}}{I_{\text{tot}}} = \frac{1}{t} \int_0^t \frac{2\omega r}{\pi r^2} \, dt' \tag{8}$$

where ω is the width of the layer, and assuming a uniform beam. Equation (8) becomes

$$\frac{I_{\text{gb}}}{I_{\text{tot}}} = \frac{2\omega}{\pi} \left\langle \frac{1}{r} \right\rangle \tag{9}$$

and so the ratio of the mean effective probe inverse-radius using EELS to that using EDX can be rather directly measured. We have

$$\frac{(I_{\text{gb}}/I_{\text{tot}})_{\text{EELS}}}{(I_{\text{gb}}/I_{\text{tot}})_{\text{EDX}}} = \frac{\langle 1/r \rangle_{\text{EELS}}}{\langle 1/r \rangle_{\text{EDX}}} \tag{10}$$

The depletion of chromium has been handled by assuming it to be equivalent to the presence of a thin layer of Fe at the grain boundary, with the concentrations of Fe and Cr elsewhere as normal. One finds experimentally that the ratio in equation (10) is 3 ± 1, where the error reflects experimental uncertainty. Calculation based on the known beam spreading suggests that the ratio of mean effective inverse-radii should be

$$\frac{\langle 1/r \rangle_{\text{EELS}}}{\langle 1/r \rangle_{\text{EDX}}} \simeq 2$$

The calculated ratio is rather lower than the experimental value. EDX has in effect analysed more volume than expected. This may be due in part to secondary and backscattered electrons falling in regions away from the incident beam and also to uncertainty in the interpretation of the spectra.

The main point is that the volume analysed with EELS is considerably smaller than that with EDX, and grain boundary depletion of Cr is much more easily detected with EELS than with EDX. Thus for *micro*analysis (with the emphasis on the prefix) EELS has been shown to be generally superior to EDX. Note, however, that for detection of grain boundary segregated species it is desirable to have both spectra simultaneously gathered, since the EDX spectrum positively identifies the presence of the segregant and enables the EELS spectrum to be analysed quantitatively.

CONCLUSIONS

EELS has been shown to be useful as a quantitative microanalytical tool, although often it has less accuracy than EDX. The results for the mass concentration ratios are generally in agreement for the two techniques to within the statistical error of about 10%. The main advantage of EELS is its ability to analyse much smaller volumes in the specimen than EDX. Hence studies of, for example, grain boundary segregation are most effectively done using EELS, together with EDX to confirm the identification of elements.

The use of EELS requires the knowledge of partial cross-sections if standards are not used. For L edges, the results using Egerton's partial cross-sections are in agreement with EDX results. For both L and M edges, partial cross-sections from equation (5) yield results in agreement with EDX. The fundamental equivalence of the electronic excitation process in energy loss and X-ray emission causes the cross-section to be simply related, and this is discussed further in the Appendix.

Further consideration of much of the above, especially concerning the volumes analysed near grain boundaries, will be given in later papers.

ACKNOWLEDGMENTS

The authors take pleasure in thanking Dr S. J. Pennycook for helpful discussions. Thanks are also due to Professor A. H. Cook, FRS, for provision of laboratory facilities, and to the Science Research Council for financial support. One of the authors (S.A.C.) acknowledges the receipt of a SRC CASE studentship with TI Ltd.

APPENDIX

Cross-sections: EELS and EDX

It is useful to recall that at the semi-classical level of treatment, the X-ray cross-sections are very simply related to the EELS cross-sections. The total cross-section for the ionization of an electron of binding energy E_i is given approximately by the Bethe expression

$$\sigma_i \propto Z_i b_i \frac{1}{E_i E_0} \ln \left(c_i \frac{E_0}{E_i} \right) \tag{A1}$$

where Z_i is the number of electrons in the shell concerned, and b_i and c_i are shell-dependent constants. With appropriate numerical values, this formula is used to estimate X-ray cross-sections (e.g. Ref. 6).

In the electron energy loss case, only a part of the total cross-section is measured, and to an adequate approximation one writes:[7]

$$\sigma_i(\alpha, \Delta) = \sigma_i \eta(\alpha)\, \eta(\Delta) \tag{A2}$$

where $\eta(\alpha)$ and $\eta(\Delta)$ are 'efficiency factors' relating to the choice of apertures and energy windows. The factor $\eta(\Delta)$ is found from

$$\eta(\Delta) = \frac{\displaystyle\int_{E_i}^{E_i+\Delta} \frac{d\sigma}{dE}\, dE}{\displaystyle\int_{E_i}^{E_0} \frac{d\sigma}{dE}\, dE} \tag{A3}$$

In the absence of prominent band structure effects, one finds that

$$\frac{d\sigma}{dE} = BE^{-s} \tag{A4}$$

For E_0 much greater than E_i, (A3) becomes

$$\eta(\Delta) = 1 - (1 + \Delta/E_i)^{1-s} \tag{A5}$$

Similarly, $\eta(\alpha)$ is found from

$$\eta(\alpha) = \frac{\displaystyle\int_0^{\alpha} \frac{d^2\sigma}{dE\, d\Omega}\, d\Omega}{\displaystyle\int_0^{\sqrt{(2\theta_E)}} \frac{d^2\sigma}{dE\, d\Omega}\, d\Omega} \tag{A6}$$

The differential cross-section is given by

$$\frac{d^2\sigma}{dE\, d\Omega} \propto \frac{1}{\theta^2 + \theta_E^2} \tag{A7}$$

and (A6) becomes

$$\eta(\alpha) = \frac{\ln(1 + \alpha^2/\theta_E^2)}{\ln(2/\theta_E)} \tag{A8}$$

The maximum scattering angle in (A6) is taken to be $\sqrt{(2\theta_E)}$, which is the position of the Bethe ridge, where classically all the scattering occurs.

Substitution of (A8) and (A5) in (A2) yields

$$\sigma_i(\alpha, \Delta) \propto \frac{Z_i b_i}{E_i E_0} \frac{\ln(c_i E_0/E_i)}{\ln(2/\theta_E)} \times \ln\left[\left(1 + \frac{\alpha^2}{\theta_E^2}\right)\right]\left[1 - \left(1 + \frac{\Delta}{E_i}\right)^{1-s}\right] \tag{A9}$$

Since θ_E, equal to $E_i/2\theta_E$, is small and c_i is of the order of unity, the quotient of logarithms in (A9) is, to a very good approximation, unity. If only losses from the same shell are to be compared, b_i is a constant, and equation (5) of the text follows.

Finally, we consider the possibility of a ratio-type expression (see equation (1)) for EELS. In the EDX case, the lack of α and Δ dependence, and the existence of the fluorescence yield and the beryllium window, all conspire to favour an empirical determination of k_{AB}. The equivalent expression in EELS is, from equations (3) and (4),

$$\frac{C_A}{C_B} = P_{AB} \frac{I_A(\alpha, \Delta)}{I_B(\alpha, \Delta)} \tag{A10}$$

where

$$P_{AB} = \frac{A_A \sigma_B(\alpha, \Delta)}{A_B \sigma_A(\alpha, \Delta)} \tag{A11}$$

Now, to derive a calibration curve for P_{AB} equivalent to that for k_{AB}, α and Δ would have to be kept constant. Whilst this could be accomplished for α, it cannot be accomplished for Δ; the useful energy window is very edge-dependent, according to the spacing of edges and to whether they are slowly-rising, whether band structure effects are prominent, and so on. In addition, there is the problem of multiple scattering for thicker specimens: the spectrometer will collect more electrons for one edge than for another, apart from the effect of differing partial cross-sections. It seems then that at this stage an experimentally determined ratio expression for EELS is not viable, and the analyses are better done theoretically. Thus, the search for better cross-sections is an important part of the development of the technique.

REFERENCES

1 P. DUNCUMB: *J. de Microscopie,* 1968, 7, 581

2 R. TIXIER and J. PHILIBERT: Proc. 5th Int. Cong. X-ray Optics and Microanalysis, eds. G. Mollenstedt and K. H. Gaukler, 1969, Springer-Verlag, Berlin, 180

3 J. I. GOLDSTEIN, J. L. COSTLEY, G. W. LORIMER, and S. J. B. REED: 1977, SEM/1977, **1**, ed. O. Johari, IITRI, Chicago, Ill., 315

4 R. F. EGERTON: 1980, private communication

5 L. M. BROWN: *J. Phys. F. (Met. Phys.),* 1981, **11**, 1

6 S. J. B. REED: 'Electron Microprobe Analysis', 1975 (C.U.P.), Ch. 12

7 M. ISAACSON and D. JOHNSON: *Ultramicroscopy,* 1975, **1**, 33

Discussion overleaf

DISCUSSION of paper by Collett, Brown, and Jacobs

D. C. JOY:
I do not agree with your estimates of accuracy for EELS quantitation performed by the 'efficiency factor' method (originally proposed by Isaacson and Johnson*). The energy efficiency from $1-[1+(\Delta/E_k)^{1-s}]$ contains the edge exponent s which must be obtained by curve fitting. The value of s changes with many factors including the value of the chosen energy window Δ. Thus choosing different windows Δ on the same edge will change the derived value of s and lead to a change in the result as the experimental parameters are altered. This is well documented (e.g. Joy, Egerton and Maher: 'Scanning electron microscopy 1979', Chicago, SEM Inc., 2, 817) and represents a severe limit to the usefulness of the technique. Secondly, the method is not applicable to M edges (or L_{23} edges for elements in the range Al through to Ca) because they display a delayed maximum in their profiles.

S. A. COLLETT:
It is clear from our paper that the efficiency factor method gives answers which are not statistically different from those obtained by EDX. There is a marked difference in attitude between those who attempt to write 'foolproof' computer programs which can be operated routinely (for example, by biologists) and those who attempt to use physical understanding to extract data under the many varied circumstances encountered in different problems. If the values of r and s are calculated using the longest sensible straight piece of the log-log plot, and the value of the energy window Δ is chosen sensibly (it will often be less than the energy range over which r and s are determined), one gets the results we quote. It seems essential to compose edges of similar type, i.e. not to form ratios from L and M edges, but only from L edges on M edges. It is important to note that in a practical case it is the ratios of elemental concentrations one wants; the area under the zero loss window does not figure in the calculation. It is also important that the specimen be thin – about 50 nm thick or less.

C. Y. F. COLLIEX:
The use of simplified efficiency factors of the type shown during the paper is very dangerous for M edges because of the strong delay in oscillator strengths for such edges. It would be necessary to check the stability of the results as a function of Δ.

S. A. COLLETT:
It is clear that there is room for both attitudes. There will be relatively routine situations in which unskilled operators will be able to accomplish quantitation using EELS. However, we do not encounter them in the type of work we do. One final comment: EDX is intrinsically more foolproof than EELS, but (alas) not for light elements.

A. J. BOURDILLON:
The beam spread is presumably due to inelastic effects and to elastic effects as in dynamical theory. Should not the latter be the same in EDX as in EELS? Is it not the more important?

S. A. COLLETT:
In the case of EELS, the finite acceptance angle of the spectrometer rejects any signals from electrons which have suffered large-angle scattering. The same is not the case for EDX.

* M. ISAACSON and D. JOHNSON: *Ultramicroscopy*, 1975, **1**, 33

Measurement of carbon in V(C,N) precipitates extracted from HSLA steels on aluminium replicas

A J Garratt-Reed

Electron energy loss spectroscopy has been used to determine the carbon content of presumed V(C, N) precipitates in a vanadium HSLA steel. Carbon extraction replication was inappropriate as the carbon in the replica would have interfered with the desired analysis. The production of aluminium replicas was therefore perfected. With appropriate care taken during specimen preparation it is possible to observe these replicas for several hours within significant contamination build-up. Within the sensitivity available (approximately 5 at.-% at present) no carbon is found in the precipitates. This result was unexpected, as the precipitation is found to be independent of the nitrogen content in the steel.

The author is at the
Center for Materials Science and Engineering,
Massachusetts Institute of Technology,
Cambridge, Mass., USA.

Carbon extraction replication has been a standard technique in transmission electron microscopy for a long period of time. Recently the technique has again become prominent with the advent of scanning transmission electron microscopy (STEM), enabling one, for instance, to perform analyses of ~50 Å precipitates using energy-dispersive X-ray analysis.[1]

When available analytical techniques were sensitive only to those elements heavier than neon, the presence of the carbon in the extraction replica proved to be of little importance. With the development of methods to analyse for light elements ($z \leqslant 11$), such as electron energy loss spectroscopy (EELS)[2] or energy-dispersive X-ray analysis with an ultra-thin window detector,[3] occasions arise when an analysis for carbon is indicated. In such cases carbon extraction replication is clearly inappropriate.

The present work was part of a wider study of the mechanical properties of high strength low-alloy (HSLA) steels containing vanadium.[4,5] It had been found that the mechanical properties correlated well with an observed distribution of precipitates which X-ray microanalysis indicated to be vanadium-rich. This correlation occurred over an extensive temperature range, in both fully austenitic steels and those partially transformed to ferrite. By analogy with the results from niobium steels[6] it was surmised that the precipitates were vanadium carbonitrides. Experiments with steels of differing nitrogen contents had indicated no measurable differences in the mechanical or precipitation behaviour as a result of changing this parameter. It was concluded that the carbon content (which was not varied in the experiments) controlled the precipitation. It was therefore desired to measure the carbon/nitrogen ratio in these precipitates as a function of other experimental parameters.

In order to avoid the problem of introducing carbon in replicas, the precipitates were extracted onto aluminium films by the following steps:

1 After mounting and polishing, the specimens were etched for 20 s in 3% nital.
2 An aluminium film approximately 500 Å thick was deposited on the specimens by vacuum evaporation. The pressure in the evaporator was kept below 10^{-5} torr to minimize the oxidation of the films. After a few tries it was found easy to judge the thickness of the film (which was not at all critical) by looking at the shadow of a folded piece of paper placed alongside the specimen in the evaporator.
3 The replicas were released by etching in 12% nital for approximately 30 s.
4 The specimens were plunged into distilled water, into which the replicas separated in small pieces. Pieces were caught on 200 mesh copper E.M. grids in the usual way.

Replicas made in this way were usually very much smaller than conventional carbon replicas, possibly because the grain structure of the aluminium weakened the film. The success rate was lower, good replicas being obtained in perhaps one in every three attempts. Nevertheless, excellent results were achievable. Figure 1 shows a low-magnification view of a replica, in which can be seen details of the surface relief of the specimen. In a high-magnification view (Fig. 2), the precipitates are visible as the dark spots, while the grain structure of the film can be seen as the low-contrast mottling in the background.

Carbon contamination can be a serious problem in analytical electron microscopy. Even in an ultra-high vacuum microscope such as the VG HB5 used in this study, thin layers of contamination on the specimen itself can migrate to the beam and build up a layer of carbon. To minimize this effect, the replicas were baked at about 150°C for a few minutes in a vacuum $<10^{-5}$ torr. It was then possible to study the samples in the microscope for extended periods without observing any contamination build-up.

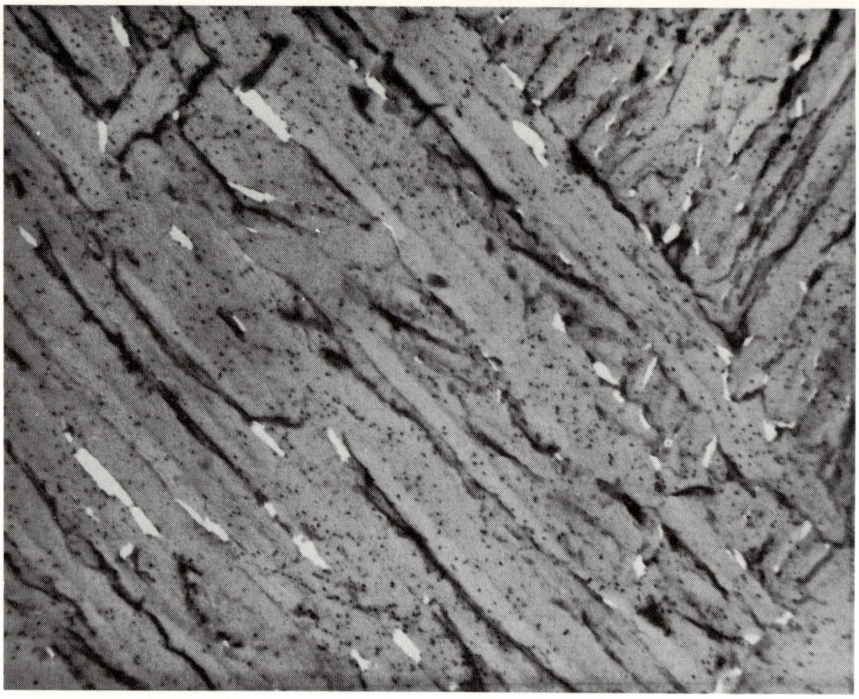

1 Low magnification electron micrograph of Al extraction replica. Precipitates (small dark spots) and surface relief (large-scale elongated features) are visible (×20 000)

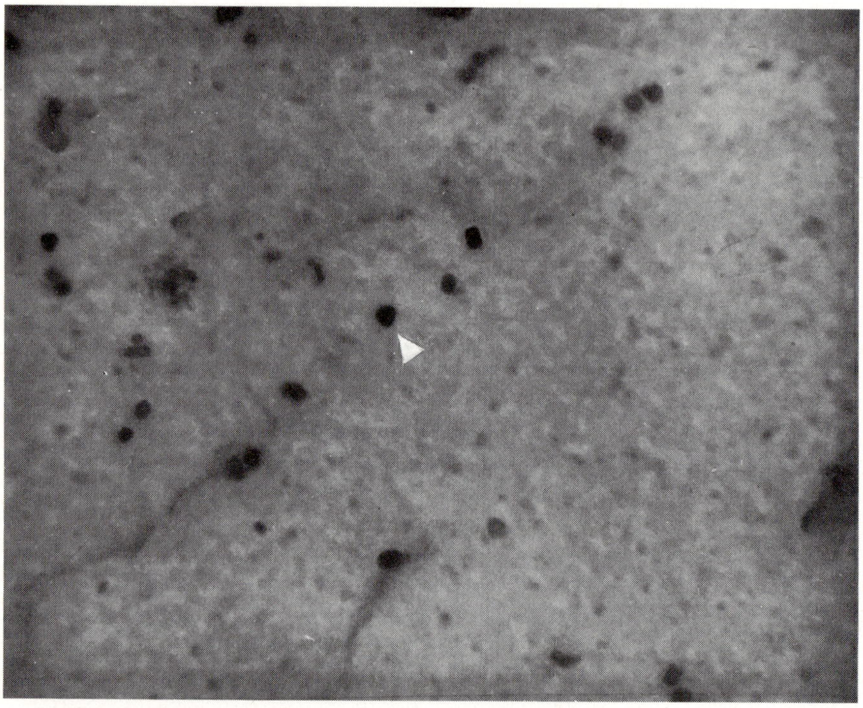

2 High magnification micrograph of area of Al extraction replica. Dark spots are precipitates of interest; background mottling indicates grain structure of Al film. Pointer indicates precipitate from which spectrum of Fig. 4 was obtained (×100 000)

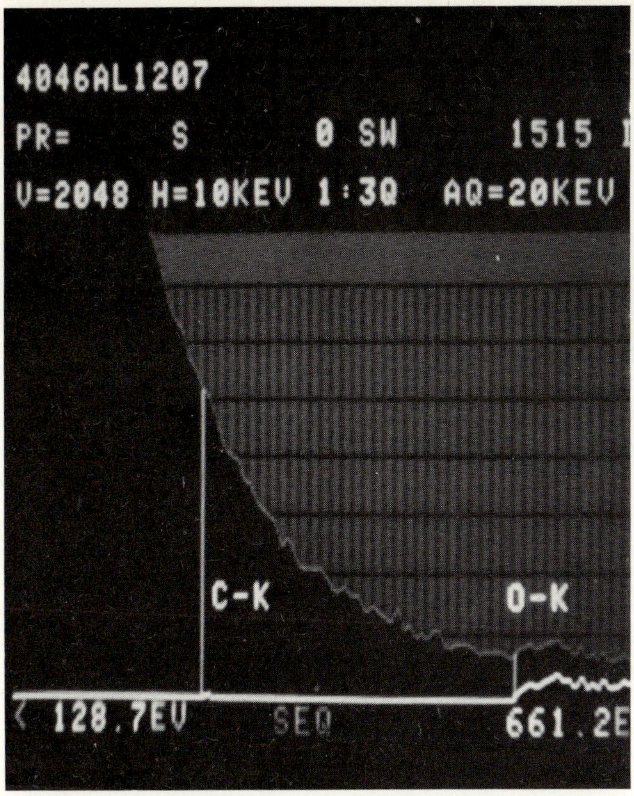

3 Electron energy loss spectrum obtained from region of Al film. Energy range encompasses loss edges corresponding to C, N, O, V. The only signal observed corresponds to oxygen

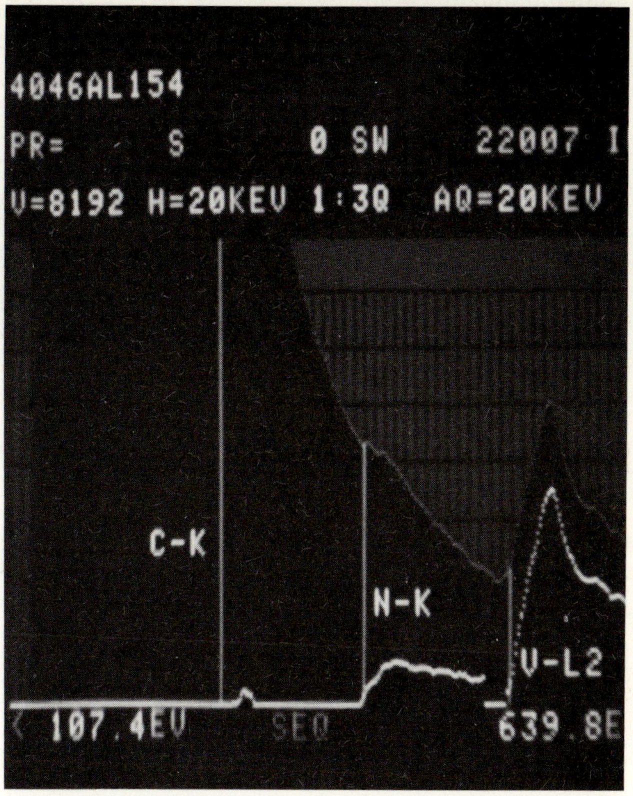

4 Spectrum similar to Fig. 3 but obtained from precipitate indicated in Fig. 2. Clear V and N signals are seen, and, in this example, a small signal at position corresponding to C (comparison of many spectra suggests latter signal due simply to system noise)

Electron energy loss measurements were performed to determine the carbon and nitrogen concentrations. The spectra were stored in the multichannel analyser memory of a Kevex 7000 X-ray spectrometer, and programs supplied by Kevex Corporation[7] were used for analysis. The illustrations of spectra are photographs directly from the display of the analyser. In these the upper line separating the grey area from the dark area is the raw spectrum; the lower white line shows the 'stripped' edges.[7]

The HB5 microscope was operated with conventional settings. The convergence of the incident probe on the specimen was about 10 mrad, and the acceptance angle of the spectrometer was about 3 mrad. The probe current was about 1.5 nA, and the probe diameter about 1.5 nm.

The sensitivity of EELS is (for very thin, <50 nm samples) measured in terms of the mass of a particular element of interest present in the volume irradiated (for constant beam current density). For maximum sensitivity, therefore, the largest available precipitates should be measured. For initial study those chosen were formed in a 0.1V–0.02N HSLA steel. The specimen was austenitized at 1250°C for 1 h, cooled rapidly to 885°C, compressed to a strain of $\epsilon = -0.3$, held at the test temperature for 3300 s, and quenched with water. The precipitates formed during the holding period, and, as can be seen in Fig. 2, had grown to a size of 15–25 nm.

Figure 3 shows the EELS spectrum from a region of the aluminium film. The only observable signal corresponds to oxygen, which would be expected to be incorporated into the film during the evaporation. Note the absence of any carbon signal. Figure 4 shows the spectrum obtained from the precipitate indicated by the pointer in Fig. 2. Strong signals are seen corresponding to the V L_{2-3} edge and the N K edge. In this example a small signal is seen at approximately the position of the C K edge. Averaging the results from many such spectra, however, it was concluded that the carbon level was below that which could be reliably detected. It was estimated that this level corresponded to one-tenth that of the observed nitrogen signal and therefore that the composition of the precipitates was $V_X(C_{<0.1}, N_{>0.9})$. It should be noted that no quantitative measurements of the vanadium concentration were made; comparison of K and L absorption edges is not straightforward, and the purpose of the work was to determine the carbon/nitrogen ratios. The value of X, therefore, is not determined here.

This result was not expected. The full implications will be discussed elsewhere.[4] It is not immediately apparent why the precipitation should be so insensitive to the nitrogen content of the steel when nitrogen appears to be the main constituent of the precipitates.

It has, however, been demonstrated that good analysis for carbon in small precipitates is achievable by the use of aluminium extraction replicas. The work described used electron energy loss spectroscopy in a rather poorly matched system. The use of a system employing a well-matched spectrometer[8] should generate better sensitivity. Similar specimens could also be examined using an ultra-thin window energy-dispersive X-ray detector.[3] In application of this technique to other materials, of course, the etching solutions and times would be adapted accordingly.

ACKNOWLEDGMENTS

The author wishes to thank Professors John B. Vander Sande and Walter Owen for their interest and discussion, and the Vanadium International Technical Committee who financed the project of which this work was part.

REFERENCES

1 J. B. VANDER SANDE and A. J. GARRATT-REED: 1980, 38th Annual Proc. EMSA, 102
2 D. M. MAHER: in 'Introduction to analytical electron microscopy', eds. J. J. Hren, J. I. Goldstein, and D. C. Joy, 1979, Plenum Press, 259
3 L. E. THOMAS: 1980, 38th Annual Proc. EMSA, 90
4 M. J. SCHMUTZ, A. J. GARRATT-REED: J. B. VANDER SANDE, and W. S. OWEN: 1981, to be published in *Metall. Trans.*
5 A. J. GARRATT-REED, M. J. SCHMUTZ, J. B VANDER SANDE, and W. S. OWEN: submitted 1981 to *Metall. Trans.*
6 S. S. HANSEN, J. B. VANDER SANDE, and MORRIS COHEN: 1980, *Metall. Trans.*, **11a**, 387
7 Kevex Corporation, Bulletin EELS, 1980
8 O. L. KRIVANEK: 1980, 38th Annual Proc. EMSA, 86

DISCUSSION of paper by A. J. Garratt-Reed ('Measurement of carbon')

P. R. SWANN:
I would like to suggest that you use silicon monoxide as an extraction replica material when analysing carbide precipitates. SiO is almost structureless and is chemically inert to most etchants.

A. J. GARRATT-REED:
Thank you for the suggestion. I had not tried SiO.

R. C. COCHRANE:
Could you tell me the processing conditions under which you produced the VN precipitates?

A. J. GARRATT-REED:
This information is contained in the paper.

T. N. BAKER:
Would you give us the bulk analysis of your material?

A. J. GARRATT-REED:
The analysis is 1.35%Mn–0.1%V–0.02%N plus some Si and C.

T. N. BAKER:
As the strengthening in HSLA steel is associated with particles <10 nm diameter, have you analysed down to this size and, secondly, would you anticipate any change in C/N ratio at sizes smaller than those you used in your illustration?

A. J. GARRATT-REED:
We have analysed smaller precipitates. As the size goes down, so does the signal. Therefore the detection limits (in terms of concentration) increase. We have not observed carbon in smaller precipitates.

We do not anticipate a change in C/N ratio as a function of size, but we would expect precipitates formed at lower temperatures to have more carbon. We have not analysed these at present.

R. C. COCHRANE:
Having heard from the previous speaker the composition of the steel, unless the solubility of VC in austenite is different on your side of the Atlantic I would expect only VN to be precipitated over the whole range of nitrogen contents, given your processing conditions.

I would be much more interested in the fine precipitates in the ferrite since it has already been shown that V precipitation kinetics are strongly influenced by transformation temperature and the state of combination of the nitrogen, either as VN or AlN.

A. J. GARRATT-REED:
I agree that VN should be precipitated. The whole point of these measurements was that we found no dependence of the precipitation on nitrogen content, and this led us to doubt the prediction. We have, in fact, confirmed the thermodynamic prediction of VN precipitation.

We have difficulty with sensitivity with smaller particles. Of course the precipitates in the ferrite are interesting, but unfortunately at present we cannot make useful measurements on them.

E. D. BOYES:
The technique of field-ion atom probe microanalysis is also very suitable for determining the composition of very small particles down to 2–3 nm diameter. It is particularly useful because it is a broad Z-spectrum technique and is relatively quantitative and easy to interpret, with the additional capability of being able to analyse directly for matrix elements included in particles. Aluminium alloys and steels have been analysed in this way.* The major limitation is the special form and small size of the specimen, but it is very suitable for homogeneous precipitation close to peak hardness. Perhaps the author could comment on the relative applicability of these two techniques to this problem.

A. J. GARRATT-REED:
I agree with the comment. We are in the process of arranging for field-ion atom probe study of these samples because we expect the technique to be very useful.

J. F. MANSFIELD:
Did you see Al peaks in the EELS spectra from your Al replicas and how did their intensities compare with those from the particle you were analysing? Would this have affected the detection of any of the other elements?

A. J. GARRATT-REED:
We did see Al peaks in the EELS measurements. The intensity was similar to the vanadium signal. There would have been a small effect on sensitivity for other elements because of the increase in the background due to the 'tail' of the Al signal (as discussed by Joy in these proceedings).

* E. D. BOYES: Proc. 9th Intl. Congr. on EM, Toronto, 1978, 104

Part 4
Instrumentation

Experiences of analytical performance of Hitachi 200 kV TEM/STEM

E Metcalfe and H T Owen

The authors present some results obtained using the Hitachi H700H TEM/STEM. The microscope is a 200 kV instrument with a 68° positive take-off energy-dispersive detector geometry which, combined with a low background kit fitted in the column to reduce spurious X-rays, allows a high X-ray performance to be obtained. Those aspects of the microscope which are significant in optimizing the high spatial resolution X-ray microanalysis of thin films, such as a high accelerating voltage and favourable column geometry, are discussed along with the limiting factors of specimen drift and contamination. The microanalytical facility has been applied to a wide variety of materials and a number of examples are presented covering carbide precipitation in steel, analysis of oxide structures, precipitation in high alloy steel and analysis of impurities in polymeric insulation material.

Dr Metcalfe is in the Central Electricity Research Laboratories (CERL) at Leatherhead, Surrey; Mr Owen is with the Electricity Research Council (ECRC) at Capenhurst, Chester, UK.

Analytical electron microscopy is increasingly becoming a general laboratory tool and is being applied to a widening range of materials problems. This paper presents some results obtained using two Hitachi H700H TEM/STEMs installed in the Electricity Council Research Laboratories and the Central Electricity Research Laboratories in 1980. The microscope is a 200 kV TEM with a scanning attachment for STEM operation. The EDS detector which is fitted to the column has a high positive take-off angle position, which, combined with a low background kit fitted in the column to reduce spurious X-rays, allows a high X-ray performance to be obtained without sacrificing the high resolution imaging capability.

Those aspects of the microscope which are significant in optimizing the high spatial resolution X-ray microanalysis of thin films are discussed along with the limiting factors of specimen drift and contamination.

The instruments have been used for structural and compositional analysis in a wide variety of materials. Four examples are presented in detail covering carbide precipitation in steel, analysis of oxide structures, precipitation in high alloy steel, and analysis of impurities in polymeric insulation materials.

ANALYTICAL MODE

200 kV operation

Apart from the advantage of increased penetration in all imaging modes (Fig. 1), it has been shown that increasing the incident beam energy improves the spatial resolution, improves the peak--to-background ratios, increases peak counts and thus improves mass sensitivity.

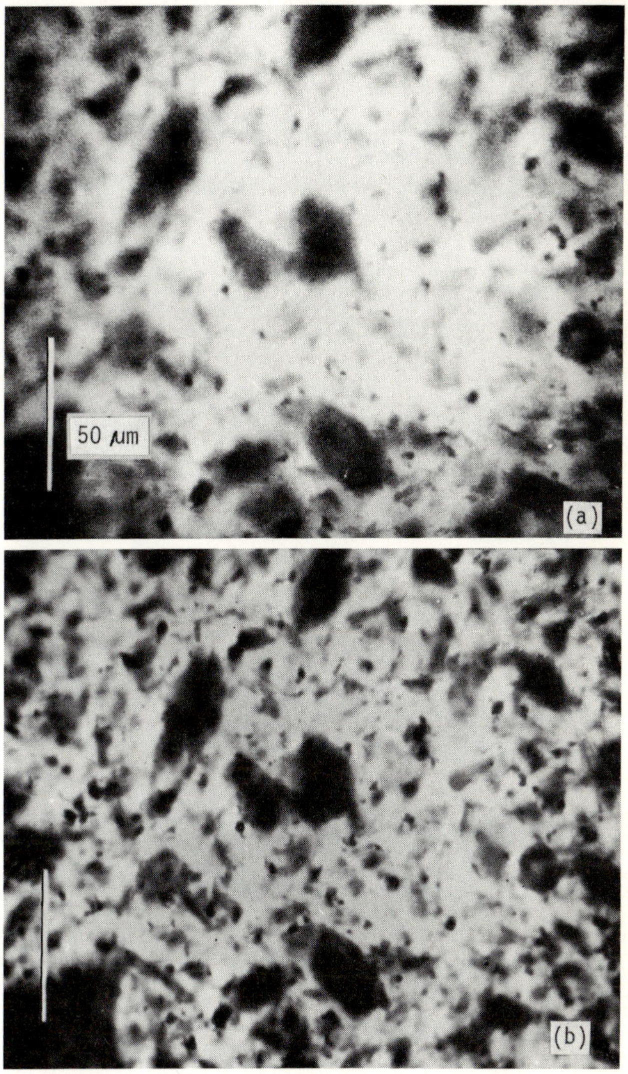

1 STEM image of silica-filled epoxy resin at (a) 100 and (b) 200 kV

TABLE 1 Peak counts and peak-to-background ratios for NiAl alloy as function of kV

kV	Al K_α	Bg	P/B	Ni K_α	Bg	P/B
75	784	112	7.03	4140	48	87
100	1202	190	6.33	7138	78	92
150	1338	216	6.2	8811	89	99
175	1401	235	6.0	9628	92	105
200	1462	243	6.0	10093	101	101

Conditions: counting period 50 s; 10 nm beam dia.; point count; 240 eV window per peak position

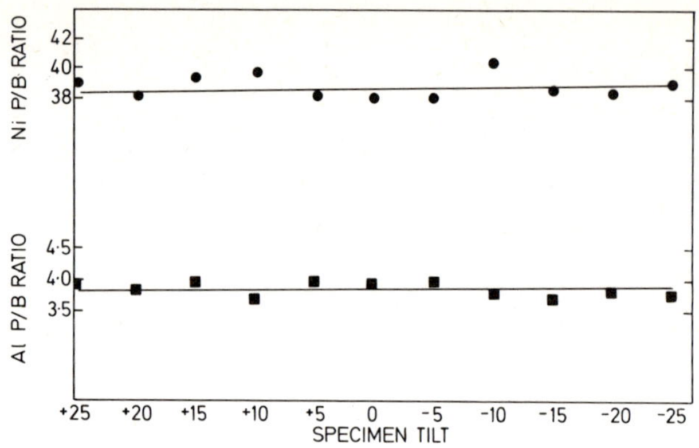

3 **Ni and Al peak-to-background ratios as function of specimen tilt**

X-ray spatial resolution is a function of the incident beam diameter and the spreading of the beam as it passes through the thin foil. This beam-broadening effect has been discussed by many workers. The simple but widely accepted model of Goldstein *et al.*[1] shows that broadening b is related to specimen composition parameters (atomic weight and number, A and Z, and density, ρ, the specimen thickness t, and the accelerating voltage E_0, by

$$b = K\left(\frac{Z}{E_0}\right)\left(\frac{\rho}{A}\right)^{1/2} t^{3/2} \qquad (1)$$

Although this single scattering model breaks down with increasing specimen thickness, results from Monte Carlo calculations of beam-spreading effects in thin foils where multiple scattering effects are taken into account[2] support the above relationship between b and E_0, and good experimental confirmation of the effect has been obtained.[3] Thus operation at 200 kV should offer advantages over lower voltages because reduced beam broadening reduces the volume from which X-rays are emitted, thus giving improved analytical resolution.

Increasing the accelerating voltage also increases the number of X-rays generated. Peak counts have been measured from a NiAl alloy as a function of kV and are shown in Table 1. Clarke[4] has discussed this effect and indicated that the increase in the number of X-ray photons can be attributed to the increased gun brightness as this is proportional to the electron energy E_0. The argument that lower accelerating voltages increase the number of X-rays generated, based on the $I/E_0 \ln E_0$ dependence of the ionization cross-section, ignores the effect of increasing gun brightness.

Theoretical considerations indicate that an increased accelerating voltage would also improve peak-to-background ratios[5] as the background due to *bremsstrahlung* radiation

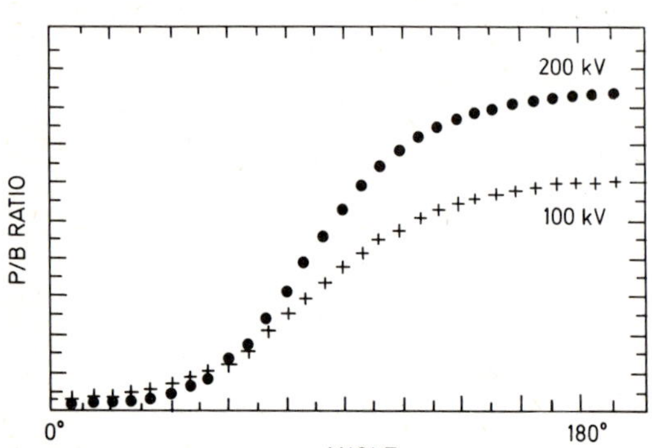

2 **Calculated peak-to-background ratio for thin Ni foil as function of detector observation angle[5]**

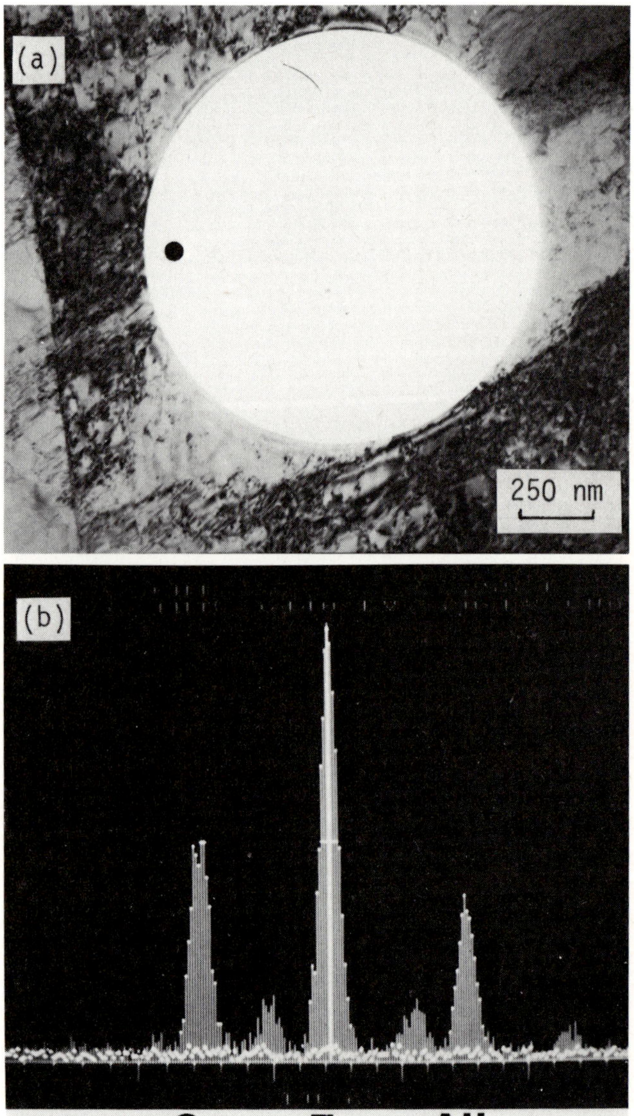

4 **In-hole count from 316 foil: (a) indicates location of beam in hole for spectrum shown in (b); (b) comparison of spectra for in-hole count (dot format) and count from thin part of foil (bar format)**

decreases. Support from this conclusion can be difficult to obtain in practice as additional scattering can occur and lead to increased background at higher kV. We have measured peak-to-background ratios on NiAl foils at different kV values and the results are also shown in Table 1. It is seen that the peak-to-background ratios for Ni are maximized at about 150–175 kV and begin to fall off at 200 kV. All of the above factors indicate that mass sensitivity during analysis should improve with increase in operating voltage.

Column geometry

Zaluzec[5] has discussed the optimum position of the detector in the STEM and has calculated the peak-to-background ratio as a function of angle measured from the forward scattering direction for various values of accelerating voltage. He has shown (Fig. 2) that the variation in peak-to-background ratio begins to level out at angles above about 140° (at 200 kV). In the H700H the detector is mounted at an angle of 158° to the forward scattering direction, thus satisfying the above optimum geometry, and in line with the specimen holder rod axis. In this position it is not necessary to tilt the specimen in order to generate an acceptable X-ray intensity at the detector. If one does tilt, for example to optimize diffraction conditions, then for a homogeneous foil the X-ray intensity varies only slightly over ±25° tilt (Fig. 3). Two further advantages of not having to tilt the specimen are that (i) there is no effective increase in specimen thickness upon tilting which would degrade spatial resolution and (ii) the background count should be lower. The emission of characteristic X-rays in a thin foil is essentially isotropic whereas the continuum is highly anisotropic with respect to the incident electron beam varying as

$$I_{\text{brem}} \propto \frac{\sin^2 \theta}{(1-\beta \cos \theta)^6} \qquad (2)$$

where θ is the angle measured from the forward direction of the beam and β is proportional to the electron velocity. If the specimen has to be tilted to high angles, say 30–45°, the *bremsstrahlung* generated can intercept an appreciable volume of the specimen and reduce the peak-to-background ratio.

The column of the H700H is fitted with a low background kit which means that all the fixed apertures are graphite and the whole of the column down to and beyond the specimen area is graphite-lined. In addition a thick molybdenum C_2 aperture is fitted as standard. The presence of spurious X-rays is usually demonstrated by the 'hole-count' effect whereby passing the primary beam through a hole in the specimen still results in the detection of a spectrum characteristic of the specimen. The ideal system would have no 'in-hole count' and in a practical system we have to define the conditions under which such measurements are made in order that they may be properly applied. A recent discussion of in-hole count measurements[5] suggests the general criterion that the intensity of each characteristic X-ray line recorded during an 'in-hole measurement' should be less than 1% of the intensity measured with the beam on the thinnest part of the specimen.

We have carried out in-hole count measurements using holes of size 1 μm in type 316 foil (Fig. 4) and can readily achieve the conditions specified above and even if the beam is placed within 100 nm of the edge of the hole we get only 2–3% of the intensity from the thinnest part of the specimen. This indicates that the column is very 'clean' with respect to stray radiation.

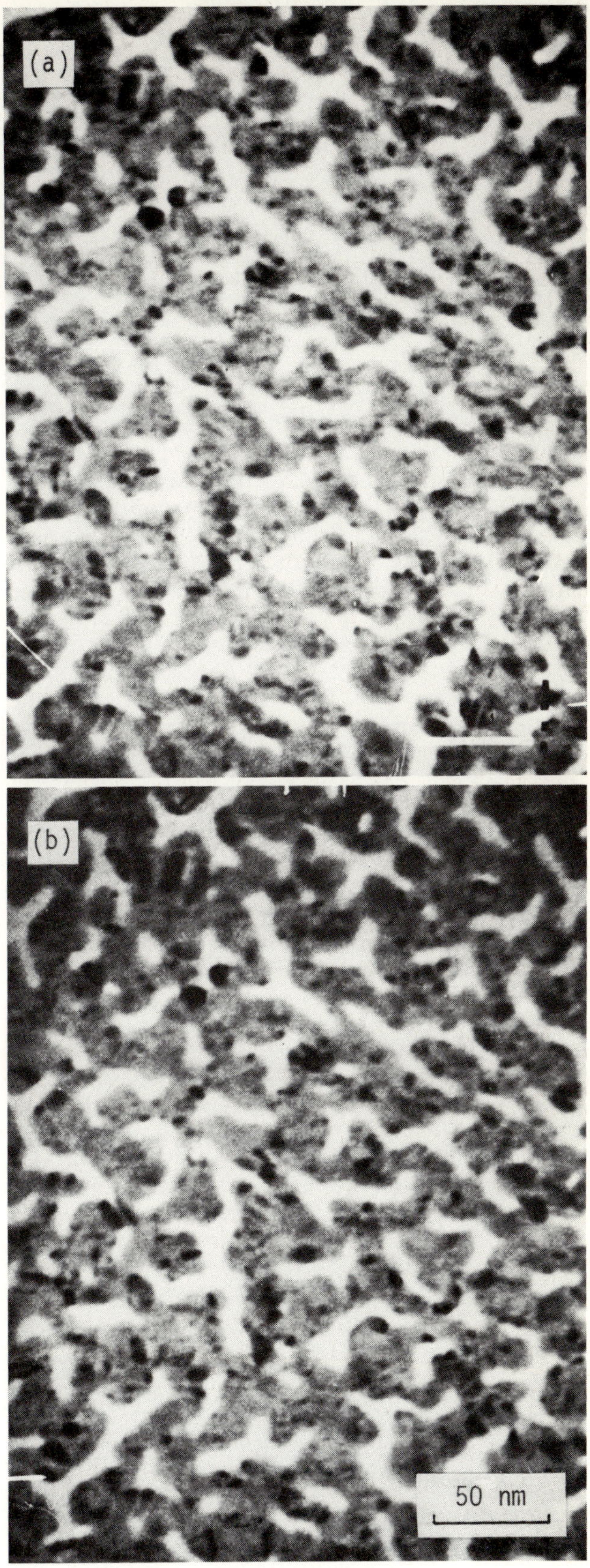

5 STEM image of evaporated gold on carbon film: (b) taken 5 min after (a) with no adjustments made to microscope in interval

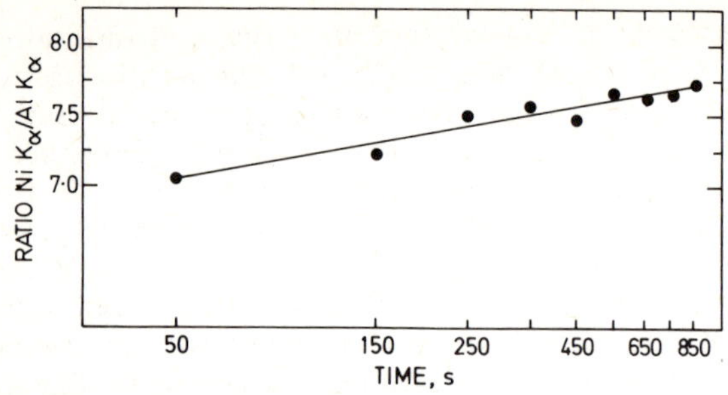

6 **Variation in Ni K_α/Al K_α ratio with accumulation of contaminant on surface**

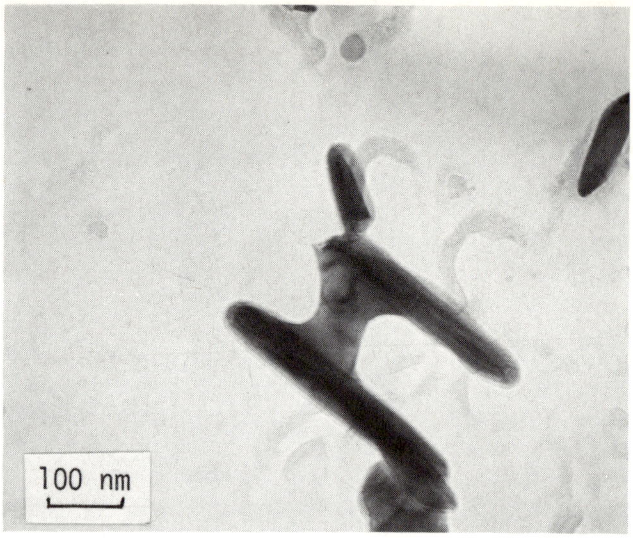

7 **TEM image of H carbide**

Contamination and drift

These are two important factors which can prejudice successful analysis of small areas in STEM. We have measured drift rate in two ways: firstly, a STEM image of a specimen of Ag deposited on a holey C film was recorded and then with no further adjustment of the microscope a second image was recorded after a 5 minute interval (Fig. 5); secondly, a contamination track was formed on the surface of a holey C film over a long time (1 h). Both tests were carried out using the standard and low background holders and gave drift rates of 1 Å s^{-1} and 4 Å s^{-1} respectively (note that these are for the worst case and drift rates for metal foils are 0.6 Å s^{-1} and 2 Å s^{-1}).

The development of STEM with its small electron probes of high current density has increased contamination rates in the irradiated area to a point where detrimental effects can occur. The most serious consequences of contamination are: (a) loss of spatial resolution as a result of beam broadening occurring within the contamination cone before the beam interacts with the specimen; (b) the production of *bremsstrahlung* X-rays when the beam interacts with the contamination cone, leading to an increase in continuum background; (c) increased excitation of backscattered electrons, resulting in secondary emission from remote areas; and (d) preferential absorption of low energy characteristic X-rays. We illustrate this last point for our system with the same alloy as used by others,[5] NiAl, and the results of intensity ratios of Ni K_α to Al K_α versus time are shown in Fig. 6. Note that the effect is quite small and can be ignored in most cases.

Contamination can arise from the vacuum system and the specimen. The H700H uses a dual pumping system in which the camera chamber and the specimen chamber are separately pumped in order to reduce contamination at the specimen, and is an improvement over earlier versions. However, in our view the process has not been taken far enough and the use of better vacuum technology and ion pumps in the specimen area, as used in some other microscopes, is a logical progression.

Several techniques have been used to reduce specimen-borne contamination, including pretreatment methods of glow discharge, ethanol immersion, beam flooding and long dwell times in the microscope prior to analysis. In our experience there is no one method which is universally acceptable for all specimens, but the use of an appropriate method can reduce contamination in most cases to acceptable levels. However, contamination from the vacuum

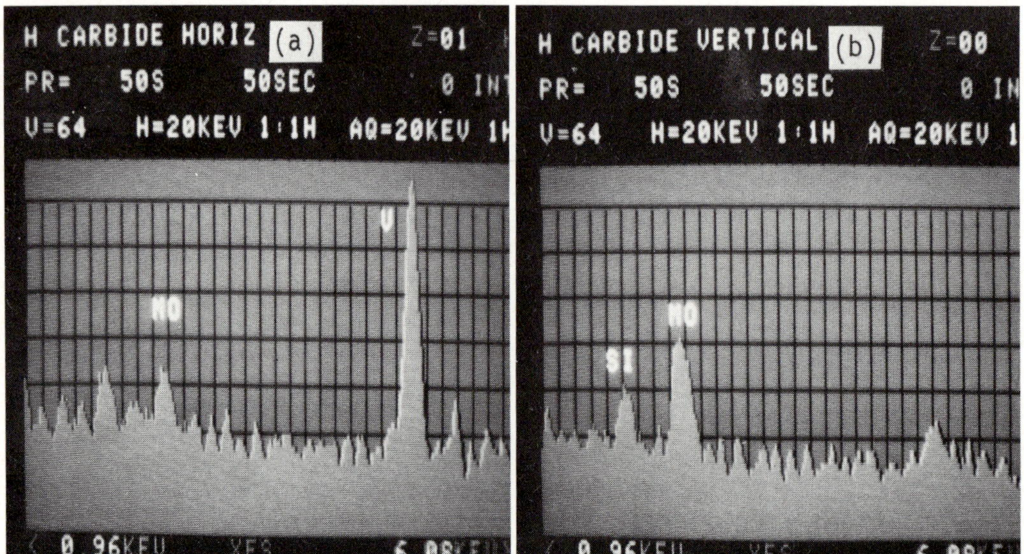

8 **Point analysis of H carbide: (a) horizontal and (b) vertical bars**

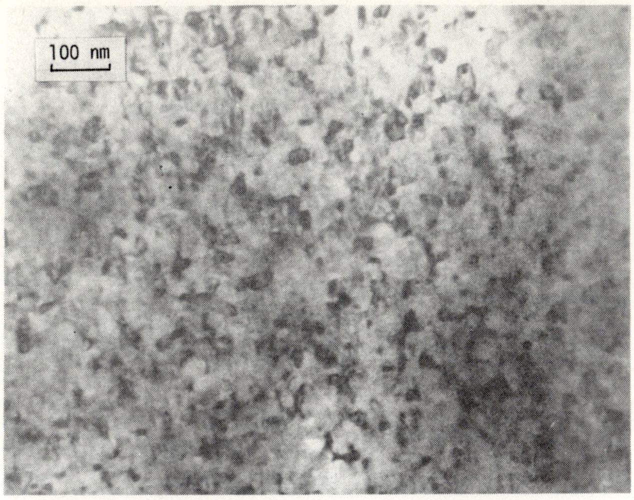

9 TEM showing details of inner layer scale formed on type 321 steel at 560°C after 20 000 h (near inner/outer layer interface)

system can still be important for we have sometimes detected Si only to find that it is in the contamination peak (see below under 'Aging and creep in alloy 800').

TYPICAL APPLICATIONS

'H' carbides in low-alloy steel

The 0.5Cr–0.5Mo–0.25V steel finds widespread application in the power generation industry, especially in heavy section pipework operating in the creep regime. The initial microstructure contains coarse-grained boundary carbides (based on Fe_3C) and a dispersion of fine V_4C_3-base carbides throughout the matrix.

During operational life of the component, usually at temperatures up to 565°C, the V_4C_3 is gradually replaced by a carbide based on Mo_2C. The latter frequently nucleates on pre-existing carbides and can result in an H-shaped carbide (Fig. 7). Examination in the STEM[6] has shown that the 'cross-bar' is V_4C_3, and the 'uprights' are Mo_2C (Fig. 8) which have nucleated on the V_4C_3 and grown at its expense, finally resulting in a pair of Mo_2C plates – the 'cross-bar' having dissolved.

It would appear that the balance of thermodynamic stabilities in this steel is such that at temperatures above 625°C V_4C_3 is the preferred carbide, whilst below 600°C Mo_2C is more stable. The initial tempering is performed at ~700°C, whilst the service is rarely above 565°C, hence the change in dispersion.

Micro-compositional segregation in oxide scales formed on austenitic steels

Where two-layered oxide scales form on steels in oxidizing environments, the inner layer is found to control the oxidation rate constituting a diffusion barrier to further gas/metal reaction. Several high temperature corrosion problems have as their cause the breakdown of this inner layer, leading to subsequent linear rates of corrosion and high associated metal loss rates.

Recent TEM studies by Rowlands and Manning[7] identified a significant structural difference between scales formed on different types of steel. Comparing inner layer scale structures on ferritic and austenitic steel, it was found that although both appear to be fine-grained using optical or SEM metallography, the scales on austenitic steels show a single crystal structure over areas large enough to correspond with the original steel grain size of 50 μm. In comparison, the grain sizes of the scales on ferritic steels are about 0.1 μm. Within the larger single crystal areas of the inner layer scales on austenitic steels, microstructural features are, however, observed on a much finer scale. Using STEM, these microstructural features have been found to be related to fine scale compositional variations. In transverse thin sections prepared normal to the oxidized surface, STEM microscopy was used to investigate the chemical composition of single crystal regions showing the fine scale contrast effects as shown in Fig. 9. This micrograph is of the inner layer oxide grown on a type 321 steel at 560°C in a power station boiler for 20 000 h. Chemical analysis using STEM on several inner layer samples of this type consistently demonstrated that the light and dark areas of the scale had significantly different compositions. The width of these chemical and contrast variations within the inner layers varied, being of 20–30 nm dimension in the first-formed parts of the scale and 150–200 nm close to the most recently formed scale close to the metal/oxide interface. Many pairs of analyses were made from adjacent light and dark areas. Compositions determined are shown in Fig. 10. Although the steel contains 12%Ni, the light areas were virtually nickel-free and had a composition corresponding to $Fe_{1.5}Cr_{1.5}O_4$. The dark phase contained considerable quantities of nickel with a mean composition of $Fe_{1.5}Cr_{0.6}Ni_1O_4$.

It was proposed that the chemical segregation discovered occurred due to diffusional rearrangement of the alloying constituents in the underlying alloy during oxidation and were not the result of spinel disproportionation during cooling.[7]

The main significance of these results at the present time is that they demonstrate that the application of STEM techniques to the study of oxidation processes can yield significant new details of oxidation processes that have not been discovered in spite of the most extensive studies over many years using optical and SEM microscopy.

Aging and creep in alloy 800

Commercial alloy 800 and experimental casts of modified alloy 800 composition have been examined[8] in order to define the compositional factors affecting precipitation strengthening and creep ductility. The principal precipitate was Ni_3Ti or $Ni_3(Ti,Al)$ gamma-prime. Direct energy-dispersive X-ray evidence was obtained in the STEM of the replacement of titanium by aluminium in the Ni_3Ti precipitate (size 5–10 nm). Figure 11*a* shows the spectra from Ni_3Ti with no spurious Al peak; Fig. 11*b* shows the presence of Al in γ' from alloy containing 0.3%Al, with in addition to the Al peak a consequent reduction in the height of the titanium peak. Also shown is a silicon peak, although

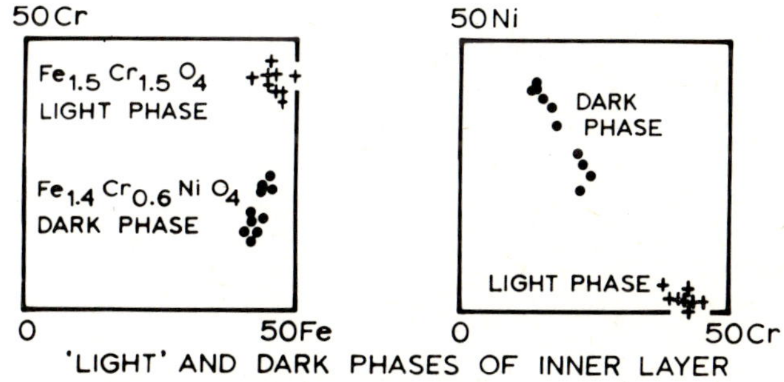

10 Spot compositions measured on inner layer scale, showing clear differentiation in composition between light and dark contrast regions

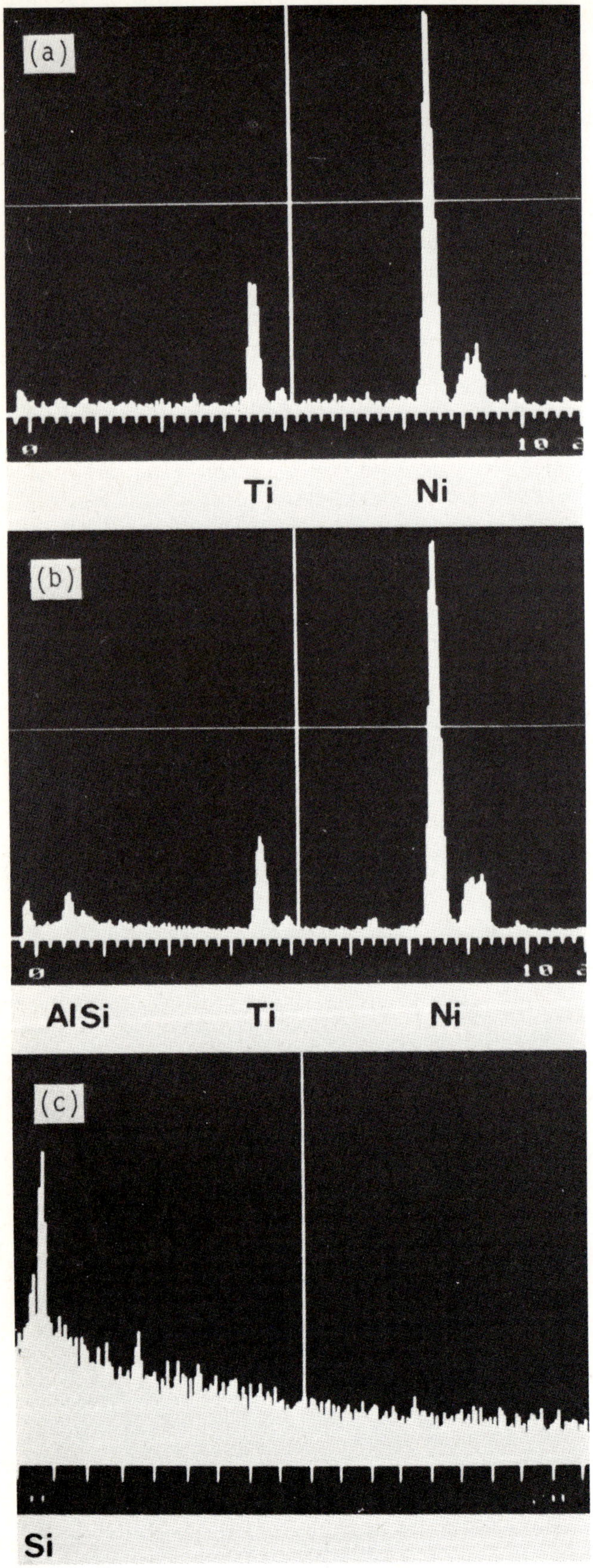

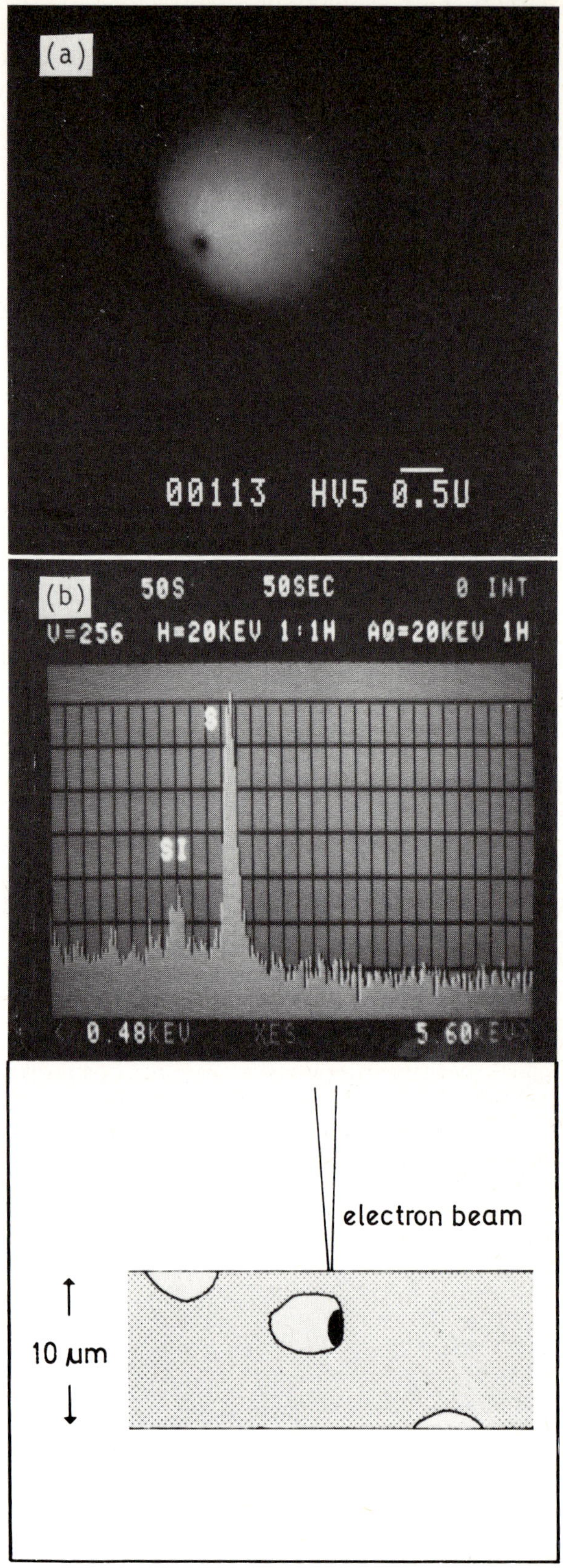

11 Analysis of γ' particles extracted from alloy 800:
(a) beam placed on cluster of precipitates from Al-free alloy
(b) beam placed on individual precipitate from alloy containing 0.3%Al
(c) beam placed on area of carbon grid remote from precipitates

12 Analysis of particle in void in polyethylene:
(a) STEM image at 200 kV; (b) analysis of particle

analysis of silicon was difficult because of its existence in the contamination peak which formed. Figure 11*a* was obtained using a broadened beam on a group of particles and no silicon was seen. If the beam is focused onto individual particles then silicon is observed (Fig. 11*b*) and if the beam is placed on the carbon grid which supports the γ' precipitates, in an area remote from them, silicon is still observed and is in the contamination peak. Thus great care is needed in the interpretation of silicon analyses.

Impurities in polymeric insulation materials

The use of polymeric insulated electric cables has been increasing over the last 15 years. However, when polyethylene is used as insulation in conditions where both moisture and electric stress are present it has been found that internal fern-like defects can grow and a large reduction in electric strength can occur. It is widely quoted that voids and particles are the growth centres.

Work on cables containing such defects[9] has enabled individual growths to be sectioned and examined by STEM. The work involves the location and analysis of voids within the foil. Thick sections (10 μm) are required because the voids (typically 1–2 μm diameter) need to be contained completely within the foil to ensure that any particles within the void are not lost. It has been found that the presence of the contamination is more important than that of voids in causing defects to grow. Figure 12 demonstrates that small particles in voids well within a thick polymer slice can be identified, for the beam broadening in such a low atomic number material is low.

DISCUSSION

During the past 8–12 months in which the two microscopes have been in commission they have been used equally in both the TEM and STEM modes. In particular, the optimizing of the EDS detector geometry combined with the 200 kV operation makes it a high performance analytical tool which, as shown above, has been applied to a wide range of problems. We normally carry out microanalysis at 200 kV with beam diameters of typically 10–15 nm and with the usual foil thicknesses of 100–200 nm can expect to be working with about 1000 cps total X-ray input count rate. Increasing kV significantly improves the spatial resolution but a decrease in probe diameter has a less significant effect on spatial resolution.[10]

No experimental work has yet been done to quantify the minimum mass fraction or minimum detectable mass in the microscope. However, indications based on hole count measurements and peak-to-background ratios point to a performance which is comparable to other systems.

In our experience the main problem of contamination is that the area of interest becomes obscured and any further analysis of an immediately adjacent area is difficult or impossible. In order to minimize contamination strict attention has to be paid to cleanliness and the setting of higher standards of operating procedures than have previously been necessary for other forms of electron microscopy. As the effects of contamination increase greatly with a smaller probe size it is better to work with the maximum probe size that the size of the features under examination will allow. This also increases the total X-ray count and therefore counting statistics, and is the normal operating regime for the microscopes.

ACKNOWLEDGMENTS

The authors would like to acknowledge those of their colleagues who have allowed use of their results, in particular Dr P. C. Rowlands, Dr M. I. Manning, Mr F. E. Asbury, Dr R. D. Naybour and Dr M. Collins. They also wish to thank the Directors of CERL and ECRC for permission to publish this paper.

REFERENCES

1 J. I. GOLDSTEIN, J. L. COSTLEY, G. W. LORIMER, and S. J. B. REED: Proc. Ann. Scanning Electron Microsc. Symp., 1977, 315
2 R. H. GEISS and D. F. KYSER: *Ultramicroscopy*, 1979, **3**, 397
3 R. HUTCHINGS, M. H. LORETTO, I. P. JONES, and R. E. SMALLMAN: *Ultramicroscopy*, 1979, **3**, 401
4 D. R. CLARKE: in 'Electron and positron spectroscopies in materials science and engineering', Academic Press, 1979, 315
5 N. J. ZALUZEC: Ninth Int. Cong. on Electron Microscopy, ed. J. M. Sturgess, publ. Microscopical Society of Canada, Toronto, 1978, **1**, 548
6 R. B. CARRUTHERS and M. J. COLLINS: these proceedings
7 P. C. ROWLANDS and M. I. MANNING: Int. Conf. High Temperature Corrosion NACE, San Diego, 2–6 March 1981
8 F. E. ASBURY: to be published, 1981
9 R. D. NAYBOUR: to be published, 1981
10 P. DOIG, D. LONSDALE, and P. E. FLEWITT: *Phil. Mag. A.*, 1980, **41**, 761

DISCUSSION of paper by Metcalfe and Owen

G. CLIFF:
Would the authors specify the manufacturer's quoted drift rate (for the Hitachi 200 kV TEM/STEM)?

E. METCALFE:
A 4 Å s^{-1} drift rate is obtained under the worst conditions, i.e. when using a holey C film in the graphite low-background holder. When a metal foil is used the drift rate appears to be about 2 Å s^{-1}. It should be noted that because of the higher X-ray yield reaching the detector in the H700H compared with the EM400, shorter counting times may be used which will reduce the effects of specimen drift.

R. A. RICKS:
Experience with an EM400T at Cambridge University has shown that a drift rate of 4 Å s^{-1} is unacceptable and measured drift rates in the above system give values of approximately 1 Å in 2 or 3 seconds. Have you any comment on this?

E. METCALFE:
The manufacturers only quote a drift rate for the high-resolution holder (0.6 Å s^{-1}) and not for the low-background holder.

Considerations in design of computer-based acquisition system for EDS and EELS

P J Statham and J M Watson

Various methods for interacting with a computer are summarized to show how the instrumentation requirements for simultaneous digital acquisition of X-ray energy-dispersive and electron energy loss spectra can be satisfied. Facilities which improve the quality of accumulated data are discussed with particular reference to a multi-spectrum acquisition system designed around a single minicomputer. The system described illustrates how experimental control can be exercised over a variety of signals from a single console, thus helping to unify the analytical capabilities of a microscope.

The authors are with Link Systems Ltd, High Wycombe, Buckinghamshire.

There are a variety of ways in which a peripheral device can request attention from a controlling computer. With special hardware, the device can access memory directly (DMA). This is the fastest method but the computer program essentially has no control over the transaction. Alternatively, a circuit can be built to allow the device to interrupt the central processing unit (CPU) and temporarily suspend the current program. After contents of the computer registers have been saved, the interrupting device may be identified and appropriate action taken before the registers are reloaded and the program is resumed. This is typically at least an order of magnitude slower than DMA but more control can be exercised when servicing the device. The simplest method is for the computer to 'poll' a device by reading in a binary status word at an input port and watching for a change in one of the bits before taking action; however, to ensure a fast response, the program cannot be doing anything else while it is waiting for the status to change. This last point is important because the prime aim in real-time computer control is to increase efficiency by making sure that the program is never idle while there are still jobs to be done. In this light, the concept of 'multi-tasking' is worth mentioning since this is a method whereby a single computer can perform several jobs apparently simultaneously by switching between tasks whenever the current task becomes temporarily idle.

Serial acquisition of electron energy loss spectra (EELS) involves counting pulses for a preset time interval, storing the result in memory and incrementing the memory address ready for the next channel of information[1-4] and simple computer control methods are adequate. However, if high count rates are to be accommodated and more than one device serviced by the same computer, more elaborate procedures are normally required. This paper considers instrumentation, control facilities and overall design philosophy for a multi-spectrum acquisition system which has been developed using the Link 860 analyser, which contains a CPU with typical minicomputer performance and 64 Kbyte memory capacity.

DATA ACQUISITION

For EDS, data rates are invariably below 50 kHz because of the limitation set by electronic noise in pulse processing. Although interrupts and even device polling could be used, DMA is preferable to leave the CPU free for other tasks. For EELS, however, data rates in excess of 1 MHz may be involved and since this is of the order of the memory cycle rate of the CPU, use of DMA could slow the computer down by at least a factor of two. By using a special interface to do the counting and reading the total accumulated count after a specified time interval, computer accesses are typically reduced to below 1 kHz and can conveniently be handled by interrupt processing. Thus, with an additional interface to provide timing information for EDS, both X-ray and EELS data can be acquired simultaneously. Coincidence counting is possible in principle and can offer advantages for some experiments[5] but when a high-resolution X-ray detector is used the relatively small collection solid angle means that most of the available EELS signal has to be sacrificed, despite an improvement in signal-to-background ratio.

Instrumentation requirements for EELS acquisition are illustrated by Fig. 1. Photomultiplier current can be digitized by an analogue-to-digital converter (ADC) or a voltage-to-frequency converter (VFC). An ADC gives a direct reading independent of the dwell time, which can be an advantage,[3] but multiple measurements or smoothing circuitry must be used to improve statistical precision and the dynamic range is limited by ADC resolution. A VFC provides a pulse train which has to be counted to obtain a number proportional to current and precision can be improved simply by extending the counting interval with no adjustment to bandwidth.[6] Both methods become unreliable at very low currents where it is preferable to count individual pulses derived from a threshold discriminator; this has the added advantage that statistical precision can be calculated directly from the accumulated data. For a Poisson distribution in time, if the mean electron count rate is n Hz, then roughly 1% of pulses arrive separated in time by less than $0.01/n$ s. Thus, to avoid losses due to pile-up, the PM tube phosphor, threshold discriminator and counting chain have to be designed for considerably

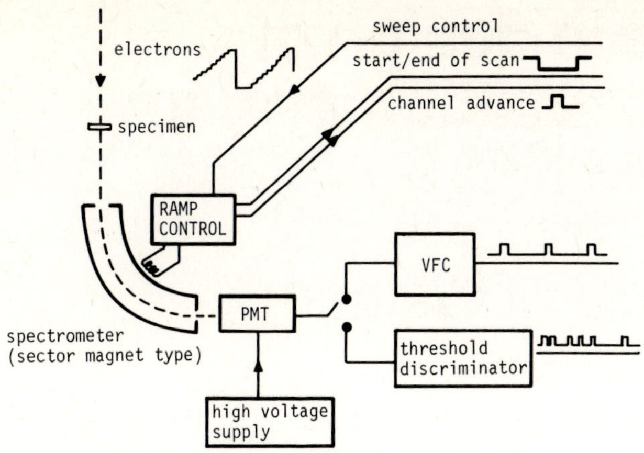

1 Control and detection requirements for serial acquisition of electron energy loss data

higher data rates than are actually recorded. By using a high-bandwidth discriminator and ECL logic, we are able to detect pulses separated by 10 ns (100 MHz) so that count rates of 1 MHz can be recorded with only 1% losses. This is nowhere near fast enough to accommodate counts in the zero-loss region at normal beam currents so a combination of pulse counting and analogue methods is desirable. Although, in principle, both systems can be run simultaneously,[6] in practice it can be difficult to adjust photomultiplier gain to avoid non-linearities due to saturation in the low-loss region while maintaining good signal-to-noise for pulse counting at higher energies. If a fast high-voltage power supply is available for the PMT, the gain can be switched automatically to avoid this problem. For EELS, counts are accumulated in a memory with 32-bit capacity; this affords a dynamic range of 10^9 which is the maximum that would normally be encountered in electron flux.[1]

By providing inputs for 'start', 'end of scan' and 'channel advance', we can acquire spectra from microscopes where the image scanning circuits are used to generate the EELS energy ramp. However, if the computer is used to control the ramp directly, several advantageous features can be implemented, as discussed below. When generating a voltage ramp with a digital-to-analogue converter (DAC), it is important to provide equal-sized voltage steps from channel to channel otherwise non-uniform sampling of the energy spectrum will produce errors, particularly in fine-structure analysis where the frequency content of the spectrum is under consideration. By using a 12-bit rather than 10-bit DAC to provide 1024 channels, channel-to-channel step variations can be kept within about 10%. Under computer control, the ramp can be directed in a direction of either increasing or decreasing energy. Thus, hysteresis can be seen as a sideways shift in the spectrum and distortion due to long-term effects of the intense zero-loss signal on scintillator and photomultiplier will be revealed.[7] The ramp can be 'frozen' at any desired position, or positioned directly on the zero-loss peak, to provide a signal for imaging. The dwell time per channel can be altered for different regions in the spectrum so that more counting time can be spent in the region of core losses where the signal is weak, rather than in the intense zero-loss region. As an extension to this facility, the sweep can be programmed so that the time (in clock pulses) taken to accumulate a preset number of counts is recorded. The statistical noise will then be constant from channel to channel so that, for example, smoothing algorithms can be more effective[8] and the reciprocal of the recorded data will correspond to the required spectrum. Where a pulse train with rate proportional to beam current is available, this can be used instead of the computer clock to cancel out drift effects. In some situations, mains interference can be cancelled out with a compensator,[7] but where this is not effective a special interface allows the computer-controlled sweep to be synchronized with the AC mains cycle; by recording with dwell times much less than 20 ms, interference that would normally cause the selected energy to oscillate and thus degrade resolution is instead spread over a wide range of energy so that energy resolution is essentially unaffected.[6]

Where multiple sweeps are being used it is possible to shift the data after each sweep before adding the data to memory.[3] The idea here is to compensate for drift by monitoring the zero loss peak, but in order to do this it is essential for the zero loss peak to be present with a well-defined maximum; saturation must not occur and the dwell time has to be chosen to ensure that sufficient counts are recorded in each sweep to avoid statistical errors in determining the zero channel.

If there is a facility to control beam position with X and Y deflection voltages, X-ray spectra can be recorded on a grid of points over the field of view and integrals from up to 18 regions of interest can be stored for every grid point.[9] The same digital mapping technique can be used for EELS provided the microscope has suitable 'de-scanning' facilities to keep the transmitted beam passing through the spectrometer as the probe is scanned across the specimen. Although this is a slow process because of the serial energy scan at each position, subsequent processing of the digitized data permits background subtraction, elemental ratioing and correction for thickness variations.[9]

EXPERIMENTAL CONTROL

Though control of a single spectrometer (EDS or EELS) is comparatively straightforward, problems arise when two or more types of data are being recorded in the same system because of differences in dynamic range and different requirements for display annotation. While recording data, it is extremely useful to be able to monitor the acquisition in progress. Whereas an oscilloscope has simple controls to change vertical and horizontal axes directly, a computer display has to be re-formatted every time a change is made and continual updating is necessary when recorded counts are changing the contents of memory. By using a multi-tasking approach, together with a priority interrupt structure and a TV display interface with limited intelligence, a system has been developed to control and monitor acquisition from multiple devices.

The conceptual layout (given in Fig. 2) centres around the video display and keypad which is used to initiate all operations. Any one of the data memories can be connected to channel A for display, adjustment of acquisition conditions such as dwell time, data processing, or output to any one of a number of peripheral devices. The contents of another memory can be compared by connecting channel B and the overlayed display can be switched on or off at will. Once the acquisition details have been defined, a single key can be pressed to start analysis in all memories that have been initialized. Similarly, all analyses currently in progress can be immediately suspended so that, if necessary, the beam can be repositioned to compensate for drift before the experiment is resumed. A second EELS channel is shown which can be used for alternative signals, such as

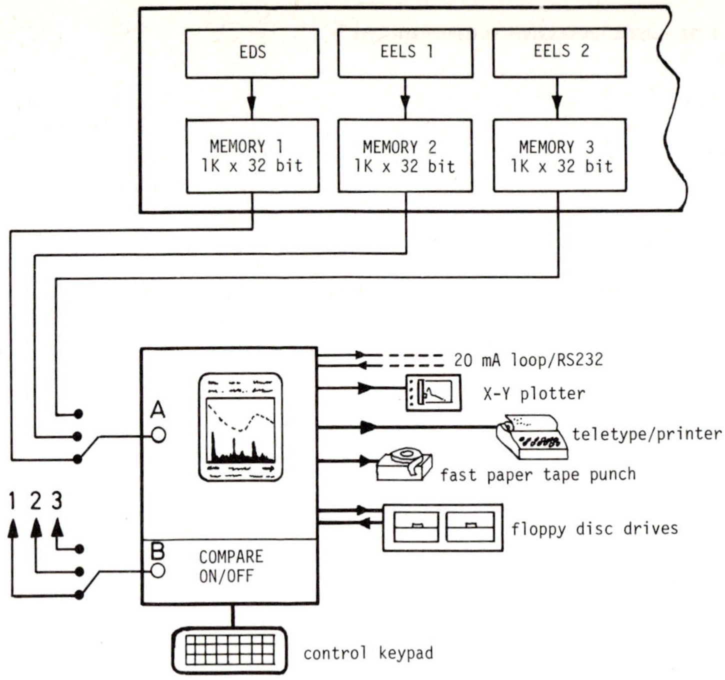

2 **Conceptual layout for multi-spectrum acquisition system based on the Link 860 computer**

from cathodoluminescence, and further memories may be added if required.

The display area can be expanded, contracted and shifted in both horizontal and vertical directions, thus offering the same flexibility and ease of use as an oscilloscope. By 'rolling' the spectrum down and then expanding the vertical scale, small structures, normally superimposed on a high background, can be revealed. A logarithmic scale permits a complete EELS spectrum to be viewed without adjusting the vertical axis and a bipolar mode can be obtained to allow negative values, as produced by differentiation of the data, to be displayed. The horizontal scale is labelled with decimal values for cursor energy and end limits. Since both zero position and energy range depend on the type of data being observed, each memory has its own calibration constants for the energy scale which can be adjusted whenever necessary. Furthermore, to accommodate differences in spectral shape and count rates, each memory retains its own vertical scaling constant for display. Thus, when transferring between X-ray and EELS displays, the horizontal axis automatically switches from keV to eV and, by making use of channel B, simultaneous acquisition of EDS and EELS spectra can be viewed on a single monitor provided the vertical scales in each memory have been suitably adjusted.

Local storage of spectra is available on floppy diskettes but it is sometimes desirable to transfer data to a different machine; since different computers rarely use the same diskette format, alternative modes for data transfer are necessary. With a serial 20 mA current loop (or RS232) link, suitable protocols can be used for communication (as, for example, have been implemented for the central IBM 360 computer at Cambridge University). Alternatively, paper tape can be punched in ASCII code suitable for direct input to a FORTRAN or BASIC program. Besides being able to copy the current display, plotting facilities include a log/log option which uses the energy calibration constants to calculate $\log E$, where E is the energy loss; such a plot reduces any curve of the form AE^{-r} to a straight line and can therefore help to reveal small edges on an otherwise curved background in EELS spectra.

With memories A and B defined by the channel connections, a series of mathematical operations (e.g. smooth A, $A \times$ constant, A–B) can be performed. More sophisticated processing such as background extrapolation and SIGMAK/SIGMAL computation[10, 11] is currently handled by a separate program which operates on diskette files, but it is planned to include such operations within the main system in the near future.

CONCLUSION

The system described illustrates how a single minicomputer can be adapted to handle a variety of signal inputs and provide the control necessary to supervise and monitor simultaneous analyses. Whereas separate, less-powerful computers could be used to perform the various tasks, the present system provides a more unified approach where analytical facilities are controlled from a single console.

REFERENCES

1 P. E. BATSON: *Ultramicroscopy*, 1978, **3**, 367
2 D. M. MAHER, P. MOCHEL, and D. C. JOY: Proc. 13th Ann. Conf. Microbeam Analysis Society, 1978, Ann Arbor, Michigan, 53A
3 R. F. EGERTON and D. KENWAY: *Ultramicroscopy*, 1979, **4**, 221
4 R. F. EGERTON: 'Scanning electron microscopy 1980', **I**, 41, 1980, SEM Inc. AMF O'Hare, Chicago, Ill.
5 D. B. WITTRY: 'Electron microscopy 1980', **3**, 14 (Proc. 7th European Congr. E.M. and 9th Int. Conf. XROM)
6 A. J. CRAVEN, A. M. MacLEOD, and R. P. FERRIER: *ibid.*, 170, 1980
7 D. C. JOY and D. M. MAHER: 'SEM 1980', **I**, 25, 1980
8 T. INOUYE: *Nucl. Instrum. Meth.*, 1971, **91**, 581
9 P. J. STATHAM and M. JONES: *Scanning*, 1980, **3**, 168
10 D. M. MAHER: ch. 9 in 'Introduction to analytical electron microscopy', Plenum Press, 1979
11 R. F. EGERTON: *Ultramicroscopy*, 1979, **4**, 169

DISCUSSION of paper by Statham and Watson

E. D. BOYES:
It is usually necessary to work with an electron visual rate image for establishing the necessary experimental conditions, selecting the probe position, etc. Is it necessary to provide dedicated hardware to be controlled by the computer for this mode?

P. J. STATHAM:
In order to generate a visual rate image, a new 'picture point' intensity value would be required every 100 ns or so to give 'broadcast quality'. Even for a coarse image, it is not practical to consider driving the TV with a simple interface or a minicomputer with instruction execution times of around 1 μs because several program instructions would be necessary to read an intensity value from memory, output to the display with deflection information and increment memory address ready for the next point. Therefore, more elaborate hardware is required which either has its own high-speed memory for holding the image or can access blocks of data in the minicomputer memory by fast DMA.

High spatial resolution microanalysis with the HB501

P E Bovey, N J Long
and I R M Wardell

The requirements for an instrument capable of high spatial resolution microanalysis are discussed and are shown to be satisfied by the HB501 scanning transmission electron microscope. Examples are given of applications which demonstrate the instrument's flexibility and the advantages of a small probe and field emission source for routine analysis.

The authors are with VG Microscopes Ltd, East Grinstead, Sussex.

GENERAL INSTRUMENTAL REQUIREMENTS

A number of quite specific instrumental requirements is implied by the term 'high spatial resolution microanalysis'. These requirements are as follows:

1 The instrument should be capable of operating in scanning mode in order to facilitate the selection of points for analysis and the acquisition of elemental line profiles and images.
2 The electron optical system must produce a probe with a diameter of less than 1 nm so that, as far as possible, the specimen and not the instrument limits the volume of material contributing to the collected data.
3 The beam current in this small probe must be sufficient to permit the acquisition of good signal-to-noise data in a reasonable time. This requirement arises out of the relatively small cross-sections for the production of X-rays and characteristic energy loss electrons.
4 The instrument should be capable of structural as well as chemical analysis at the sub-nanometer level for a more complete characterization of the specimen. This means that provision of microdiffraction as well as EDX and ELS facilities is necessary.
5 Instrumentally induced specimen contamination must be eliminated, otherwise high spatial resolution microanalysis becomes impossible. Microdiffraction patterns are rapidly destroyed by contamination growth and energy loss spectroscopy becomes quite insensitive.

The above requirements are satisfied by the HB501 scanning transmission electron microscope. The instrument is fitted with a field emission source which enables beam currents of up to 10^{-9} A to be formed into a probe 0.5 nm in diameter. The probe-forming optics allow the probe size to be varied above and below this value as required but for most analytical purposes this condition is convenient since it permits simultaneous high resolution imaging and analysis (EDX, ELS and microdiffraction). The electron optical column and vacuum system of the HB501 have been designed using ultra-high vacuum techniques and materials with the result that a pressure of 10^{-9} mbar is routinely achieved in the specimen chamber and instrumentally induced contamination is absent.

The instrument has been applied to a wide range of materials problems, some of which are now described.

PRECIPITATION IN FERRITIC STEEL

Figure 1 shows a precipitate obtained by carbon extraction replica from a ferritic steel. In order to refine the grain structure of the steel aluminium nitride had been introduced during high temperature heat treatment. When the steel was subsequently tempered molybdenum and carbon were found to deposit heteroepitaxially on aluminium-rich particles to produce the composite precipitate of Fig. 1. The HB501 was used to chemically characterize the precipitate and to confirm or otherwise the presence of nitrogen in the aluminium-rich region.

Using the same electron optical conditions as for imaging the precipitate was analysed by both EDX and ELS

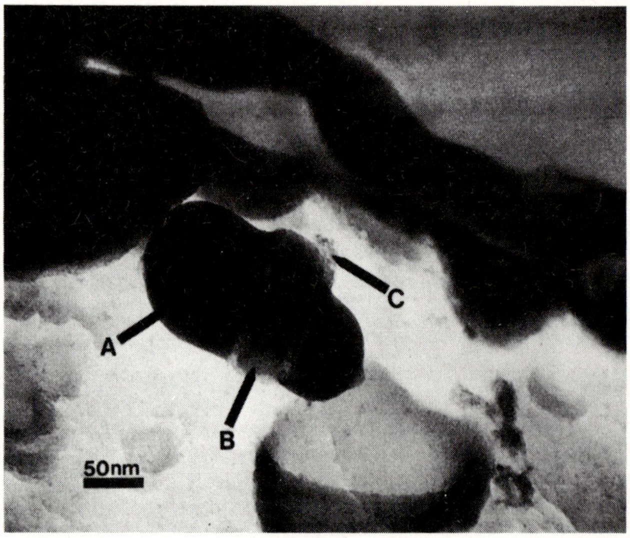

1 **Micrograph showing precipitate analysed**

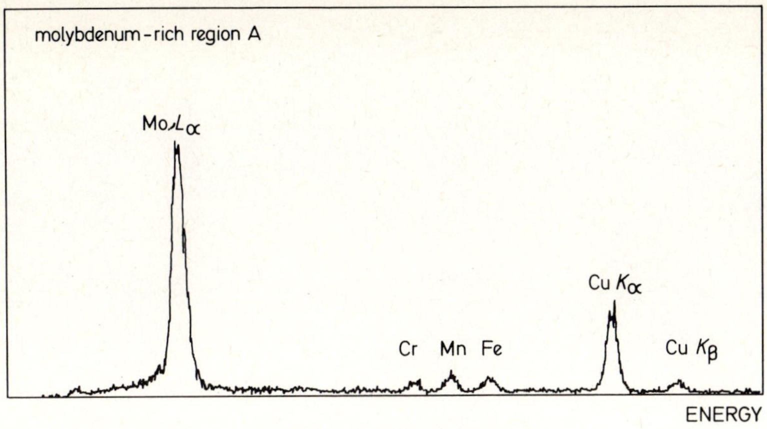

2 **X-ray spectrum from molybdenum-rich region *A***

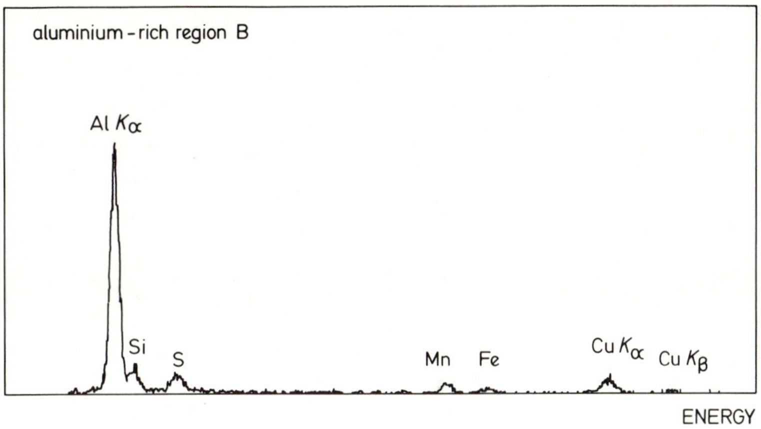

3 **X-ray spectrum from aluminium-rich region *B***

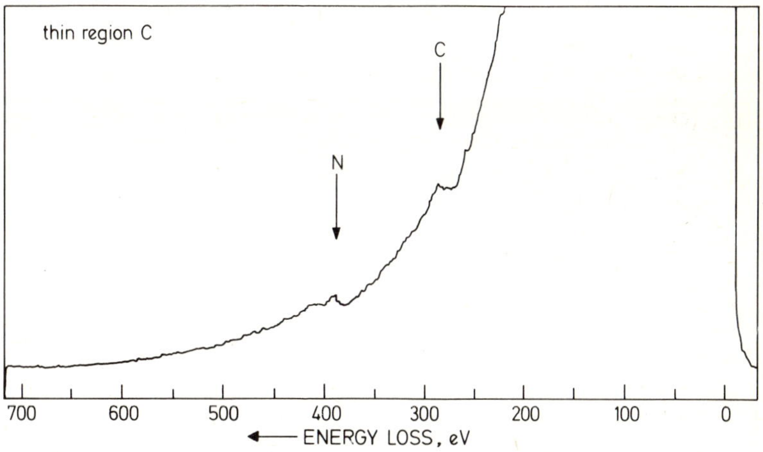

4 **Energy loss spectrum from thin region *C***

systems. Figure 2 shows that region *A* was molybdenum-rich while Fig. 3 shows that region *B* was aluminium-rich. The ELS spectrum of Fig. 4 from area *C* confirmed that nitrogen was present in the central aluminium-rich region of the precipitate. Also evident in Fig. 4 is a carbon peak which persisted when the probe was shifted off the precipitate and onto the carbon support film. This indicates that the film rather than the precipitate is the origin of the carbon signal. The measurements illustrate a number of important features of the HB501:

(1) The value of multiple analytical techniques is evident since they enable the EDX results to be complemented with light element analysis using the ELS system. This leads to a more complete characterization of the specimen.
(2) Use of the same instrumental conditions for both analysis and high resolution imaging greatly simplifies the acquisition of high quality data. The ability to image the area under analysis (whether by EDX or ELS techniques) gives increased confidence in the results, particularly when very small areas are involved.

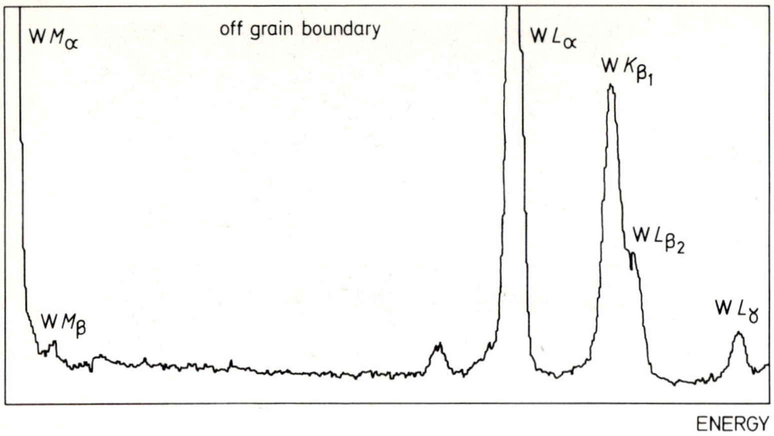

5 X-ray spectrum from region of tungsten carbide specimen away from boundary

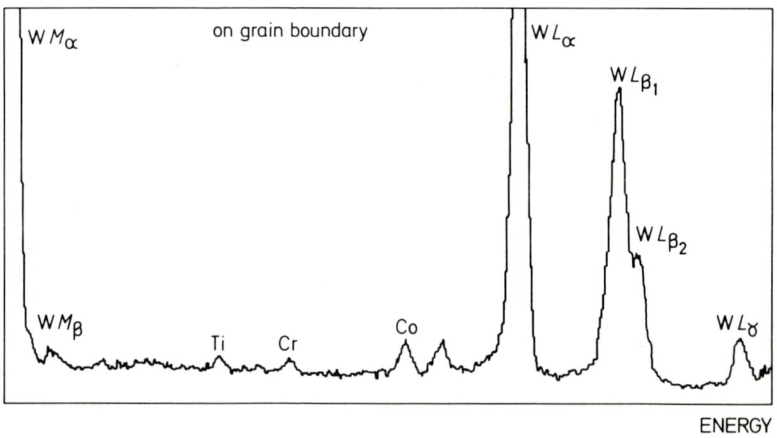

6 X-ray spectrum from grain boundary in tungsten carbide specimen

(3) In order to detect easily the nitrogen peak above background it was necessary to select an area less than 10 nm in size from the thin region *C* for ELS analysis. In the thicker parts of the specimen such as area *B* multiple scattering significantly reduced the peak-to-background ratio. This emphasizes the necessity of high spatial resolution for analysis as well as imaging even for cases where the overall particle size is large.

GRAIN BOUNDARY SEGREGATION IN TUNGSTEN CARBIDE

Grain boundary segregation is an important and widely investigated phenomenon in materials science. Because of the difficulty of preparing ultra-thin specimens of most materials, investigations are usually carried out in regions that are several tens of nanometres in thickness.

It is often argued that, as a result of electron scattering and resultant beam broadening, the value of a small incident probe diameter is questionable for this type of work. However, the fact remains that for boundaries where the segregants are highly localized the smaller the probe the greater will be the sensitivity for segregant detection. This is because, in the case of a small probe, a higher proportion of the incident beam strikes segregant atoms so that a greater fraction of the total X-ray signal comes from these atoms.

This is illustrated in Figs. 5 and 6 which show X-ray spectra from the matrix and a grain boundary respectively in a tungsten carbide specimen. The grain boundary was oriented such that it was parallel to the incident beam direction. Although it was suspected that cobalt segregated to the boundaries in this material, there was no difficulty in detecting the cobalt using the HB501. Also it was clearly seen that titanium and chromium were also present in the boundary as seen from the Ti and Cr peak heights which are significant with respect to the background count level. These elements were difficult to detect in a conventional tungsten hairpin filament TEM where the higher *bremsstrahlung* hides these weaker peaks in the background. These elements are important in this material since they are added in trace quantities, specifically to modify grain boundary migration behaviour. This result demonstrates the practical benefits of the HB501 and its field emission source in a typical materials science application requiring both high spatial resolution and a clean (i.e. low background) X-ray acquisition capability.

ELECTRONIC BAND STRUCTURES OF HIGHLY ANISOTROPIC MATERIALS

There is considerable interest in the band structure of materials which are highly anisotropic such as intercalated layer compounds and, in particular, graphite intercalates. There are two main reasons for this interest: the technological potential for materials with very anisotropic high conductivity; and the desire for a fundamental understanding of the behaviour of these materials.

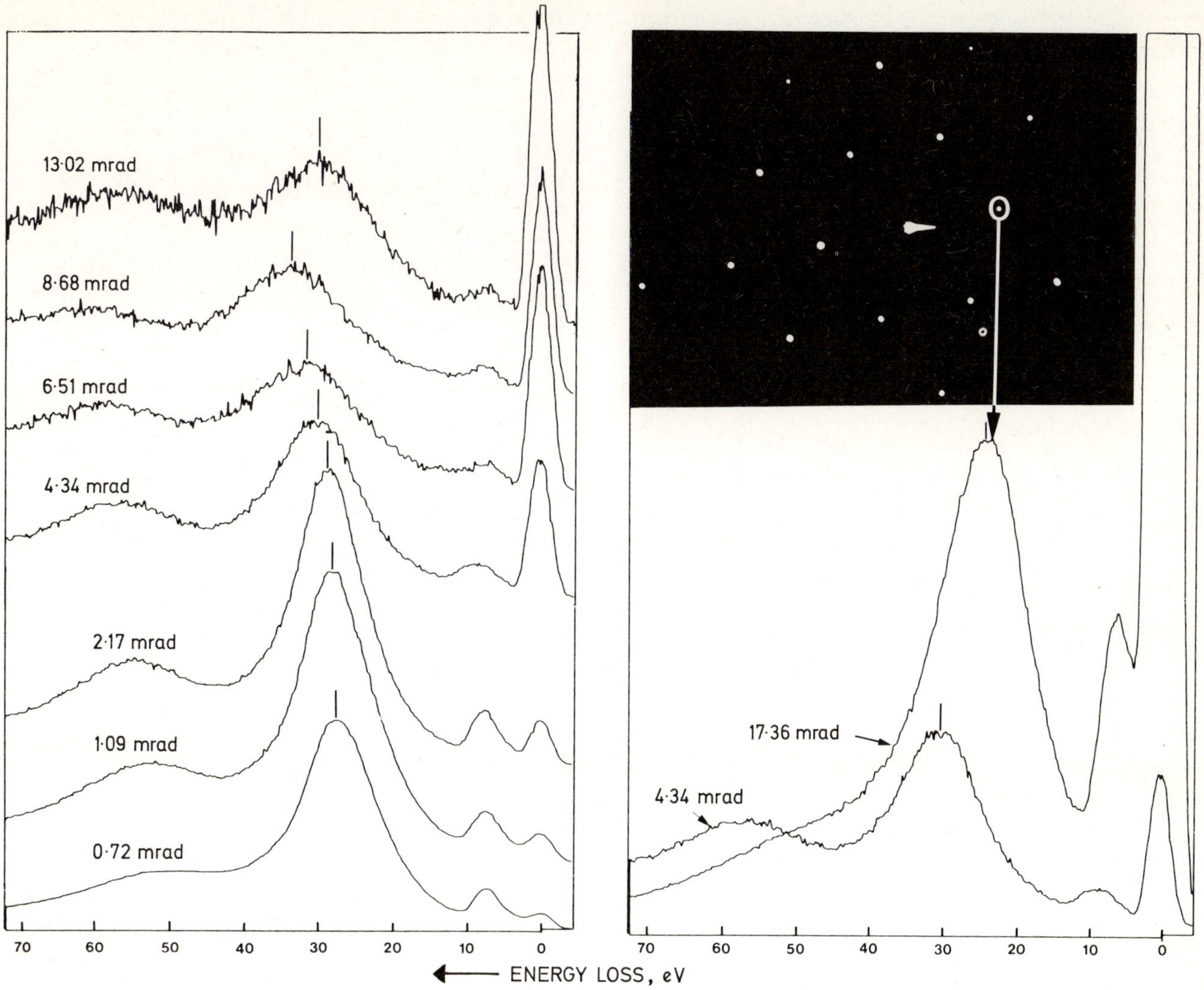

7 Energy loss spectra from single crystal graphite as function of scattering angle. (*Inset*) selected area diffraction pattern, used to orientate crystal, indicates 17.36 mrad spectrum corresponding to $10\bar{1}0$ Bragg reflection

Although there have been several theoretical calculations of the band structure there is a lack of experimental results to test quantitatively these predictions. The energy loss spectrometer on the HB501 can be used to derive conveniently and accurately the complex dielectric function and hence the interband transition energies of, for example, intercalated graphite crystals as a function of direction through the crystal. Only the low loss end of the spectrum is needed for these measurements since this corresponds to the energy range of the electronic transitions.

Figure 7 shows a set of measurements typical of those required to make such an analysis. The specimen under examination is single crystal graphite and, as the diffraction pattern indicates, the crystal is oriented such that its *C*-axis is parallel to the electron beam.

Once the crystal is oriented correctly it is a straightforward matter to record energy loss spectra from selected points in the diffraction pattern corresponding to a range of scattering angles defined by, say, θ and ϕ. In the example shown the measurements refer to a range of values of θ at constant ϕ with the maximum value of $\theta = 17.36$ mrad corresponding to the $10\bar{1}0$ Bragg reflection as indicated. It is clear that the positions and relative intensities of the peaks in the spectra are a strong function of the scattering angle.

When a high energy electron passes through a specimen the probability that it suffers an energy loss $\Delta E = \hbar\omega$ and that it is scattered through a particular angle depends upon the dielectric function of the specimen $\epsilon(\omega, \theta, \phi)$. Thus, in particular, it is possible to derive the imaginary part of ϵ as a function of scattering angle and energy loss by appropriate analysis of the spectra in Fig. 7. Peaks in Im (ϵ) correspond to critical point transitions between electronic bands and these observed values may then be quantitatively compared with the result of band structure calculation.

Important features of the HB501 illustrated by the above results are:

1 Energy loss spectra, electron micrographs and diffraction patterns can be obtained using the same detector. It is therefore easy to select and specify the scattering angle and to monitor diffraction patterns occasionally in order to check the orientation of the specimen.

2 The field emission source of the HB501 enables a small collection angle to be used whilst maintaining a satisfactory count rate. Thus the anisotropy of ϵ may be measured with a high angular resolution.

3 The field emission source also enables small areas of the sample to be analysed with a good signal-to-noise ratio. This avoids the necessity of preparing specimens containing large single crystals.

4 The true UHV environment of the HB501 means that surface contamination is not a problem.

ACKNOWLEDGMENTS
The authors are grateful to Dr J. Titchmarsh, UKAEA Harwell, for cooperation in obtaining Figs. 1–4 and to Dr M. Loretto, University of Birmingham, for his help in obtaining Figs. 5 and 6. They would also like to thank Professor S. Solin, Michigan State University, USA, for helpful discussions and assistance in obtaining Fig. 7.

REFERENCE
D. M. HWANG, M. UTLAUT, M. S. ISAACSON, and S. A. SOLIN: *Phys. Rev. Lett.*, 1979, **43**, 882

Elimination of X-ray hole count in HB5 microscope

A J Garratt-Reed

VG Microscopes has delivered recent HB5 and HB501 microscopes with a 'virtual objective aperture' with the purpose of reducing the hole count (X-ray signal generated while the electron beam is in a hole in the specimen). The early instrument at MIT has recently been so retrofitted and evaluation indicates that on this instrument, at least, the hole count is now so low as to be absolutely insignificant in any real analytical situation.

The author is in the Center for Materials Science and Engineering, Massachusetts Institute of Technology, Cambridge, Massachusetts, USA.

In any analytical electron microscope it is clearly imperative that the only ionizing radiation reaching the sample should be that in the focused electron probe. Otherwise, X-rays will be generated from regions of the specimen remote from the probe and will confuse and limit the usefulness of the data, both in terms of quantitative analysis and spatial resolution.

Spurious radiation can be generated either by the scattering of primary electrons on collision with the walls of the system, lens polepieces or apertures, or by the generation of high energy X-rays (both characteristic and *bremsstrahlung*) after these electron collisions.

The presence of such spurious radiation can be detected by passing the probe through a hole in the specimen and recording the resultant X-ray spectrum, which is consequently termed the 'hole count'. Ideally, no X-rays should be recorded. In the VG HB5 scanning transmission electron microscope at MIT, in the worst cases this hole count was about one-third of the signal detected from usable areas of specimens.[1] The effect on quantitative analysis is twofold. Firstly, the quasi-constant hole count will tend to mask changes in the composition at different points on the specimen, and, because the relative intensities of X-ray lines of different energies are not the same in the hole count as in the electron-generated signal (for at least some of the hole count is generated by X-rays with very different ionization characteristics from the electrons) there is an apparent change of composition with thickness. Secondly, the spurious radiation bombards the specimen holder (in this case made of graphite) and, while not generating detectable characteristic X-rays, it does generate a significant *bremsstrahlung* continuum, seriously degrading the signal-to-noise ratio for trace elements in the spectrum. It is possible to record the hole count and subtract it from the spectrum taken at each point to produce corrected data. Unfortunately, statistically subtraction is identical to addition, so this process degrades the signal-to-noise level still further, as the background noise is increased by the subtraction.

VG Microscopes has sold recent HB5 and HB501 instruments with a 'virtual objective aperture' designed to reduce the magnitude of the hole count. It is readily shown on the HB5 that the majority of the hole count is associated with the objective aperture. If the microscope is set into the selected area diffraction mode – the area selected being in a hole in the specimen – and the probe is directed through the centre of the objective aperture, no X-ray signal is generated. Further, if the probe is scanned over the objective aperture (as happens when a selected area diffraction pattern is displayed on the bright field CRT) with the detector set for X-ray mapping, the X-ray window including the entire specimen, the result is as shown in Fig. 1. Here the specimen was Inconel 600 (Ni–15%Cr–9%Fe), and each

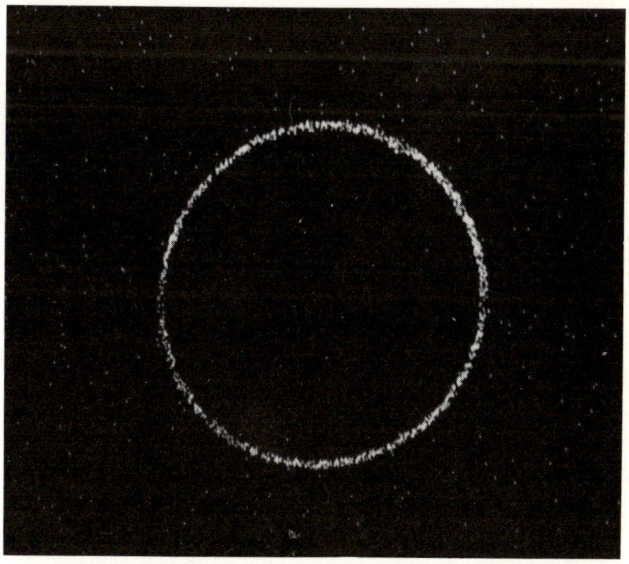

1 **X-ray map with microscope in selected area diffraction mode. Spots indicate position of beam (in front focal plane of objective lens) whenever X-ray photon from the specimen was detected**
Note that X-ray generation was particularly intense as the beam impinged on the edge of the objective aperture (the bright ring), was less intense but detectable as the beam was intercepted by the aperture disc, and was not observed at all as the beam passed through the aperture and on through a hole in the specimen. In this case the specimen was a 3 mm disc of Inconel 600, thinned by jet polishing to produce a small hole surrounded by an electron-transparent region

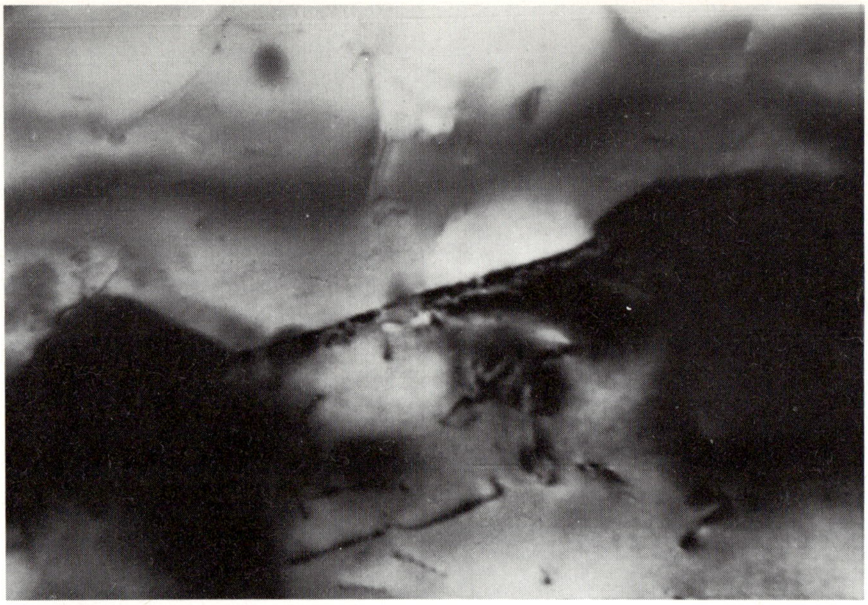

2 Micrograph of grain boundary in heat-treated Inconel 600 from which profiles of Fig. 3 were obtained (×100 000)

white spot indicates the detection of one X-ray photon. The outline of the objective aperture (100 μm diameter) is marked by a high intensity of detected X-rays. While the beam is impinging on the platinum a low intensity X-ray signal is observed, but while the probe passes through the aperture no X-rays at all are detected. It can then be concluded that most of the hole count results from X-rays generated by the beam striking the objective aperture disc and being transmitted through the platinum to the specimen, and electrons scattered at the edge of the aperture to reach the specimen at varying positions. This effect is exacerbated by the fact that the objective aperture is close to the specimen (about 2 mm).

Optically a plane conjugate to the objective aperture plane exists between the gun and the condenser lens. An appropriately sized aperture placed at this position is identical, in its effect upon the transmitted beam, to one placed at the conventional position of the objective aperture. Spurious radiation generated here, however, is readily blocked by using the field-limiting aperture. This is the rationale of the virtual objective aperture.

To test the efficacy of this modification, which has been retrofitted to the HB5 at MIT, a specimen of Inconel 600 which had been solutionized and then annealed at 700°C for 5 h was studied. In this specimen, severe depletion of chromium at the grain boundaries is observed. Figure 2 is a micrograph of the grain boundary chosen for measurement, while Fig. 3 shows the chromium concentration profile across the boundary. In this case no sensitivity correction has been applied; the data plotted are N_{Cr}/N_{Tot} where N_{Cr} is the number of counts in chromium K_α peak, and N_{Tot} is the total number of counts in the chromium, nickel, and iron K_α (in both cases measured after subtraction of the continuum background). The data in Fig. 3 were measured with both the conventional objective aperture (open circles) and the virtual aperture (closed points). As can be seen, the results differ significantly. The data taken with the real objective aperture were corrected by subtracting the hole count. The profile so generated was then identical (within the statistical fluctuations of the numbers of counts measured) with that measured with the virtual objective aperture, and is not plotted in the figure. The amount of work involved, of course, was significantly more than twice that involved in generating the profile using the virtual objective aperture, An attempt to make a similar correction to data taken from the ferritic region of a pearlitic steel, where the chromium concentration was <1 at.-%, failed because of the poor signal-to-noise ratio.

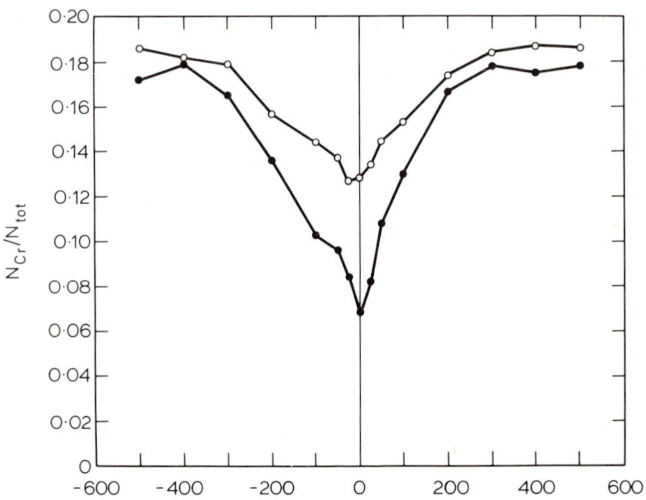

3 Profiles of chromium X-ray counts as proportion of total X-ray signal as function of position across grain boundary shown in Fig. 2; measurements were taken with both real objective aperture (○) and virtual aperture (●)

Measurements in this material were only possible after the virtual aperture was fitted.[2]

The electron probe (current $\simeq$1.5 nA) was passed through a hole in the specimen, with the virtual objective aperture in use. The X-ray count rate was less than one per second in the range 0–20 keV. With the probe incident on a thin carbon film the X-ray signal was several tens of counts per second in the same range.

It is thus concluded that in the original HB5 design the hole count was generated at the objective aperture and was objectionably intense. The hole count may be reduced by the use of the virtual objective aperture so that it is at least an order of magnitude below the signal generated by the interaction of the electron beam with a thin carbon film. Thus in any analytical situation it will be negligible compared with effects generated within the specimen itself and can safely be ignored.

ACKNOWLEDGMENTS

The author wishes to thank Dr E. L. Hall and Professor J. B. Vander Sande for helpful discussion; Professor G. R. Piercy and Dr B. Robertson of McMaster University for making available facilities to perform some of the experiments reported here; The Kearns Group (now V. G. Instruments Inc.) for travel facilities; and the National Science Foundation for financial support under block grant no. DMR-78-24185.

REFERENCES

1 G. S. WAS: ScD thesis, Massachusetts Institute of Technology, Cambridge, Mass., June 1980
2 G. D. W. SMITH, A. J. GARRATT-REED, and J. B. VANDER SANDE: these proceedings

COMMENT on paper by A. J. Garratt-Reed ('Hole count in HB5')

L. M. BROWN:

May I make a comment on the history of the virtual objective aperture? My recollection is that it was conceived and designed in outline by Dr A. J. Craven in the autumn of 1977, a few months before he went from our lab to Glasgow. He suggested that our HB5 should be modified to receive the aperture when it received its X-ray detector. We too vouch for the effectiveness of the virtual objective aperture!

Microanalysis with an FEG STEM

E D Boyes, T G E Davey-Winter, P J Harris, and R T Kerr

The Oxford FEG (field emission gun) STEM has a number of design features which are advantageous for microanalysis, such as a 35° take-off angle, the use of a gun/condenser aperture in a virtual objective mode, and of a digital scan generator; a completely clean ion-pumped vacuum system is used to minimize contamination and thermal field emission is being investigated to improve the long-term stability necessary for quantitative analyses, with rapid video fluctuations being compensated electronically. The authors use analyses of particle and of segregation to discuss the advantages of the higher current density probes and to indicate possible directions for further improvements. In particular they explore the potential usefulness of in-situ analyses of monolayer segregation at grain boundaries which can be imaged in their metallurgical context.

The authors are in the Department of Metallurgy and Science of Materials at the University of Oxford.

The Oxford STEM is an inverted column microscope with a combination of mechanical and electrical alignments for the triode field emission gun (FEG), two magnetic lenses with separate stigmators, a 5 Å image point resolution at 80 kV and facilities for X-ray microanalysis and electron micro-diffraction. The microscope is generally similar to other instruments of this type but appreciably shorter in length, especially in the critical gun/interface/condenser lens area where there is an interchangeable and alignable aperture which can be used in a virtual objective mode.[1] The microscope has been designed and largely constructed in Oxford.

FIELD EMISSION GUN (F E G) IN MICROANALYSIS

There are advantages in the superior quality of the scanning images produced with an FEG. Not only are these required for small ($\leqslant$5 nm) particles but they also permit more precise orientation of planar features such as some grain boundary segments, in order to optimize the resolution, and hence sensitivity, in microanalysis.

Although there are somewhat conflicting reports[2,3] on beam spreading within the sample there would appear to be combinations of small specimen thickness and/or low atomic number (Z) materials where the probe size with a conventional gun may well be the limiting factor. Illustrated in Fig. 1 are some examples for aluminium and copper using a revised analytical approach[4] which more closely approximates the full Monte Carlo calculations and experimental results. From these results it will be seen that the thickness of copper has to approach 1000 Å before the beam spreading at 80 kV is similar to the CTEM probe size for microanalysis of 100–200 Å and that for aluminium this critical thickness would be doubled. Foils of these thicknesses would be adequate for analysis of heterogeneous effects as well as homogeneous precipitation near peak hardness in many materials, and the smaller probes have advantages for these applications.

There would be considerable advantages in being able to detect and analyse monolayer type coverage at grain boundaries in their metallurgical context, as has been achieved for segregated zones.[3] In some cases the segregation may be an independent embrittling agent suitable for Auger analysis of fracture surfaces, but in others it may only lead to an increased stress corrosion or similar susceptibility which is only accessible *in-situ*. A very simple analysis of the geometry involved suggests that a minimum effective probe size will be required for this application. The sensitivity (S) is defined as (mass of sample)/(matrix included under the probe plus extraneous X-ray source equivalents, if any). For a very thin sample the idealized analysed volume will be given by $0.25\,\pi d^2 t$ where d is the probe diameter (plus allowance for beam spreading) and t the specimen thickness. It can be shown that for a segregated zone parallel to the beam and width w for $w < d$ the sensitivity is given by:

$$4\rho_1 w d t / \rho_2 \pi d^2 t$$

or

$$S \propto 1/d$$

where ρ_1 and ρ_2 are the respective densities.

The limits of sensitivity will have to be carefully established experimentally but in principle it should be possible to approach single monolayer coverage with a suitable FEG system.

At 80–100 keV rather thin specimens would be required but it can be seen from Fig. 1 that a higher voltage would better exploit the advantages of the FEG in this application.

X-RAY DETECTION AND TAKE-OFF ANGLE

A positive take-off angle is important for the following purposes:

(a) to maximize the collected ratio of characteristic to *bremsstrahlung* X-rays (P/B);

(b) to reduce absorption of lower energy X-rays in the specimen and fluorescence of it and the specimen holder; and

(c) to increase the choice of specimen tilts that may be used for efficient collection of X-rays in order to

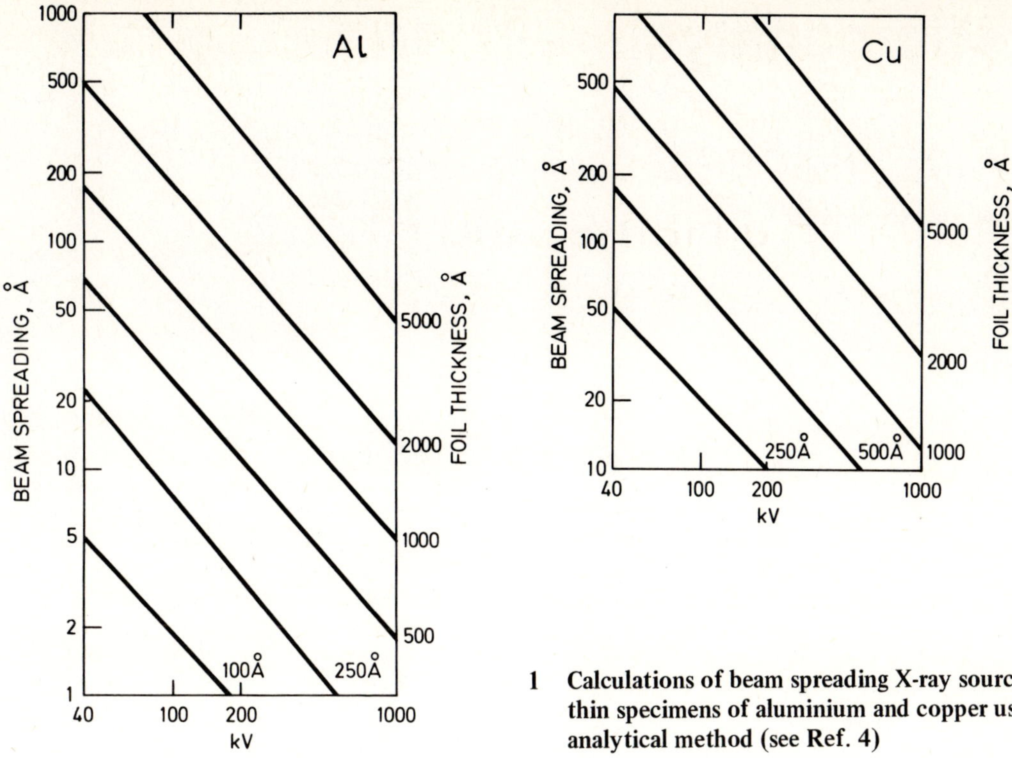

1 Calculations of beam spreading X-ray source size in thin specimens of aluminium and copper using modified analytical method (see Ref. 4)

(i) set-up image diffraction contrast conditions;
(ii) minimize the projected width of features to be microanalysed, especially recrystallized grain boundaries with planar segments; and
(iii) permit microanalysis at or close to zero tilt which minimizes the electron beam path length and hence beam spreading effects in the specimen.

The Oxford STEM operates with a high take-off angle (γ) of 35°. This is a compromise in lens design in which image performance is considered to be less important. Zaluzec[5] suggests that there is little improvement in P/B ratio at angles above 45° (135° in his notation) and high energy backscattered electrons may be more of a problem at high angles. A conventional top-entry ±45° double tilting goniometer cartridge with a light alloy nose cap is used and X-rays can be collected without tilting the specimen and at up to at least 5° of reverse tilt but there are some systematic influences of geometry and of incident beam vector. The adsorption by contamination is tackled with (a) an entirely clean pumped vacuum system and (b) a new clean air lock in which the specimen and holder may be pretreated under high vacuum conditions prior to loading into the stage.

ELIMINATION OF F E G HOLE-COUNT

None of the original FEG STEMs was outstanding for X-ray microanalysis and the advantages, particularly of sensitivity, to be expected from using the high current density of a field emission gun (FEG) therefore remained to be fully exploited. The original Oxford instrument had an excellent sensitivity of $\sim 1 \times 10^{-18}$ g for microanalysis of small particles widely dispersed on a carbon film and particularly so when a beryllium grid was used. However, the performance was much inferior if the particle analysed was embedded in a metal foil the outer regions of which were thick in terms of the electron penetration distance. In other words there was a high hole count.

Under unfavourable circumstances the hole count could be unacceptably high, due mainly to the need to use the objective aperture B in Fig. 2 to control the probe convergence and hence the C_s-limited probe size. This was the only separately alignable beam aperture essentially because of the difficult access to the liner tube construction and to that of the complex commercial interface between the gun and the rest of the column. Since this aperture was immediately adjacent to the specimen – about 4 mm in front of it – any forward scattering of electrons or hard X-rays interacted with the thicker specimen regions and even the light alloy holder to produce background counts. These could be considerable since this aperture intersected a large total current and the signal from very thin specimen regions was therefore degraded. The information could not be retrieved readily by spectral processing due amongst other things to the very limited throughput capability (~4500 c/s) of the EDS pulse processing instrumentation. Other instruments would appear to have some of the same characteristics.[6]

The lower part of the column has been replaced by a better aligned, screened and mechanical arrangement incorporating a custom-built UHV adjustable and interchangeable aperture mechanism at the FEG exit (A, Figs. 2 and 3). It can also be electrically floated to monitor the probe current continuously. This aperture is now used to set a C_s-limited probe of some 20-30 Å diameter and in all respects to act as a virtual objective aperture of 50-100 μm equivalent size at B but now some 250 mm away from the specimen. Of course the actual aperture size depends on the demagnification used. The narrow lower bore (4 mm dia.) of the objective lens shields the outer regions of the specimen-holder and further controls

2 Column of Oxford FEG STEM showing positions of pre-specimen apertures. From table upwards these are: gun-condenser or virtual objective aperture (*A*); selected area aperture (*D*); regular objective aperture (*B*); and collector aperture (*C*) incorporated in post-specimen airlock

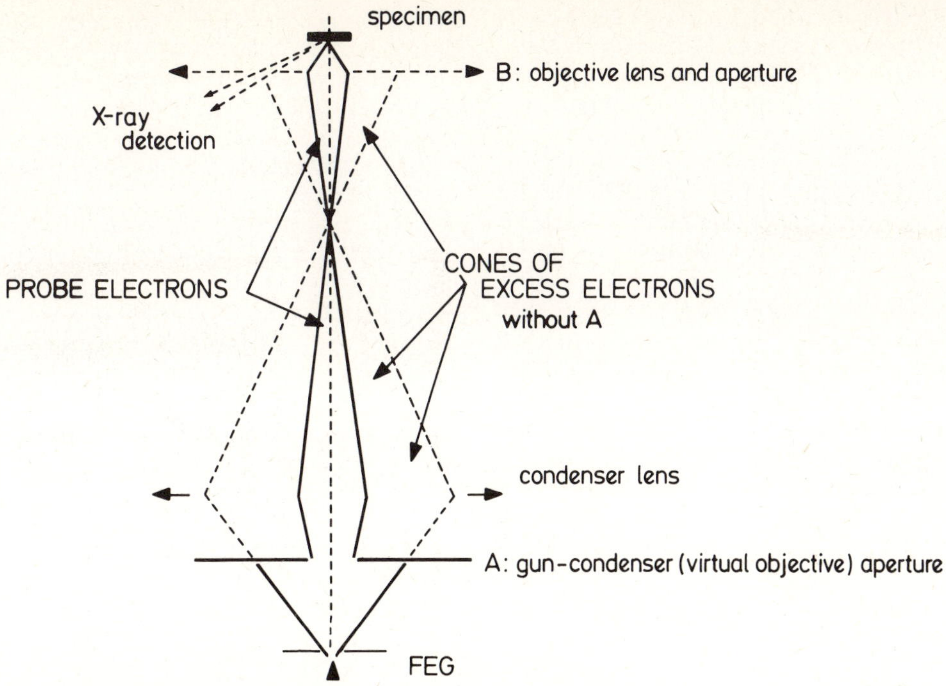

3 Schematic diagram of role of gun-condenser or virtual objective aperture mode

extraneous electrons and column X-rays. The results are outstandingly good with the hole count being completely eliminated. The improvement under equivalent conditions is at least a factor of ×300. Some apparently random background of 0–6 counts per second remain due to random events[16] independent of whether the detector is connected to the microscope. This FEG STEM and similarly modified commercial instruments have now been converted from a previous hole-count situation significantly worse than the conventional TEM/STEM instruments to a much better performance. A conventional objective aperture is used to further reduce the probe aberrations for imaging purposes, to obtain diffraction contrast and to reduce the size of the beams in the microdiffraction patterns.

MULTIPLE POINT ANALYSES

X-ray maps and even line scans of analogue intensities are of somewhat limited utility in the high resolution microanalysis of thin foils. Firstly the statistics may be poor since the total counts are low compared to an EPMA WDS system and secondly there may be non-uniformly thick regions of thin foil specimens. If the available time and counts are concentrated into a limited number of points in a defined array the precision of the chemical analyses can be improved at the expense of spatial resolution. This may be preferable for examining potential sources of stress corrosion susceptibility related to the local chemical compositions. For example in Stawström's model it would be most important to know whether the chromium concentration dips below 12% in a high-Cr alloy.

In addition there may be thickness variations, perhaps associated with chemical composition, which would invalidate intensity profiling even when background stripping was performed. A possible solution is an array of point analyses which can be processed to provide relative compositions in different regions of the specimen. They should be essentially independent of thickness, crystallographic and excitation variations. Ratio line scanning also has some of these advantages.[8]

A preliminary hardware implementation of this concept is achieved with a programmed digital scan generator. A series of points can be designated along a line (Fig. 4) across the visual rate electron images and the analysis probe is subsequently switched along the sequence of positions with a spectrum being recorded at each point. The line of points may be rotated with the usual analogue scan rotation function, for example to align it normal to a grain boundary, since to achieve this digitally is complex.[9] A developed version will be computer coupled.

PARALLEL MICRODIFFRACTION

Obtaining microdiffraction patterns by scanning reciprocal space has advantages of amplification, CRT display and ready quantitation.[10] However, where high pattern resolution is required, very few of the available electrons contri-

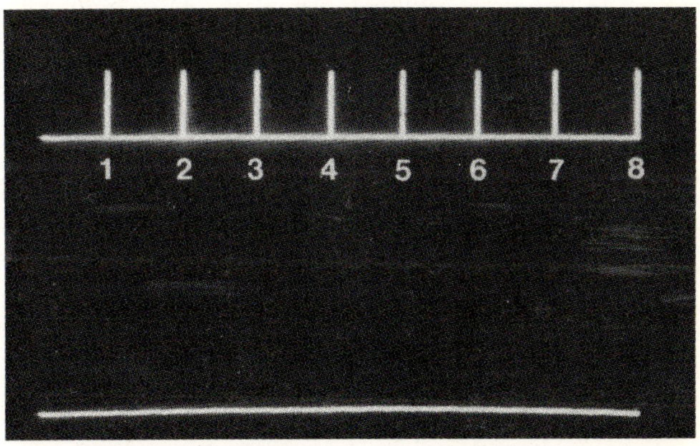

4 A digital scan generator is used to define a sequence of multiple analysis points used for profiling segregation more quantitatively

bute to the final results and as a consequence there can be significant S/N limitations.

To scan the present ±100 milliradian range of microdiffraction pattern with an acceptable 1 mrad resolution would require sampling the area in increments of only 0.01% with a similar detection efficiency for the primary electrons from the specimen. A scintillator/PMT detection efficiency of 20% would reduce the detected signal to 0.002%. A scan time of 50 s with a typical $\sim 1 \times 10^{-10}$ A high resolution probe produces only ~100 electrons ($\sqrt{N} = 10\%$) per sampled element even if all the probe electrons contribute uniformly to the pattern, which is usually an optimistic assumption. The numbers explain the poor quality scanned microdiffraction patterns often obtained and in many cases the parallel detection system using photo film and a TEM/STEM with a conventional gun has considerable advantages of sensitivity and pattern quality, but not of spatial resolution. The clean semi-UHV ion pumped vacuum system in the specimen chamber of our STEM ($\sim 5 \times 10^{-9}$ torr) makes it difficult to introduce photo film for recording and a displayed pattern is also required, for example, for setting-up dark field conditions. An internal phosphor screen display with an external camera for recording is therefore used (Fig. 2). Another advantage of this arrangement, with a hole through the assembly for the bright field or other axial beam, is the continuous availability of the instantaneous microdiffraction concurrently with the image.[11] The sensitivity of the system is further improved by current experiments with a low light level (SIT) TV camera (LLTV). A primary detected quantum efficiency (DQE) for electrons of at least 0.1 (10%) is expected. The improvement in quality is so large as to be important qualitatively as well as quantitatively. The system also provides the possibility of alternative ways of obtaining the selected area diffraction patterns without the area increase associated with the pre-specimen scanning previously required.

The possibility of very short exposure of 0.04–1 s with this system is particularly important with agglomerations of small particles typical of catalysts; where the specimen is subject to contamination; or where there may be problems with instabilities due to charging or heating under the beam. An accurate probe positioning system is provided by the digital scan generator (Fig. 4) in conjunction with the relatively high image resolution of ~5 Å. Recording is not delayed by the need to lift screens etc. with increased specimen exposure. A digital video frame-store system is available for on-line or subsequent processing to enhance the quality, reduce noise and provide quantification and geometry correction facilities on slow scan and TV rate images and recordings.[12]

CONCLUSIONS

The requirements of a microscope for microanalysis[1,13] are rather different and in some ways more critical than for the highest resolution.[14,15] In general the column geometry is more important and the electronics should be capable of considerable flexibility. It is suggested that optimum results at 80–100 kV are finally beginning to be approached.

ACKNOWLEDGMENT

The Science Research Council is thanked for supporting this project under GR/A 60718.

REFERENCES

1 E. D. BOYES *et al.*: EMAG 79 (1979), Inst. Physics Conf. Series 1980, **52**, 433
2 J. I. GOLDSTEIN, J. L. COSTLEY, G. W. LORIMER, and S. J. B. REED: SEM 77, 1977, 315
3 J. B. VANDER SANDE, T. F. KELLY, and D. IMESON: Proc. 28th Ann. EMSA Meeting 1980, 356
4 R. T. KERR, J. M. TITCHMARSH, and E. D. BOYES: to be published in Proc. Microbeam Analysis Soc. (MAS), 1981
5 N. J. ZALUZEC: Proc. 9th Intl. EM Congr., 1978, **1**, 548
6 H. L. FRASER: *ibid.*, **1**, 552
7 C. STAWSTRÖM and M. HILLERT: *J. Iron Steel Inst.*, 1969, **207**, 77
8 D. R. CLARKE: Proc. 37th Ann. EMSA Meeting 1979, 486
9 A. JONES: SEM 78, IITRI 1978
10 C. W. B. GRIGSON: *Rev. Sci. Intr.*, 1965, **36**, 1587
11 J. M. COWLEY and J. C. H. SPENCE: *Ultramicroscopy*, 1979, **3**, 433
12 E. D. BOYES, P. R. KELLAR, M. J. GORINGE, and G. DIXON-BROWN: Proc. 38th Ann. Meeting EMSA 1980, 226
13 J. A. VENABLES and A. JANSSEN: *Ultramicroscopy*, 1980, **5**, 297
14 A. V. CREWE and J. WALL: *J. Molec. Biol.*, 1970, **48**, 375
15 E. D. BOYES *et al.*: EMAG 79 (1979), Inst. Physics Conf. Series 1980, **52**, 445
16 *This residual 'hole count' was shown to be independent of the beam intensity down to zero current. X-rays generated in the gun, at 2×10^{-11} torr pressure, may have been the origin but alternative explanations are possible. Care is required to ensure that very low count rates are faithfully recorded by the electronics.* (P. STATHAM: Proc. NBS Workshop on ED X-ray Spectrometry held at NBS, 1979.)

COMMENT on paper by Boyes *et al.*

D. C. JOY:

You might like to know that we are successfully observing and recording diffraction patterns from our HB501 using a GE CCD-array TV camera focused via a telephoto lens onto a rare-earth YAG phosphor. Because the new YAG phosphors are so bright it is not necessary to use a low-light TV system, and we obtain images of excellent S/N at normal beam currents. The TV images are strobed into an APPLE computer which over a period of 2 seconds stores and displays a 256 × 256 × 8 bit image.

Digital X-ray mapping from thin foils

P D Hunneyball, M H Jacobs, and T J Law

The authors describe a new technique for acquiring quantitative compositional area maps from thin foils. Conventional X-ray distribution maps are of limited value since the intensity is dependent not only on composition levels but also foil thickness and orientation. The digital contour mapping technique, used in conjunction with both energy-dispersive and wavelength-dispersive X-ray spectrometers, has evolved from the interfacing of Elscint's programmable processing system and element distribution processor to EMMA-3. The theoretical basis for the method is described and the sequence of data processing is explained. Digital contour images composed of a matrix of 128 × 126 picture cells (pixels), each of which can accumulate up to 256 X-ray counts, are illustrated. The number of counts in each pixel is displayed as a colour, or a grey level, within a range of 16 equal levels. Examples described are Cu–75Au, Al–18Zn and Fe–17Cr.

The authors are in the TI Research Laboratories at Hinxton Hall, Saffron Walden, Essex.

This paper is concerned with the technique and data processing required to acquire quantitative compositional area maps from thin foils. For the first time it has been possible to convert raw X-ray data from pulse-recorded area scans into quantitative contour maps. This was achieved by adding an electron probe scanning system to EMMA and processing the X-ray pulses from a thin specimen with an Elscint Proxan programmable data processing system. Both energy-dispersive spectrometry (EDS) and wavelength-dispersive spectrometry (WDS) have been investigated. The technique has evolved from developments in electron probe microanalysis of bulk samples in the authors' laboratory of which a very brief history is relevant.

Conventional 'dot mapping', in which single X-ray pulses from the surface of a bulk sample are recorded as bright spots during an area scan, is a popular method of building up an element distribution map, sometimes via the superposition of several frames. Dot mapping is at best semi-quantitative, but it is nonetheless frequently useful. Quantitative contour mapping recently was made possible[1] by interfacing the electron probe microanalyser to Elscint's Element Distribution Processor (EDP) coupled with the Proxan data handling and floppy-disc recording systems. This enables the acquisition of digital images consisting of a matrix of 128 × 126 (=16 128) picture cells ('pixels') each of which can accumulate up to a maximum of 256 X-ray counts. The final number of counts in each pixel is displayed as a colour within a range of 16 equal colour levels which span the visible spectrum from blue to red, or, alternatively, as 16 graded grey levels (as shown in the illustrations). The raw X-ray data are easily and quickly corrected by the system's Nova 3 minicomputer to yield calibrated digital contour maps. Images acquired via WDS from bulk samples are superior vis-à-vis EDS because of the much lower background count.[1]

Conventional dot maps may also be obtained from thin specimens in a transmission electron microscope equipped with electron probe scanning and EDS or WDS. Such maps are useful for discriminating between the distributions of particle types on an extraction replica. They are of much less value, and may be misleading, when applied to thin foils because of the sensitive dependences of X-ray intensity on specimen thickness and crystal orientation as well as composition level. The thickness problem is illustrated in Fig. 1 which shows a wedge-shaped foil on Al–18Zn which contained a grain boundary. The sample had been solution-treated at 793 K for 1 h followed by 453 K for 18 h to promote precipitation of zinc particles at grain boundaries (not seen in this field of view) with associated depletions in zinc levels either side of the boundary. Figure 1*a* shows the TEM image, with the wedge increasing in thickness along the boundary from *A* to *B*. The corresponding dot map (1*b*) reflects the thickness gradient but gives no evidence of zinc depletion adjacent to the grain boundary. Figure 1*c* shows the zinc contour map generated by the method described in this paper; darker grey levels correspond to lower zinc levels, and it will be seen that the influence of the wedge has been eliminated and the zinc depletion unscrambled via appropriate data processing.

INSTRUMENTATION

EMMA was interfaced to the Elscint digital acquisition and processing equipment as shown schematically in Fig. 2. The slow scan unit was used to drive EMMA's probe via the beam shift coils in synchronism with the EDP, line time was 0.04 s and frame time was 10 s. EMMA was operated at 100 kV, and typical probe currents were 5 nA for EDS and 50 nA for WDS. The Kevex silicon detector was of 80 mm^2 active area and of high sensitivity (some 50 times greater than WDS at same probe current). Image accumulation times were ~200 s (20 frames) for EDS and ~1000 s (100 frames) for WDS. The WDS interface unit (Fig. 3) was designed to enable simultaneous acquisition and storage of X-ray pulses from EMMA's two crystal spectrometers. This was accomplished via appropriate comparator and delay

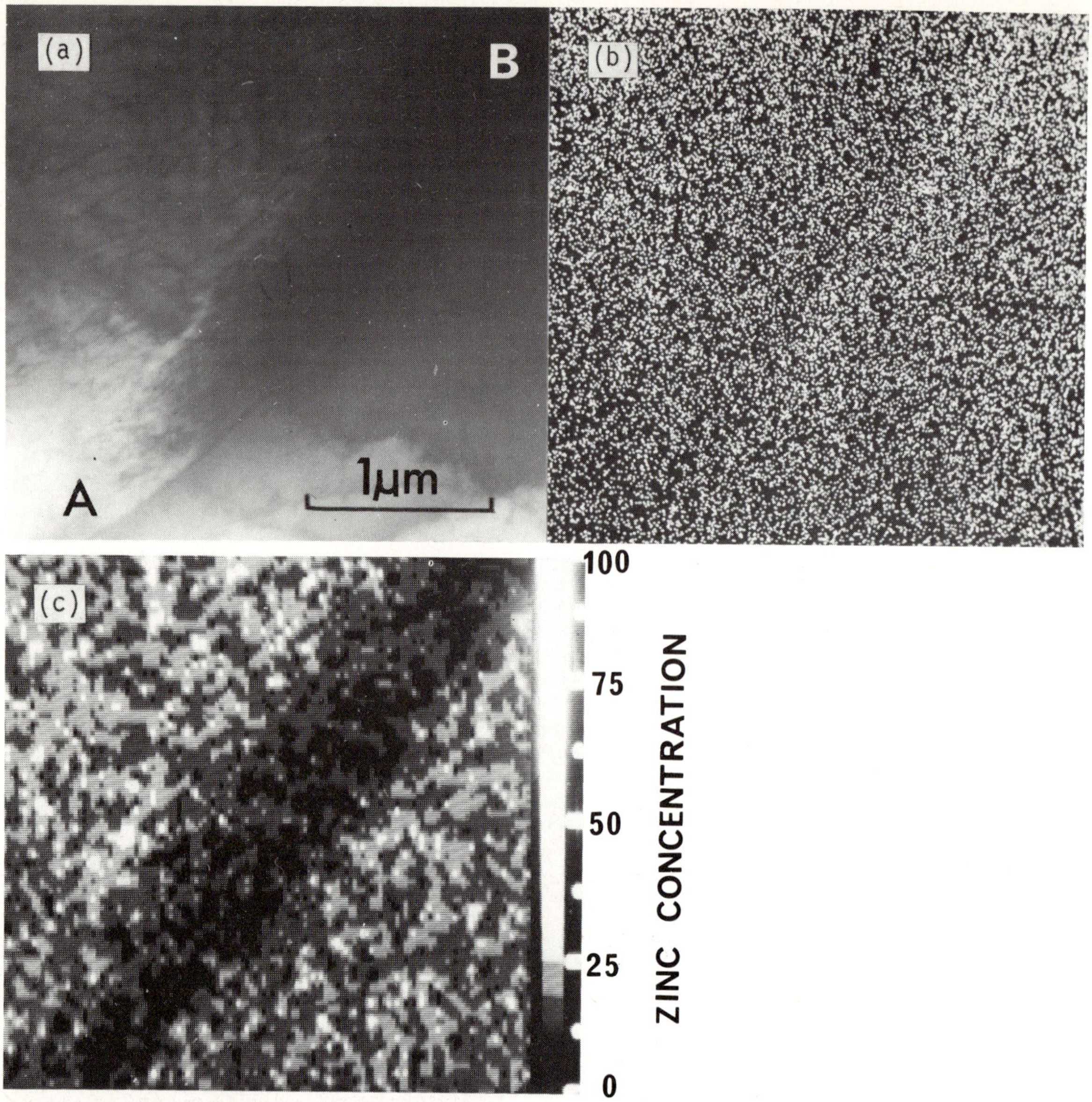

1 **Wedge-shaped thin foil of heat-treated Al–18Zn: (a) TEM image showing wedge increasing in thickness along grain boundary from *A* to *B*; (b) conventional dot map employing Zn K_α X-rays; (c) digital contour map showing Zn distribution (note elimination of influence of wedge and exposure of zinc depletion adjacent to boundary)**

circuits such that correct pulse discrimination and channel allocation was achieved.

For EDS analysis of binary alloys the 128×126 pixel matrix was divided into four quarters (see Fig. 4) with simultaneous recording in two quarters of two separate peak counts, while their corresponding backgrounds were recorded in the other two quarters. Figure 4 shows chromium and titanium peak and background distributions from extracted carbides – an application to which we will return later. For WDS analysis the two peak distributions were recorded first, and then the crystal spectrometers were detuned to record the background distributions. Once acquired, the data recorded in each quarter were transferred for permanent record to a floppy disc and for subsequent processing.

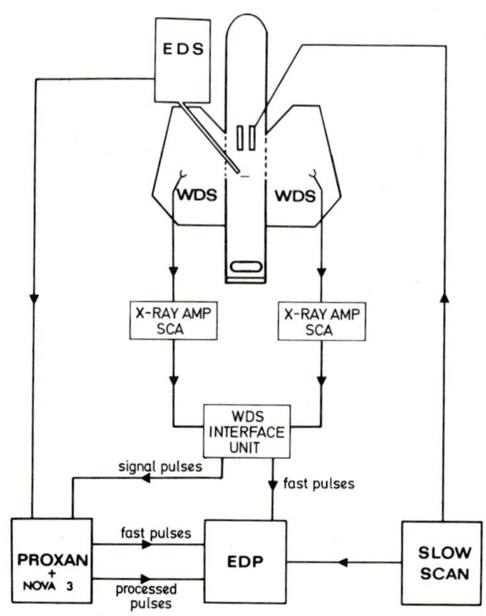

2 **Block diagram showing interfacing between EMMA and Elscint's Proxan and Element Distribution Processor**

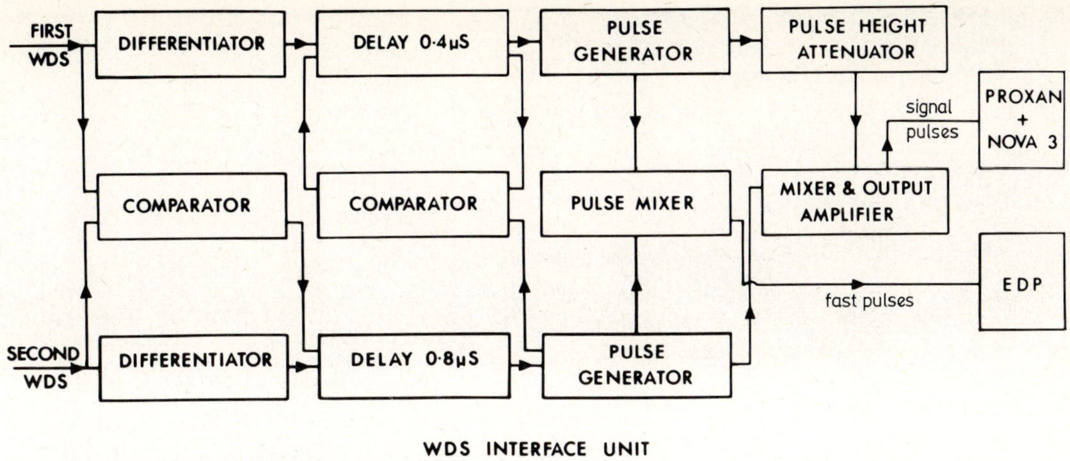

3 Block diagram of WDS interface unit for simultaneous acquisition and storage of X-ray pulses from EMMA's two crystal spectrometers

THEORY

The following theory is developed for a binary alloy sample containing elements *a* and *b* of compositions C_a and C_b. Binary alloys are the only materials studied experimentally so far, although the theory and practice is easily expanded to multicomponent systems.

Consider the four quarters of pixel arrays illustrated in Fig. 4, each of which contains 64 x 63 picture elements. Let *l* and *m* designate the co-ordinates of an identical pixel in each quarter. $P_a^{l,m}$ and $P_b^{l,m}$ are peak counts accumulated in two of the quarters, while $B_a^{l,m}$ and $B_b^{l,m}$ are background counts accumulated in the remaining two quarters. Let

$$I_a^{l,m} = P_a^{l,m} - B_a^{l,m} \qquad (1)$$

and

$$I_b^{l,m} = P_b^{l,m} - B_b^{l,m} \qquad (2)$$

From the early theory of thin foil X-ray microanalysis[2,3] the concentration ratio associated with the pixel *l*, *m* is

$$\frac{C_a^{l,m}}{C_b^{l,m}} = k_{a,b} \cdot \frac{I_a^{l,m}}{I_b^{l,m}} \qquad (3)$$

where $k_{a,b}$ is a constant dependent upon the efficiencies of X-ray generation and detection. For a binary alloy

$$C_a^{l,m} + C_b^{l,m} = 1 \qquad (4)$$

Rearrangement of equation (3) yields

$$C_b^{l,m} = \left(\frac{1}{k_{a,b}} \cdot \frac{I_b^{l,m}}{I_a^{l,m}}\right) C_a^{l,m} \qquad (5)$$

Substitution of equation (5) into equation (4) gives

$$C_a^{l,m} + \left(\frac{1}{k_{a,b}} \cdot \frac{I_b^{l,m}}{I_a^{l,m}}\right) C_a^{l,m} = 1 \qquad (6)$$

from which

$$C_a^{l,m} = \frac{k_{a,b} \cdot I_a^{l,m}}{k_{a,b} I_a^{l,m} + I_b^{l,m}} \qquad (7)$$

Similarly

$$C_b^{l,m} = \frac{I_b^{l,m}}{k_{a,b} \cdot I_a^{l,m} + I_b^{l,m}} \qquad (8)$$

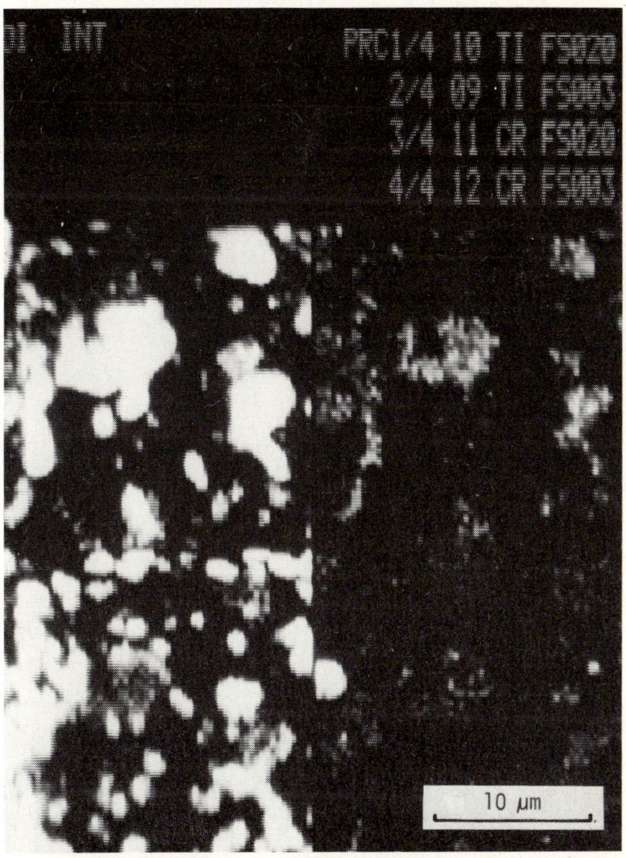

4 Simultaneous digital display of four WDS images from carbide particles extracted from Nimonic alloy:

top left image from Ti K_α peak
top right image from Ti K_α background
bottom left image from Cr K_α peak
bottom right image from Cr K_α background

Consequently the fully quantitative compositional map for element *a* is given by $C_a^{l,m}$ with $l = 1$ to 64 and $m = 1$ to 63, i.e. a digital matrix of compositional data.

PRACTICE

Data processing may be listed as the following seven steps:

1. Acquire peak images in two separate quarters
2. Acquire background images in the remaining two quarters
3. Subtract peak images from corresponding background images
4. Multiply one background-corrected image by appropriate *k* factor
5. Add images to obtain $k_{a,b} \cdot I_a + I_b$
6. The final ratios corresponding to equations (7) and (8) are performed via a normalizing routine which displays the concentration map on a scale between 0 and 100 wt-% (each concentration level, be it grey or colour, represents 2.56 wt-%)

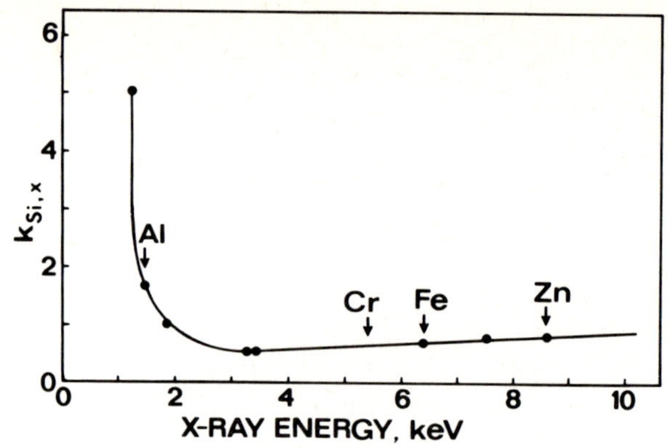

5 Calibration curve for K_α X-ray lines analysed by Kevex EDS used in this work: $k_{a,b}$ given by $k_{Si,a}/k_{Si,b}$ where *a* and *b* are the two elements of interest

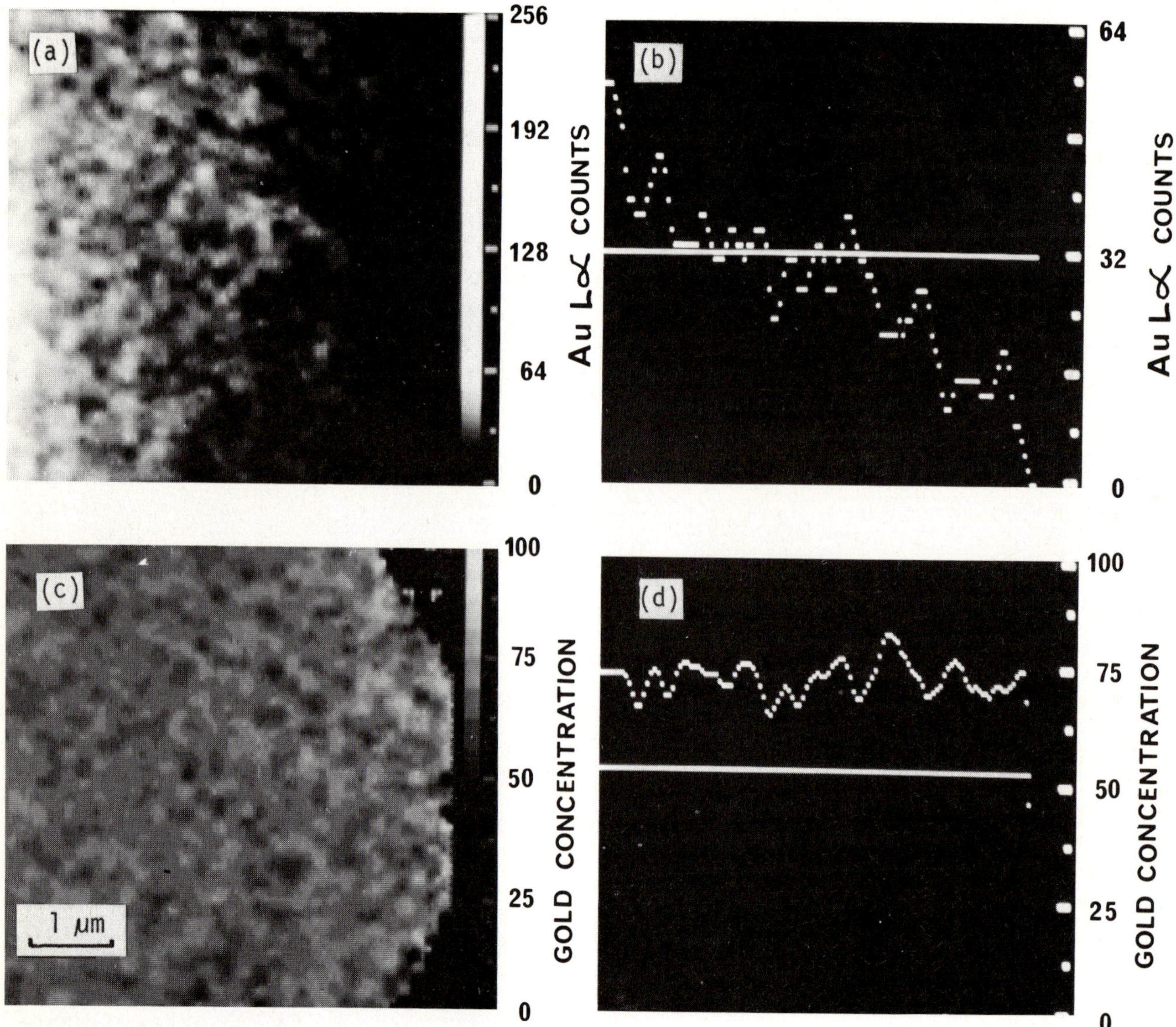

6 Microanalysis for Au in wedge-shaped foil of Cu–75Au: (a) background-corrected Au L_α data showing decrease in counts down the wedge from left to right; (b) decrease in counts from *A* to *B*; (c) quantitative contour map for Au; (d) concentration is seen to be approximately uniform with mean concentration of 75%

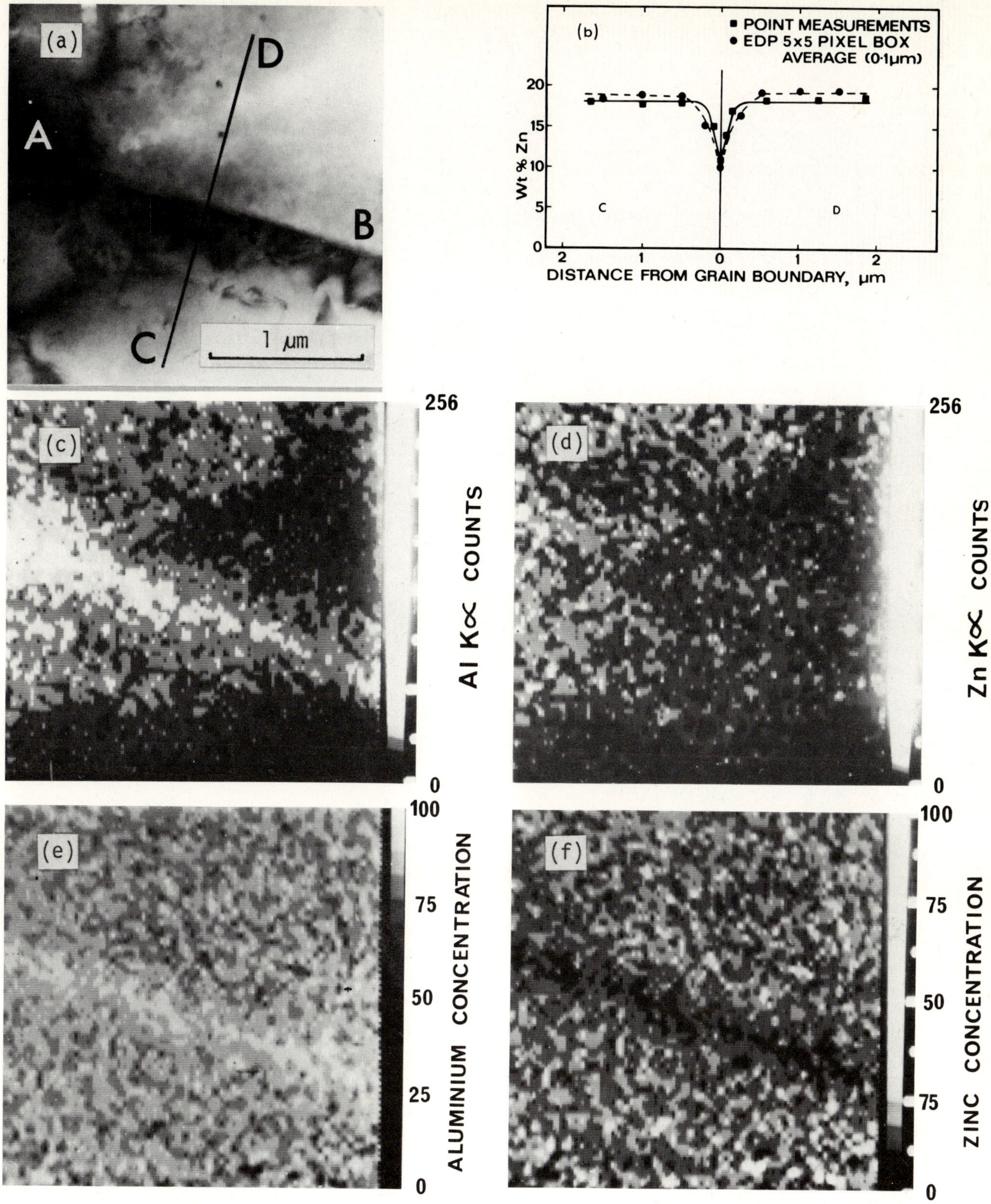

7 Microanalysis of Zn and Al in wedge-shaped foil of Al–18Zn: (a) TEM image showing grain boundary running from thick to thin region along *A* to *B*; (b) point analyses across boundary between *C* and *D* (*see text*); (c) and (d) background-corrected X-ray raw data for Al and Zn respectively; (e) and (f) fully processed Al and Zn compositional maps respectively

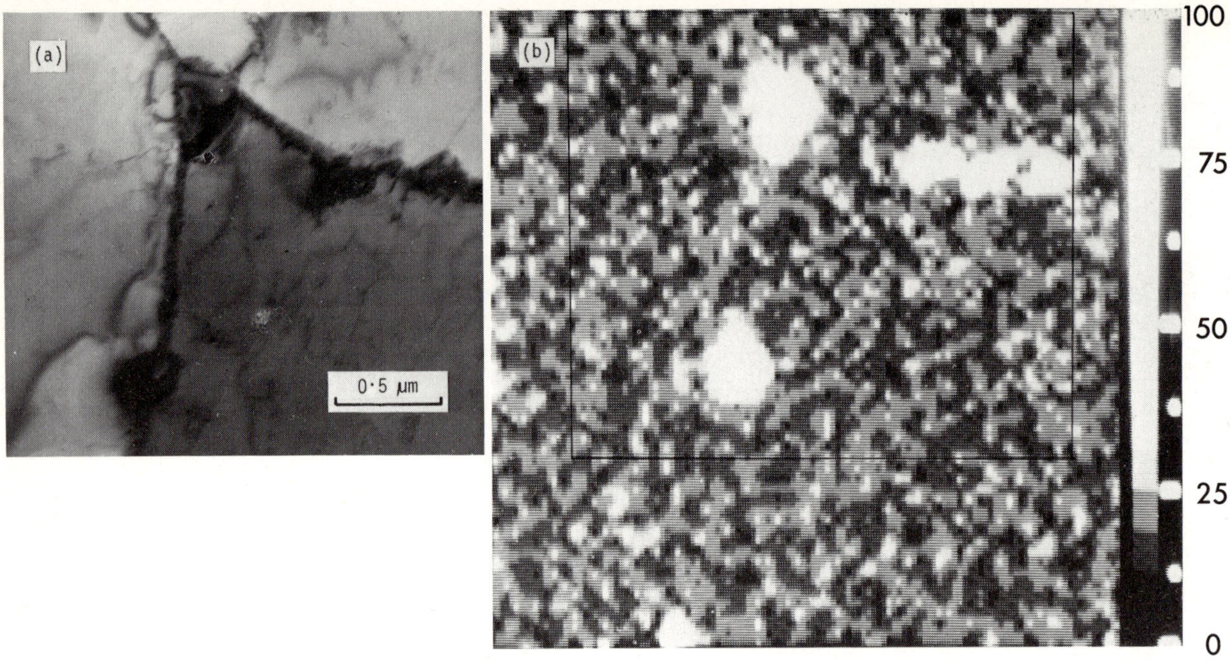

8 Foil of type 430 stainless steel (Fe–17Cr–0.1C): (a) TEM image showing large chromium carbonitride particles on grain boundaries at triple point; (b) Cr distribution map (area corresponding to (a) is outlined)

7 Display final images on full screen for photography.

The k factor is obtained by prior calibration in the normal way. For EDS we have employed the Cliff–Lorimer[4] method, and the appropriate K_α X-ray line calibration for our detector is shown in Fig. 5. For WDS equivalent thin foil standards can be employed, or reference to bulk standards.[5]

Comparisons between EDS and WDS for this work showed that the advantages of the much higher count rate from EDS more than offset its marginally higher background which, for thin foils, is inherently low for both EDS and WDS.

APPLICATIONS

The choice of applications was constrained by two factors:

(i) to binary alloys for convenience
(ii) to alloys where the appropriate X-ray k factors were whole numbers (the system in its present form cannot multiply stored data by a fractional number).

Examples selected were:

Cu–75Au	$k \simeq 3$ for Cu K_α, Au L_α on WDS
Al–18Zn	$k \simeq 2$ for Al K_α, Zn K_α on EDS
Fe–17Cr	$k \simeq 1$ for Fe K_α, Cr K_α on EDS

Cu–75Au

Sheet material was annealed and then electropolished to produce a wedge-shaped foil. The specimen was analysed by WDS via Cu K_α on LiF crystal and Au L_α on LiF crystal. The results for Au are illustrated in Fig. 6. The background-corrected raw X-ray data are shown in Fig. 6*a*, and the decrease in count rate from A to B down the wedge is evident, as shown graphically in Fig. 6*b*. The final processed image is given in Fig. 6*c*, where it will be seen that the Au concentration is uniform down the wedge, as reinforced by the line profile (Fig. 6*d*) which indicates a mean concentration of 75%. The fluctuations in composition about the mean are not completely understood; they are partly due to poor counting statistics, but the scale of the patchiness suggests some artifact originating from the specimen.

Al–18Zn

As a further example following that shown in Fig. 1, it is instructive to compare conventional point analysis with digital contour mapping. The area of foil studied is shown in Fig. 7. The TEM image, Fig. 7*a*, again shows a grain boundary running from a thick to a thinner region from A to B. Conventional point analyses across the boundary between C and D indicated zinc depletion adjacent to the boundary (Fig. 7*b*). Figures 7*c* and *d* show the background-corrected X-ray raw data for Al and Zn, with no strong indication of Zn depletion in the neighbourhood of the boundary. Figures 7*e* and *f* show the fully processed Al and Zn compositional maps; Zn depletion and Al enhancement are clearly visible in the vicinity of the boundary. The Zn profile across the boundary in Fig. 7*f* is plotted for comparison in 7*b*. At each point of interest the data were averaged over a 5×5 pixel box which is equivalent to a region 0.1 μm across (comparable with the spot size employed for the conventional point analyses). Both profiles in Fig. 7*b* are similar.

Fe–17Cr

A sample of type 430 stainless steel was solution-treated at 1323 K for 30 min and then heat-treated at 923 K for 284 h. The TEM image of the foil, Fig. 8*a*, shows the expected large particles of chromium carbonitride along the grain boundaries. The Cr concentration map is shown in 8*b*. A depletion in Cr at the boundaries in between particles was not detected. If such a depletion did exist, then it was below the spatial resolution of our instrumentation at ~0.1 μm.

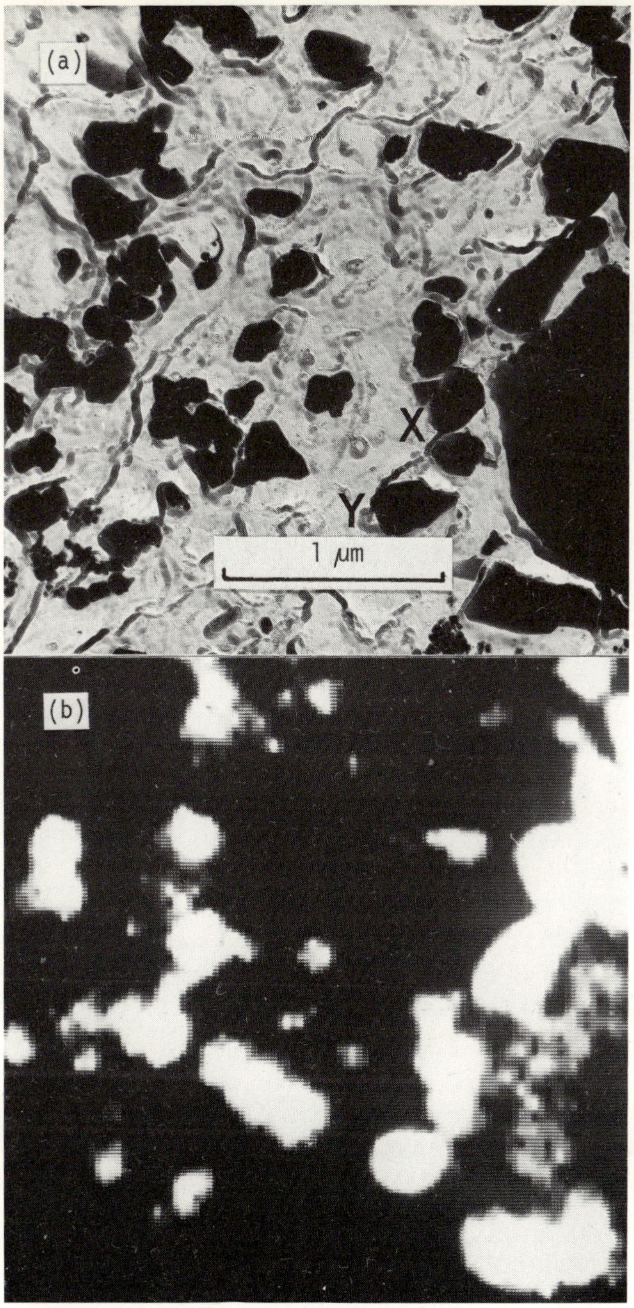

9 Extraction replica containing chromium carbides and titanium carbides from Nimonic alloy: (a) TEM image of area with number of small chromium carbide particles, but only one large TiC particle at bottom right; (b) chromium K_{α} raw X-ray distribution map (note that particles *X* and *Y* are just resolved on X-ray image)

DISCUSSION AND CONCLUSIONS

The results reported here demonstrate that digital compositional mapping from thin foils via EDS or WDS X-ray analysis is a practicable proposition. It is regrettable that we have not been able to illustrate here in print the vast enhancement in visual impact associated with colour mapping compared with grey level mapping, for it is the authors' opinion that colour increases enormously the perception of the information contained within a map.

The spatial resolution of the X-ray maps is limited by the pixel size and the size of the electron probe. A much smaller high-intensity probe would enable higher magnifications to be fruitfully employed. If the probe size exceeds the effective pixel diameter on the image, then it is the probe which limits the spatial resolution. This is illustrated in Fig. 9, which shows extracted chromium carbide particles from a Nimonic alloy. The separation between the two particles *X* and *Y* is equal to 0.1 μm (i.e. the probe size), but these are only just resolved on the X-ray image.

It is concluded that the addition of digital mapping to high resolution analytical TEMs and STEMs would add a further dimension to the quality of information usefully extracted from the instrument. It seems likely that, with appropriate instrumentation, electron energy-loss maps are also a real possibility for the future.

REFERENCES

1 P. D. HUNNEYBALL, M. H. JACOBS, T. J. LAW, and R. B. NEWBERY: 'Electron Microscopy 1980', **3**, 36–37, The Netherlands, 7th European Congr. Electron Microscopy Foundation
2 P. DUNCUMB: *J. de Microscopie,* 1968, 7, 581
3 J. PHILIBERT and R. TIXIER: *J. Phys. D. (Appl. Phys.),* 1968, **1**, 685
4 G. CLIFF and G. W. LORIMER: *J. Microsc.,* 1975, **103**, 203
5 M. H. JACOBS and J. BABOROVSKA: Proc. 5th European Congr. Electron Microscopy, 136–137, 1972, London and Bristol, Inst. Physics

DISCUSSION of paper by Hunneyball, Jacobs, and Law

D. C. JOY:
Could you not include non-integer k_{ab} factors by multiplying both terms in the manipulation expression by integers (i.e. 2 and 3 to give 1.5)?

M. H. JACOBS:
Yes, in principle you can do this. However, since each picture element can contain counts only up to 256 any multiplication by large numbers necessarily means that the initial X-ray data must be low in number of counts and therefore of poor statistical accuracy. Although it is not

fitted to our system, there is available a facility which enables multiplication of the data by any number from 0.10 to 10.

P. J. STATHAM:
Since your hardware has a capacity of 256 counts per pixel, the apparent fluctuations in concentration to which you draw attention must surely be on the limit of statistical significance. With a uniform field of 100 counts per pixel, pixel-to-pixel apparent intensity fluctuation would frequently exceed 10% relative error and any smoothing of the data would correlate the fluctuation and give apparent oscillations in concentration.

M. H. JACOBS:
A 9-point smoothing routine is available on the equipment. This was not used in any of the results presented in our paper. Clearly, such a smoothing routine has to be used with caution, since it tends to even out concentration variations.

R. G. FAULKNER:
Can the display system be modified to supply quantitative information about volume fractions of specific composition particles?

M. H. JACOBS:
Yes, we have software to do this; any colour or colour range can be selected and processed to give an area fraction. The accuracy is limited by the picture element size relative to the features of interest.

G. W. LORIMER:
The apparent Au-rich region observed in your Au–Cu alloy could surely be statistical noise. If you examined the same area twice and compared the images this would confirm or disprove this possibility. Have you tried this?

M. H. JACOBS:
No. This would be a worthwhile experiment to carry out.

Part 5
Convergent beam diffraction and other techniques

Quantitative microanalysis: a comparison of techniques

J A Venables

Quantitative microanalysis aims to characterize solid samples on a microscopic scale as completely as possible. Such characterization can be classified as sample shape and topography, chemical composition, positional order (i.e. crystallographic structure, orientation and defects in this structure), and electronic states. The major techniques in this area, which are based on new forms of 'analytical' electron microscopy, have developed along somewhat different lines for bulk samples (based on SEM techniques) and for thin film samples (based on CTEM/STEM techniques). This paper compares these two approaches to characterizing (i) topography and surface structure using secondary and transmitted electrons; (ii) chemical analysis using Auger electron spectroscopy, X-rays and energy loss electrons; (iii) crystallographic order using back-reflected and transmitted electrons; and (iv) electronic state information from cathodoluminescence, electron beam induced conductivity and surface work function. Surface-sensitive techniques are highlighted throughout. Techniques based on probes other than electrons are also being improved in sensitivity and spatial resolution. Techniques which may well become competitive with microanalysis using electron probes are discussed briefly.

The author is in the School of Mathematical and Physical Sciences, University of Sussex, Brighton.

The papers in this volume describe a range of microanalytic techniques, their quantification, and their application to materials problems. Of these techniques, it is clear that scanning transmission electron microscopy (STEM) is becoming widely used, especially in conjunction with energy-dispersive X-ray analysis (EDX), and increasingly with electron energy loss spectroscopy (EELS) and/or convergent beam diffraction (CBD). A few other papers describe the 'ultimate' analytic technique, the combination of a field ion microscope (FIM) with analytical facilities (the atom probe), which has demonstrated single-atom sensitivity and which also shows very encouraging agreement with STEM-EDX analysis.

In the present paper it is not proposed to dwell on the above techniques but to describe in outline other microanalytic techniques not dealt with elsewhere in this volume. Emphasis is given to surface-sensitive techniques and to the complementary advantages of bulk sample (largely based on the scanning (SEM) principle) and thin foil sample analysis used in conventional transmission (CTEM) or STEM. The concentration on SEM techniques could be criticized on the grounds that these are known not to yield the 'highest' spatial resolution, due to the extent of the electron cascade and the smearing effects of backscattered electrons. On the other hand, there is a strong case for considering microanalysis at an 'appropriate' resolution for the problem being studied, and for contrasting the three-dimensional (3D) information often obtainable by SEM with the 2D projection typically obtained using TEM techniques. SEM techniques are often more appropriate for problems spanning a large range of sizes in a single sample, and specimen preparation is generally easier than for TEM.

Because of limited space, this paper concentrates on giving references to other papers where the topics are discussed in more detail, and does not present any pictures. The following sections discuss in turn: topography and surface structure; chemical analysis, especially Auger electron spectroscopy (AES) and scanning Auger microscopy (SAM); diffraction techniques and imaging; and electronic state information. Techniques based on probes other than electrons, whose spatial resolution is being improved in the sub-micron range, are discussed in the sixth and last section.

TOPOGRAPHY AND SURFACE STRUCTURE

Both TEM and SEM can be surface-sensitive at the monolayer level under suitable conditions, and a detailed review of surface-sensitive electron microscopy techniques is given elsewhere.[1] So it would seem that there is no problem in principle of observing arbitrarily complex surface topography. However, in practice, SEM observation is often far superior to TEM, partly because of the 3D 'effects' given by the backscattered and secondary electron signals, which mimic our own perception of 'shadows' obtained with visible light on everyday objects. These effects are particularly strong at glancing incidence, as we have seen in various studies of crystal growth.[2]

By contrast, TEM information can often be disappointing and difficult to extract. Monolayer level information is available at high resolution, as in the work of Lehmpfuhl and Uchida[3] on MgO or Cherns on Au.[4] But the observability of surface steps is very sensitive to diffraction conditions, and there is often ambiguity as to which surface these steps are on. Under well controlled conditions, excellent high-resolution, surface-sensitive (sub-monolayer) information can be obtained from clean surfaces by CTEM, as is shown by the work of the Tokyo group on the Au (111) surface super-lattice, its removal by sub-monolayer Ag, and on other sub-monolayer deposits.[5]

Small particles have often been observed by STEM but their appearance is rather uninformative and belies their complex 3D shape. In order to abstract information about their shape and crystallography both multiple dark field imaging and detailed analysis of weak-beam fringes have been developed.[6] High resolution ($\lesssim$1 nm) SEM would be interesting for these applications, and the use of low-loss imaging at glancing incidence is a step in this direction.[7] Such pictures often give a remarkably different picture of surface topography from normal CTEM images, the surface appearing to be much rougher.[8]

The secondary electron signal is sensitive to effects at the monolayer level, largely via the work-function change. This is demonstrated by our recent work on Cs monolayers on W[9] where the work function change is $\simeq$3 V, but it can also be seen where the change is much smaller (e.g. Refs. 2 and 10). This information can be supplemented by AES and reflection electron diffraction (RHEED) which are also sensitive, and quantifiable, at the sub-monolayer level.

CHEMICAL ANALYSIS VIA AES AND SAM AND COMPARISON WITH EDX AND EELS

Reviews of the Auger electron (AES and SAM) techniques have been given over the last few years, to which the reader is referred.[1,11,12] There are many examples of SAM which are useful at quite low ($\simeq$5 μm) lateral resolution, such as the surface diffusion work of Butz and Wagner.[13] Using SEM/SAM techniques in a field emission gun higher spatial resolution ($\lesssim$30 nm) has been obtained[14] and used on various samples, including grain boundary and surface diffusion in Ag/Au[15] and composition profiles in $Ga_xAl_{1-x}As$ laser structures.[16] For planar structures such as these laser devices, taper profiling is very advantageous and has resulted in 1% compositional accuracy with 2.5 nm depth resolution, using a beam with only 5 μm lateral resolution.[17] Clearly, use of field emission machines with the taper geometry could result in even better depth resolution, with a limit of $\gtrsim$0.5 nm determined by the escape depth of the Auger electrons.

The lateral resolution and the quantitative accuracy of analysis are influenced by electron backscattering. The signal level is increased by a factor $(1 + \beta\eta)$, where η is the electron backscattering factor, and β is found to be $\simeq$2; thus for high atomic number (Z) elements with $\eta \simeq 0.5$–0.6, around half the Auger electrons are produced by backscattered electrons. Further accurate measurements and theories of these effects are required for improved quantification, although ratio techniques[18] may minimize the effects of backscattering in relatively homogeneous samples.

The effects of backscattering on resolution have been investigated in a preliminary way with qualitatively predictable results.[19] Edge profiles contain a central part which has the resolution of the probe size, with wings due to backscattered electrons. These wings are narrower for high-Z elements, even though more Auger electrons are produced by backscattering, and in practice resolution is presently limited as much by signal-to-noise ratio (S_n) considerations.

Improvements in SAM resolution can come about by increased detection efficiency of analysers, by improved field emission columns, and by better data processing; changes in this direction are encouraging[20] and will lead to better S_n values. With such improvements it is hoped to achieve ~2–3 nm SAM resolution with several nA probe currents, although with bulk samples the contributions of backscattered electrons over a 50–200 nm radius would still be present.

This backscattered contribution can be greatly reduced by the use of thin film samples, with a consequent improvement of spatial resolution and signal-to-background ratio, as explored in a preliminary way by Wiedman and Seiler[21] and emphasized by Wittry[22] in a comparison of microchemical techniques. We estimate that SAM may then become comparable or even superior to STEM-EDX or EELS techniques, with a detection limit around 10 atoms in favourable cases. In particular the volume analysed will be very small ($2^2 \times 0.5$ nm) since the mean free path of the Auger electrons determines the sampling depth, rather than the foil thickness for EDX or EELS. Already the volume sampled by these other techniques is limited by the foil thickness and beam spreading, as discussed in other papers in this volume. On suitably thin films, EDX has detected ~30 atoms,[24] and EELS, which can be further developed via parallel recording, may attain even greater sensitivity because of its superior detection efficiency. At the higher resolution level, of course, radiation damage in beam-sensitive materials may be a prior limitation for all three techniques, as pointed out by Bauer and Seiler;[23] but metallic and semiconductor materials do seem to be suitably resistant to beam damage in general.

Despite all such improvements, the three chemical analysis signals, AES, EDX and EELS, which are all caused by the same initial inelastic (ionization) event, will remain weaker signals than either the secondary electron or the elastic (diffracted) electron signal. They are thus often best used in a complementary mode, as point, line-scan or restricted area analyses, to identify features in micrographs taken with one of the stronger signals.

ELECTRON DIFFRACTION, CRYSTALLOGRAPHIC ANALYSIS AND IMAGING

Thin film crystallographic and defect analysis using transmission electron diffraction (THEED) and dark- and bright-field microscopy have been a mainstream activity of materials research for the last 25 years. With the advent of STEM and convergent beam diffraction (CBD), such crystallographic analyses can be obtained from individual particles as small as a few nm across; the use of higher order Laue zone (HOLZ) lines makes accurate lattice parameter measurements ($\lesssim$1 part in 10^3) possible; such lines can be split by individual dislocations, making Burgers vector characterization possible on a very fine scale;[25] symmetry analyses of the patterns go a long way towards absolute structure determination and, even without detailed structure analysis, enable phase identification on a microscopic scale to be made very readily.

Reviews of such developments have been given by Steeds.[26] Thus CBD is an excellent example of a quantitative microanalytic technique, which combines high precision with high signal strength, based on the strong forward peaking of electron diffraction with the efficient parallel recording of the photographic plate. In dedicated ultra-high vacuum STEMs, not containing photographic plates, the diffraction pattern has to be recorded via a suitable optical system[27] and despite recent improvements further technique developments are needed in this area to achieve comparable precision.

When bulk samples are used, the crystallographic signal is camouflaged by the non-crystallographic effects of the electron cascade, though information is still present which can be used to extract crystal structure and orientation. Electron channelling patterns (ECP), electron backscattering patterns (EBSP) and X-ray Kossel patterns use such crystallographic information; the techniques have been compared elsewhere.[28] The X-ray technique produces the most accurate lattice parameter values, but the EBSP technique can be seen to come from the smallest volume ($\lesssim 20 \times 100$ nm), and can be analysed to give orientation relationships to better than ±0.5°. A related technique using a small-angle backscattered electron detector can give crystallographically enhanced images.[29] Electron channelling 'micrographs', exploiting small variations in the angle of incidence relative to the lattice, have been in use for some time.[30] Energy filtering of backscattered electrons has been used to reduce unwanted (non-crystallographic) information from the cascade. With such a (field emission) system, individual sub-surface defects such as dislocations and stacking faults have been observed in bulk samples in the SEM,[31] using such channelling techniques.

At glancing incidence (i.e. at smaller normal component of electron energy) diffraction effects from the surface layers become dominant and inelastic cascade effects less severe. Reflection diffraction (RHEED) is highly surface-sensitive at the sub-monolayer level and has the added advantage of being very sensitive to 3D protrusions (such as crystalline islands) and of being compatible with high (lateral) spatial resolution (see Refs. 1, 2, 27, 35). Recent developments of RHEED by Ino and coworkers[32] have included the recording of higher-order Laue zones (in a fashion analogous to CBD developments) and a projection which displays the diffraction pattern in an analogous manner to low-energy diffraction (LEED).

More recently the corresponding dark-field reflection microscopy has been developed by the Tokyo[32] and Arizona groups.[34] Using an ultra-high vacuum specimen environment and in-situ sample cleaning, spectacular surface pictures have been obtained, especially of Si(111) surface steps, emergent dislocations terminating such steps, the 'step' roughening transition, surface phase transitions, and sub-monolayer adsorbed phases on this surface.[33] Similar scanning pictures of oxidized copper have been obtained, but not yet at as high resolution.[34] In these techniques very flat surfaces are needed, and much remains to be done to quantify the techniques and to compare scanning and conventional imaging. More detailed reviews are given elsewhere.[1,35,36]

The use of glancing incidence illumination for chemical analysis also seems fruitful, and varying this angle is a way to sample different depths. EDX, AES and RHEED have been used in various combinations to sample depths in the range 1-30 nm;[37,38] such methods need to be developed further, backed by a computational scheme which can readily handle varying incidence angles.[39]

ELECTRONIC STATE INFORMATION

Once the topography, chemical composition and crystalline order of the sample are known, the next questions concern chemical or electronic state information, electrical transport, etc. This information is often particularly important for device applications, for example in the semiconductor or chemical industries. As microanalysts we are interested in producing such information but also in knowing where (spatially) it comes from; this means that techniques which are known to be sensitive to electronic states have to be 'localized', and that those signals which can be stimulated by an electron beam will be especially useful.

Many of these techniques exist in principle, though most are not widely available, having been developed in a small number of laboratories and not yet commercialized. Perhaps the most widely used are electron beam induced conductivity (EBIC)[40] and cathodoluminescence (CL).[41] These are extremely powerful, especially for semiconducting and insulating materials, since they are able to locate recombination centres (e.g. dislocations and precipitates) and to observe the small energy level shifts associated with recombination in different places. Optical resolution of wavelength shifts of 1 nm at 500 nm, for example, corresponds to an energy resolution of 5 meV. A particularly impressive example of STEM-EBIC-CL techniques is shown by the work of Petroff,[42] who has applied this combination instrument to studies of recombination centres in GaAs and other semiconducting compounds.

An energy resolution of a few meV is very difficult to achieve with direct electron techniques; it can only be done using energy filtering on both input and output beams, which, of course, limits intensity severely. Such work has been pursued vigorously in a few laboratories[42,43] and applied to studies of valence band states, plasma losses, molecular vibrations, and surface states. In somes cases[43] typical electron microscope energies are used, whereas for surface studies[44] beam energies $\simeq$5 eV are common. The first type have the capability of developing into truly microanalytic techniques, especially if field emission sources are used, which already have good (if not sufficient) energy resolution before filtering.

Using other electron spectroscopic techniques it may be possible to achieve relatively high energy resolution without placing such restrictions on the incident beam. We have used low-energy secondary electron spectroscopy to measure the local work function of samples, with $\lesssim$40 meV energy and $\lesssim$0.4 μm spatial resolution.[45] This 'work-function mapping' detects adsorbed monolayers with very high sensitivity and considerable precision. For the case of an adsorbed Cs monolayer on W, with a work function shift of 3 V, the total secondary electron signal is $\simeq$50% higher from the Cs layer, but all this extra signal occurs in the range up to 3 eV above secondary electron onset. The signal can therefore be detected with great ease, and a feedback mechanism has been constructed to track this onset energy.[9,45] This type of technique can therefore be seen as an initial stage towards examining surface states in general on a microscopic scale.

Surface and chemical state information may also be obtainable from small energy shifts in other electron spectroscopies, most notably AES and photoelectron spectroscopy (ultraviolet (UPS) and soft X-ray (XPS), or collectively electron spectroscopy for chemical analysis (ESCA)). AES lines involving valence transitions shift by a few volts, depending on chemical state, for example from 92 V for Si to 79 V for Si in SiO_2. UPS lines are, however, better for this purpose, because their intrinsic line widths are smaller; electron energy distribution curves with ultraviolet excitation have become the standard means of exploring band structure and surface states. The corresponding microscopy is photoemission electron microscopy. Although this has been commercially available for some time,[46] it has not yet been developed in a microanalytic direction. This is one of the possible future techniques discussed in the next section.

OTHER MICROANALYTIC TECHNIQUES

Although the main techniques of microanalysis are based on using electron probes, there are others which use photons, ions or even neutral atoms. An idea of the vast range of available techniques is given in diagrams presented by Wittry[22] and others. Electron, photon, ion or neutral atom emission (or desorption) are also possible bases for techniques, without an obvious external probe. The field ion microscope and the atom probe are the most spectacular examples of such techniques, as discussed elsewhere in this volume;[47] the field emission microscope has also been widely used.[48]

Using photons, soft and hard X-rays offer interesting analytical possibilities below 1 μm resolution, but are difficult to focus. Focusing using zone plates is under consideration, using the high X-ray flux available from synchrotron sources.[49,50] Comparisons with electrons show that X-ray illumination may be preferable in causing less damage under certain conditions;[50] photoelectron microscopy has been advocated[1,51] as an analytical tool with spatial resolution in the 1-30 nm range. Particular advantages would be that specific element core levels can be excited and that exit depth and chemical state information can be obtained by energy analysing the emergent electrons.[1]

Alternative approaches, using high-power X-ray sources in-situ, have been used to generate either contact exposures (i.e. shadow images) of resists with $\simeq$10 nm resolution,[52] or scanning images of thin targets using the photoelectrons emitted.[53] While all of these approaches may not necessarily be capable of the highest spatial resolution, it seems likely that they will be developed to the point where we can use our extensive knowledge of X-ray absorption, fluorescence and photoemission on a microscopic scale.

Ion techniques have been developed extensively, using both large (MeV) accelerators and smaller dedicated instruments. Microbeam attachments have recently been added to several MeV accelerators around the world,[54] with spatial resolution approaching 1 μm in certain cases, using MeV proton-induced X-ray emission (PIXE). Advantages of the technique include much better X-ray trace element sensitivity approaching 1 ppm[54,55] due to the lower *bremsstrahlung* background, use of Rutherford backscattering for depth dependent analysis, nuclear reactions for isotope detection, and less lateral spreading of the beam in the target ($\lesssim$25 nm at 1 μm depth for 1 MeV protons in Si[56]).

Resolutions less than 1 μm require higher-brightness ion sources, and field ion (H^+ and H_2^+)[57] and liquid metal (especially Ga^+)[58] sources have been constructed and compared.[56] Resolution $\lesssim$1 nm is hoped for; a particularly detailed discussion of recent progress in scanning transmission ion microscopy (STIM) is given by Levi-Setti and Fox.[57] It is shown that protons both suffer increased scattering and cause increased damage, relative to electrons of the same energy. Information per unit of damage is comparable for the two particles, with protons maybe having an advantage for very thin foil samples but not otherwise.

At the highest resolution it certainly seems that protons (and heavier ions) will be uncomfortably destructive. This destruction is, of course, used to advantage in secondary ion mass spectrometry (SIMS) or in its imaging version, the ion microprobe. These techniques have remarkable sensitivity, especially for trace elements, because of the almost non-existent background signals obtainable in mass spectrometry. Present spatial resolution in commercial equipment[59] is around 1 μm, though dependent on the particular ion species and concentration.

At currently obtainable detection sensitivities (sputtering yield $\simeq$2, ion fraction $\simeq$0.01 and collection efficiency 0.1), a current of 10^{-12} A into a 10 nm probe (around the limit of field ion sources) would be removing $\gtrsim$2 μm/s, while detecting 100 ion/s from an atomic concentration of 1%,[57] which is certainly destructive! Comparable figures for 1 μm lateral resolution are 10^{-10} A, 20 nm/s and 10^4 ion/s for 1% concentration, which is much more manageable. Quantification of SIMS has proved difficult in general, but for trace elements of given atoms in a well defined matrix (e.g. typical semiconductor problems such as dopants in Si or GaAs) the method correlates well with other techniques (e.g. Rutherford backscattering) at high concentration, and extrapolates reproducibly down to 10^{13} atom/cm^3 or 1 part in 10^9 – a quite remarkable achievement.[60]

ACKNOWLEDGMENTS

The author is grateful to several colleagues for providing information about techniques they are using, and to Dr D. J. Fathers for a critical reading of the manuscript.

REFERENCES

1 J. A. VENABLES: *Ultramicroscopy*, 1981, in press; *see also Ref. 16*

2 K. HARTIG, A. P. JANSSEN, and J. A. VENABLES: *Surf. Sci.*, 1978, **74**, 69; J. A. VENABLES, J. DERRIEN, and A. P. JANSSEN; *ibid.*, 1980, **95**, 411; P. AKHTER, J. DERRIEN, A. P. JANSSEN, and J. A. VENABLES: Inst. Phys. Conf. Ser. **52**, 1980, 383 (EMAG '79)

3 G. LEHMPFUHL and Y. UCHIDA: *Ultramicroscopy*, 1979, **4**, 275

4 D. CHERNS: *Philos. Mag.*, 1974, **30**, 549

5 T. TANASHIRO *et al.*: Proc. 4th Int. Conf. Solid Surfaces (Cannes), 1980, **1**, 683; K. TAKAYANAGI and G. HONJO: *ibid.*, 1980, **1**, 267; K. YAGI *et al.*: *Surf. Sci.*, 1979, **86**, 174

6 K. HEINEMANN, M. J. YACAMÁN, and H. POPPA: *Phys. Status Solidi (a).*, 1979, **54**, 675; *J. Cryst. Growth*, 1979, **47**, 177, 187

7 O. C. WELLS: *Appl. Phys. Lett.*, 1971, **19**, 232; O. C. WELLS, A. N. BROERS, and C. G. BREMER: *ibid.*, 1973, **23**, 353

8 W. KRAKOW and A. N. BROERS: Proc. EMSA, 1980, **38**, 162

9 P. AKHTER and J. A. VENABLES: *Surf. Sci.*, 1981, **103**, 301

10 M. G. WELLS and D. A. KING: *J. Phys. C (Solid State Phys.)*, 1974, **C7**, 4053

11 J. A. VENABLES and A. P. JANSSEN: Proc. 9th Int. Cong. Electron Microscopy (Toronto), 1978, **3**, 280; A. P. JANSSEN and J. A. VENABLES: SEM '79, 1979, **2**, 259; Inst. Phys. Conf. Ser. **52**, 1980, 357 (EMAG '79)

12 A. VAN OOSTROM: *Surf. Sci.*, 1979, **89**, 615

13 R. BUTZ and H. WAGNER: *Surf. Sci.*, 1979, **87**, 69, 85

14 J. A. VENABLES, A. P. JANSSEN, C. J. HARLAND, and B. A. JOYCE: *Philos. Mag.*, 1976, **34**, 395

15 A. P. JANSSEN, J. A. VENABLES, J. C. M. HWANG, and R. W. BALLUFFI: *Philos. Mag.*, 1977, **36**, 1537

16 J. A. VENABLES *et al.*: *J. Microsc.*, 1980, **118**, 351

17 A. VAN OOSTROM, L. AUGUSTUS, W. NIJMAN, and L. LESWIN: *J. Vac. Sci. Tech.*, 1980, **17**, 40; J. M. WALLS, D. D. HALL, and D. E. SYKES: *Surface and Interface Analysis*, 1979, **1**, 204

18 A. P. JANSSEN, C. J. HARLAND, and J. A. VENABLES: *Surf. Sci.*, 1977, **62**, 277

19 A. P. JANSSEN and J. A. VENABLES: *Surf. Sci.*, 1978, **77**, 351; M. M. EL-GOMATI, A. P. JANSSEN, M. PRUTTON, and J. A. VENABLES: *Surf. Sci.*, 1979, **85**, 309; J. KIRSCHNER: *Appl. Phys.*, 1977, **14**, 351

20 J. A. VENABLES and A. P. JANSSEN: *Ultramicroscopy*, 1980, **5**, 297; J. A. VENABLES and G. D. ARCHER: Proc. 7th Eur. Conf. Electron Microscopy (The Hague), 1980, **1**, 54; D. J. FATHERS, C. J. HARLAND, G. D. ARCHER, and J. A. VENABLES: work in progress

21 P. WIEDMAN and H. SEILER: *Optik*, 1977, **49**, 295; *ibid.*, 1976, **44**, 505

22 D. B. WITTRY: Proc. 7th Eur. Conf. Electron Microscopy (The Hague), 1980, **3**, 14

23 H. E. BAUER and H. SEILER: Proc. 7th Eur. Conf. Electron Microscopy (The Hague), 1980, **3**, 214

24 H. DEXPERT, E. FREUND, and J. LYNCH: this volume

25 R. W. CARPENTER: 'Microbeam analysis 1980' (ed. D. B. Wittry), 1

26 J. W. STEEDS: in 'Introduction to analytical electron microscopy' (eds. Hren, Goldstein, Joy), 1979, 387; *Proc. EMSA*, 1980, **38**, 188; Inst. Phys. Conf. Ser. **52**, 1980, 197 (EMAG '79); this volume

27 J. A. VENABLES, C. J. HARLAND, and A. P. JANSSEN: Inst. Phys. Conf. Ser. **41**, 1978, 313; J. M. COWLEY: Inst. Phys. Conf. Ser. **52**, 1980, 239 (EMAG '79); *Proc. EMSA*, 1979, **37**, 372; *Ultramicroscopy*, 1979, **3**, 433; *Chem. Scripta*, 1978–79, **14**, 33

28 J. A. VENABLES, C. J. HARLAND, and P. AKHTER: *Proc. EMSA*, 1980, **38**, 184; D. J. DINGLEY, SEM 1981, in press; L. REIMER, W. PÖPPER, and W. BRÖCKER: SEM '78, 1978, **1**, 705

29 C. J. HARLAND, P. AKHTER, and J. A. VENABLES: *J. Phys. E. (Sci. Instrum.)*, 1981, **E14**, 175; Proc. 7th Eur. Conf. Electron Microscopy (The Hague), 1980, **1**, 252

30 *See for example* D. C. JOY, D. E. NEWBURY, and P. M. HAZZLEDINE: SEM '72, 1972, **1**, 97; D. L. DAVIDSON: SEM '77, 1977, **1**, 431; *for the equivalent case in magnetic materials see* D. J. FATHERS and J. R. JAKUBOVICS: *Physica*, 1979, **86-88B**, 1343; D. J. FATHERS, D. C. JOY, and J. P. JAKUBOVICS: Proc. 8th Intl. Conf. Electron Microscopy (Canberra), 1974, **1**, 630

31 P. MORIN, M. PITAVAL, M. BESNARD, and G. FONTAINE: *Philos. Mag.*, 1979, **40**, 511; *Scanning*, 1979, **2**, 217; 'Microbeam Analysis 1980' (ed. D. B. Wittry), 22

32 S. INO: *Jpn J. Appl. Phys.*, 1977, **16**, 891; Y. GOTOH and S. INO: *ibid.*, 1978, **17**, 2097

33 N. OSAKABE, Y. TANASHIRO, K. YAGI, and G. HONJO: *Surf. Sci.*, 1980, **97**, 393; *Jpn J. Appl. Phys.*, 1980, **19**, L309; Proc. 4th Intl. Conf. Solid Surfaces (Cannes), 1980, **2**, 1007; *Surf. Sci.*, 1981, **102**, 424

34 P. E. H. NEILSEN and J. M. COWLEY: *Surf. Sci.*, 1976, **54**, 340; *Ultramiscroscopy*, 1975, **1**, 145

35 G. G. HEMBREE and J. M. COWLEY: SEM '79, 1979, **1**, 145; *Oxidation of Metals*, 1979, **13**, 331

36 K. YAGI: *Proc. EMSA*, 1980, **38**, 290

37 S. INO, T. ICHIKAWA, and S. OKADA: *Jpn J. Appl. Phys.*, 1980, **19**, 1451; D. F. MITCHELL, P. B. SEWELL, and M. COHEN: *Surf. Sci.*, 1977, **69**, 310 and previous papers

38 S. J. BRETT and R. D. DOHERTY: *Mater. Sci. Eng.*, 1978, **32**, 255; C. LEA, S. J. BRETT, and R. D. DOHERTY: *Scr. Metall.*, 1979, **13**, 45

39 D. J. FATHERS and P. REZ: SEM '79, 1979, **1**, 55

40 *See for example papers in J. Microsc.*, 1980, **118**, pp 255 ff

41 *See for example* S. M. DAVIDSON and A. RASUL: SEM '77, 1977, **1**, 225, 233; S. J. PENNYCOOK, L. M. BROWN, and A. J. CRAVEN: *Philos. Mag.*, 1980, **A41**, 589; E. M. HÖRL; Proc. 7th Eur. Conf. Electron Microscopy (The Hague), 1980, **3**, 2

42 P. M. PETROFF, R. A. LOGAN, and A. SAVAGE: *J. Microsc.*, 1980, **118**, 255; SEM '78, 1978, **1**, 325

43 J. SILCOX: Proc. 9th Intl. Conf. Electron Microscopy (Toronto), 1978, **3**, 259; B. SCHRÖDER, J. GEIGER, and H. W. MÜLLER: *ibid.*, 1978, **1**, 534; M. ESSIG and J. GEIGER: Proc. 7th Intl. Conf. Electron Microscopy (The Hague), 1980, **3**, 64

44 H. IBACH, H. HOPSTER, and B. SEXTON: *Applications of Surf. Sci.*, 1977, **1**, 1; H. FROITZHEIM: in 'Electron spectroscopy for surface analysis' (ed. H. Ibach), Springer, 1977, 205

45 A. P. JANSSEN, P. AKHTER, C. J. HARLAND, and J. A. VENABLES: *Surf. Sci.*, 1980, **93**, 453

46 B. BURNETT, K. E. PARKER, and A. J. BAKER: 'The Balzers "Metioscope" model KE3', Proc. 8th Intl. Conf. Electron Microscopy (Toronto), 1978, **1**, 612

47 G. D. W. SMITH, A. J. GARRATT-REED, and J. B. VANDER SANDE: this volume

48 R. GOMER: *Crit. Rev. in Solid State and Materials Sciences*, 1978, 8, 119; J. P. JONES: in 'Chemical physics of solids and surfaces' (ed. M. W. Roberts and J. M. Thomas), 8; 1980, Royal Society of Chemistry, London

49 Projects under way at Daresbury Synchrotron Radiation Facility and at other facilities abroad

50 M. ISAACSON and M. UTLAUT: *Optik*, 1978, **50**, 213; D. SAYRE, J. KIRZ, R. FEDER, D. M. KIM, and E. SPILLER: *Ultramicroscopy*, 1977, **2**, 337

51 Various papers by L. WEGMANN, W. ENGEL, and O. H. GRIFFITHS over the last 15 years

52 E. SPILLER, R. FEDER *et al.*: *Science*, 1976, **191**, 1172; *ibid.*, 1977, **197**, 259

53 C. J. HOVLAND: *Appl. Phys. Lett.*, 1977, **30**, 274; Proc. 7th Intl. Vac. Cong. (Vienna), 1977, 2363; J. PERRIN, D. MOUZE, and J. CAZAUX: Proc. 8th Intl. Vac. Cong. (Cannes), 1980, **1**, 291; J. CAZAUX: *Mater. Sci. Eng.*, 1980, **42**, 45; *J. Phys.*, 1977, **38**, L473

54 G. J. F. LEGGE: 'Microbeam Analysis 1980' (ed. D. B. Wittry), 70; F. BOSCH *et al.*: *Nucl. Inst. Meth.*, 1978, **149**, 665

55 F. FUJIMOTO: Proc. 7th Eur. Conf. Electron Microscopy (The Hague), 1980, **3**, 130

56 F. W. MARTIN: *Nucl. Inst. Meth.*, 1978, **149**, 475

57 R. LEVI-SETTI and T. R. FOX: *Nucl. Inst. Meth.*, 1980, **168**, 139

58 L. W. SWANSON, G. A. SCHWIND, and A. E. BELL: SEM '79, 1979, **1**, 45; G. RINGO and V. E. KROHN: *Nucl. Inst. Meth.*, 1978, **149**, 735

59 *For example* Cameca 'Ion microprobe' model IMS-3F

60 K. WITTMAACK: *Nucl. Inst. Meth.*, 1980, **168**, 343; J. B. CLEGG: *Surf. and Interface Analysis*, 1980, **2**, 91; R. G. WILSON and V. R. DELINE: *Appl. Phys. Lett.*, 1980, **37**, 793; K. WITTMAACK and J. B. CLEGG: *ibid.*, 1980, **37**, 285

Microanalysis by convergent beam electron diffraction

J W Steeds

This contribution is concerned with the special advantages of convergent beam electron diffraction for microanalysis as compared with more frequently used techniques such as energy-dispersive X-ray analysis and electron energy loss spectroscopy. The method basically relies on the ability to perform accurate crystallographic analysis on small particles and then to make use of the wealth of crystallographic data which is available.

The author is in the Physics Department at the University of Bristol, UK.

Convergent beam electron diffraction (CBED) has a long history starting with its discovery in 1939 by Kossel and Möllenstedt. However, it is only in recent years that microscopes have been developed which permit microanalysis with high spatial resolution. These instruments are also capable of yielding CBED results of unparalleled quality and utility. Suddenly, a much overlooked aspect of electron microscopy has become a powerful analytical tool available to many research workers. By accepting from the outset the dynamical nature of the diffraction, it emerges that far from being a rather inexact relative of X-ray or neutron diffraction, electron diffraction has become a rigorous and accurate crystallographic technique in its own right. Aspects of electron diffraction which were previously regarded as weaknesses now turn out to be strengths. Moreover, and perhaps most surprisingly, much of the information which can be obtained by CBED is readily extracted without the need for calculations based on dynamical diffraction theory. As there have been several recent reviews on CBED[1-3] it suffices here to give a brief discussion of the principles before illustrating the effectiveness of the technique by a number of examples.

BASIC PRINCIPLES

The reasons why CBED is so much more potent than selected area diffraction are related to the well-known difficulties of specimen preparation for transmission electron microscopy. The unavoidable bending and thickness variation within the selected area mean that the diffracted intensities are averaged in a very complicated, specimen-dependent way over the two parameters which most strongly affect them. Since the typical extinction lengths for electron diffraction are a few tens of nm and the angular width of large-angle scattering in the electron microscope is typically 10^{-4} rad, it can be appreciated that the specimen requirements for a 1 μm selected area are unreasonably stringent. However, for CBED with probe sizes of 50 nm or less, most thin specimens will yield adequate results.

Terminology can be a problem in CBED. The term 'microdiffraction' is often used in related contexts, but the authors generally limit themselves to the spacings and angles between diffraction maxima. It might be wise therefore to accept this usage and to reserve the term 'CBED' for those experiments where the details of intensity information within the diffracted discs is being exploited. A marginal case occurs when the relative intensities of featureless discs are used to deduce symmetry information. Some workers use the term 'Kossel-Möllenstedt [or K-M] patterns' to refer to CBED where the diffracted discs of different order do not overlap, while patterns with overlapping discs are referred to as 'Kossel' patterns.

Implicit in the above discussion has been the idea that CBED patterns are composed of discs of intensity. In CBED a probe is focused onto an area of interest with the second condenser lens and the CBED pattern is generated in the back focal plane of the objective lens or a conjugate plane in the lens system of the microscope. Position in the CBED pattern defines an angle of incidence with respect to the optical axis of the microscope. The larger the convergence angle in the illumination focused on the specimen, the greater the disc size. The disc size can therefore be controlled by choice of the second condenser (C_2) aperture. Each disc is a map of intensity variation, direct or diffracted, for a range of angles of incidence with respect to the optic axis (and a crystal axis if it is lined up with the optic axis). The intensity for a given direction of incidence in the illuminating cone is to be found in each of the diffracted discs in precisely the equivalent position of the direct path in the central disc (Fig. 1).

For small spacings between the diffraction maxima, it is necessary to perform CBED with small C_2 apertures, and the resulting pattern becomes what has been tentatively called (above) a microdiffraction pattern. Disc spacings and angles can be measured quite accurately from such patterns

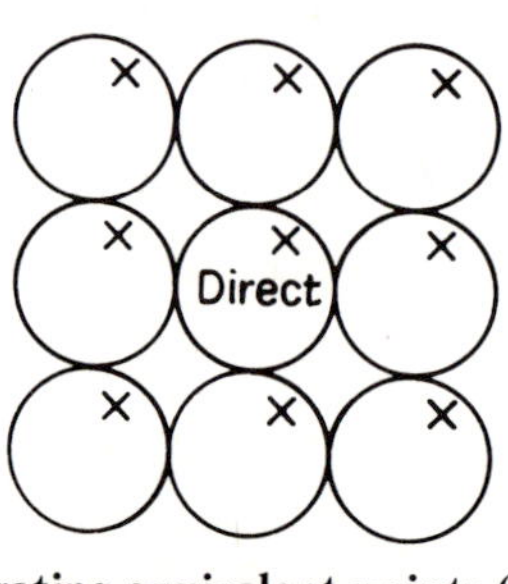

1 Diagram illustrating equivalent points (×) in each of discs of CBED diffraction pattern

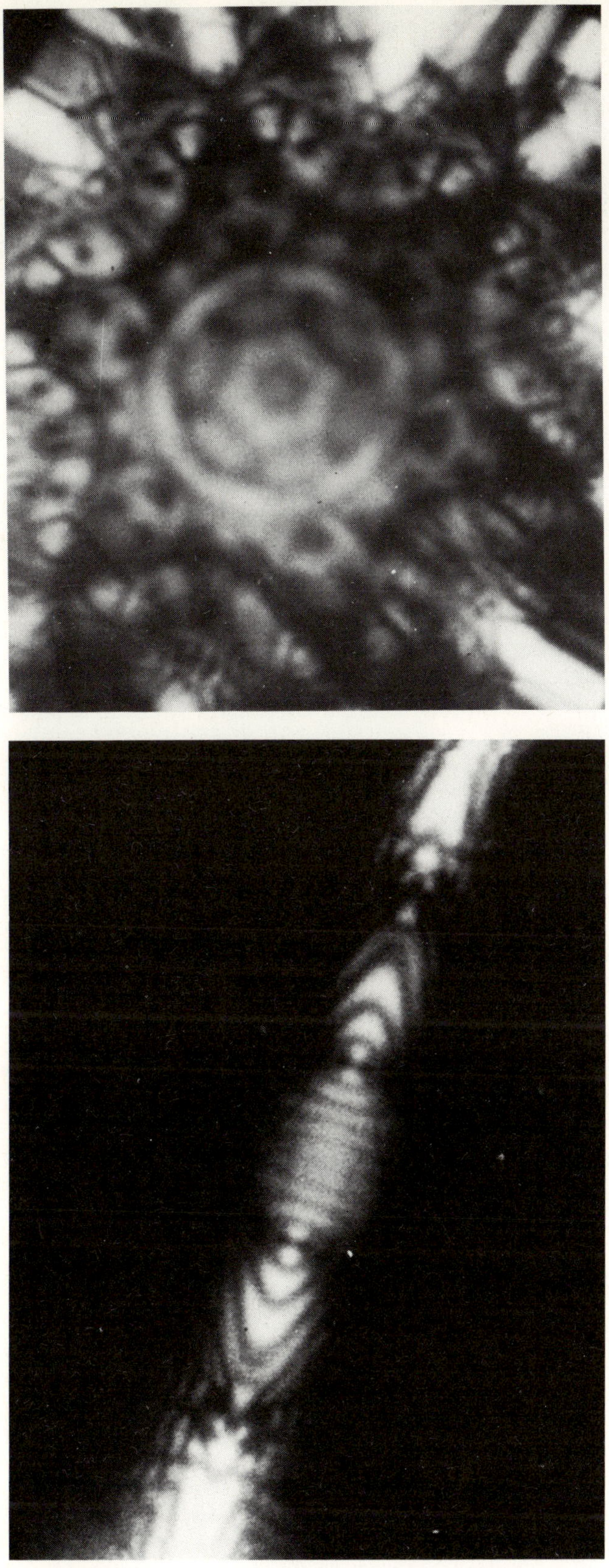

a

b

2 **Double rocking CBED zone axis patterns: (a) direct beam [0001] Mo_2C; (b) 220 beam [111] Ge**

so long as a number of criteria are satisfied. Firstly, it is necessary that there is a sharp image of the C_2 aperture in the back focal plane of the objective lens (the responsibility for this lies essentially with the manufacturer). Secondly, the probe size must be large enough for the Bragg angles to be well-defined. Thirdly, the C_2 aperture must be clean, otherwise charging effects at its perimeter will cause problems. In this situation, it is possible to regain the more detailed information typical of CBED by using a double rocking technique devised and implemented by Dr Eades.[4] A cone of large angular convergence is produced in time by rocking a narrow, essentially parallel beam about the point of interest with the scan coils of a STEM unit. The same STEM unit may then be used to 'un-rock' the beam after passing through the specimen onto an axial STEM detector further along the microscope column in a plane conjugate with the diffraction plane. The resulting CBED pattern, for any chosen order of reflection, is displayed on the CRT of a STEM unit. Other orders of reflection are always at angles to the chosen order fixed by the Bragg reflection condition and, by suitable adjustment of the microscope lenses, can be made to miss the detector collecting the chosen order. Different offsets can be used to display any chosen direct or diffracted disc (Fig. 2).

Most CBED experiments have so far been performed with probe sizes in the range 10–50 nm and using tungsten hairpin or LaB_6 filaments. Under these circumstances the illumination is extremely incoherent and the cone of electrons converging on the sample may be considered as composed of independent plane waves which do not interfere with each other. Recently, experiments have been performed in Arizona in Professor Cowley's group using a much finer coherent probe having a size of about 0.5 nm. Under these circumstances exciting and completely different sorts of CBED patterns are produced. Since different directions of incidence then interfere with each other, the patterns are sensitively dependent on the focus setting of the objective lens. Moreover, since the probe sizes are comparable with or smaller than the unit cell size, the patterns vary with translations comparable with the unit cell dimensions. Other novel aspects of these patterns are that the intensity maxima are not necessarily at the centres of the CBED discs and one disc may have double maxima. Clearly the interpretation of such patterns is going to be difficult but it certainly offers a challenging goal for those interested in highest possible resolution of microanalysis by diffraction. Interesting questions of local symmetry in such situations are being explored by Dr Spence[5] of the same group.

To return to more conventional, incoherent CBED, there are two important features of CBED which should be mentioned because of their importance in microanalysis. Through CBED the three-dimensional nature of electron diffraction has been thoroughly appreciated and exploited for the first time. The CBED pattern is composed of discs in the zero layer and then rings of discs in each of the successive layers of the reciprocal lattice (the so-called Laue zones; see Fig. 3). A two-dimensional film produced in the electron microscope is capable of revealing the reciprocal lattice in three dimensions (3D) so long as the microscope permits a large enough angular view. The implications of this development for investigation of crystalline materials are far-reaching. Two aspects of this wealth of information which are particularly useful are the sharp higher-order Laue zone (acronym 'holz') lines often visible in the central disc of diffraction patterns from thick crystals and the symmetry properties of the patterns. It is here that the

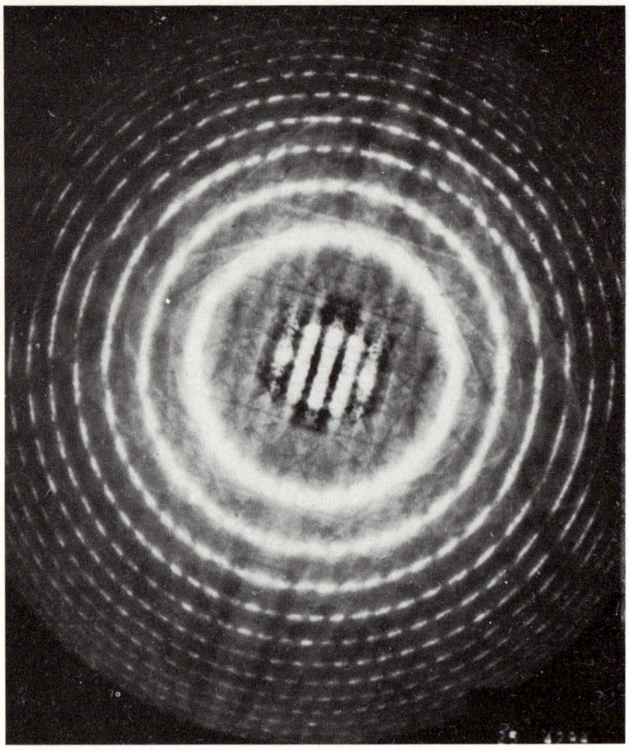

3 Short camera length CBED pattern of $NbSe_3$ showing many Laue zones; to reduce thermal diffuse scattering the sample was cooled to ~ 30 K

previously supposed weaknesses of electron diffraction turn out to be great strengths.

As has been discussed elsewhere,[1,7] the holz lines give an accurate measure of lattice parameter variation or strains within the sample. Indeed, they are sometimes invisible precisely because they are so sensitive that strains in the specimen, which might not otherwise have been noticed, are sufficient to broaden them, making them undetectable. The results are independent of camera length and achieve precisions of up to 2 in 10^4, a result which should be contrasted with the familiar inaccuracy of d-spacings from conventional selected area diffraction patterns. The symmetry of the patterns is related to the crystal point group.[6] It is precisely because of the dynamical nature of electron diffraction that this is true. In the weak scattering limit, Friedel's Law holds, and the 32 crystal point groups are reduced to the 11 Laue groups. Electron diffraction is therefore superior to X-ray or neutron diffraction in several respects. It only requires tiny isolated crystals; regions can be chosen for diffraction experiments which are demonstrably defect-free; a three-dimensional view of the reciprocal lattice (sometimes with thousands of reflections) can be obtained from carefully aligned crystals with an exposure of a few seconds' duration; and the crystal point group can often be determined from a single pattern. Even the so-called 'problem of double diffraction' can be turned into a positive advantage. Many kinematically forbidden reflections from perfect crystals appear by multiple diffraction routes and when these reflections are studied in carefully oriented samples by CBED, they frequently show a characteristic central line of missing intensity. When these dynamical absences occur in patterns with 3D diffraction information, it is possible to decide by inspection whether the symmetry element producing this effect is a screw axis perpendicular to, or a glide plane parallel to, the incident beam. Space group determination becomes a routine and rigorous exercise.

SOME PRACTICAL POINTS

It is a surprising fact that specimens which were expected to give good holz lines sometimes revealed none at all while others which looked most unlikely to give any results produced excellent holz patterns. There are a number of points to attend to if holz lines are not observed. Often fine lines cannot be seen on the fluorescent screen but are clearly visible in negatives. For crystals with large unit cells the angular view is restricted and better results may well be obtained at a different operating voltage. Strains resulting from carbon contamination, surface films created during electropolishing or ion etching can be further causes of failure. For isolated small particles lacking good thermal connection with the specimen grid, beam heating could explain the lack of lines. Finally, planar or point disorder, which may not be evident at first sight, can greatly attenuate large-angle scattering. There are three techniques for increasing the strength of holz lines which are worth bearing in mind. By cooling the specimen or reducing the operating voltage, 3D diffraction effects are enhanced at a particular zone axis. Cooling also greatly reduces or completely halts the process of carbon contamination. Finally, an alternative, and generally lower-symmetry, zone axis can be investigated. For strong 3D diffraction, it is generally advantageous to choose a direction with closely spaced reciprocal lattice planes (small first-order Laue zone). Thickness variation within the illuminated area may cause broad features within CBED discs to become indistinct, but will not affect holz lines significantly.

The symmetry of CBED patterns is that of the crystal as a whole. It can be seriously affected by planar faults parallel to the surface of the specimen and therefore not easily detected. It can also be affected by interfacial boundaries, strong tilting of the specimen, and steep wedge angles. For the identification of precipitates and inclusions, it is hence generally preferable to work with extraction replicas where matrix strains and interfaces are removed. There are many other advantages to the use of extraction replicas. The number of particles extracted onto a replica vastly exceeds the small number available in the thin area of an electropolished sample. Thus, the extraction technique greatly facilitates generalization of the results to macroscopic properties. In addition, particles can be selected from particular small areas of interest such as from cracks in a heat-affected zone near a weld. This task is essentially impossible for thin film microscopy. Finally, the EDX spectra from extracted particles serve as ideal standards whereas the spectra from particles contained in a matrix almost invariably include some unwanted matrix information. For symmetry determination it is essential that the CBED discs do not overlap, so that for large unit cells quite small C_2 apertures have to be used. Precise alignment of the crystal on its axis is essential and may be facilitated by lateral shift of the C_2 aperture once the alignment is within a degree or so.

For measurement of the spacing of the reciprocal lattice planes along the beam direction, overlap of the various diffraction orders in the h.-o.L. zones is a distinct advantage. This point is illustrated in Fig. 4. The fine structure often observed within the discs of a holz ring results from dynamical effects which are discussed elsewhere.[7,8] Let us assume, as is sometimes the case, that there is a single bright line within each holz disc. This line lies approximately

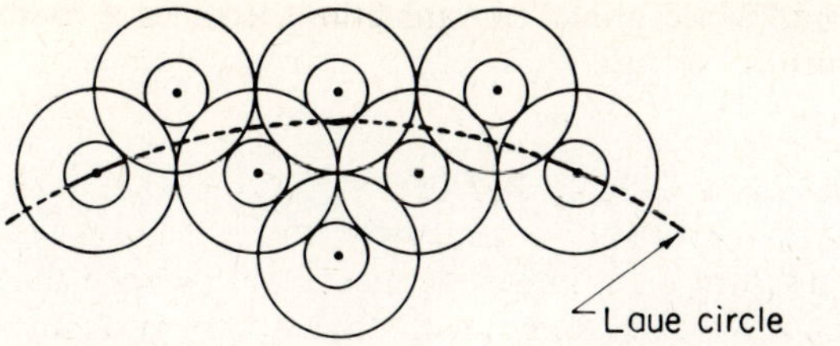

4 For small discs only solid portions of Laue circle would show; for large overlapping discs Laue circle would be continuous

where the circle, cut off from the Ewald sphere by the appropriate layer of the reciprocal lattice, intersects the disc centred on each of the reciprocal lattice points in the layer. For large discs, which overlap each other, a continuous ring is generated in that layer. Each of the Laue zones is defined by a separate ring and these rings give the most precise measure of reciprocal lattice plane spacings. Various refinements are possible. There may be distortion in the diffraction pattern especially at large angles from the optic axis. By using a small C_2 aperture revealing the individual reflections in each layer, the reflections may be indexed and a calibration graph constructed which relates real and ideal scattering angles. An example is shown in Fig. 5. The pattern formed from overlapping discs may be digitized and computer routines used to fit circles to the data, locate the centre of the pattern and hence deduce experimental values for the diameters of the Laue circles. Further refinement would take account of dynamical effects. Even without this our experiments indicate that an accuracy of 1% is possible.

CASE HISTORIES

(i) Precipitates and inclusions in 316 steel

During examination of long-term aged samples of this steel, we have been able to distinguish carbides, borides and oxides as well as diamond grit left over from mechanical polishing. The work was performed on particles extracted onto C replicas. Extraction may be performed either on specially prepared surfaces or from fracture surfaces. An example of an SEM-extraction replica comparison for a fracture surface of 316 steel is shown in Fig. 6. The carbides discovered were $M_{23}C_6$ and η-carbide, and these were readily distinguished by their different space groups (*Fm3m* and *Fd3m*) and by the totally different diffraction effects at each of the 15 or so zone axes of these materials which we studied. As these phases are both cubic and have similar cell constants, it could be difficult to distinguish them by conventional electron diffraction. Once these phases had been distinguished unambiguously, EDX spectra were taken and considerable chemical differences were found between them. The $M_{23}C_6$ always had a very high Cr content while the η-carbide contained a lot of Mo, quite a large quantity of Si and a trace of V. In view of the large Si content in this η-carbide structure and the enormous range of compounds known to have this structure (including oxides, nitrides and silicides[9]) it is actually uncertain whether this phase contained any carbon at all.

The boron content of the steels was of considerable interest and once their CBED patterns and EDX spectra had been established, the two borides which we discovered turned up quite regularly. One of these borides, with formula M_2B, had a very high Cr content and could therefore easily be confused with $M_{23}C_6$ in the absence of diffraction evidence. Although an orthorhombic phase (space group *Fddd*), it is pseudohexagonal and was generally multiply-twinned on a very fine scale. The resulting pseudohexagonal patterns led to considerable confusion until finally a small region of good crystal was discovered. The second boride, M_3B_2 (space group *P4/mbm*) was the more common precipitate in the steels which we studied. The [001] CBED patterns showed characteristic dynamic absences (Fig. 7) as do also the σ-phase precipitates at this axis, but the σ-phase space group ($P4_2/mnm$) and cell constants are different. EDX analysis showed that the M_3B_2 phase had a very high Mo content and a trace of V. In these respects it was similar to the η-'carbide' but its Ni content was much lower than that in η-'carbide', and there was no silicon. Two nitrides were found in the samples. One of these, AlN (space group $P6_3mc$), was relatively rare; the other, Cr_2N (space group probably *P31m*), was quite common, especially near the surface of the material, which had been in service for a long time. The quality of the Cr_2N patterns was poor, indicative of disorder. This phase is closely related to Mo_2C which we investigated quite thoroughly; a summary of the results is presented below.

In addition to heavily disordered silicaceous slag inclusions, four oxides were encountered in the 316 steel. One of these, Al_2O_3 (space group $R\bar{3}2/c$) occurred very frequently, presumably because it had been used for polishing. Its characteristic diffraction effects became very familiar. Less frequently, spinel inclusions were encountered (approximate composition Cr_2MnO_4, space group as for η-carbide). In addition, two forms of TiO_2 were regularly encountered (rutile, space group $P4_2/mnm$ and anatase, space group $I4_1/nmd$). Although these two forms of TiO_2 are both tetragonal, the different crystal structures could

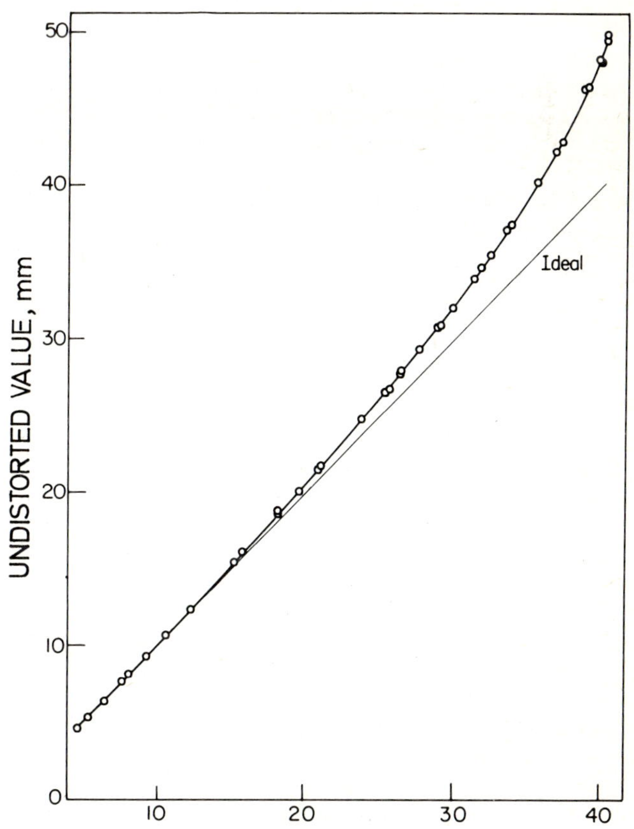

5 Calibration graph for correcting distortions in diffraction plane (EM400, 280 mm camera length)

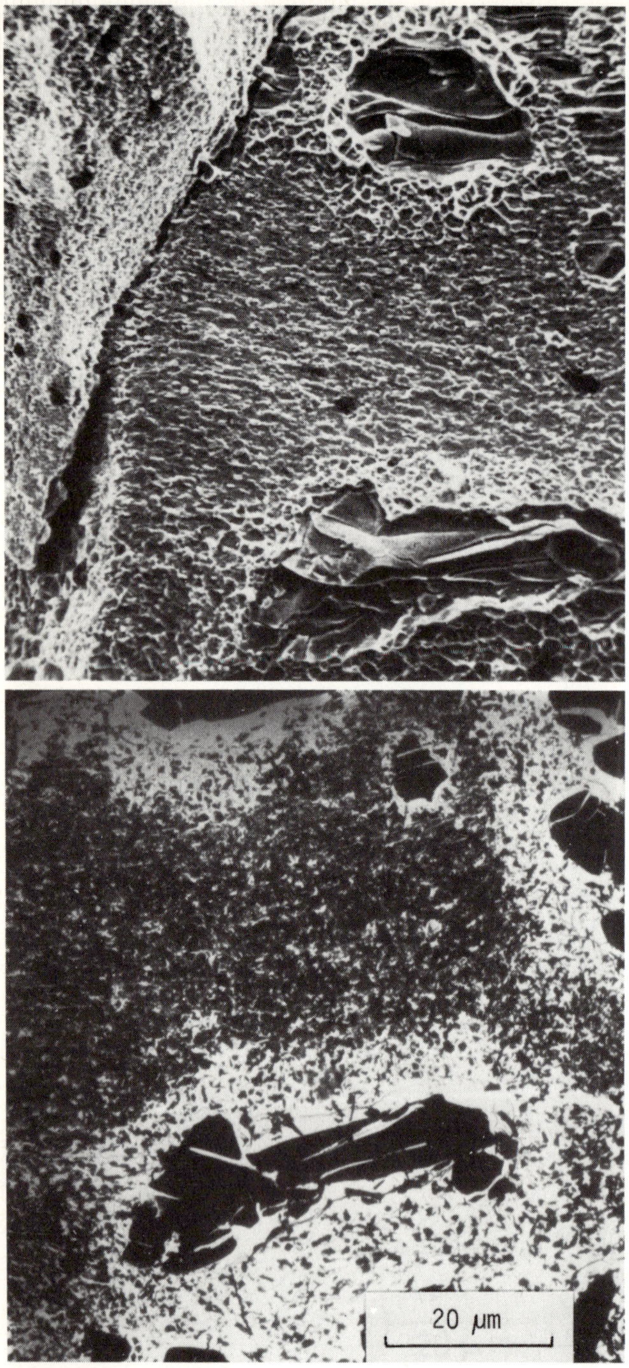

a

b

6 **(a) Scanning electron micrograph of fracture surface of long-term aged 316 steel; (b) extraction replica from surface illustrated in (a)**

7 **[001] CBED pattern from M_3B_2 precipitate extracted from sample of 316 steel.** Alternate reflections in two lines of discs passing orthogonal to each other through centre reveal dark bands (lack of intensity) which are a characteristic indication of kinematically forbidden reflections (dynamic absences)

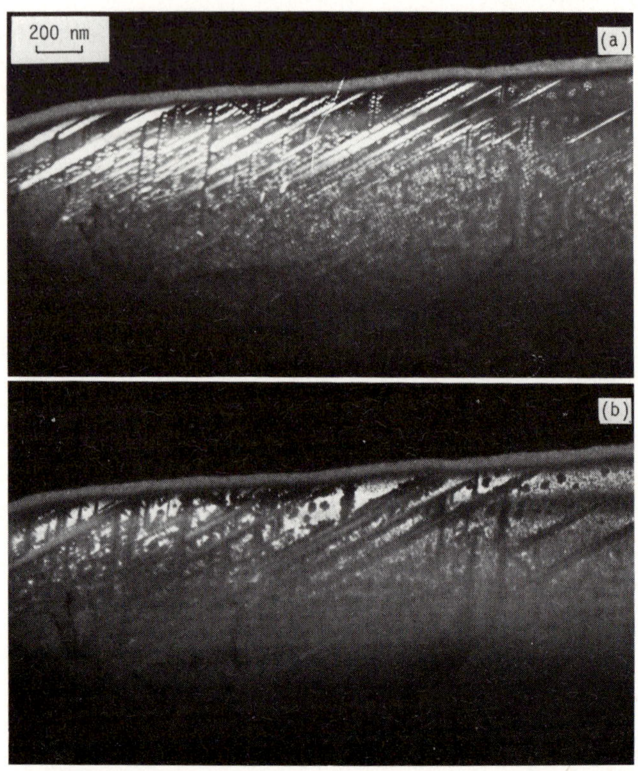

8 **Dark field micrographs in weak satellites at (a) 1/4 $(11\bar{2}0)$ and (b) 1/3 $(11\bar{2}0)$ reflections of specimen of Mo_2C_{1-x}**

be distinguished easily at more than 18 different crystallographic axes.

Diamond grit, when extracted onto the replicas, was recognized by the holz line patterns giving the cell constant of the diamond cubic lattice.

(ii) Mo_2C

Our interest was first aroused in Mo_2C when studying grain boundary carbides in molybdenum specimens. The precipitates revealed weak, very closely spaced linear structure but we were unable to account for this behaviour at that time. Large quantities of Mo_2C were produced in an undergraduate project[10] by reacting together appropriate quantities of Mo and C in an electron beam furnace. Specimens were prepared for transmission electron microscopy by ion bombardment and revealed a very interesting microstructure. It seems the melt was slightly carbon-deficient and after the solidification of a considerable amount of increasingly C-deficient Mo_2C_{1-x}, eutectic growth of laths of Mo and strongly C-deficient Mo_2C_{1-x} finally occurred. Large grains of the first material to solidify, with reasonable stoichiometry, were found by dark field electron microscopy to have a very complex microstructure. The $11\bar{2}0$ reflections of the hexagonal substructure (space group $P6_3/mmc$; $a_H = 0.30$ nm, $c_H = 0.47$ nm) were divided into both thirds and quarters by weak spots arising from the carbon ordering. Dark field micrographs in the 1/4 spots for satellites from each of three different $11\bar{2}0$ reflections, for a region oriented close to [0001] of the hexagonal cell, revealed long narrow domains elongated along the three different directions at 60° to each other (Fig. 8*a*). These domains were composed of stoichiometric Mo_2C (space group *Pbcn*; $a \simeq c_H$, $b \simeq 2a_H$, $c \simeq a_H\sqrt{3}$) in the three possible senses of elongation of the slightly distorted hexagonal cell of the orthorhombic phase. Dark field microscopy in the 1/3 spots lit up the regions between the three sets of elongated domains (Fig. 8*b*).

The Mo_2C_{1-x} in the eutectic laths, which were about 1 μm wide, revealed a weak linear structure similar to that originally observed in molybdenum specimens and sometimes broad faults with regular weak fringes like stacking faults. These regions mostly either had 1/3 spots in $11\bar{2}0$ systematic rows or else no additional reflections at all. It may be that they are a hexagonal phase ($a \simeq a_H\sqrt{3}$, $c \simeq c_H$) similar to that existing for nitrogen-deficient Cr_2N or Fe_2N. However, convergent beam diffraction at the [0001] zone axis of the hexagonal only cell revealed that the 1/3 $(11\bar{2}0)$ satellites were very much weaker than is the case for Cr_2N, and further that the satellites were of different intensity for each of the three $\{11\bar{2}0\}$ rows. It is, therefore, more likely that carbon-deficient Mo_2C_{1-x} forms an orthorhombic phase ($a \simeq 3a_H$, $b \simeq a_H\sqrt{3}$, $c \simeq c_H$). The linear structure may either represent boundaries between the different orientations of this orthorhombic cell or else be antiphase boundaries. The stacking faults appeared to mark the boundaries between the ordered Mo_2C_{1-x} and disordered hexagonal material. Work is in progress to clarify the details of this microstructure.

(iii) ηCu_6Sn_5

A somewhat different example is the long-standing problem of the long-range ordered structure of Cu_6Sn_5 described by Bernal[11] as an ordered NiAs phase with both *c* and *a* values five times that of the NiAs substructure and the source of much subsequent controversy. When one considers the number of solder joints made in the 50 years since Bernal's paper, it is perhaps surprising that this problem should remain unsolved. It has re-emerged recently in the context of solder connections between silicon device chips, and it has been established that the η phase forms at very low temperatures when copper and Pb/Sn solder are in contact. Identifying this phase by microdiffraction of these contacts proved difficult in the absence of good crystallographic data and so some samples were prepared in Bristol by direct interaction of appropriate amounts of Cu and Sn in sealed quartz ampoules.[12] The chief phase to emerge has, in fact, a monoclinic cell with approximate dimensions $a = 2.1$ nm, $b = 0.73$ nm, $c = 0.99$ nm, $\beta = 149°$ and a probable space group of *C2/c* or *Cc*. The relationhip with the NiAs sub-cell may be represented as

$$[100]_m \equiv [1\bar{2}10]_{NiAs}$$
$$[010]_m \equiv [10\bar{1}0]_{NiAs}$$
$$[\bar{2}0\bar{5}]_m \equiv [0001]_{NiAs}$$

Some of the reflections of the NiAs sub-cell are divided by five in the monoclinic phase, for example $(1\bar{1}0)_m \equiv (1/5)(21\bar{3}0)_{NiAs}$ (the NiAs subcell has approximate dimensions $a = 0.42$ nm, $c = 0.5$ nm). The phase was always found to be multiply-twinned by rotations of 60 or 120° about the pseudohexagonal '[0001]' axis. However, another large cell was also discovered in the samples which were prepared. Some of the diffraction patterns from this phase were at first sight similar to some of the diffraction patterns of the monoclinic phase. The present indications are that this phase is trigonal with space group *R3c* or $R\bar{3}c$ and cell constants $a_{trig} \simeq 3a_{NiAs}$, $c_{trig} \simeq 6c_{NiAs}$. This phase was also twinned on a very fine scale, the twins being rotated by 60° about the trigonal axis.

CONCLUSIONS

The methods outlined above for identifying phases in complex fine-scale microstructures have been used in a wide range of problems. Some other recent examples of their application are as follows:

(1) identification of phases associated with cracks in electron beam welds of 718 Nimonic alloys[13]
(2) identification of stable and metastable phases produced by crystallization of glassy CuZr films[14]
(3) determination of the phases formed between Y_2O_3 and Al in oxide dispersion-strengthened alloys[15]
(4) determination of phases in the CaO/ZrO_2 system.[16]

From a rather wide application of diffraction techniques to phase identification as an aid to microanalysis, it has emerged that there are clear advantages in comparison with energy-dispersive X-ray analysis (EDX) and electron energy loss spectroscopy (EELS). The major restriction is that the material to be analysed has to be crystalline. The chief advantage, as compared with EDX, is the lack of any limitation with regard to light elements. Another advantage is the certainty of the source of the information (there is no problem with confusion from spurious signals). In short, CBED is particularly useful for complex, multi-element, multiphase systems where it is required to establish with certainty each of the separate phases present. Thereafter, EDX fingerprints may be recorded and used for subsequent identifications.

The direct advantage with respect to EELS is that there is no restriction to thin samples (less than 50 nm thickness), no problem about light elements in the presence of heavy ones, and no overlapping peak problem for multi-element systems.

Other obvious advantages of the diffraction method are the gathering of information about crystal types, crystal defects and point or planar disorder.

ACKNOWLEDGMENTS

It is a pleasure to thank Drs K. Cooke, L. Stoter, and N. S. Evans and Messrs P. Norton, B. Kendrick and S. E. Hills for the opportunity to draw freely from the outcome of their research projects.

Some of the work reported here was supported by a CASE studentship with CEGB Berkeley (Dr P. Marshall) and by extra-mural research contracts from UKAEA Harwell which are gratefully acknowledged. Further support in the provision of equipment came from the Science Research Council.

REFERENCES

1 J. W. STEEDS: 'Introduction to analytical electron microscopy', ch. 15, 387; eds. J. J. Hren, J. I. Goldstein, and D. C. Joy, Plenum Press

2 P. GOODMAN: 'Scanning electron microscopy', vol. 1, 53; 1980, eds. J. M. Johari and A. M. F. O'Hare

3 G. LEHMPFUHL: 'Electron microscopy 1978', vol. III, 304; 1978, ed. J. M. Sturgess, Microscopical Society of Canada

4 J. A. EADES: *Ultramicroscopy*, 1980, **5**, 71

5 J. C. H. SPENCE: *Optik*, 1981 (in the press)

6 B. F. BUXTON, J. A. EADES, J. W. STEEDS, and G. M. RACKHAM: *Phil. Trans. R. Soc.*, 1976, **A281**, 171

7 P. M. JONES, G. M. RACKHAM, and J. W. STEEDS: *Proc. R. Soc. Lond.*, 1977, **A354**, 197

8 J. W. STEEDS: 'High voltage electron microscopy 1980', vol. 4, 96; 1980, eds. P. Brederoo and J. van Landuyt, Leiden

9 H. H. STADELMAIER: 'Developments in the structural chemistry of alloy phases', 141; 1969, ed. B. C. Giessen, Plenum Press

10 A. B. KENDRICK and S. E. HILLS: 3rd year undergrad. project report, University of Bristol 1979

11 J. D. BERNAL: *Nature*, 1928, **122**, 54

12 P. F. NORTON: MSc thesis, University of Bristol

13 R. VINCENT and J. W. STEEDS: Proc. 38th Annual Meeting Electron Microscopy Society of America, 332; 1980, Claitors Pub. Div.

14 A. F. MARSHALL, J. W. STEEDS, D. BOUCHET, S. L. SHINDE, and R. G. WALMSLEY: to be published in Proc. 39th Annual Meeting of EMSA, 1981

15 R. AYER and J. W. STEEDS, to be published in Ref. 14

16 A. HEUER, L. SCHOENLEIN and J. W. STEEDS: 1981, to be published

DISCUSSION of paper by J. W. Steeds

R. A. RICKS:

Concerning the use of convergent beam electron diffraction techniques to measure the carbon concentration in retained interlath austenite films in martensitic steels: do you find any problems with strain effects which are inherent in the microstructure?

J. W. STEEDS:

Yes. However, we were eventually successful in obtaining electropolished samples where the retained austenitic regions projected out well beyond the ferrite laths at the edge of a thin specimen and were then free of strain. Incidentally, by no means all of these boundary regions between the ferrite were austenitic; many were in fact composed of small packets of twinned martensite, with twin spacings of 3–5 nm and packet sizes less than 100 nm. Adjoining packets in a boundary region were twinned on different {112} planes.

C. Y. F. COLLIEX:

What is the origin of the difference of theoretical level required for the interpretation of microdiffraction patterns from a rather large area (10 nm) and a very small one (<1 nm)?

J. W. STEEDS:

Diffraction with a fine (<0.5 nm) coherent probe is considerably more difficult to interpret than CBED with, let us say, a 10 nm probe. Quite apart from the difficulties of orienting tiny crystals accurately onto zone axes, the perimeters of convergent beam discs are no longer sharply defined so that accurate plane spacing and angles cannot be determined. If one wishes to interpret the detail within the discs then many factors must be taken into account such as the defocus of the objective lens, the transfer function, the probe profile, the specimen size and shape. In the case of imperfect crystals all the distorted atomic positions have to be specified. Such calculations are not difficult to perform but they are time-consuming and the effects of all the relevant parameters have to be investigated.

L. M. BROWN:

May I first say how important these convergent beam techniques are, and how impressed I am by the recent progress in their interpretation. They are clearly the most important of the three microanalytical outputs, although of course all three are needed: microdiffraction, EDX, and EELS.

I do not consider that the interpretation of 1 nm probes is any more difficult than of 30 nm probes, provided the information sought is symmetry or lattice parameter. The high-resolution instruments are less well developed than the conventional ones, particularly in their more restricted maximum angle of deflection. However, if inhomogeneous samples (i.e. dislocation cores, single interstitials., etc.) are to be analysed, then complex mathematical calculation is required.

Study of grain boundary precipitation in Fe – 20%Cr – 25%Ni – 4%Al alloy using STEM/EDS and convergent beam electron diffraction

R A Ricks

In the intermediate temperature range (e.g. 700°C) the precipitation of the equilibrium NiAl phase in an Fe–20%Cr–25%Ni–4%Al occurs after the formation of the metastable, coherent Ni_3Al (γ') precipitate. This results in the formation of a homogeneous dispersion of NiAl after long aging times, since the Ni_3Al precipitates enhance the nucleation of NiAl. However, if the alloy is aged at temperatures above the γ' solvus the precipitation of NiAl occurs heterogeneously on boundaries and dislocations. Under these conditions the growth is diffusion-controlled and can be studied using EDS without the interference of neighbouring precipitates. The aim of this investigation is to study the growth of these grain boundary precipitates using STEM/EDS and correlate the results obtained with a parallel study using convergent-beam electron diffraction. The results have shown that the growth of these precipitates may be enhanced if the abutting austenite decomposes to ferrite to produce a duplex precipitate.

The author is in the Department of Metallurgy and Materials Science, University of Cambridge.

Aluminium and titanium are common additions to austenitic stainless steels since they are known to increase the strength of such alloys by the formation of a fine dispersion of γ' (Ni_3(Ti, Al)) precipitation.[1–4] However, if these materials are subject to either long-term aging treatments (e.g. during creep) or high temperature heat treatments, the γ' precipitation can be replaced by the equilibrium phase η (Ni_3(Ti,Al)) or β (Ni(Al, Ti)). This transformation is generally detrimental to the corrosion and mechanical properties of these alloys[5] and it is therefore important to understand the characteristics of this overaging phenomenon.

The present work is part of a wider investigation into the decomposition of supersaturated austenite in aluminium-containing austenitic steels. The results are concerned with a study of the growth of the ordered bcc NiAl precipitates in a high-purity Fe-20%Cr-25%Ni-4%Al alloy using both STEM/EDS and electron diffraction techniques. The investigation has shown that under certain conditions the formation of NiAl precipitates is accompanied by the decomposition of the adjacent regions to ferrite, with a further transformation of this ferrite to σ phase possible after longer aging times.

EXPERIMENTAL

The alloy used in this investigation was prepared from high-purity base elements to give a nominal composition (wt-%) of Fe-20%Cr-25%Ni-4%Al. The alloy was melted in an argon arc furnace and subsequently homogenized at 1200°C for 3 days under a partial pressure of argon. The ingot was cold swaged to 3 mm diameter rod prior to a solution treatment at 1200°C followed by a quench into iced brine to obtain a fully homogenized austenitic microstructure. All aging treatments were performed in muffle furnaces preset at the required temperature, which was checked using an electrical thermocouple type thermometer.

Specimens for electron microscopy were slit from the heat-treated rod under conditions of flood lubrication prior to thinning in a twin jet electropolisher with an electrolyte consisting of 10% perchloric acid, 20% glycerol, and 70% ethanol, using a potential of 35 V at room temperature. The specimens were thoroughly washed in clean methanol and examined in a Philips EM400 TEM/STEM equipped with a Link 860 EDS. A commercial thin foil correction programme was used which compensated for the effects of both X-ray detection efficiency and absorption of X-rays within the specimen. Previous studies[6] have indicated that when using this specimen preparation technique the effects of solute-rich surface films were negligible.

RESULTS

At temperatures above ~800°C the formation of the γ' (Ni_3Al) precipitate is suppressed and the equilibrium NiAl precipitate is formed, both on grain boundaries and as Widmanstätten intragranular precipitates. This paper is concerned solely with the grain boundary precipitation phenomena.

After aging at 800°C for 30 min, small precipitates are readily observed on the austenite grain boundaries (Fig. 1). These precipitates were found to obey a Kurdjumov-Sachs[7] orientation relationship with one abutting austenite grain[8] while growth occurred predominantly into the other austenite grain. Figure 2 shows a compositional profile up to the growth interface of a precipitate similar to those shown in Fig. 1. The concentrations of both Ni and Al are reduced immediately ahead of the interface, implying that

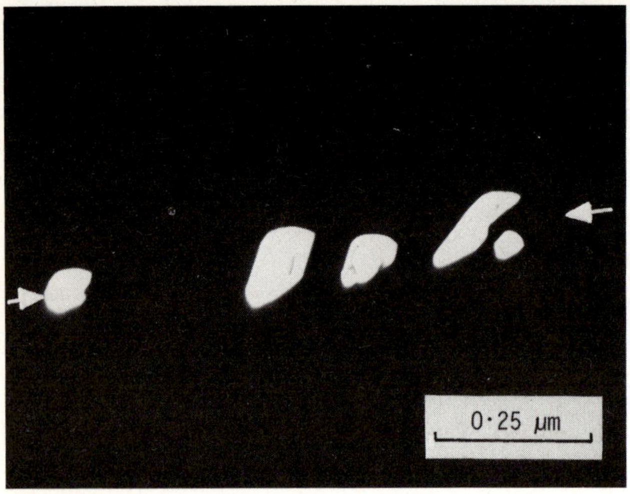

1 Centred dark field transmission electron micrograph showing precipitation of NiAl on austenite grain boundaries; aged at 800°C for 30 min

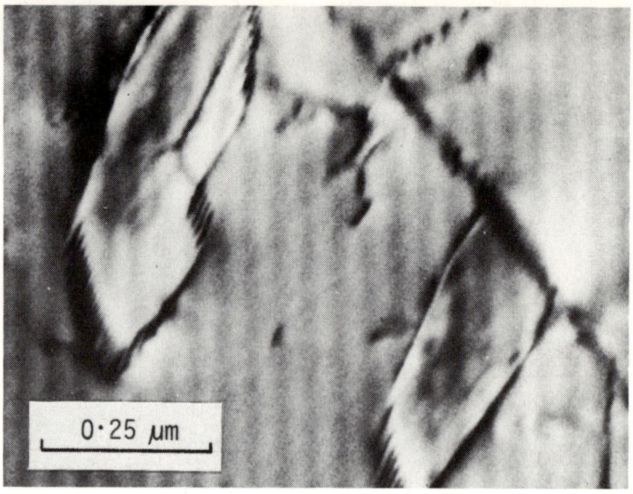

3 STEM image showing grain boundary precipitation after aging at 800°C for 1 h

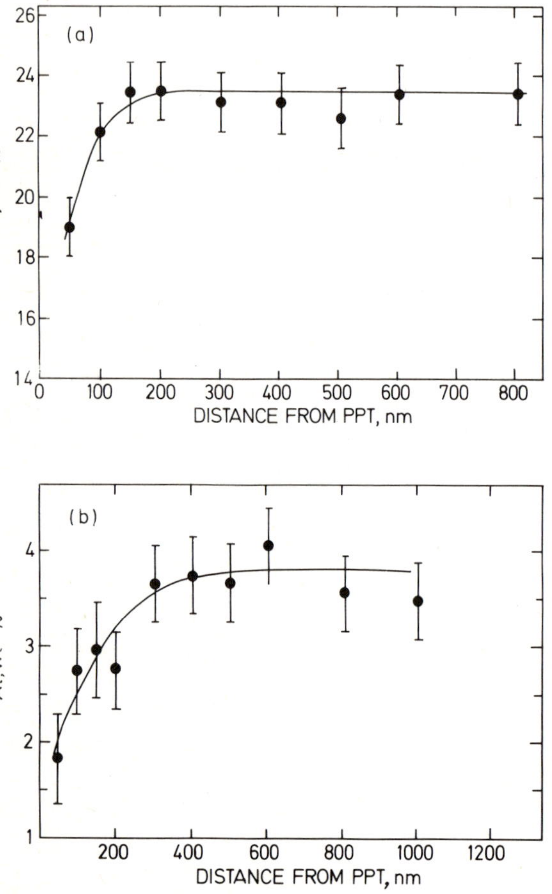

2 EDS composition profiles in austenite up to small grain boundary NiAl precipitate, showing volume diffusion fields of Ni (a) and Al (b)

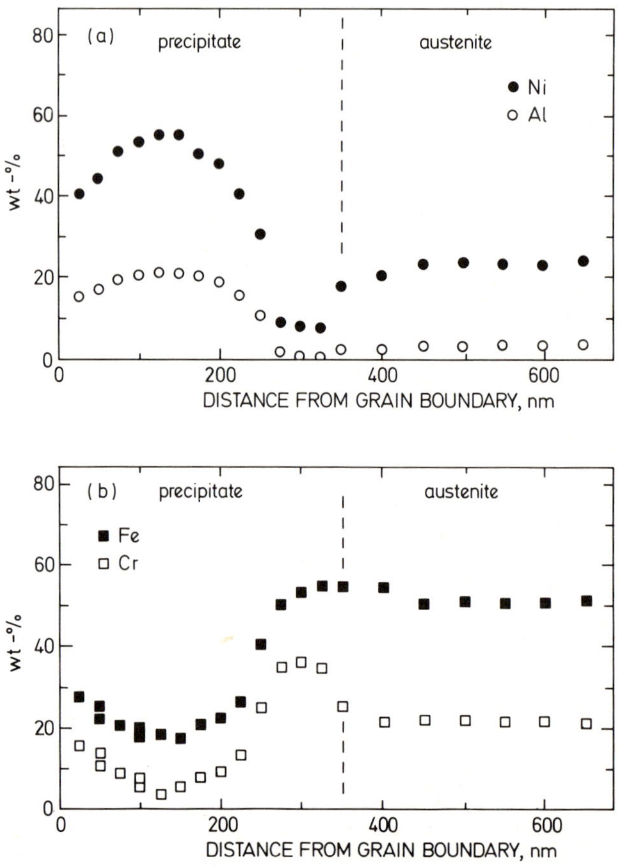

4 Composition profiles along length of small precipitate (in Fig. 3) showing Ni and Al levels (a) and Cr and Fe levels (b); for discussion see text

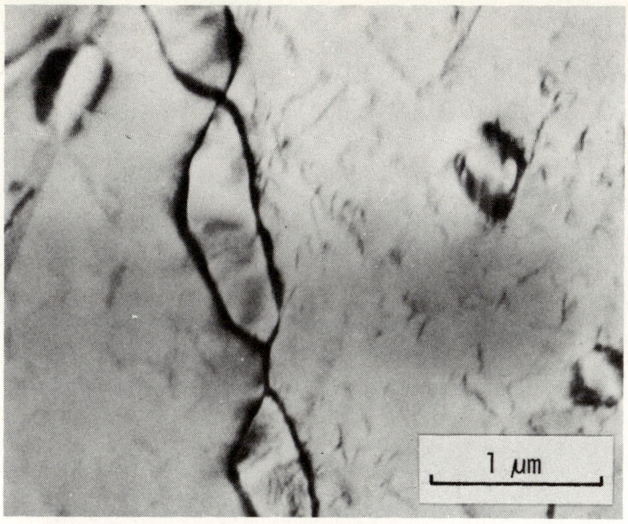

5 **STEM image showing copious grain boundary precipitation after aging at 800°C for 3 h**

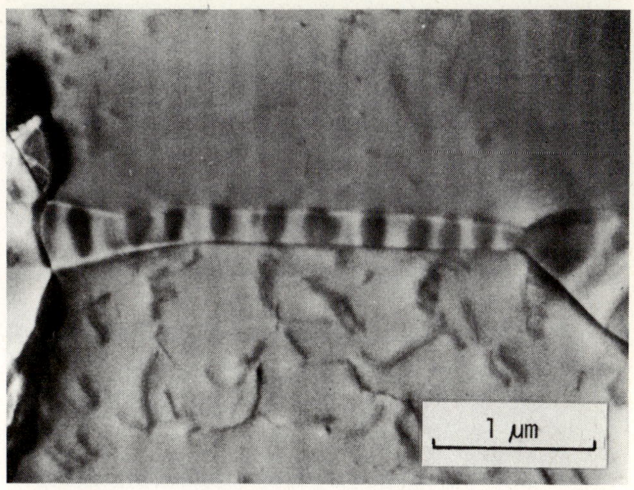

7 **STEM image showing contrast between (Ni + Al)-rich (dark) and (Fe + Cr)-rich portions of duplex grain boundary precipitate**

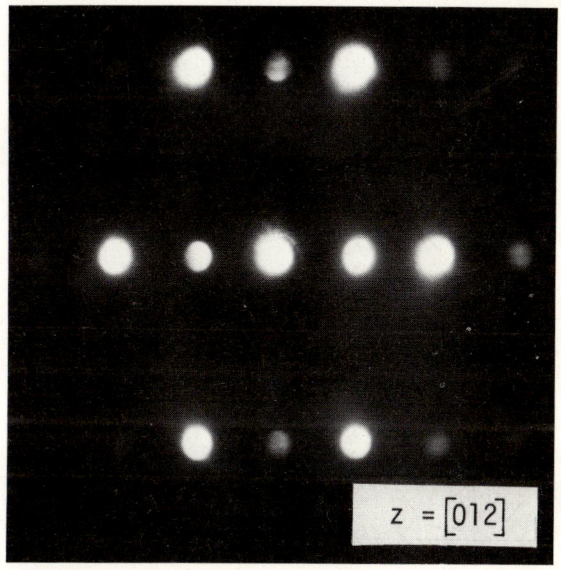

8 **Electron diffraction pattern from (Ni + Al)-rich region of duplex grain boundary precipitate**

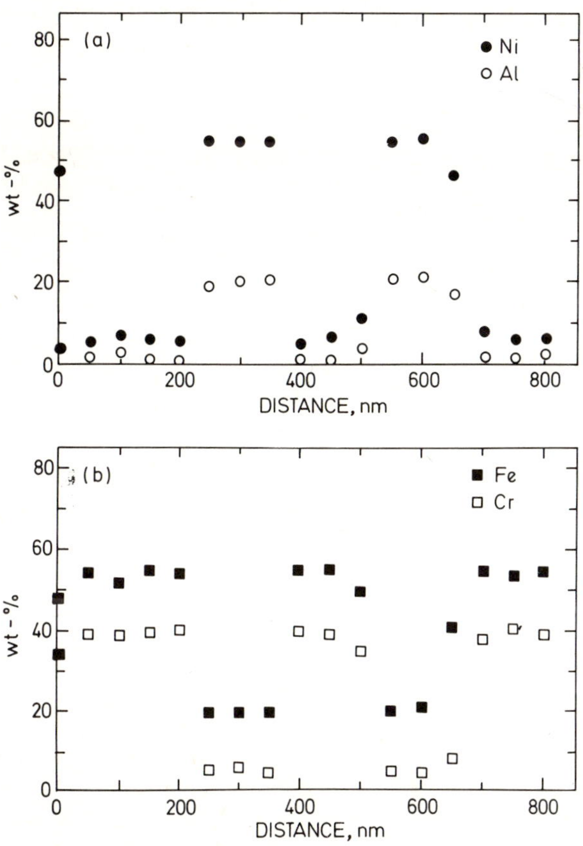

6 **Composition profiles along length of grain boundary duplex precipitates for Ni and Al (a) and Cr and Fe (b); note abrupt composition changes between β (NiAl) and ferritic regions**

the growth of the precipitates is controlled by volume diffusion. The composition of the grain boundary precipitates was found to be (wt-%): 54 ± 3 Ni, 18 ± 2 Al, 20 ± 2 Fe, 8 ± 2 Cr.*

At a more advanced stage of growth, dislocations in the interface between the precipitates and the austenite are observed (Fig. 3), whilst the precipitates have developed specific interfacial facets. The results of a compositional determination along the length of the small precipitate shown in Fig. 3, away from the austenite grain boundary, are shown in Fig. 4. These results indicate the following:

(a) although the precipitate nucleates rich in both Ni and Al, as growth proceeds the Ni and Al levels drop and the precipitate becomes enriched in Fe and Cr in the vicinity of the austenite/precipitate interface;

* It should be realised that the recorded Fe and Cr levels may well be enhanced by the effects of X-ray production from the matrix due to beam broadening at these small precipitate sizes.

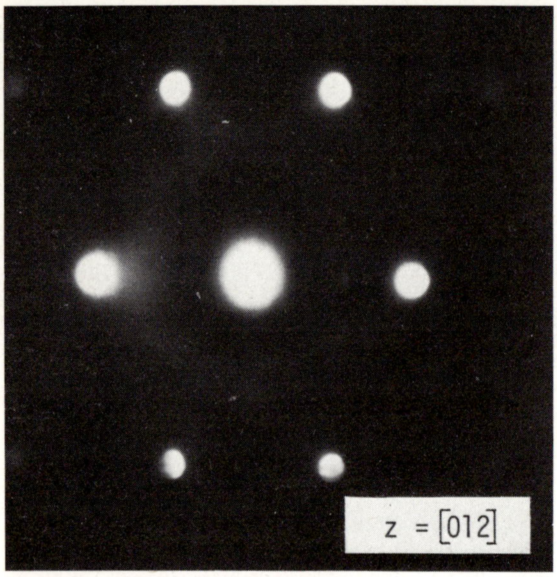

9 Electron diffraction pattern from (Fe + Cr)-rich region of duplex grain boundary precipitate

(b) ahead of the interface there is a slight drop in the Cr and Fe levels below those of the bulk.

These results suggest that, although the grain boundary precipitates nucleate as the equilibrium NiAl phase, as growth continues the precipitates become enriched in Cr and Fe to the point where the precipitate can no longer be this β phase. Since no interphase boundary could be observed between the (Ni + Al)-rich and (Fe + Cr)-rich regions of the precipitate shown in Fig. 3, it was deduced that the latter regions were isomorphous with the NiAl and were most probably ferrite (see later).

After longer aging times at 800°C copious precipitation was observed on the austenite grain boundaries as shown in Fig. 5. Analytical determinations of these grain boundary precipitates show that the composition along them fluctuates in a manner illustrated in Fig. 6, with (Ni + Al)-rich portions and (Fe + Cr)-rich portions. Under certain diffracting conditions these compositionally dissimilar portions are clearly differentiated using STEM imaging as shown in Fig. 7.

The application of convergent beam electron diffraction to the (NiAl)-rich regions of the precipitates produced diffraction patterns consistent with the ordered bcc structure of the β NiAl phase, as shown in Fig. 8. Diffraction patterns obtained from the (Fe + Cr)-rich regions confirmed that they were bcc in crystal structure (Fig. 9) and were most probably the Cr-rich α' phase. These results are also illustrated in Fig. 10 which shows a transmission electron micrograph of a duplex precipitate colony in bright field (10*a*) and in centred dark field modes using a NiAl superlattice 001 reflection (10*b*) and a combined ferrite/NiAl 002 reflection (10*c*). These results will be discussed in the following section.

After aging times of one week at 800°C the precipitates were found to grow into the austenite grains, eventually producing large colonies as shown in Fig. 11, consisting of alternate (Ni + Al)-rich portions and (Fe + Cr)-rich portions. Figure 12 shows typical X-ray spectra obtained from these different regions.

After much longer aging times the ferritic portions of the duplex precipitate colonies could transform to a Cr-rich

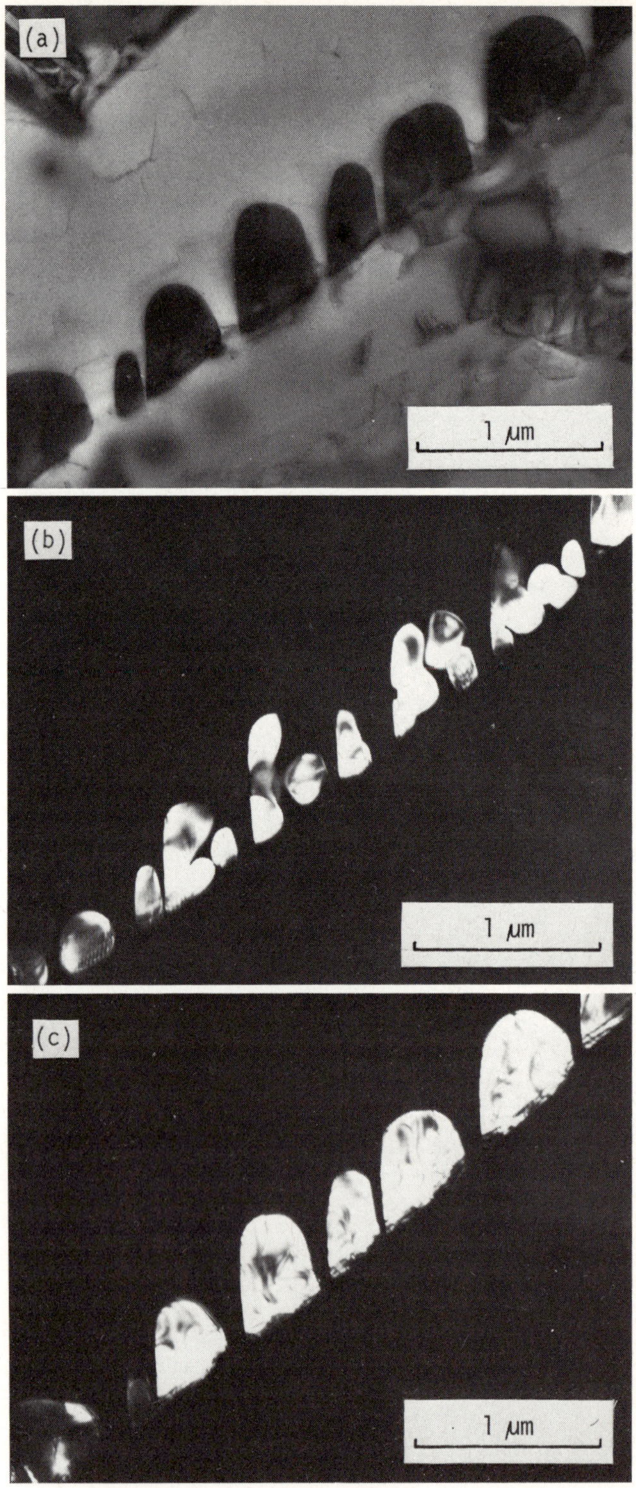

10 Transmission electron micrographs of duplex grain boundary precipitates after aging at 800°C for 3 h: (a) bright field; (b) centred dark field using 001 NiAl reflection; (c) centred dark field using 002 combined ferrite/NiAl reflection

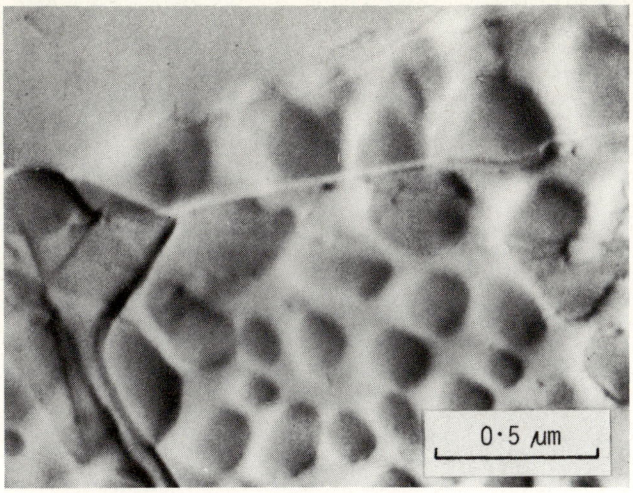

11 STEM image of large duplex NiAl/ferrite colony growing into austenite after aging at 800°C for 1 week

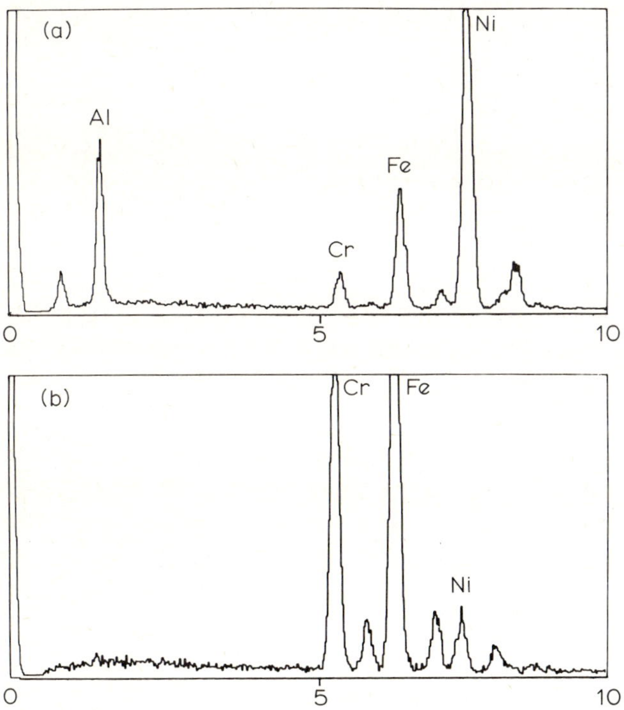

12 EDS spectra obtained from NiAl regions (a) and ferritic regions (b) of duplex precipitate colonies shown in Fig. 11

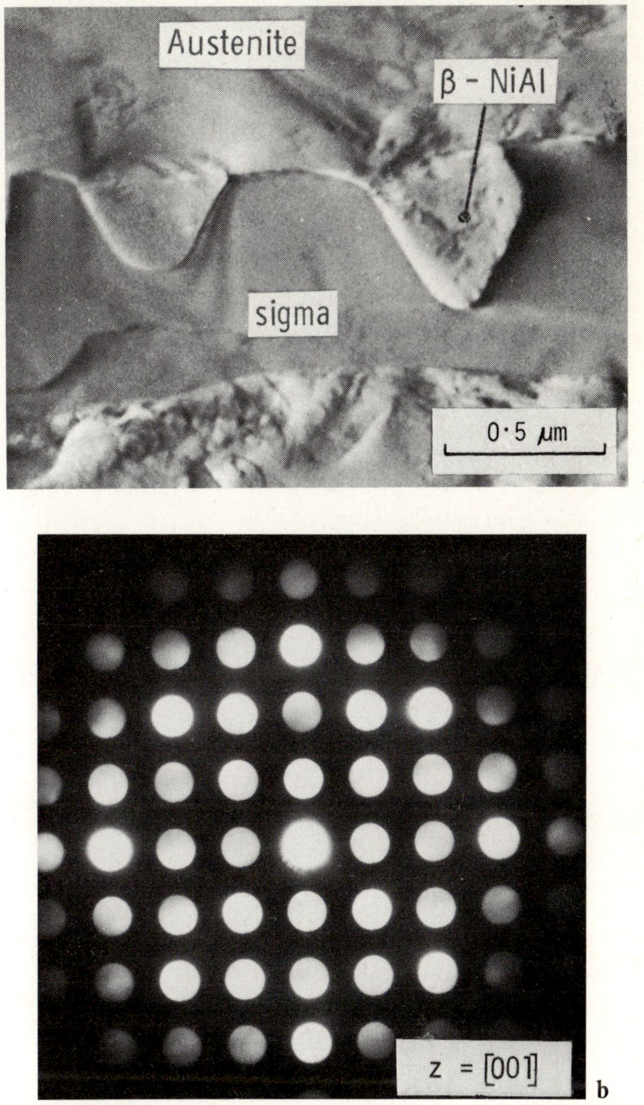

13 (a) STEM image of duplex precipitate colony formed after aging at 800°C for 2 weeks showing formation of sigma phase from the original ferritic regions; (b) electron diffraction pattern from sigma phase regions shown in Fig. 13*a*

phase. Figure 13 shows a bright field scanning transmission electron micrograph (13*a*) of a duplex grain boundary precipitate together with an electron diffraction pattern from the original ferritic portion (13*b*). The analysis of the diffraction pattern indicates that the change in phase is fully consistent with the formation of σ phase from the ferrite, whilst EDS showed the composition of this phase to be (wt-%): 17 ± 3 Fe, 78 ± 5 Cr, 5 ± 1 Ni, <0.5 Al, which varies considerably from the original ferrite composition (see Fig. 6). The remaining NiAl regions of Fig. 13*a* were found to have the following composition (wt-%): 34 ± 2 Fe, 44 ± 3 Ni, 10 ± 1 Cr, 12 ± 2 Al, which, again, varies from the original NiAl composition (Fig. 6), especially with respect to Fe, which implies that the NiAl becomes enriched with the Fe which is rejected by the σ phase.

DISCUSSION

The results outlined in the previous sections have shown that the initial growth of grain boundary NiAl precipitates is dependent on the volume diffusion of Ni and Al to the precipitate. During subsequent growth of these precipitates the Fe and Cr levels increase *within* the precipitate, which indicates that the isomorphous growth of ferrite onto the existing NiAl has taken place. The gradual change in solute levels (Fig. 4) at this stage indicates that there is no actual phase boundary between the ferrite and the NiAl, and transmission electron microscopy has confirmed this. However, after longer aging times full partitioning occurs and two distinct precipitate types are observed.

This growth mode may be explained in terms of the relatively slow diffusion rates of the solute elements in the alloy. Thus the growth of a grain boundary precipitate can be enhanced if alternate regions of (Ni + Al)-rich β phase and (Fe + Cr)-rich ferrite are formed instead of partitioning the Cr and Fe to the austenite and obtaining Ni and Al from the austenite by volume diffusion. Clearly the former mechanism allows a more rapid growth rate to be achieved.

In this manner the observed duplex NiAl/ferrite precipitate colonies could be formed.

One consequence of this reaction is that the resultant ferritic regions become favourable sites for σ phase formation since they have a high Cr concentration. Thus despite the presence of 4%Al in this alloy, which in concentrations above 1.5% is known to suppress σ phase formation,[9] the precipitation of duplex NiAl/ferrite colonies provides regions which are depleted in Al and Ni, and sufficiently enriched in Cr to allow the subsequent transformation to σ to occur. The EDS results show that, once formed, the σ phase becomes further enriched in chromium whilst the remaining, originally NiAl, regions become enriched in Fe.

ACKNOWLEDGMENTS

The author would like to thank Professor R. W. K. Honeycombe for the provision of laboratory facilities and to acknowledge the provision of a maintenance award by the SRC. Discussions with Dr P. R. Howell are gratefully acknowledged.

REFERENCES

1 F. G. WILSON and F. B. PICKERING: *Acta Metall.*, 1968, **16**, 115
2 J. M. SILCOCK and N. T. WILLIAMS: *J. Iron Steel Inst.*, 1966, 1100
3 L. K. SINGHAL and J. W. MARTIN: *J. Iron Steel Inst.*, 1967, 947
4 B. R. CLARK and F. B. PICKERING: *J. Iron Steel Inst.*, 1967, 70
5 R. W. K. HONEYCOMBE: 'Steels, microstructure and properties', 1981, 223–225, London, Edward Arnold
6 R. A. RICKS, P. D. SOUTHWICK, and P. R. HOWELL: *J. Microsc.*, 1981, in press
7 G. KURDJUMOV and G. SACHS: *Z. Phys.*, 1930, **64**, 325
8 R. A. RICKS: unpublished research

DISCUSSION of paper by R. A. Ricks ('G.b. precipitation')

J. W. STEEDS:
I do not think the convergent beam electron diffraction pictures you presented are from σ phase. The zero layer pattern looks more like what we would expect from χ phase or the 'η-carbide' structure.

R. A. RICKS:
The diffraction pattern presented in this paper [Fig. 13*b* on previous page] may be indexed in terms of the σ phase structure. However, in the light of Dr Steeds's comment, further work is now in progress to determine unambiguously the structure of this phase.

Precipitates in neutron-irradiated austenitic stainless steel

J M Titchmarsh and T M Williams

The precipitate phases present in FV548, a niobium-stabilized austenitic steel, subsequent to neutron irradiation in a fast reactor, have been identified using analytical electron microscopy. The irradiation dose, temperature and initial material condition were varied. In addition to those types of precipitate found in thermally aged, unirradiated material, the irradiated samples often contained one or more phases rich in Ni and Si, both at grain boundaries and at irradiation-induced voids, depending upon the material condition and temperature. The factors governing the formation of this irradiation-enhanced precipitation and the relevance of the precipitation to void-swelling are discussed.

The authors are in the Metallurgy Division at AERE Harwell, Oxfordshire.

The transmission electron microscope is well established as an instrument for the examination of complex alloys in experiments aimed at relating microstructures to bulk materials properties. Advances made during recent years in analytical electron microscopy with high spatial resolution has extended the potential of the instrument in areas of investigation which were previously untenable, for example the identification of very small, randomly-oriented particles. Mixed populations of such precipitates are a commonly observed feature in materials which have been subjected to high temperature neutron- or ion-irradiation.

The steels or nickel-base alloys commonly used in the manufacture of fast reactor core components are prone to a phenomenon referred to as 'void swelling' (see Refs. 1–4). Vacancies and interstitials are produced in equal numbers when atoms are displaced from their lattice sites during irradiation, but in the presence of point defect sinks such as dislocations which are known to absorb interstitials preferentially, a surfeit of vacancies can exist. These vacancies may aggregate to form 'voids', with the result that the overall dimensions of a component may increase. Clearly, since volume or shape changes are undesirable in fast reactor components manufactured to close tolerances it is important to choose alloy compositions and conditions to minimize the effect.

Irradiated materials are subject to the phenomenon of 'non-equilibrium' solute segregation caused by the flux of point defects to sinks.[5] It has recently been appreciated that the precipitation which occurs under irradiation is influenced by this non-equilibrium segregation and that copious and sometimes unexpected precipitation occurs at point defect sinks such as grain boundaries and voids. Furthermore, since the swelling behaviour of an alloy is composition-dependent (e.g. see Ref. 6) precipitation may affect the properties because it results in changes in the composition of the matrix in which it occurs. It is therefore important to study and understand the precipitation behaviour of irradiated alloys.

In this paper we describe our study of the relationship between precipitation, irradiation temperature, irradiation dose and initial material condition for the case of FV548, a niobium-stabilized austenitic stainless steel. This steel is one of the candidate materials from which certain fast reactor core components might be fabricated.[7] Some of these components would be expected to have a working lifetime of up to two years at temperatures up to 700°C and a maximum displacement dose approaching 100 displacements per atom (dpa). The void swelling behaviour of this and other austenitic alloys has been studied for a number of years but it is only now that a more detailed understanding of the interrelationship between alloy composition, precipitation and void swelling is becoming possible as a result of the use of analytical electron microscopy.

EXPERIMENTAL

The compositions of the two casts of FV548 examined in this study are given in Table 1. Cast 01388 was solution-treated for 0.5 h at 1150°C and irradiated to a total dose of 16 dpa at five temperatures, namely 466, 561, 591, 667 and 738°C. Cast VA447 was prepared in three conditions: solution-treated (ST) for 0.5 h at 1150°C, ST + 20% cold-worked and ST + aged for 3 h at 850°C. Samples in each of these conditions were irradiated to 30 dpa at temperatures of 380, 470 and 600°C. Neutron irradiation was performed in the Dounreay Fast Reactor by encapsulating 2.3 mm disc samples in a stainless steel container which, in turn, was sealed in a second container filled with inert gas to ensure adequate heat transfer when immersed in the liquid sodium coolant. The temperature was monitored by thermocouples at each end of the containers.

TABLE 1 Chemical composition of casts of FV548 and laboratory casts *A* and *B*

Cast	C	Mn	Si	Ni	Cr	Mo	Nb	Fe
VA447	0.11	1.14	0.35	11.8	16.5	1.44	0.92	bal.
01388	0.08	1.06	0.45	11.5	16.3	1.40	1.02	bal.
Alloy A	0.02	1.5	6.0	12.0	17.0	2.5	1.7	bal.
Alloy B	0.22	1.5	6.0	12.0	17.0	2.5	1.9	bal.

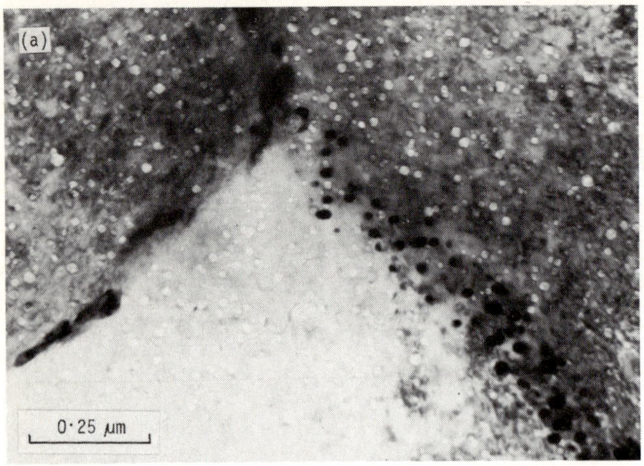

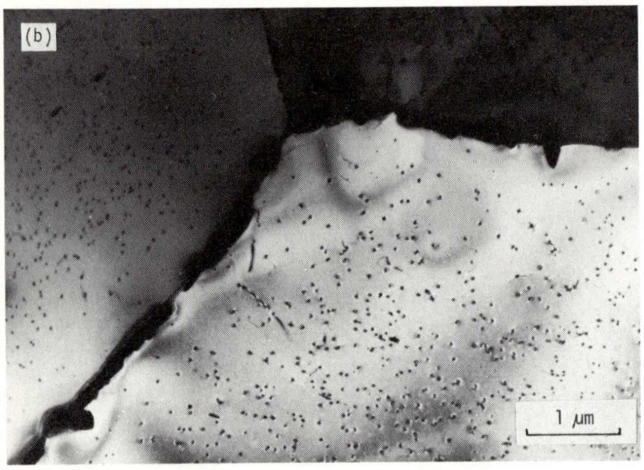

1 **Transmission electron micrographs: (a) fine-scale grain boundary precipitation in material solution-treated, aged and irradiated at 380°C; (b) much coarser grain boundary precipitation in material solution-treated and irradiated at 667°C**

TEM samples were prepared directly from the discs using standard electropolishing procedures and examined in a Philips EM400 fitted with a probe-forming objective lens and qualitative EDX system. A standard, double-tilt specimen holder was used because of the need to orient the foils accurately to obtain useful microdiffraction patterns as well as the X-ray spectra from selected regions. EDX spectra could be obtained only by tilting the specimen to high angles from the horizontal plane and such an orientation rarely enabled the simultaneous recording of a useful microdiffraction pattern and spectrum. Diffraction patterns, therefore, were recorded prior to any additional tilting necessary to record a spectrum.

RESULTS

General microstructural features

Details of the void swelling behaviour for the two casts have been published elsewhere[8] and will only be summarized briefly here. At comparable irradiation temperatures the void swelling increased with displacement dose in the solution-treated material. Measurable void swelling occurred only at irradiation temperatures up to 470°C for the doses studied although isolated, larger voids were observed after irradiation up to 600°C. The swelling was greatest in the ST+aged material and least in ST+CW material, where measurable swelling occurred only at 380°C. Following irradiation at 380°C, material in the three initial conditions each contained a high dislocation density and fine intergranular precipitation. The latter was identified, using selected-area diffraction and dark-field imaging, as NbC oriented with the matrix. The number densities of the precipitate decreased with increasing irradiation temperature, but the mean size of NbC particles increased. Material from cast 01388 irradiated at 667 and 738°C contained very few dislocations but the mean NbC particle size increased to 20–30 nm. All specimens contained a few large, round particles (up to 5 μm) of NbC, presumably not dissolved during solution treatment.

The grain boundaries in all specimens were decorated with precipitates, the mean dimensions of which increased from a few tens of nanometres after irradiation at 380°C, up to a micron or more at 738°C (Fig. 1). Where void swelling had occurred, many of the voids were associated with precipitate particles (Fig. 2). These particles were not generally crystallographically aligned with respect to the matrix. The mean particle size increased in proportion to the mean void diameter (i.e. with irradiation temperature) but even at 600°C were only a few tens of nanometres. In addition, particles of a similar size but not associated with voids were also present. After irradiation at temperatures $\geqslant$600°C specimens contained intragranular precipitates, both isolated and in small groups, with dimensions typically hundreds of nanometres.

Identification of precipitates

The ease of identification of individual precipitate particles in random orientation with respect to the matrix depended upon the dimensions of the particles and the thickness of the matrix. Both microdiffraction patterns and EDX spectra became increasingly difficult to interpret as the particle size decreased. A lower bound of approximately 50 nm was estimated as being the size of the smallest particles which could be identified with confidence. Hence, the nature of many of the particles at grain boundaries in specimens irradiated at 380°C could not be determined, and identification of the intragranular precipitation in specimens irradiated at this temperature was totally unsuccessful. At higher temperatures, however, particles at grain boundaries and at the larger voids in the intragranular regions of the foil were identified successfully.

The identification proceeded in the following way. Large, isolated grain boundary particles were oriented to

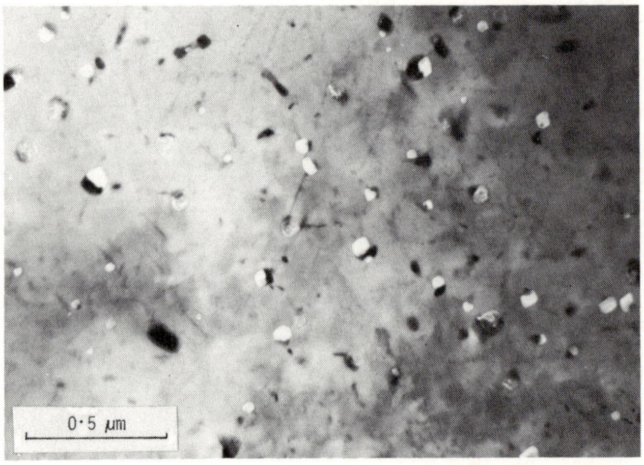

2 **Transmission electron micrograph showing high density of intragranular voids, often associated with precipitate particles, in material solution-treated, aged and irradiated at 470°C**

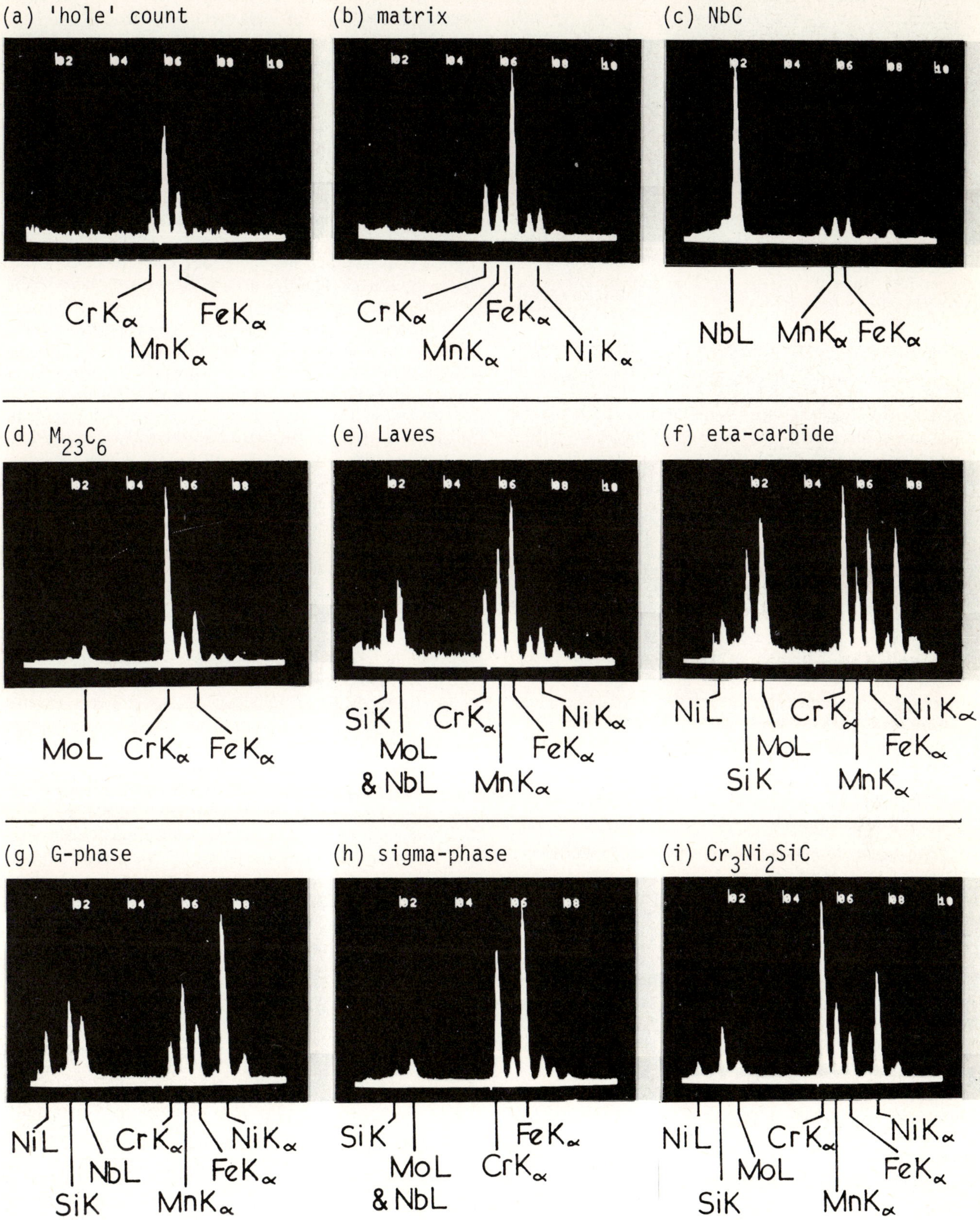

3 **EDX spectra: (a) typical hole-count with strong Mn K_α peak arising from residual activity of whole specimen; (b) matrix; (c) NbC; (d) $M_{23}C_6$; (e) Laves phase; (f) eta-carbide; (g) G-phase; (h) σ-phase; (i) Cr_3Ni_2SiC. Spectra in (d)–(i), recorded from grain boundary particles, include minimal contributions from matrix**

give microdiffraction patterns containing low-index zone axes. Where feasible, several patterns were recorded from an individual particle tilted to different orientations, the data measured from these patterns being sufficient to identify the type of particle when compared with reference crystallographic data. Face-centred cubic structures were distinguished from diamond cubic by tilting to the ⟨100⟩ zone axis and observing the presence or absence of 200-type reflections. An EDX spectrum was also recorded using a probe with a diameter small enough to minimize any matrix excitation. The ratios of the heights of the various elemental peaks then served as a 'fingerprint' to identify other particles of the same type. Some variation in peak height ratios was found as a function of irradiation temperature for some of the types of precipitate present, but such variations were found to be too small to affect the accuracy of particle identification.

All the spectra were observed to contain Mn *K* peaks, these X-rays being produced by the continued decay of the Fe^{55} isotope produced throughout the specimen during irradiation. A 'hole'-count spectrum highlights this feature (Fig. 3*a*). Subtraction of this background from spectra showed that the true concentration of Mn was negligible in all the particle types present. Typical EDX fingerprints from grain boundary particles are shown in Fig. 3.

The smaller grain boundary and intragranular precipitates were often identified using the fingerprints alone. Diffraction patterns were recorded, whenever possible, to confirm identification. The major contribution to the spectra from these particles often came from the surrounding matrix; to overcome this problem, a number of identification rules were deduced by adding a matrix spectrum and a pure spectrum, from each particle type in turn and in varying proportions, to determine which combinations might confuse identification. These rules were as follows:

(i) NbC spectra could not easily be misinterpreted, appearing as the matrix with additional Nb *K* and Nb *L* peaks and very low Si *K*.
(ii) $M_{23}C_6$ was distinguished by the increased Cr *K* peak and lack of Si, the latter characteristic preventing possible confusion with σ-phase.
(iii) Laves ($Fe_2(MoNb)$) had a Ni content similar to that of the matrix, a slightly increased Cr content and noticeable Si *K* and (Nb *L* + Mo *L*) peaks. The Si *K*/(Nb *L* + Mo *L*) ratio was typically 0.5 or less.
(iv) Eta-carbide (M_6C) had greatly increased Si, Nb + Mo, Cr and Ni concentrations compared with the matrix. It was distinguished from Laves by having a Si *K*/(Nb *L* + Mo *L*) ratio of 0.75 to 1.0 and a Ni/Fe ratio larger than in the pure matrix or Laves-plus-matrix mixtures.
(v) G-phase ($Ni_{16}Nb_6Si_7$) had an increased Ni content relative to the matrix. The Si *K*/(Nb *L* + Mo *L*) ratio was ⩾1 and this prevented confusion with any of the other types.
(vi) Sigma-phase had a significantly increased Cr content and confusion was possible only with $M_{23}C_6$ (see above).
(vii) Cr_3Ni_2SiC is a second form of M_6C, chemically different from the eta-carbide described above. It showed Cr, Ni and Si peaks larger than the pure matrix, but little Mo or Nb. It was distinguished from σ and $M_{23}C_6$ by the Ni/Fe ratio being greater than in the matrix.

In specimens irradiated at 470°C and below, particles were found to give spectra showing increasing Ni and Si alone, suggesting that at least one additional phase was present. Limited diffraction data showed that this was not $Ni_3Si(\gamma')$, which has been observed in irradiated 316 steel,[9] but was possibly Ni_2Si. The carbon content of particles identified as carbides was assumed, and some carbon might have been substituted by other light elements.

Similar types of precipitate were generally observed both within grains and at grain boundaries in those specimens where both classes were large enough to identify. The results of the precipitate analyses are presented in Table 2. Included in this table are additional results from unirradiated material, ST and aged at 600°C for 1000 h.

DISCUSSION

When complex austenitic alloys (such as FV548 steel) are aged, a variety of carbides and intermetallic phases are formed, depending on the aging time and temperature and the solubility relationships as the alloy attempts to reach thermodynamic equilibrium.

The types of phase which form clearly depend upon the concentrations of the various chemical species, the atomic diffusion rates and the ease of precipitate nucleation. Particles find easy nucleation sites at boundaries and crystal imperfections. The presence of a high point-defect concentration introduced by neutron (or heavy ion) irradiation can increase the diffusion rates of the solvent and various solute atoms and, in addition, the diffusion of point defects to sinks can induce a flux of solute atoms to or from point defect sinks. This leads to localized variations in composition and, subsequently, the formation of phases which are not normally present after aging at the same temperature without irradiation.[10,11] While all the phases identified have previously been reported in thermally-aged austenitic steels of various types, thermal aging of FV548 does not usually result in the precipitation of G-phase or either of the two M_6C phases described here.

All the types of precipitate which were apparently enhanced by neutron irradiation at temperatures where significant void swelling was observed, namely Cr_3Ni_2SiC, eta-phase, G-phase and the unidentified Ni–Si phase, were rich in both Ni and Si. The two latter phases contained very little chromium so that prolonged irradiation under conditions where these two phases are formed must ultimately lead to silicon and nickel depletion of the matrix. It has

TABLE 2 Phases identified in irradiated FV548

Case	Material condition	Neutron dose, dpa	Irradiation temperature, °C	Phases present
01388	ST	16	466	Cr_3Ni_2SiC; unidentified
			561	$M_{23}C_6$. eta, G
			591	$M_{23}C_6$; eta
			667	$M_{23}C_6$; Laves
			738	$M_{23}C_6$; Laves, σ
VA447	ST	30	380	unidentified
			470	eta; Cr_3Ni_2SiC
			600	$M_{23}C_6$; eta
VA447	ST +aged 850°C	30	380	G + unidentified
			470	$M_{23}C_6$; G
			600	$M_{23}C_6$; G
VA447	ST +CW	30	380	unidentified
			470	$M_{23}C_6$; eta
			600	$M_{23}C_6$. eta, Laves
VA447	ST aged at 600°C	–	–	$M_{23}C_6$

NbC was observed in all specimens

been claimed that such nickel depletion has led to the transformation of austenite to ferrite in neutron-irradiated type 316 steel[12] although we have not observed such a transformation in FV548 steel. Nevertheless, as suggested earlier, the loss of silicon from the matrix is a detrimental aspect since silicon is effective in reducing the void swelling while it remains in solution. Furthermore, whether or not such a transformation to ferrite occurs, there is evidence that void-swelling tends to increase as the nickel concentration in austenitic alloys decreases, so that the loss of nickel into precipitates must also be regarded as undesirable. Since increased silicon levels tend to increase the capacity for nickel to be removed from the matrix of an alloy, it follows that the addition of silicon, until recently regarded as an effective way of reducing void-swelling, is almost certainly detrimental in the long term.

On the other hand, although loss of silicon from the matrix during irradiation seems inevitable, the formation of precipitates such as eta-carbide, which contains substantial quantities of a range of elements, is not detrimental since their formation leaves the austenite matrix composition little changed.[11] Our work suggests that one way of minimizing the precipitation of undesirable phases and encouraging less deleterious precipitation in FV548 steel is to ensure maximum solubility of carbon by adequate solution treatment before irradiation.

Although isolated G-phase particles were found in solution-treated FV548 specimens, the presence of this phase after irradiation was clearly enhanced by the prior aging treatment at 850°C (see Table 2). The formation of this phase rather than the eta-carbide phase found in equivalent unaged material is considered to be due to the lack of free carbon available for carbide nucleation and growth during irradiation, the prior aging treatment greatly reducing the carbon in solution by formation of NbC particles. As irradiation proceeds the Ni, Nb and Si concentrations near point defect sinks increase and it becomes energetically favourable for the formation of a silicide (G-phase) when little carbon is available, or eta-carbide rich in Si, Ni, Nb, Cr, Fe and C, when sufficient carbon is available.[11]

In order to test this hypothesis pellets of two model alloys were prepared containing the major elements in FV548 but with the Si content increased to 6 wt-% to ensure that the Si solubility was exceeded at all aging temperatures, and with the Nb content increased to make the formation of Nb-rich phases more likely. Model *A* contained only 0.02 wt-% C and model *B* 0.22 wt-% C (the full compositions are listed in Table 1). Thus model *A* simulated the probable matrix composition in ST + aged FV548 near a vacancy sink during irradiation, and model *B* the situation in ST FV548. The pellets were solution-treated and samples from each aged in argon at 650 and 750°C for 1000 h. The precipitates in these samples were identified using the method outlined previously.

No carbide particles were found in alloy *A* in the unaged condition whereas alloy *B* contained large NbC and $M_{23}C_6$ particles. All the aged samples contained coarse particles of σ-phase and Laves-phase. Alloy *B* was found to contain a second distribution of small $M_{23}C_6$ precipitates and a nickel–silicon-rich eta-carbide phase. Alloy *A*, however, contained no eta-phase but a nickel–niobium G-phase. Thus, the formation of the G-phase and eta-phase was dependent upon the available carbon and this result supports the idea that, while irradiation can locally modify the composition adjacent to sinks, the types of induced precipitation can be influenced by careful control of specific bulk elemental concentrations and thermal history.

It is clear that the analytical electron microscope has an essential role to play in revealing the complex microstructural processes which occur during neutron irradiation of complex alloys. Inherent difficulties in the analysis of small particles embedded in a matrix will always be present. Clearly, the availability of a microscope with a high-brightness electron gun would enable more information to be obtained concerning the precipitation in material irradiated at the lower temperatures. However, the semi-quantitative and qualitative analyses described here using a conventional microscope with a thermal tungsten filament have been adequate to give important insight in an area of great technological importance.

CONCLUSIONS

1 Neutron-irradiated FV548 contained several types of precipitate particles, the types of particle present in any one specimen being dependent on the irradiation temperature and the initial condition of the material.

2 Those types of precipitate not normally found in unirradiated material contained high proportions of silicon and nickel, the removal of which from solution is deleterious with respect to the void-swelling behaviour.

3 Material irradiated after aging at 850°C contained copious quantities of G-phase, while equivalent unaged material contained eta-carbide. This occurred because aging greatly reduced the amount of free carbon in solution.

4 Laboratory melts were produced and given suitable heat treatments to simulate successfully the situation adjacent to point defect sinks in material irradiated in the aged and unaged conditions.

REFERENCES

1 Proc. BNES Conf. on 'Voids formed by irradiation of reactor materials', Reading, 1971
2 Proc. Int. Conf. on 'Radiation effects in breeder reactor structural materials', Scottsdale, Arizona, 1977
3 Proc. Int. Conf. on 'Behaviour of metallic materials and fast reactor core components under irradiation', Ajaccio, Corsica, 1979
4 Consultant Symposium on 'Physics of irradiation-produced voids', ed. R. S. Nelson, AERE Harwell, 1974, AERE Report R7934
5 T. R. ANTHONY: 'Radiation-induced voids in metals', eds. J. W. Corbett and L. C. Ianniello, AEC Symp. Ser. 26, Conf 710601, 1972, 630
6 W. G. JOHNSTON, T. LAURITZEN, J. H. ROSOLOWSKI, and A. M. TURKALO: Proc. ASM Seminar on Radiation Damage, 1976, 227
7 D. R. HARRIES: Ref. 2, p. 27
8 T. M. WILLIAMS and D. R. ARKELL: Ref. 3, p. 171
9 C. CAWTHORNE and C. BROWN: *J. Nucl. Mater.*, 1977, **66**, 201
10 T. M. WILLIAMS, J. M. TITCHMARSH, and D. R. ARKELL: *J. Nucl. Mater.*, 1979, **82**, 199
11 T. M. WILLIAMS and J. M. TITCHMARSH: *J. Nucl. Mater.*, 1979, 87, 398
12 H. R. BRAGER and F. A. GARNER: Symposium on 'Irradiation phase stability', Pittsburgh, Pa, Oct. 1980

Discussion overleaf

DISCUSSION of paper by Titchmarsh and Williams

J. W. STEEDS:
Whereas I agree with your general approach to phase identification I would like to counter to some extent the impression you leave of the complexity of diffraction analysis. A new and unknown phase of reasonable dimensions (~ 100 nm) can normally be identified by CBED in one or at most two sessions on the microscope. When a reliable EDX fingerprint has been established I would, like you, use EDX to identify it thereafter (if possible).

J. M. TITCHMARSH:
Your point is taken. If a new phase is encountered then microdiffraction techniques must be employed to establish identity of that phase. A realistic balance must be maintained between the time to identify individual precipitates with 100% certainty and the time required to identify hundreds of precipitates, by fingerprinting, with the possibility of occasional error.

Modifications to state of precipitate order which accompany passage of migrating interfaces

M P Shaw, R C Ecob, A J Porter, and B Ralph

Convergent-beam electron diffraction and energy-dispersive spectrometry have been used to study the ordering of precipitate phases in a commercial nickel-base superalloy and in a binary copper–4 wt-% titanium alloy. It is shown that the ordering scheme within the precipitates varies with the transformation route, and the observations are discussed in terms of the kinetics of precipitate growth. The highest degree of ordering is achieved when neither interfacial nor diffusional control is operative, such as when growth occurs behind a slowly moving boundary.

The authors are in the Department of Metallurgy and Materials Science, University of Cambridge.

The properties of moving grain boundaries are not as yet fully understood. While it has been established that enhanced diffusivity and solubility are associated with such defects (e.g. Refs. 1–6), it is not yet clear exactly how observed changes in metallurgical microstructures are effected. In the present paper it is demonstrated that the rate of boundary movement combined with the driving force for phase transformation or morphological modification can determine precipitate ordering in a fashion not observed in transitions associated with static grain boundaries.

It has already been established that the passage of a recrystallization interface through the microstructure of the wrought nickel-base superalloy Udimet 720 causes changes to the ordering of the intermetallic phase, γ'.[7,8] That investigation involved the use of convergent-beam electron diffraction (CBED) and revealed that γ' in conventionally heat-treated material has cubic symmetry and no evidence for a departure from the LI_2 based unit cell normally associated with γ' has been found in such material. In γ' discontinuously produced at a recrystallization interface, however (e.g. Ref. 9), cubic symmetry is not observed. On the basis of the reduction in point symmetry of [111] zone axis HOLZ patterns from $3m$ (cubic) to m, and the fact that extra superlattice reflections over and above those associated with LI_2 are not observed, an $L6_0$ tetragonal unit cell has been proposed for such γ' (Ref. 8). Tetragonality is produced by extra ordering on the *A* sub-lattices of the LI_2 cell, and does not involve any change in unit cell volume.

The purposes of the present paper are four-fold. Firstly, it reports the results of a full point group determination performed by CBED to confirm the crystal symmetry previously proposed. Secondly, it seeks to examine by energy-dispersive X-ray spectrometry (EDS) any chemical partitioning effects associated with changes in γ' ordering brought about by recrystallization. Thirdly, observations which indicate comparable changes in ordering in a binary Cu–Ti alloy as a result of the passage of a cellular precipitation interface are also presented. Finally, the properties of moving boundaries with respect to these observations are made the subject of discussion.

CONVERGENT-BEAM ELECTRON DIFFRACTION

Symmetry

Only a limited number of metallurgical studies have been made using CBED to date and therefore a brief discussion of the information which may be obtained from convergent beam diffraction patterns (CBDPs) will now be presented. For a full treatment of CBED, the reader is referred to the comprehensive review by Steeds.[10] He and his co-workers, in a series of papers (for example Refs. 11 and 12), have shown that the distribution of diffracted intensities is related to the crystal structure of the phase under observation, both via the intrinsic interaction of incoming electrons with the structure of a crystal, and also by the generation of higher order Laue zone (HOLZ)* lines in the discs of the zero layer diffraction pattern (see under next heading). Buxton and co-workers[11] have shown that consideration of CBDP symmetry, combined with systematic tilting experiments, can lead to full point group determination. In addition, some features of CBDPs can be related to space group symmetry (e.g Ref. 13). The present authors have used a simple geometrical interpretation of holz lines to study rather subtle changes in precipitate symmetry, as well as relative lattice parameter variations[7,8] and a brief review of this technique is given.

Holz lines

These form a part of the fine structure of the discs in a CBDP and they arise from the interaction of excited reflec-

* Hereafter shown as acronym 'holz' (likewise 'folz' later in text).

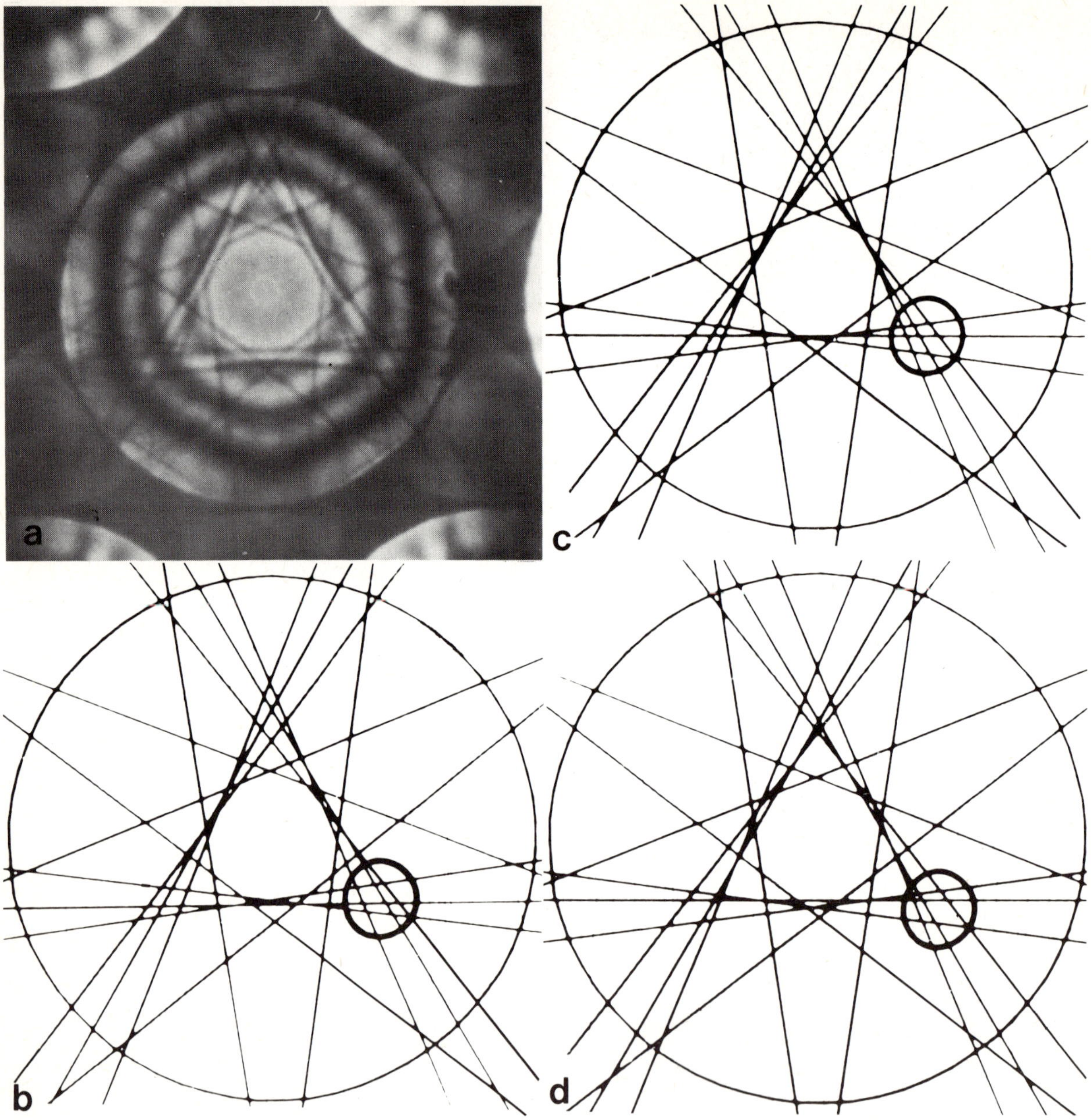

1 **(a) [111] holz pattern taken from γ matrix of uncrystallized Udimet 720; (b) simulated pattern $\lambda = 3.44$ pm, $a = 0.355$ nm; (c) $a = 0.3556$ nm; (d) $a = 0.356$ nm.**
(c) makes closest match; both (b) and (d) are easily distinguishable from experimental pattern by consideration of circled regions; in these and all subsequent simulations, 'condenser apertures' are of arbitrary size

tions in upper Laue zones with the zero layer diffraction discs. Diffracted information from higher order Laue zones involves relatively large Bragg angles, and it is this fact that makes holz lines so sensitive to small variations in interplanar spacing.

The sensitivity of holz line positions means that small symmetry variations can be recognised relatively easily, when such variations give rise to changes in unit cell dimensions. However, there are limitations on the accuracy with which measurements of holz line positions can be made; these are primarily associated with their finite width. For this reason, the approach used in the current investigation involves close examination of the size and symmetry of the polygons formed by the high angle intersection of indexed holz lines (see later). Furthermore, it has been found that the sensitivity of the technique can be increased by comparing experimental CBDPs with computer-simulated patterns. The comparison of experimental and simulated patterns and the matching of the two is believed to be the most effective method of optimizing the information available. The way in which experimental patterns are simulated is the subject of the following section.

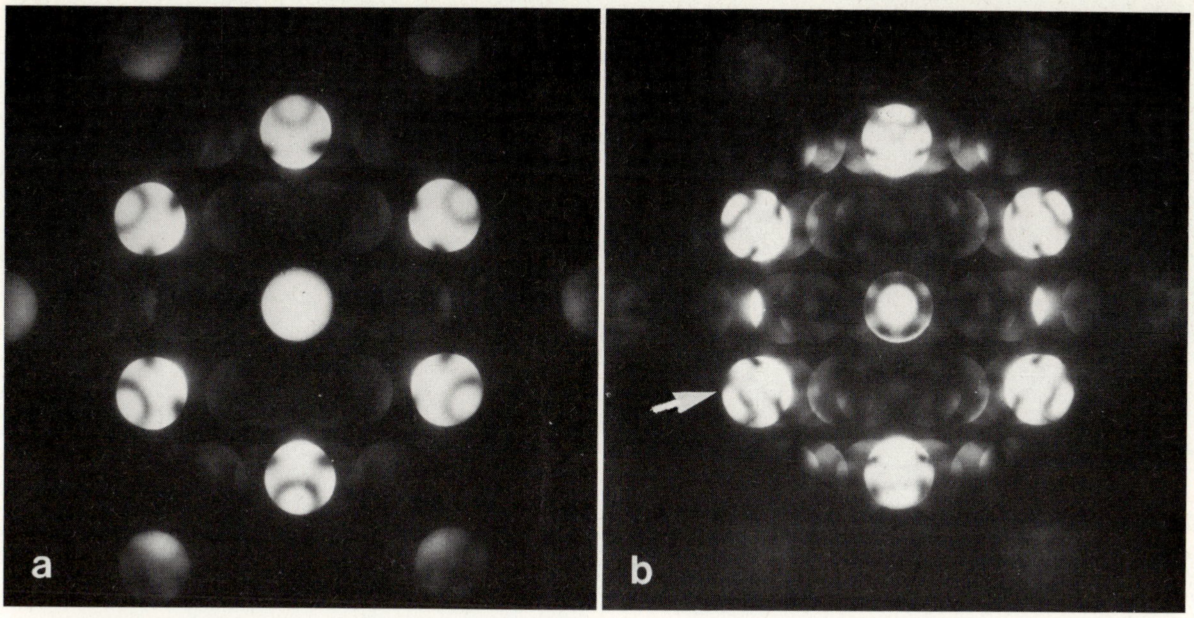

2 **[010] CBDPs from cellular β plates in copper–4 wt-% titanium: (a) isothermally transformed 3 h at 760°C; point symmetry *mm*2; (b) isothermally transformed 1 h at 740°C; point symmetry *m*; horizontal mirror is absent, as indicated by arrowed region of pattern**

The simulation technique

This technique has been described in detail elsewhere.[7] Essentially, the geometrical treatment employed allows computation of the distance of each holz line from the origin of the diffraction pattern and of its orientation. The folz (first order Laue zone) ring is formed by the intersection of the first Laue layer with the Ewald sphere, and the defect holz line associated with a particular folz reflection is separated from the ring by twice the Bragg angle. Both the angular radius of the folz ring and the Bragg angle depend critically on the electron wavelength; since the latter is not accurately known the wavelength and the lattice parameter are treated as non-independent variables. The orientation of each line is calculated from the projected direction of the folz reflection.

These calculations are performed on a high resolution graphics unit (Tektronix 4051) and the simulations may be computed and displayed in around 20 s each. Figure 1 shows the results of a search for the appropriate lattice parameter in the case of a ⟨111⟩ zone axis pattern. It is clear in Fig. 1 that while the absolute movements of the lines in the simulated patterns are not easily measurable, the changes in the polygons (such as those marked) are appreciable and they reflect the extreme sensitivity of holz patterns to very small changes in lattice parameter.

Note, however, the lack of agreement between certain areas of simulated and experimental patterns. Holz lines intersecting at low angles can hybridize and hence quantitative comparisons are limited to areas containing high angle intersections.

Experimental details

The work presented in this paper was performed on a standard Philips EM400 microscope possessing TEM/STEM capability. It should perhaps be appreciated that there are microscope refinements which have been designed specifically to assist the production of high quality CBDPs (specimen cooling, continuous voltage adjustment, etc.) but these were not used in the present work. Standard twin-jet specimen polishing techniques were used to prepare the samples investigated.

Although CBDP work can be carried out on typical thin foils commonly used for imaging purposes, there are certain specimen limitations that need to be appreciated if the information from the technique is to be maximized. Since it is holz lines which lead to quantitative analysis, the particularly important variables are outlined as follows:

(a) The visibility of holz lines is dependent upon foil thickness. The critical thickness for visibility is between 3 and 6 extinction distances and this means that information can be obtained from relatively thick foils when the operating voltage is 120 kV. However, if it is necessary to examine thinner regions the microscope voltage can be altered to modify the extinction distance.
(b) Specimen quality can play a vital role in determining holz line visibility. In particular, contamination and foil bending are deleterious.
(c) Final tilting of the beam to an exact on-zone orientation can be achieved using displacement of the condenser aperture. However, excessive enthusiasm in this area can lead (owing to microscope aberrations) to distortion of holz patterns.

In the section which follows, quantitative information deduced from holz lines is of central importance. Other features of CBDPs are also used, however, in order to determine point group symmetry.

RESULTS

Copper–4 wt-% titanium

Phase transformations in the copper–titanium system and the structure of the metastable β′ and equilibrium β phases have been described in detail elsewhere (e.g. Refs. 14–16). It is important to note, however, that both phases have the same nominal composition, Cu_4Ti. The present results are

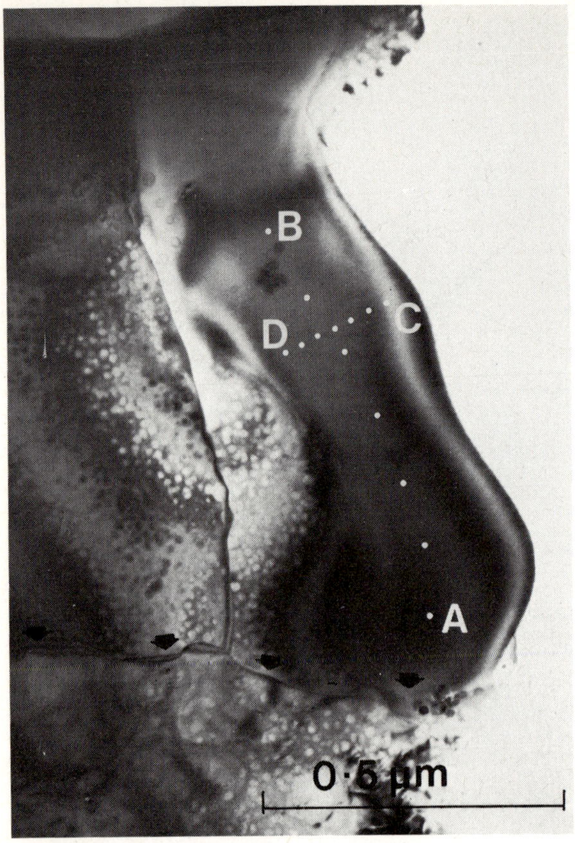

3 **Bright field TEM micrograph of cellular γ' precipitate in partially recrystallized Udimet 720; positions marked correspond to stations of electron probe during EDS analyses; recrystallization interface arrowed**

concerned with apparent variation in the point group of the cellular β product depending on transformation conditions.

Figure 2*a* shows an [010] zone axis CBDP taken from β isothermally transformed at 760°C above the metastable β' solvus. Here, a mixture of Widmanstätten and cellular β is observed, the latter reaction being rather slow (around 3 h to full transformation). The particular CBDP shown was taken from a cellular β plate, but identical patterns have been observed to arise from Widmanstätten precipitates. It is evident that the pattern displays 2*mm* whole pattern symmetry and this is consistent with the β phase having point group *mmm*. While no holz lines are present, it is clear that symmetry information is contained in the CBDP.

In addition, cellular β has been observed to belong to the *mmm* point group when specimens are quenched and tempered at 690°C. Here, metastable β' is produced initially, but the cellular transformation product replaces this in around 1 min.

However, at intermediate temperatures below the β' solvus but where the cellular transformation is slower, CBDPs such as that of Fig. 2*b* are observed where the whole pattern point symmetry is reduced to *m*, implying 2*mm* point group, and thus a different atomic ordering within the β phase. While the unit cell dimensions are unchanged, there is actually more ordering within β produced under these intermediate transformation conditions.

Udimet 720

Examination of Udimet 720 was performed on material heat treated as for commercial use (e.g. Ref. 9) cold rolled, and then partially recrystallized at 1100°C.

TABLE 1 EDS analyses of γ' in Udimet 720 (wt-%)

Element	γ' in commercially heat-treated material (cubic)	Discontinuously produced γ' (tetragonal)
Ni	69.3±1.6	69.4±1.1
Co	10.0±0.8	10.1±0.5
Al	4.2±0.7	4.4±0.5
Ti	11.0±0.3	10.7±0.4
Cr	4.8±0.4	5.0±0.4
Mo	0.7±0.5	0.5±0.5

Errors are given as standard deviation in mean composition obtained from a number of analyses.

(a) EDS analysis

Typical compositions for γ' in the unrecrystallized regions of the foil, and for discontinuously produced γ' formed during recrystallization are presented in Table 1. It is clear that no measurable γ' composition change is associated with the change in γ' morphology resulting from recrystallization. A typical discontinuous γ' particle, cube/cube related with the matrix of the recrystallized grain surrounding it, is shown in Fig. 3. EDS analyses have been taken along and across this particle – i.e. with the probe tracking as indicated on Fig. 3. The results of these analyses are shown graphically in Fig. 4, and no statistically significant changes in γ' composition along or across the precipitate are detected. Hence, the changes in ordering of γ' previously reported[8] and discussed more fully below are not associated with any measurable change in the partitioning of constituent elements.

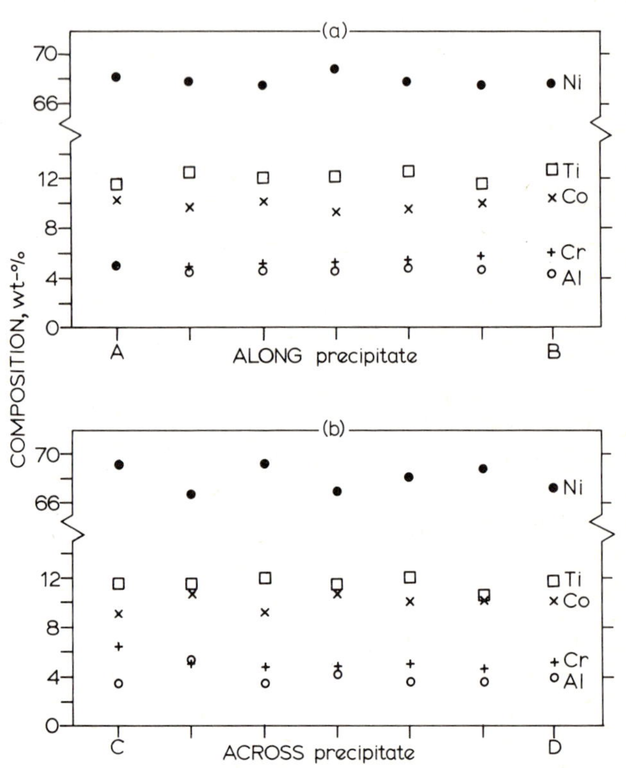

4 **(a) Composition variation of γ' particle of Fig. 3 plotted against distance in *AB* direction; errors as for Table 1; (b) as for (a) for variations in *CD* direction**

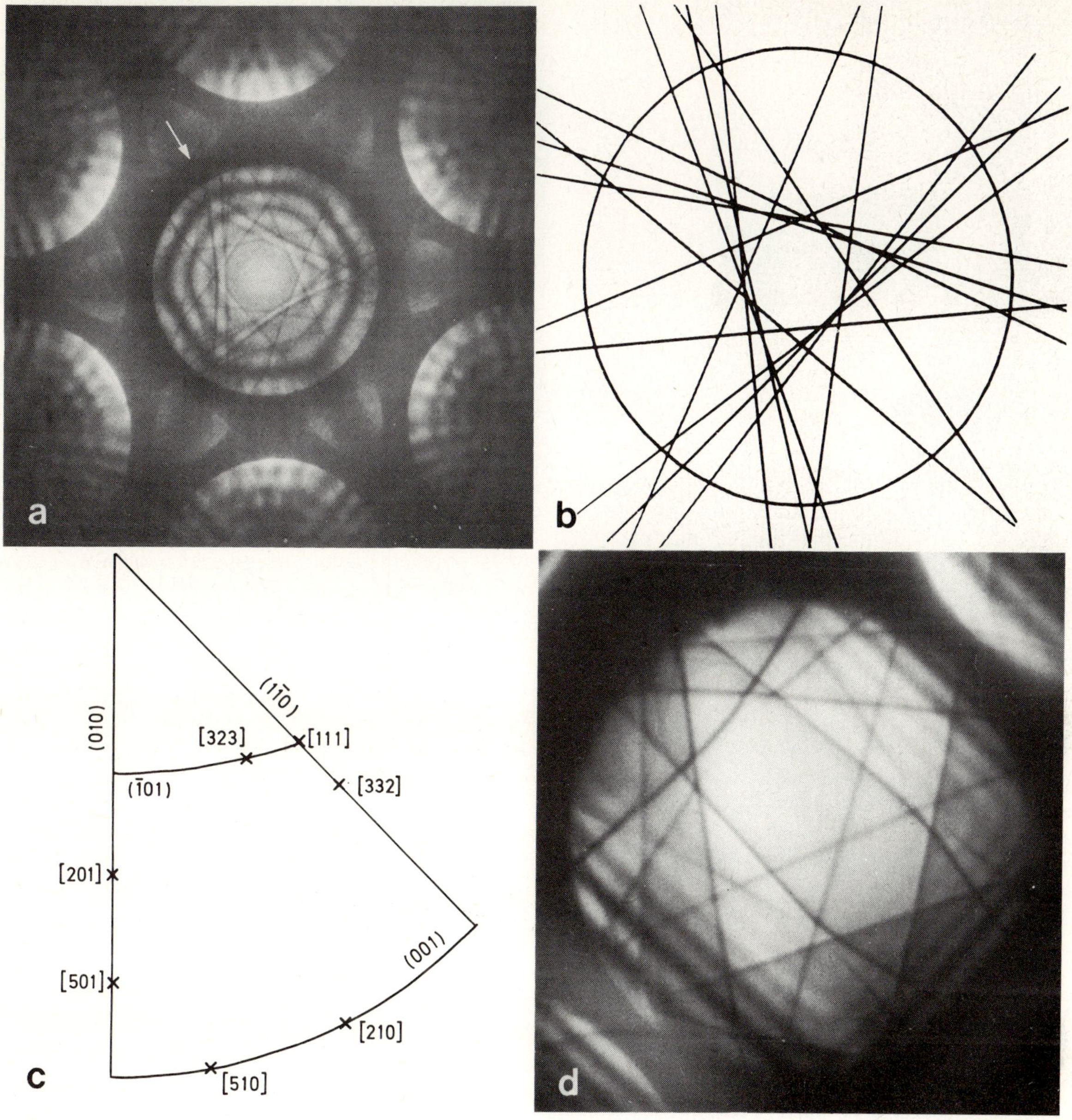

5 (a) [111] holz pattern taken from cellular γ' in partially recrystallized specimen of Udimet 720 – by comparison with Fig. 1*a*, point symmetry has been reduced from 3*m* to single mirror plane (arrowed); (b) simulated [111] holz pattern considered to match (a) – λ = 3.44 pm, *c* = 0.3554 nm, *a* = 0.3563 nm; (c) stereogram showing indexing scheme for zone axes of CBDPs examined during point group determination – note that mirror retained in (a) is (1$\bar{1}$0); (d) [332] holz pattern, which again displays (1$\bar{1}$0) mirror plane, is in correct orientation with respect to (a)

(b) CBED analysis

Examination of [111] zone axis patterns in discontinuously produced γ' reveals that the 3*m* point symmetry associated with cubic material is replaced in this case by *m* symmetry in the holz pattern (Fig. 5*a*). The mirror is assigned indices (1$\bar{1}$0), and the simulated pattern judged to give best match is shown in Fig. 5*b*. The zone axes of some of the patterns examined in the production of a full point group determination of such a γ' particle are shown schematically in the stereogram of Fig. 5*c*. The [332] pattern (Fig. 5*d*) also possesses the (1$\bar{1}$0) mirror. Examination of the [210] pattern reveals that the (001) mirror plane is also retained, as is the (010) mirror in the [201] pattern. These patterns are shown in Fig. 6, together with the relevant simulations, which were produced using lattice constants identical to those used in the [111] simulation (tetragonal, c/a = 0.9975). It is clear from examination of the indicated areas of the [210] and [201] patterns that these patterns are not related by a [100] tetrad axis. Lastly, examination of the [323] pattern of Fig. 7, which has no holz symmety, confirms the absence of the ($\bar{1}$01) mirror associated with the LI_2 unit cell.

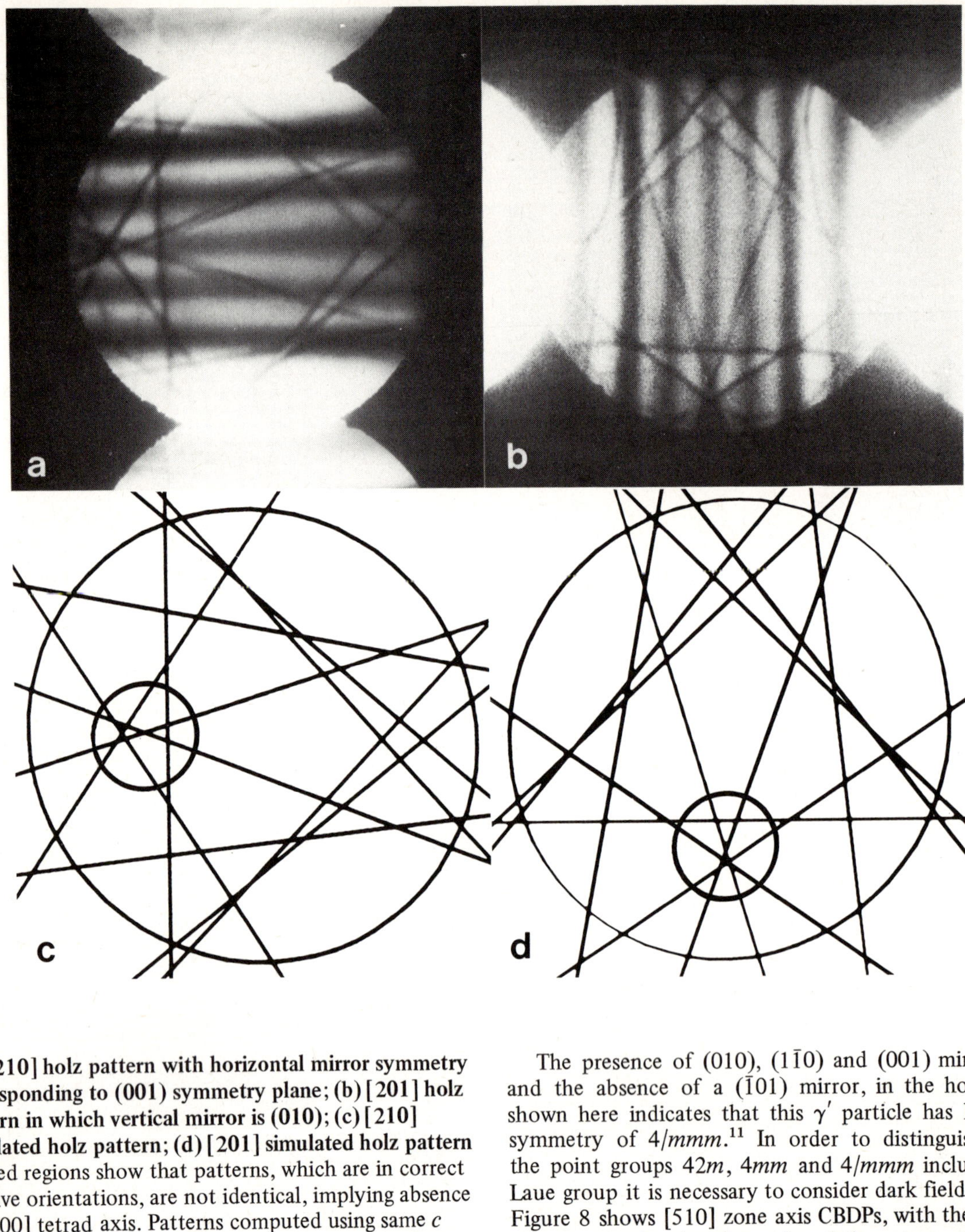

6 **(a) [210] holz pattern with horizontal mirror symmetry corresponding to (001) symmetry plane; (b) [201] holz pattern in which vertical mirror is (010); (c) [210] simulated holz pattern; (d) [201] simulated holz pattern** Ringed regions show that patterns, which are in correct relative orientations, are not identical, implying absence of [100] tetrad axis. Patterns computed using same *c* and *a* values as Fig. 5*b*. Both simulated patterns closely match their respective experimental patterns, so that postulated lattice parameters give rise to fully consistent set of simulations

The presence of (010), $(1\bar{1}0)$ and (001) mirror planes, and the absence of a $(\bar{1}01)$ mirror, in the holz patterns shown here indicates that this γ' particle has Laue group symmetry of $4/mmm$.[11] In order to distinguish between the point groups $42m$, $4mm$ and $4/mmm$ included in this Laue group it is necessary to consider dark field symmetry. Figure 8 shows [510] zone axis CBDPs, with the condenser aperture displaced either side of the mirror plane in the holz pattern. This figure indicates mirror symmetry, thus suggesting that the relevant point group symmetry is $4/mmm$. It can thus be stated that the unit cells previously proposed,[8] which involve extra ordering on the *A* sites of the LI_2 structure, are consistent with this more complete determination of crystal symmetry.

DISCUSSION

The results presented above both from a copper–titanium alloy and from Udimet 720 show that the degree of order in the precipitate phase concerned varies with the mechanism of its formation. Classically, order–disorder transformations occur when the enthalpy and entropy terms balance in the expression for the free energy change associated with the transition. In the cases presented here, such equilibrium considerations do not predominate. In the β phase, fully ordered precipitates occur in a regime in the middle of the transformation range, with partially ordered β produced at both higher and lower temperatures. In the

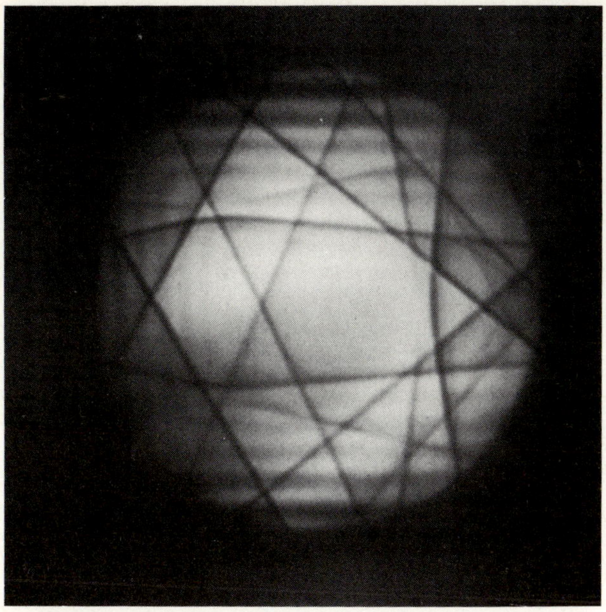

7 [323] holz pattern showing no symmetry, and confirming absence of $(10\bar{1})$ mirror. Trace of $(10\bar{1})$ is horizontal

γ' case, discontinuously produced precipitates in recrystallized regions are fully ordered by contrast to those formed prior to recrystallization at a very similar temperature, via homogeneous nucleation and growth.

The explanation proposed for these phenomena is therefore founded on kinetic arguments. It is well known that homogeneous precipitation in some aluminium alloys, for example, proceeds through several non-equilibrium stages because the lower interfacial energies of metastable precipitates lead to a faster rate of decomposition of the supersaturated solid solution. The thermodynamics of the interphase interfaces involved modify the kinetics of the nucleation process. However, precipitation on grain boundaries of the equilibrium phase is sufficiently fast for not all of the intermediate stages to be required.[17] Equivalently, modification of the thermodynamics of the alloy system by the presence of a grain boundary may lead to such areas of the microstructure effectively lying above the metastable solvi.

Analogously, it is suggested in the present case that although the fully ordered stated is presumably thermodynamically more favourable, the metastable partially ordered state is produced because it allows faster kinetics for precipitate *growth*.

Precipitate growth is commonly understood to be limited by one of two factors – the rate of incorporation of solute atoms into the particle (interface control) or the rate of diffusion of solute to the particle (diffusion control).*

Consider the growth of an intragranular γ' precipitate. In Udimet 720 the γ/γ' misfit is very small and the particle is fully coherent even at a diameter of $>1\,\mu$m, suggesting that the growth is subject to interface control. The lower

* A third rate-controlling process, that of the nucleation of growth ledges, is not considered to be of relevance here because the consumed precipitates are three-dimensionally coherent, and the growing precipitates extend behind a migrating grain boundary.

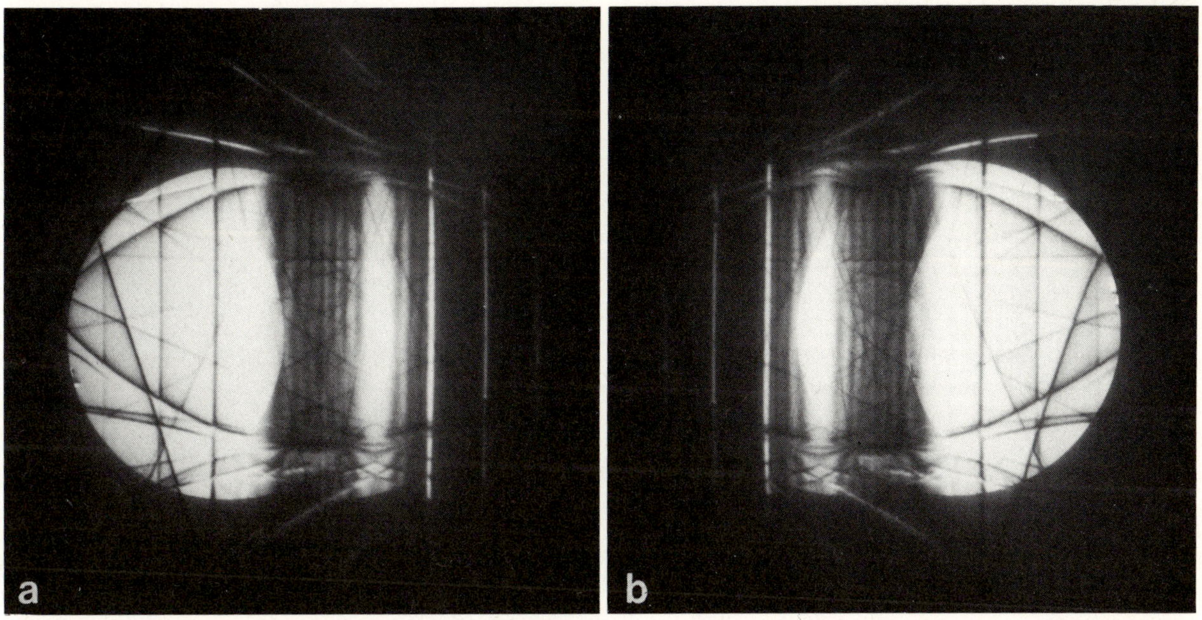

8 (a) Tilted [510] CBDP, centred near 002 so as to show dark field holz pattern; (b) as (a) centred near $00\bar{2}$. Two dark field patterns are mirror-related, confirming reality of (001) symmetry plane (vertical). Point group is thus $4/mmm$

the degree of order within the γ', the less is the degree of coordinated atomic movement across the interface required for growth; therefore the overall rate of decomposition of the supersaturated solid solution is expected to be faster.

When a γ' precipitate is growing in a discontinuous fashion, however, the only interface in motion is the incoherent boundary forming part of the recrystallization front. In the absence of the homophase regions of the latter, γ' growth would be expected to be volume diffusion controlled. In reality all the diffusion of solute atoms occurs in the migrating recrystallization front, and is therefore assumed to be fast.

The rate-determining step, however, during growth of a cellular colony or a eutectoid colony is still usually taken to be interfacial diffusion. Thus the precipitate particle can only extend as fast as solute atoms are supplied to it. However, if the rate of solute supply is actually faster than the potential growth rate of the particle then diffusion control is no longer operative.

It is suggested that it is precisely this situation which sometimes obtains during the discontinuous growth of γ' behind a recrystallization front in Udimet 720. Here, the driving force for boundary migration is low, and the migration is impeded by the presence of a high density of large γ' precipitates in the consumed grain. There is thus a very high rate of solute supply combined with a slow growth rate, allowing sufficient time for the production of the fully ordered thermodynamically stable phase, i.e. tetragonal rather than cubic γ'.

There are two further observations which corroborate the above hypothesis. The first of these is that the composition of γ' in the recrystallized and unrecrystallized regions being invariant suggests that all the various solute atoms have sufficient time for redistribution in the moving boundary – i.e. that their diffusion is not rate-limiting. If it were, then the faster diffusing solutes would be expected to be enriched within the growing γ' during the early stages of reprecipitation. Variations in composition within a growing γ' particle might then be expected to occur, but are not observed. The other observation is that when recrystallization occurs more quickly, γ' in the consumed grain does not reprecipitate but is by-passed, and is incorporated into the growing grain in its original orientation (e.g. Ref. 9). This suggests that only under conditions such that diffusion is fast with respect to the rate of boundary migration is tetragonal γ' observed.

Consider now the cellular precipitation in the copper-4 wt-% titanium alloy. Here the partially ordered phase is observed after fast growth of the colony into a grain containing very fine metastable β' at 690°C. At 740°C, however, the driving force is lower since not only have β and β' the same nominal composition, but the latter has coarsened and become semicoherent before being consumed. In this coarser state, it exerts a larger restraint on the growth front of the cellular colony (e.g. Ref. 18). It is therefore suggested that, in parallel to the superalloy case, a slow cellular growth rate, which is not diffusion-limited, leads to full ordering in the β phase.

At the highest transformation temperatures, above the β' solvus, cellular growth must again be diffusion-controlled, since there are no obstacles to boundary migration. The relatively high solute atom mobility, combined with a slow reaction, might be expected to lead to full ordering (point group $2mm$), but only mmm β is observed. In order to explain this apparent anomaly, it is necessary to postulate that at the highest temperatures the $mmm \rightarrow mm2$ ordering temperature is exceeded. Since the structures are similar so must be their enthalpies, but the configurational entropy change for the formation of the $mm2$ structure is larger than that for mmm and therefore the ordering temperature must be lower.

If the above explanations for the present observations are given credence, there is one further question which must be addressed. Why is no transition from the partially to the fully ordered state observed in either the β or γ' phases, when held in the temperature regime over which fully ordered precipitation occurs? Again, it is expected that such a reaction would be thermodynamically favourable; however, as mentioned above, the structural differences between the states are very subtle, and it is postulated that no suitable mechanism is available to accomplish the transition under such a small driving force. Cooperative atomic interchanges would be required not just on the scale of an individual unit cell, but over volumes at least as large as that giving rise to the diffraction information on which the present arguments are founded.

ACKNOWLEDGMENTS

The authors are grateful to Professor R. W. K. Honeycombe for the provision of laboratory facilities and to Dr W. M. Stobbs for stimulating discussions. Financial support from the Science Research Council, Inco Research and Development Centre, and the United Kingdom Atomic Energy Authority is gratefully acknowledged.

REFERENCES

1 M. HILLERT and G. R. PURDY: *Acta Metall.*, 1978, **26**, 333
2 K. SMIDODA, C. GOTTSCHALK, and H. GLEITER: *Met. Sci.*, 1979, **13**, 145
3 A. R. JONES and B. RALPH: *Acta Metall.*, 1975, **23**, 355
4 J. W. CAHN and R. W. BALLUFFI: *Scr. Metall.*, 1979, **13**, 499
5 C. Y. BARLOW and B. RALPH: Proc. Risø Int. Symp. Recrystallization (eds. N. Hansen, A. R. Jones, T. Leffers), 165–171; 1980, Risø National Laboratory, Denmark
6 A. J. PORTER and B. RALPH: *ibid.*, 147–153
7 R. C. ECOB, M. P. SHAW, A. J. PORTER, and B. RALPH: *Philos. Mag.*, 1981, in press
8 A. J. PORTER, M. P. SHAW, R. C. ECOB, and B. RALPH: *Philos. Mag.*, 1981, in press
9 A. J. PORTER and B. RALPH: *J. Mater. Sci.*, 1981, **16**, 707–713
10 J. W. STEEDS: 'Introduction to analytical electron microscopy', 387–422; 1979, New York, Plenum Press
11 B. F. BUXTON, J. A. EADES, J. W. STEEDS, and G. M. RACKHAM: *Phil. Trans. A*, 1976, 1301
12 P. M. JONES, G. M. RACKHAM, and J. W. STEEDS: *Proc. R. Soc.*, 1977, **354**, 197
13 P. GOODMAN: 'Electron diffraction 1927–1977', p. 116, Inst. Physics, Bristol and London
14 R. C. ECOB, J. V. BEE, and B. RALPH: *J. Microscop.*, 1979, **116**, 141
15 R. C. ECOB, J. V. BEE, and B. RALPH: *J. Microscop.*, 1980, **119**, 153
16 R. C. ECOB, J. V. BEE, and B. RALPH: *Metall. Trans.*, 1980, **11A**, 1407
17 R. B. NICHOLSON: 'Phase transformations', 269–312, 1970, ASM
18 M. F. ASHBY: Proc. Risø Int. Symp. Recrystallization (eds. N. Hansen, A. R. Jones, T. Leffers), 325–336, 1980, Risø National Laboratory, Denmark

DISCUSSION of paper by Shaw, Ecob, Porter, and Ralph

D. B. WILLIAMS:
I question your method of determining the rate-controlling process in the discontinuous reaction by measuring the composition of γ'. Previous investigations of discontinuous precipitation have demonstrated interface control by observing discrete composition changes *across* the advancing interface, and diffusion profiles in the *matrix* phase behind the advancing interface.

A. J. PORTER:
I emphasized during my presentation that EDS results do not constitute our only evidence for saying that diffusion in the recrystallization interface is not the rate-controlling process for boundary migration. Pinning of the interface by γ' particles – some of which are not dissolved during recrystallization – determines the rate at which the process can occur. Varying the prior dispersion of γ', which substantially alters the pinning process, while not markedly affecting interfacial diffusion, changes the rate at which recrystallization proceeds.

The lack of any change in the distribution of 'cooling' γ' in the matrix as one moves away from the recrystallization interface indicates that no substantial diffusion of γ'-forming elements occurs in the matrix behind the advancing interface.

R. D. DOHERTY:
Were the discontinuous, tetragonally ordered precipitates single domains? If so, was there any consistent relationship between the *c* axis and the direction of movement of the recrystallization front?

A. J. PORTER:
We only ever observe single domain tetragonal particles. The present paper discusses the extra ordering of such precipitates in terms of there being 'time and space' in the recrystallization interface for the thermodynamically most favourable unit all to be formed, and we would therefore expect it also to be favourable for a growing particle to continue an existing domain rather than to begin a new one.

We have not found a consistent relationship between the *c* axis and the direction of movement of the recrystallization interface. Such a relationship would impose severe restrictions on the regions of a recrystallization interface at which discontinuous re-precipitation of γ' could occur. The most obvious relationship one could propose, for example, is with the *c* axis of the tetragonal cell parallel to the direction of movement of the recrystallization interface. Not only will very few regions of a randomly oriented recrystallization interface satisfy this criterion, given that the γ' particle has a fixed orientation relationship with the rest of the recrystallized grain, but it is also likely that the direction of movement of any part of the interface will change as it migrates.

Comparison between atom probe and STEM microanalysis

G D W Smith, A J Garratt-Reed, and J B Vander Sande

The atom probe and the field emission STEM are the two microanalytical techniques which currently give the highest spatial resolution. Both techniques function in a regime where independent verification of the quantitative accuracy of the results is difficult or impossible since other, more traditional, methods of analysis cannot be used. It is therefore desirable to cross-check the results obtained from the atom probe with those obtained from field emission STEM X-ray analysis. This paper reports the first work of this kind. Three problems of interest to physical metallurgy have been studied. (1) The phase chemistry of a nickel-base superalloy, IN939, has been examined and comparative results obtained for the composition of the gamma, gamma-prime MC carbide and $M_{23}C_6$ carbide phases. (2) The role of silicon in retarding the rate of tempering of martensitic steels has been investigated, and results obtained for the partitioning of silicon between the epsilon carbide phase and the martensitic matrix. (3) The redistribution of chromium between the ferrite and cementite phases during the pearlite reaction in steels has been measured at temperatures well below the thermodynamic 'no-partitioning' temperature. In each of these investigations, care was taken that specimens from the same batch of material should be studied in both instruments. The results permit a critical comparison to be made between the two techniques and indicate a number of areas in which complementary information may be obtained by using a combination of both instruments.

G.D.W.S. is in the Department of Metallurgy and Science of Materials, University of Oxford; A.J.G.-R. and J.B.V.S. are in the Department of Materials Science and Engineering, Massachusetts Institute of Technology, Cambridge, Mass., USA.

The objective of the work described in this paper was to compare and contrast the microanalytical capabilities of atom probe mass spectrometry and field emission STEM X-ray analysis with regard to quantitative accuracy, spatial resolution, and versatility. During the past decade, much effort has been expended to refine each of these analytical methods. Both techniques function in a regime of high spatial resolution where independent verification of the results is difficult or impossible, since other more traditional methods of analysis cannot be used. It is therefore most desirable to cross-check the analytical results obtained from the two techniques, using identical specimen materials in each case. Only in this way can the relative strengths and weaknesses of each technique be fully assessed. It is believed that the present paper represents the first systematic work of this kind to be reported. Specimens were selected from a number of project areas in which one or other of the authors' research groups are engaged. The analyses presented are thus representative of practical investigations in physical metallurgy. The specific problems chosen for study were designed to be taxing for each analytical technique although they do not, of course, cover the whole spectrum of possible topics to which either instrument could be applied.

EXPERIMENTAL TECHNIQUES

Atom probe

The time-of-flight atom probe used in this work has been described in detail elsewhere.[1] The instrument consists essentially of two parts: a field ion microscope (FIM) section in which an atomic resolution image of the specimen is obtained, and a mass spectrometer section in which the chemical identity of ions field-evaporated from the specimen surface is determined. A small aperture hole in the FIM imaging screen allows ions to enter the mass spectrometer section and defines the area from which the chemical analysis is obtained. The time-of-flight principle is used to identify the ions. In this method, ions are removed from the specimen surface at well-defined instants by the application of a train of nanosecond rise-time high voltage pulses, superimposed on the normal d.c. imaging voltage in the FIM. The ions are accelerated, and pass from the FIM section into a drift tube at the end of which is situated a particle detector of single-ion sensitivity. The mass-to-charge ratio (m/n) can then be estimated for each ion from the following equation:

$$m/n = 0.19297\,Et^2/l^2 \qquad (1)$$

where E is the total potential through which the ion is accelerated (kV), t is the flight time to the detector (μs), and l is the flight path (m).

Flight times are measured using standard digital techniques, the reference frequency in this case being provided by a crystal-controlled oscillator operating at 159.625 MHz. Up to four parallel timing channels are available to allow for the possible evaporation of multiple ions on a single

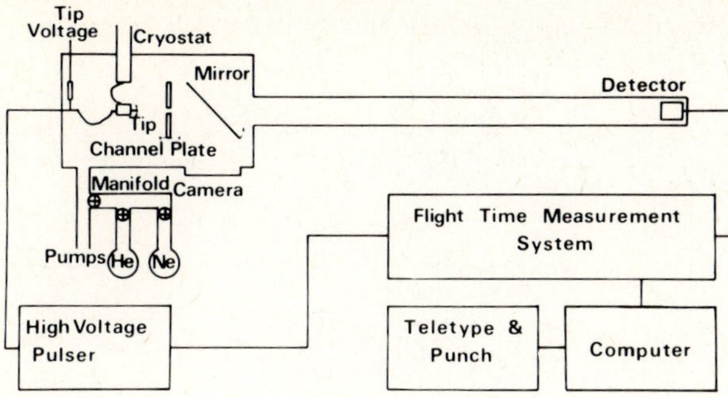

By courtesy of 'Surface and Interface Analysis' (Heyden & Sons)

1 Schematic: time-of-flight atom probe

high-voltage pulse. The digital data are fed into a PDP11/10 minicomputer which controls the experimental cycle, calculates and stores the mass spectra, and provides the operator with an ion-by-ion output of the composition of the specimen. The whole system is shown schematically in Fig. 1.

At first sight, a particular advantage of the atom probe technique is the simple, quantitative interpretation which is possible for the data on chemical composition. However, considerable care must be taken to establish the correct conditions for stable operations. For example, if the vacuum is poor, one or more chemical species on the specimen may be selectively etched away by reaction with chemically active gas molecules. Also, if the standing voltage on the specimen is too high, some species may be preferentially lost from the surface by d.c. field evaporation during the time interval between the application of successive high voltage pulses. In either case, errors will arise in the resultant chemical analysis, which is obtained only from ions desorbed during the period of the high voltage pulse. In the case of new or unfamiliar specimen materials, it is generally found desirable to run preliminary experiments in which the apparent chemical composition is measured as a function of the ratio of the pulse voltage to the d.c. standing voltage (the 'pulse fraction'). A certain minimum pulse fraction, usually 15% or more, is needed in order to obtain consistent results. When suitable standard materials of known composition are available, these procedures can be extended to determine the absolute quantitative accuracy of the atom probe technique. Some examples are given later in the present paper.

Typical operating conditions used in the present work are summarized as follows:

(a) system base pressure below 5×10^{-10} torr
(b) all analyses carried out in a vacuum of better than 1×10^{-9} torr, i.e. with the image gas removed
(c) specimen cryostat cooled with liquid nitrogen
(d) average evaporation rates of between 0.1 and 0.001 ions per pulse
(e) pulse repetition rates of 50–100 Hz
(f) pulse fractions of at least 15%, depending on specimen material.

In addition to the time-of-flight atom probe described above, use was made of an imaging atom probe of the kind first described by Panitz.[2] This instrument is shown schematically in Fig. 2. The specimen is located at the centre of curvature of a double curved chevron channel plate assembly, with a phosphor screen and fibre optics face-plate behind it. By time-gating the voltage supply to the channel plate assembly, following the application of a high voltage pulse to the specimen, the phosphor screen can be used to display an image of the spatial distribution of any selected chemical species over the whole of the specimen surface.

STEM

The field emission STEM used in this work was a V.G. Microscopes HB5 commercial instrument fitted with an energy-dispersive X-ray analysis system and an electron energy loss spectrometer. Full details of this instrument are given elsewhere.[3] The electron source is a cold tungsten field emitter. There is a two-lens electron probe forming system, by which typically 2 nA of 100 kV electrons can be formed into a probe approximately 1.5 nm diameter on the specimen. The vacuum in the specimen chamber is $\sim 2 \times 10^{-9}$ torr, which minimizes any problems of specimen contamination. The X-ray detector is a slightly modified Kevex Si (Li) detector, with a Kevex 7000 multichannel analyser and microcomputer system for spectral manipulation. The modification to the detector consists of the addition of a thin mylar film across the beryllium window, to act as an electron trap. One effect of the mylar is to reduce the sensitivity of the detector to low-energy X-rays ($< \sim 2$ kV) compared to that of the standard detector.

Specimens in the STEM consist either of thin sections of bulk material or of carbon films supporting particulate material. Whilst it is possible to image specimens many tens to several hundred nanometres thick, effects such as beam spreading, specimen fluorescence, and absorption become of increasing importance in modifying the resolution and sensitivity of X-ray analysis.

Interpretation of energy-dispersive X-ray data requires a great deal of care. The usual way to consider the effects of X-ray production is to employ the sensitivity factor method due to Cliff and Lorimer,[4] embodied in the equation:

$$\frac{C_A}{C_B} = \frac{I_A K_A}{I_B K_B} \qquad (2)$$

where C_A/C_B is the ratio of the concentrations of the elements A and B in a specimen; I_A and I_B are the integrated X-ray counts in chosen lines due to each element; and K_A and K_B are sensitivity factors. Often the ratio K_A/K_B is used as a relative sensitivity factor denoted by K_{AB}, relating the sensitivities of the two elements. In this

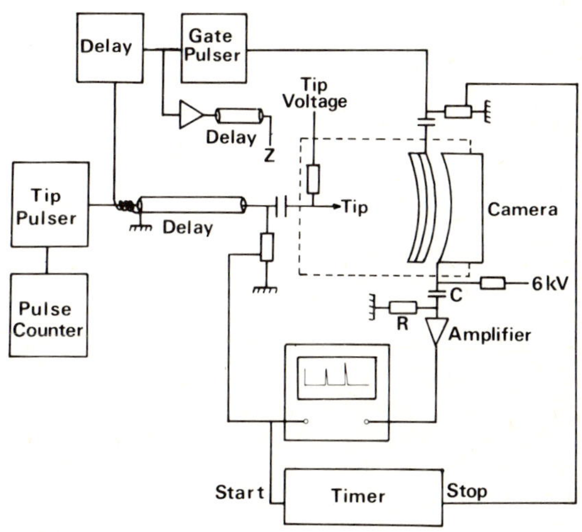

By courtesy of the Institution of Metallurgists

2 Schematic: imaging field desorption atom probe

TABLE 1 Composition of alloy IN939 (at.-%)

Cr	Co	Al	Ti	W	Nb
24.5	18.2	4.0	4.4	0.6	0.6
Ta 0.4	Zr 0.06	C 0.7	B 0.05	Ni balance	

TABLE 2 Compositions of alloys used as standards for calibration of γ and γ' phases in alloy IN939 (at.-%)

γ standard	Ni–19.5Co–27.0Cr
γ' standard	Ni–20.1Al–6.1Ti

Compositions determined by electron probe microanalysis

case, some element in the specimen is chosen as the reference, and the relative sensitivity factors for each of the other elements are determined from measurements on specimens of known composition. It is thus possible to deduce the composition of the sample from the X-ray intensities. Strictly, the factor K is affected only by the beam voltage and the transitions giving rise to the X-rays. It is convenient, however, to consider also the effect of the detector geometry and the absorption in the detector window, and to include them in the value of K, the disadvantage of this course being that the values of K become instrument-dependent.

If the X-rays were generated in an infinitely thin specimen in free space, then the measured X-ray signals would correspond exactly to the generated X-ray intensities. Such, however, is not the case, and it is necessary to consider (a) the modification to the X-ray intensities as they travel through adjacent material and selective absorption occurs, and (b) generation of spurious X-rays from remote parts of the specimen and the microscope as a result of spurious radiation (electrons and X-rays) generated in the microscope column or as a result of interaction of the main electron beam with the specimen itself.

It has been shown that in the HB5 fitted with the so-called 'virtual objective aperture', the contribution of spurious radiation generated by the microscope is negligible compared with that generated by the interaction of the beam with the specimen.[5] Further, if the microscope is not so equipped, this 'hole count' can simply be recorded by passing the beam through a hole in the specimen, and subtracting the resulting X-ray spectrum from that recorded while the beam is incident on a feature in the specimen. The main disadvantage of this latter procedure is the large increase in the background noise level.

For uniform thin foils, correction equations have been developed to allow for the effects of selective absorption.[6] In such cases fluorescence is usually of less importance. However, in multi-phase materials the situation is not so simple. In general the X-ray path through each phase is not known, and the distance from the X-ray source to different fluorescing phases is not well characterized. Thus accurate correction of the X-ray intensities is not usually possible in such cases. To some extent the problems may be overcome by the use of extraction replica techniques, but these cannot be employed in cases where it is required to measure the compositions of, say, precipitates and matrix in immediately adjoining regions. In such instances it is of some assistance to use specimen geometries in which the extent of interaction between the various phases is minimized. Some examples of this approach are given below.

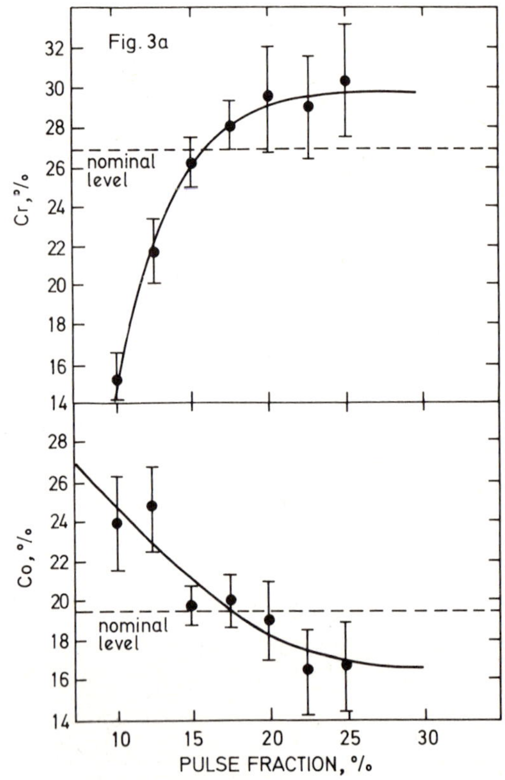

3a **Atom probe analysis of reference Ni–Co–Cr alloy as function of pulse fraction**

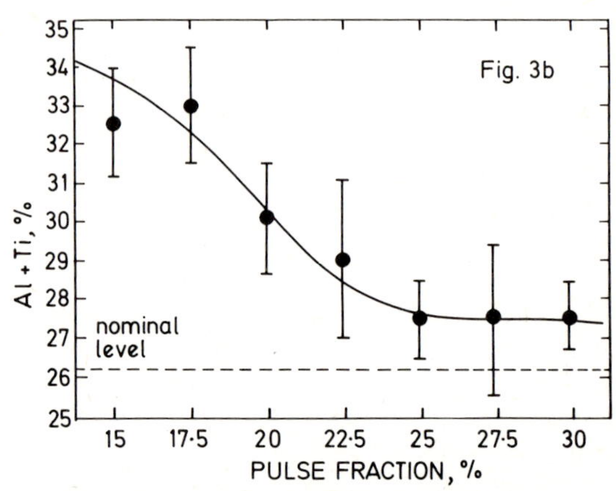

3b **Atom probe analysis of reference Ni–Al–Ti alloy as function of pulse fraction**

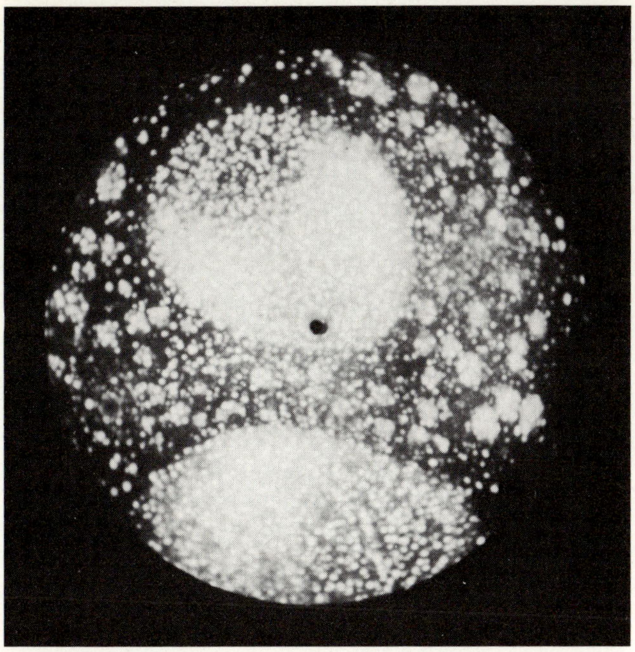

By courtesy of the Institute of Physics (from Ref. 9)

4 Neon field ion micrograph of alloy IN939 showing coarse and fine gamma-prime precipitates (19 kV)

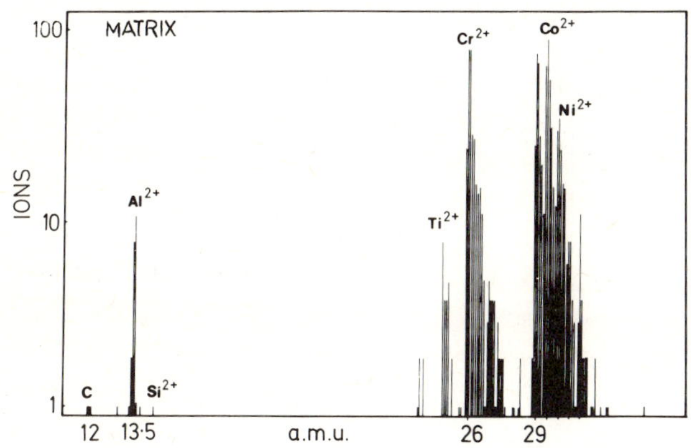

By courtesy of the Institute of Physics

5 Atom probe mass spectrum of IN939 gamma phase

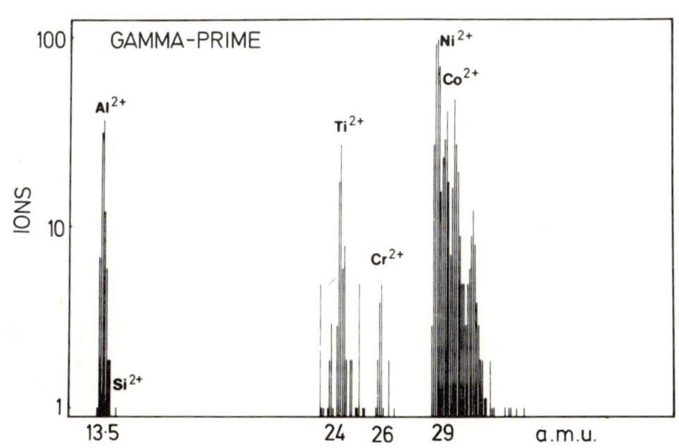

By courtesy of the Institute of Physics

6 Atom probe mass spectrum of IN939 gamma-prime phase

EXPERIMENTAL RESULTS

Phase analysis of nickel-base superalloy IN939

The measurement of phase compositions in nickel-base superalloys forms an essential part of the predictions of long-term phase stability in these materials, particularly with respect to resistance to formation of sigma phase. Alloy IN939* forms one of the new class of high-chromium nickel-base superalloys, developed especially for use in marine and industrial environments where high corrosion resistance is required.[7,8] It is therefore particularly important to carry out phase analysis on this new material. Preliminary atom probe experiments were conducted by Beaven *et al.*[9] and a much more extensive study is currently being completed by Delargy.[10] The full composition of the alloy is given in Table 1.

Atom probe analysis of reference standard materials was necessary in order to define the correct operating conditions for the analysis of major phases in this alloy. Two standard alloys were chosen, having compositions related to those expected for the gamma and gamma-prime phases respectively (see Table 2). Analysis of these alloys showed pronounced pulse fraction effects, which are illustrated in Fig. 3. In order to obtain self-consistent microanalytical results, pulse fractions approaching 25% are necessary in each case. This imposes stringent requirements on vacuum conditions, since at low d.c. holding fields there is an increased likelihood of corrosion of the specimen surface by chemically active residual gases in the vacuum system. The small discrepancy (up to 2%) observed between the nominal composition of the reference alloys and the atom probe results obtained at high pulse fractions may possibly be ascribed to this effect. When specimens of the Ni–Al–Ti alloy were pre-baked and out-gassed before insertion in the atom probe, the agreement between nominal and measured compositions was significantly improved, and this procedure was adopted as standard for the superalloy specimens.

A field ion micrograph of a fully heat-treated specimen of IN939 is shown in Fig. 4. The heat treatment involves a double aging cycle, which produces a duplex distribution of coarse (~100–150 nm diameter) and fine (~5–10 nm diameter) gamma-prime particles in a gamma matrix. The gamma-prime precipitates image brightly in the FIM, against a background of the slightly darker gamma phase. The aperture hole connecting the FIM imaging section of the atom probe to the mass spectrometer section may be seen as a small black disc in the centre of the image. Partial mass spectra of the gamma matrix, and coarse gamma-prime precipitates, are shown in Figs. 5 and 6 (the refractory metal peaks being omitted for the sake of clarity). The Al, Ti and Cr peaks are well separated. The ability to determine aluminium contents, in particular, in a relatively high atomic number matrix, represents a considerable advantage of the atom probe technique. The separation of Ni and Co is more difficult, since the monoisotopic ^{59}Co lies between the ^{58}Ni and ^{60}Ni species, and deconvolution of the mass spectra is possible only when there is sufficient Co present to give a well-defined intermediate mass peak. This usually requires at least 5% Co concentration, which is fortunately the case for both gamma and gamma-prime phases in this particular alloy. The atom probe analyses of major phases are summarized in Tables 3 and 4. Data have been included

* IN939 is a registered trademark of International Nickel Ltd.

TABLE 3 IN939 : analysis of gamma matrix

	FE STEM		Atom probe	
	(a)	(b)	(a)	(b)
Ni	40.9	40.4	41.4	41.1
Co	23.1	22.6	24.0	24.4
Cr	32.9	32.9	31.1	30.8
Al	~1.6	~1.8	1.7	1.7
Ti	1.0	1.3	0.7	0.6
W	~0.2	~0.5	1.0	1.2
Ta	–	~0.25	0.1	0.2
Nb	~0.3	~0.25	–	–

All concentrations in atomic percent. Analyses are presented of two areas studied by each technique. FE STEM results contain contribution from fine secondary gamma-prime precipitates dispersed in gamma matrix.

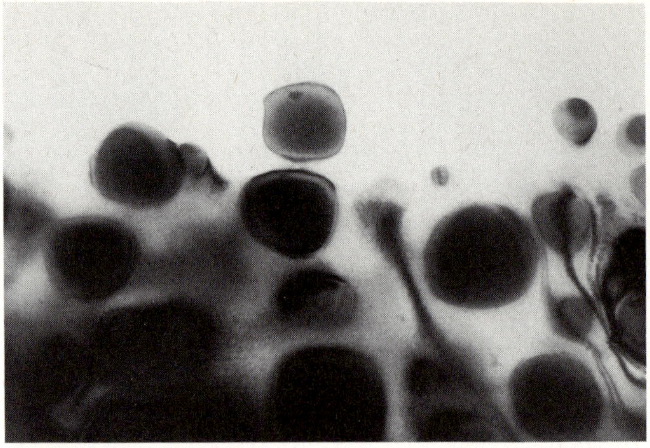

7 **Field emission STEM micrograph of IN939** (× **50 000**)

TABLE 4 IN939 : analysis of gamma-prime

	FE STEM		Atom probe	
	(a)	(b)	(a)	(b)
Ni	65.5	63.6	60.2	61.2
Co	8.6	8.3	11.5	10.4
Cr	2.7	2.0	1.8	2.0
Al	~10.7	~12.8	12.1	12.5
Ti	11.5	11.8	12.7	12.3
W	~0.4	~0.4	0.6	1.0
Ta	~0.2	~0.6	1.1	0.6
Nb	~0.4	~0.5	–	–

All concentrations in atomic percent. Analyses are presented of two areas studied by each technique.

from analyses of two samples of each phase to indicate the degree of consistency in results which is attained.

The field emission STEM X-ray analysis of this alloy was also preceded by analysis of the reference standard materials, in order to determine relative sensitivity factors for the major alloying elements (calculated with respect to nickel as the reference element). An immediate difficulty was encountered, in that the absorption correction factor for aluminium is large and strongly thickness-dependent. Since accurate measurement of foil thickness is extremely difficult in the case of a material as complicated as a superalloy, it was decided to employ an approximate correction for aluminium, based on the signal intensity obtained from a region of foil subjectively estimated to have a thickness of about 100 nm. Other correction factors were found to be much less dependent on foil thickness, and were also much closer to unity than in the case of aluminium.

A STEM micrograph showing the gamma and gamma-prime phases is shown in Fig. 7. It was found advantageous to analyse gamma-prime precipitates located in protruding positions at the foil edges, in order to minimize the problem of overlap of X-ray signals from the two phases. A problem remained in the analysis of the gamma matrix, however, in that the signal due to the small volume fraction (~8%) of fine secondary gamma-prime precipitates could not be separated from that of the matrix. The analytical results tabulated in Tables 3 and 4 thus include a small contribution from these fine particles. Bearing in mind the difficulties in applying corrections to the aluminium X-ray intensity, the degree of agreement between the results from the two instruments is exceptionally good.

Some analyses of carbide phases were also carried out by both atom probe and field emission STEM X-ray techniques, and are summarized in Tables 5 and 6. More difficulties arise in obtaining accurate quantitative results in these cases, since no suitable reference standard materials are available for calibration purposes. Bearing this in mind, the agreement in the results is extremely good in the case of the MC-type carbide phase. The analysis of the $M_{23}C_6$ phase reveals a wider discrepancy, but this is probably due to the fact that the only particle of this kind encountered during atom probe analysis came from material which has been subjected to long-term exposure at elevated temperatures, and may thus have formed as a result of the decomposition of a primary MC carbide rather than being formed directly from the matrix during the initial heat treatment of the alloy. The higher refractory metal content of this particle tends to support this interpretation.

TABLE 5 IN939 : analysis of MC-type carbides

	FE STEM	Atom probe
Ti	62.4	68.3
Nb	12.8	12.0
Ta	14.3	10.3
W	6.3	8.0
Zr	n.d.	0.7
Cr	2.2	0.7

FE STEM data expressed in terms of relative integrated peak intensities; atom probe data in atomic percent. In each case, only abundances of metallic elements are recorded.

TABLE 6 IN939 : analysis of $M_{23}C_6$-type carbides

	FE STEM	Atom probe
Cr	95.5	85.6
Co	2.2	2.3
Ni	1.4	5.4
W	0.42	3.2
Ti	0.24	3.2
Nb	0.12	–
Ta	0.10	0.1
Al	0.02	0.2

FE STEM data expressed in terms of relative peak intensities; atom probe data in atomic percent. In each case, only abundances of metallic elements are recorded.

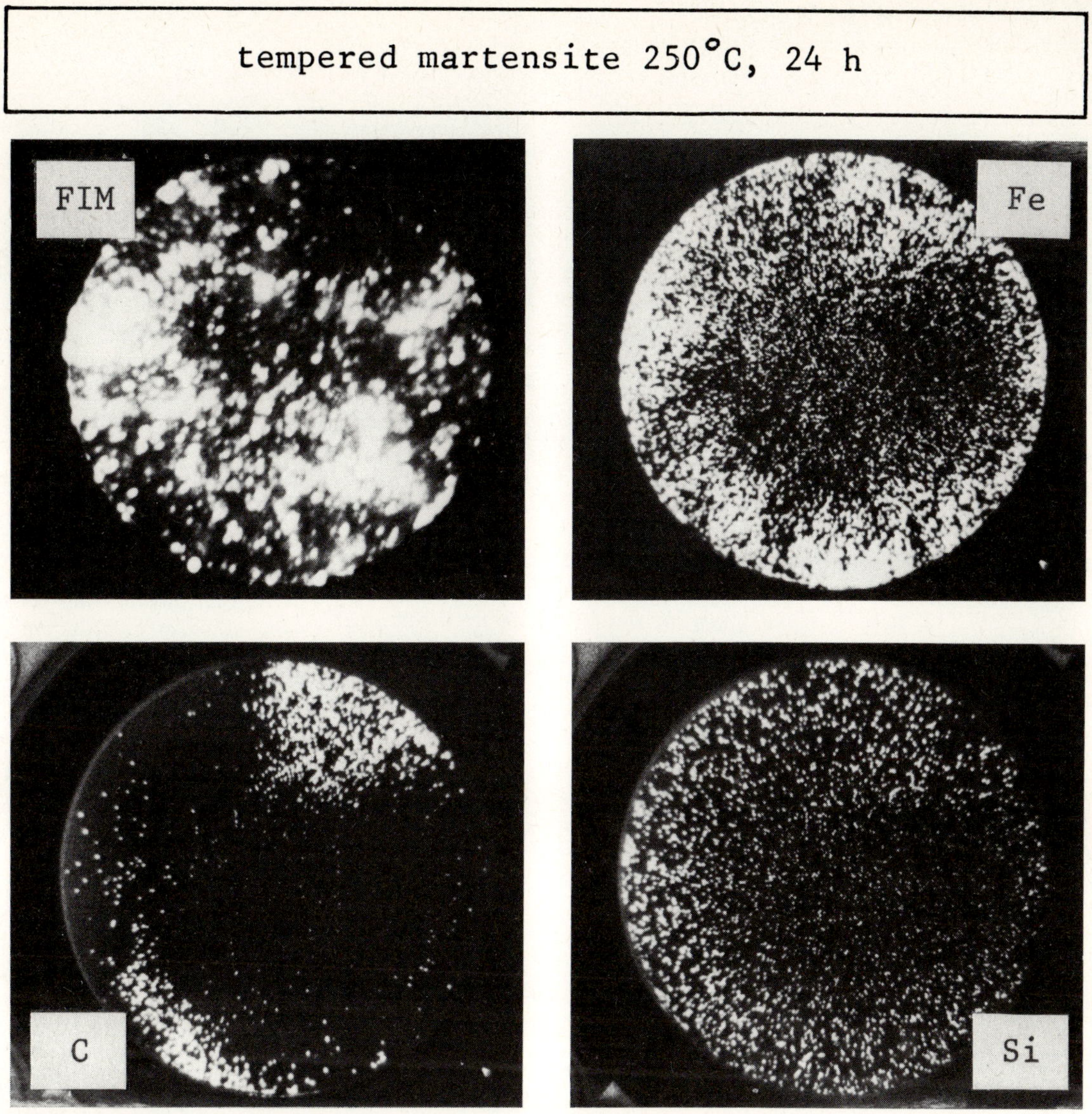

8 Neon field ion micrograph of silicon steel, and gated desorption images of iron, carbon and silicon

Silicon distribution in tempered martensitic steel EN45

It has long been recognized that the addition of a small percentage of silicon to a martensitic steel can produce a marked retardation in tempering kinetics. The effect is exploited commercially in a number of steels designed for use in large springs, drive shafts, anti-roll bars and other heavy duty components. The ability of these steels to be tempered to higher temperatures than usual without undue loss of hardness results in an outstanding combination of strength and toughness. The mechanism by which silicon affects the tempering process has been the subject of much dispute, however. One view has been that silicon may diffuse into the epsilon carbide phase during the first stage of tempering and stabilize it against the transition to cementite. In support of this view, Reisdorf[11] and Gordine and Codd[12] have carried out electron microprobe analyses of extraction replicas which seemed to indicate that the silicon content of epsilon carbide in such steels can be as high as 30%. An entirely different view of the process has been taken by other workers. For example, Owen[13] has

9 Field emission STEM micrograph of silicon steel (× 100 000)

TABLE 7 Composition of silicon steel (at.-%)

C	Si	Mn	Cr	Ni	Cu	P	S
2.69	3.46	0.88	0.20	0.22	0.22	0.04	0.04

shown from kinetic measurements that the rate of growth of the epsilon carbide phase is virtually unaffected by silicon content, whereas the rate of growth of cementite is substantially retarded. He therefore suggests that the critical factor in the kinetics is the rejection of silicon ahead of growing cementite plates in the third stage of tempering, rather than the stabilization of the epsilon carbide phase. In order to distinguish between these two models, it is clearly necessary to measure the distribution of silicon with much greater precision and spatial resolution than was possible in the earlier electron microprobe work.

The atom probe suffers from one potential disadvantage in this application. The singly charged silicon ion Si^{+} would occur predominantly at mass-to-charge ratio 28, which is the same as that of the main iron species, Fe^{++}. Fortunately, experiments by Miller and Smith[14] have shown that silicon in solid solution in iron- and nickel-base alloys field-evaporates predominantly as the doubly charged species Si^{++}. It appeared likely that a similar pattern of behaviour should exist in a carbide phase which field-evaporates at a similar field strength to iron, but it is not possible to establish this point unambiguously. Therefore a particularly strong case existed in this instance for carrying out measurements both by atom probe and STEM X-ray microanalysis.

The steel used for this investigation was a high-silicon, high-carbon spring steel, EN45, whose full composition is shown in Table 7. The material was austenitized at 1050°C, water quenched, and tempered for 24 h at 250°C to produce a relatively coarse dispersion of epsilon carbide particles.

Earlier atom probe studies of a binary Fe–Si alloy by Miller and Smith[14] had shown that a pulse fraction approaching 20% was necessary to produce an accurate quantitative analysis of such a material. The atom probe investigation of the tempered martensite was carried out by Barnard[15] using both the conventional time-of-flight instrument and an imaging atom probe. A gated desorption sequence of a carbide-containing specimen, recorded in the imaging atom probe, is shown in Fig. 8. It is evident that the high local concentration of carbon in the epsilon carbide phase is not accompanied by any significant enrichment of silicon level. Quantitative analysis carried out in the time-of-flight atom probe confirms this point. The mean silicon content determined for the carbide phase was 2.96 ± 0.55 at.-%. For the martensitic matrix the corresponding value was 3.93 ± 0.31%, thus indicating that, if anything, the silicon level was slightly depleted in the carbide.

A micrograph of the same material, taken in the field emission STEM, is shown in Fig. 9, which illustrates some of the difficulties in the electron microscopy of this material. The foils tend to etch irregularly, and there are often traces of an amorphous deposit on the foil surface. It was, however, possible to take a series of X-ray spectra from carbide and matrix regions, and to compare the relative intensities of the silicon K_α lines in each case (normalized to the iron K_α line). Care was taken to select particles in thin areas of the foil and oriented parallel to the electron beam. No significant difference could be detected between the silicon levels of the two phases, in broad agreement with the atom probe results.

Some additional STEM X-ray analysis was carried out on an homogeneous Fe–5.8at.-%Si alloy, in order to try and determine the relative sensitivity factor for silicon in iron. No consistent figure could be obtained, the apparent silicon content always appearing to increase towards the thinnest regions of the foil. This result, taken together with the observation of amorphous deposits on the surface of foils of the tempered martensite, suggests a possible reason for the discrepancy between the present work and earlier electron microprobe investigations. The procedure used to produce the carbon extraction replicas may well remove an amorphous later of silicon hydroxide from the specimen surface along with the carbides which it was intended to examine. This would greatly affect the subsequent X-ray analysis results.

Partitioning of alloying elements in a pearlitic steel

It is well known that small additions of alloying elements such as chromium, manganese or molybdenum can have a profound effect in retarding the rate of the pearlite reaction in carbon steels. At the highest reaction temperatures, the retardation can be explained in terms of the thermodynamic requirements for the substitutional alloying element to partition preferentially to one or other of the product phases in order for the reaction to proceed. However, a strong kinetic effect is still observed at lower reaction temperatures, where no partitioning of substitutional elements is thermodynamically necessary. It is therefore of interest to determine the distribution of alloying elements after reaction at lower temperatures, in order to obtain information about the factors which control reaction kinetics in this temperature regime. Early electron microprobe work by Razik *et al.*[16] using the EMMA-4 instrument at Manchester to study thin foils of pearlitic steels containing chromium and manganese, indicated that no redistribution of these elements occurred below the thermodynamic no-partitioning temperatures (~700°C). However, atom probe work by Miller and Smith[17] revealed evidence of redistribution of these elements at reaction temperatures as low as 550°C, and a detailed atom probe investigation by Williams[20] has shown evidence of extensive redistribution of chromium, and to a lesser extent of manganese, at a reaction temperature of 597°C. Subsequent electron microprobe work using EMMA-4 by Al-Salman and Ridley[19] and Chance,[20] this time employing an extraction replica technique instead of thin foil analysis, has confirmed and extended the atom probe data. It is therefore a matter of particular interest to see whether thin foil X-ray analysis, using a field emission source, can reproduce the atom probe results.

The composition of the steel used in this work is shown in Table 8. The steel was austenitized for 1 h at 900°C, and then isothermally transformed to pearlite in a salt bath held at 597°C. The reaction time was 2 min, and the samples were then water-quenched. The minimum pearlite spacing obtained by this procedure was approximately 100 nm. A field ion image of a specimen of this material is shown in Fig. 10, together with a diagram illustrating the crystallography of the specimen. A cementite plate, approximately 12 nm thick, can be seen lying across the centre of the imaging region. Experiments by Miller *et al.*[1] have shown that a pulse fraction of 10% is sufficient for the quantitative

TABLE 8 Composition of chromium steel (at.-%)

C	Cr	Mn	Si	Ni
2.72	0.90	0.65	0.50	0.20
Cu	Mo	S	P	Fe
0.13	0.02	0.04	0.03	balance

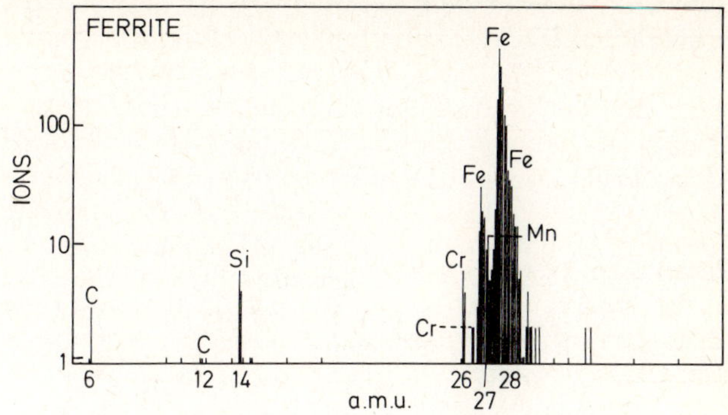

11 Atom probe mass spectrum of ferrite phase of chromium steel

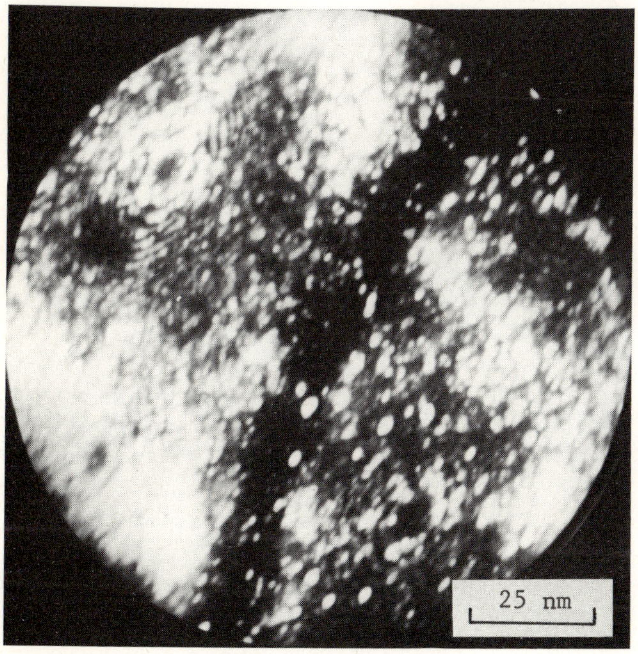

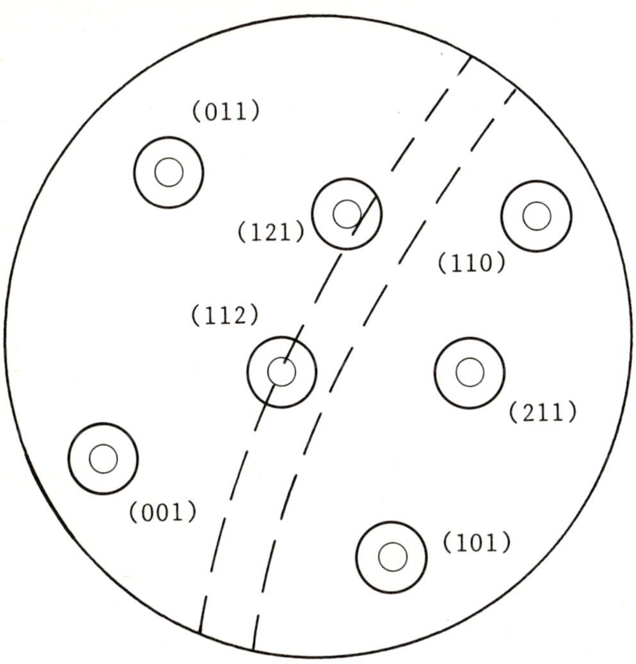

10 Neon field ion micrograph and indexing for chromium steel, showing cementite plate (12.5 kV)

atom probe analysis of chromium and manganese in steels, but in this work a pulse fraction of 15% was chosen, to permit a better measurement of silicon as well. Atom probe mass spectra taken from the centre of the ferrite and cementite phases respectively are shown in Figs. 11 and 12. It is evident that the chromium content is markedly lower in the ferrite than in the cementite phase, whilst silicon is enriched in the ferrite. The manganese content of the cementite phase is slightly above that of the ferrite, but estimation of this element is difficult, as the monoisotopic ^{55}Mn lies between the ^{54}Fe and ^{56}Fe isotopes.

The quantitative atom probe data from these regions are summarized in Table 9, together with data taken from areas of ferrite lying within 10 nm of a ferrite/cementite interface. It is evident that further depletion of chromium and enrichment of silicon occurs in the ferrite close to the carbide/matrix interface, presumably due to the limited distances over which diffusion can occur during the available reaction time. A more detailed atom probe composition profile across the ferrite/cementite interface is illustrated in Fig. 13, and shows that just within the cementite phase the local chromium concentration is very strongly enriched, to a level of approximately 5%.

A STEM micrograph of a similar pearlitic steel specimen is shown in Fig. 14. For analysis purposes, lamellae were chosen to be as closely parallel to the beam as possible. The distributions of chromium and manganese are plotted as a function of position in Fig. 15, the position and direction of the trace being as indicated on the micrograph of Fig. 14. The chromium and manganese X-rays are not severely attenuated by iron, and in any case only the relative changes in concentrations were required. Therefore

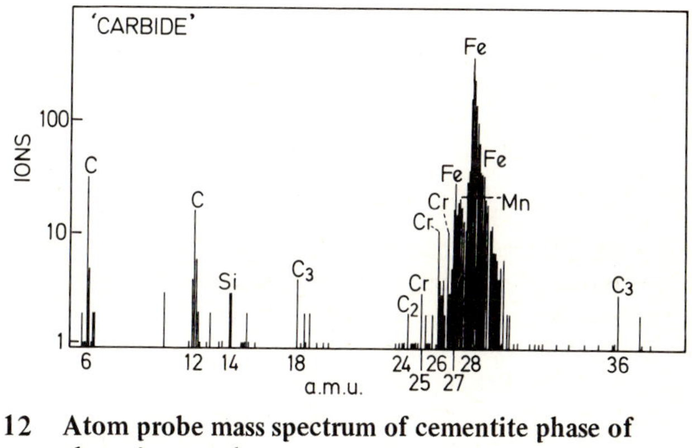

12 Atom probe mass spectrum of cementite phase of chromium steel

no sensitivity factors were measured, and the data plotted are the integrated K_{α} counts for the elements concerned, divided by the total signal from the specimen. The manganese K_{α} line overlaps the chromium K_{β} line, so the accuracy of the measurements is less for manganese than for chromium. The concentration ratios of chromium and manganese between the various regions of the specimen, as determined by both atom probe and STEM techniques, are summarized in Table 10. The agreement between the two sets of results is generally very satisfactory in the case of chromium, although the STEM measurements do not reveal the full extent of the depletion in chromium level on the ferrite side of the ferrite/cementite interface, or the corresponding local enrichment of chromium on the cementite side of the interface. The agreement in the case of manganese is poor, and indicates the practical limitations of both experimental techniques in this particular case. The observation of fine-scale chromium redistribution in this steel using the field emission STEM is in marked contrast to the failure to detect this effect in the EMMA-4 instrument,[16] and illustrates some of the advantages of the high

TABLE 9 Atom probe analysis of chromium steel (at.-%)

	Ferrite (centre)	Ferrite within 10 nm of interface	Cementite (centre)
Cr	0.59±0.12%	0.18±0.13%	2.16±0.16%
Mn	0.68±0.12%	0.81±0.27%	1.09±0.11%
Si	0.61±0.12%	0.9±0.28	0.07±0.03%

All concentrations expressed in atomic percent

TABLE 10 Concentration ratios of Cr and Mn in pearlitic steel

	Atom probe	STEM
Cr(cementite)/Cr(ferrite centre)	3.7	4.5
Cr(cementite)/Cr(ferrite near interface)	12.0	8.7
Mn(cementite)/Mn(ferrite centre)	1.6	~3.5
Mn(cementite)/Mn(ferrite near centre)	1.4	~5.3

Atom probe ratios based on atomic concentration measurements; STEM ratios derived from normalized relative peak intensities

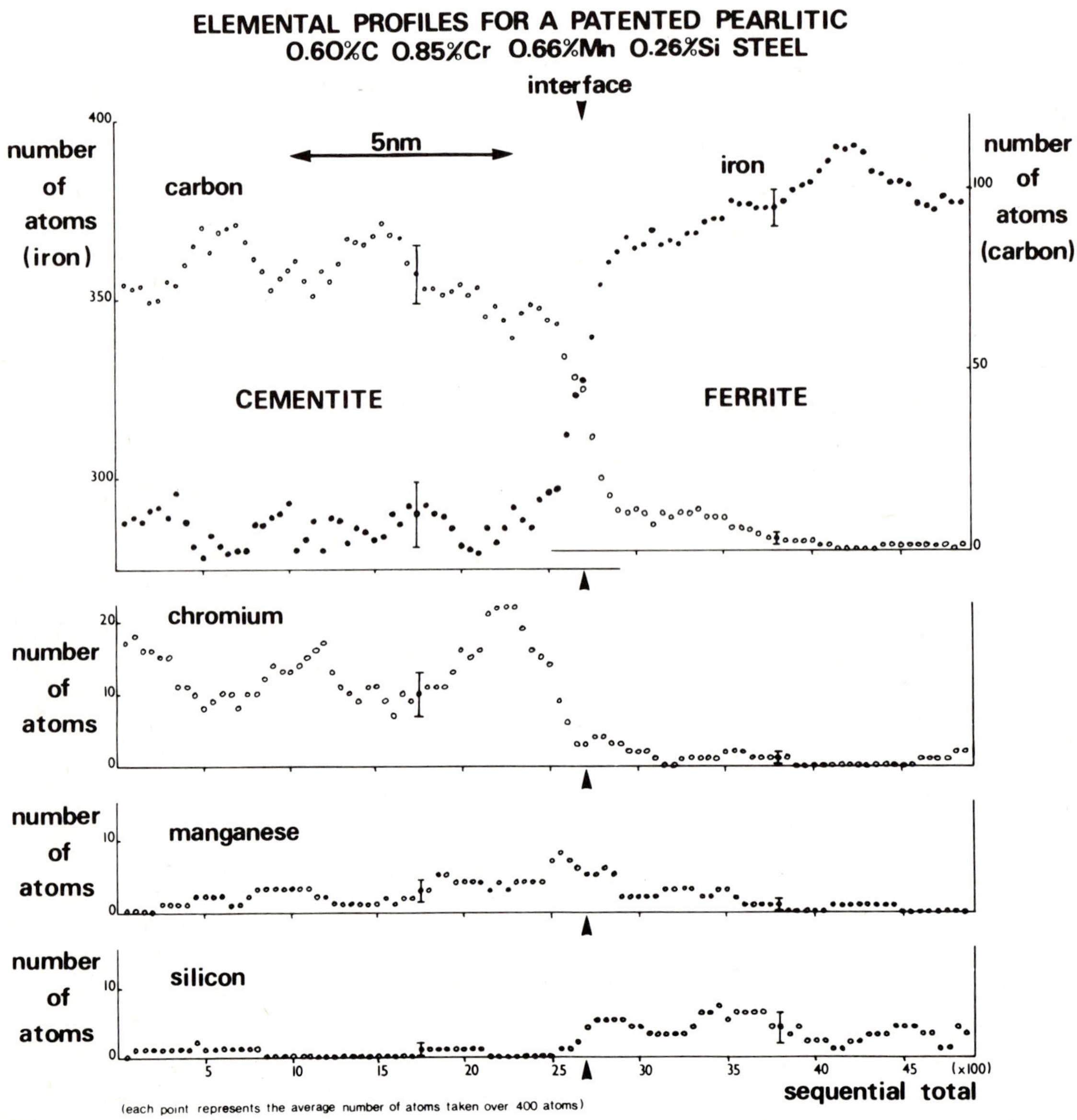

By courtesy of the Institution of Metallurgists

13 Atom probe composition profile across cementite/ferrite interface in chromium steel

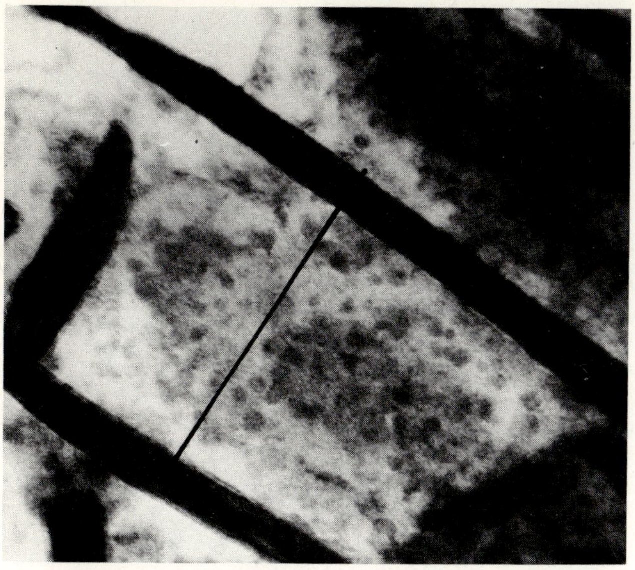

14 Field emission STEM micrograph of chromium steel (× 200 000)

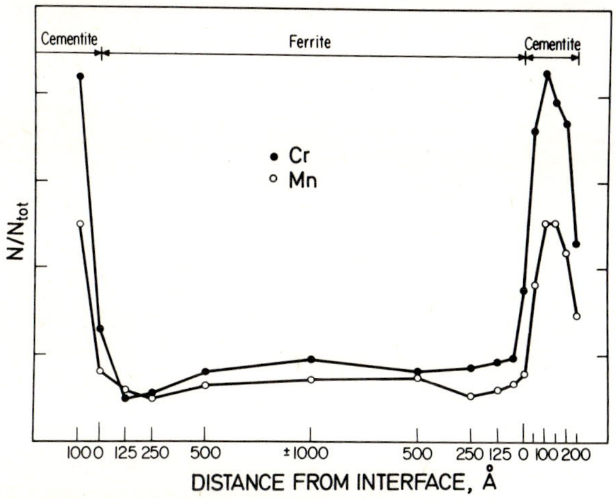

15 Relative intensities of Cr and Mn X-ray signals from points along line indicated in Fig. 14

intensity electron source and electron optics of the HB5. However, even the HB5 cannot reveal the full details of the redistribution process, this being possible only by atom probe analysis.

DISCUSSION

It is convenient to discuss separately the spatial resolution, quantitative analysis capability and versatility of the two techniques.

Spatial resolution

(a) Atom probe

The spatial resolution of the atom probe may be considered in two components, the 'depth resolution' (i.e. in a direction perpendicular to the surface of the tip) and the 'lateral resolution', in the plane of the surface being analysed. The lateral resolution is determined mainly by the size of the aperture in the image screen assembly, divided by the magnification of the image. The minimum size of this aperture is governed by the required statistics in the data. Simply, the larger the aperture, the more atoms are recorded. It would be possible to use a variable geometry slit assembly to match the aperture to the area of interest, but this would present formidable engineering difficulties and has not yet been done. A fixed aperture (physically, a hole in the channel plate) is used, and the magnification is adjusted by changing the distance from the tip to the plate. Typically, an aperture corresponding to a region 2–4 nm in diameter on the specimen is used. The lateral spatial resolution may be degraded, however, in regions where non-uniform curvature of the tip (because, for instance, of second-phase particles) distorts the local electric field, leading to aberrations in the trajectories of the desorbed ions.

If the atomic planes are nearly parallel to the local surface of the tip, the atoms tend to evaporate from the edge of the plane, giving the effect of the plane 'peeling back' from the edge. If the hole in the channel plate is sufficiently small that it does not extend from the edge of one plane to the next, then single atomic plane depth resolution is possible. In practice, this possibility is only useful if the tip can be pre-aligned so that the plane of interest (an interface, for example) is set parallel to the surface before analysis begins.

(b) STEM

The factors affecting spatial resolution in the STEM are the size of the incident electron probe and the thickness of the sample. Whilst the image resolution is governed principally by the incident probe diameter (ample signal being available for good signal-to-noise ratios and aperturing and energy analysis effectively eliminating electrons scattered out of the direct beam), X-ray analysis involves working with much poorer signal-to-noise ratios leading to the employment of larger beam currents, and hence larger beam sizes, and with the possibility of beam-spreading in the sample and fluorescence. In some cases, the form of the specimen renders these problems insignificant, e.g. analysis of 6 nm particles on a carbon replica,[21] and in others corrections or deconvolutions are possible, e.g. segregation at grain boundaries in ceramics over distances <10 nm.[22] In general, however, each specimen must be considered separately, the need for correction determined and the data treated accordingly.

Quantitative analysis

(a) Atom probe

Once the appropriate pulse fraction for a particular material has been determined the quantitative analysis capability of the atom probe is generally excellent. Very often statistical fluctuation in the number of counts recorded is the limiting problem. Consider, for example, a single atomic layer: if the specimen is adjusted for ideal conditions for single atomic layer detection, and if an aperture corresponding to 4 nm on the specimen (at the large end of the typical range) is used, then ~400 atoms will be analysed. If half of these are of element A and half of element B, then the total number of atoms of each is 200. As the standard deviation of this distribution is 10, then the analysis result would give $50 \pm 2.5\%$ of each element. Another difficulty comes from the limited mass-resolution of the time-of-flight spectrometer. If enough counts are obtained, deconvolution of adjacent mass peaks can be attempted, but otherwise it becomes very difficult to make quantitative measurements of small amounts of element $Z+1$ in the presence of a large concentration of element Z, as is shown in this paper by the attempts to measure Mn and Co in the presence of Fe and Ni. A similar difficulty arises if a multiply charged ion of one species has the same mass-to-charge ratio as a singly charged ion of another element, as with the potential problem of silicon measurements in carbides in the tempered silicon steel.

(b) STEM
Quantitative analysis in the STEM is considerably more difficult than in the atom probe. A wide variety of problems, some instrumental and some specimen-related, affect the STEM X-ray measurements. Detailed descriptions will be found elsewhere.[6, 23] These problems include: variations in tip current; generation of spurious radiation in the microscope leading to the so-called 'hole count'; varying sensitivity of the X-ray detector to X-rays of different energies; low signals because of small angles subtended at the specimen by the detector; signal-to-noise limitations because of *bremsstrahlung* background; spurious peaks because of detector anomalies (mainly escape peaks and sum peaks); specimen drift; differential polishing of different elements in an alloy; specimen contamination; reabsorption of emitted X-rays by the specimen; fluorescence of remote parts of the specimen by radiation generated by the specimen; and spreading of the beam by the sample. Some of these factors are fixed in a given instrument, and are therefore incorporated in empirically-determined k factors, and some apply to all signals uniformly and hence only affect the counting statistics. Specimen drift can be corrected at intervals during an analysis, whilst contamination can usually be prevented by cleaning the specimen in an appropriate way. The hole count can either be measured and subtracted, or may be insignificant.[5] During the analysis, therefore, the three most important variables are re-absorption of the emitted X-rays, fluorescence, and beam spreading. Each of these is minimized by making measurements on the thinnest possible regions. If the thickness is uniform and can be determined by, for example, convergent beam diffraction,[24] and the specimen is homogeneous, then corrections can be made to allow for absorption, Usually, neither of these conditions is met, and then it is only possible to determine the likely range of the correction. In the results presented here, the only attempt to make allowance for these effects was in an effort to measure in similar thicknesses of foil in both the superalloy and the superalloy standard specimens. Considering this, the agreement between the atom probe and STEM X-ray results is excellent. Fluorescence undoubtedly prevented better measurements from being made in the pearlitic steel samples, as the cementite was much thicker than the ferrite and would have intercepted a large amount of the scattered radiation whilst the beam was incident on the ferrite close to the interface. Chromium in the cementite would then have been excited and contributed to the measured X-ray data. Again, in STEM analysis, each specimen – and even each data point – must be individually evaluated and appropriate corrections or allowances made for departures from ideal conditions.

Versatility
(a) Atom probe
The atom probe is limited to studying sharp tips of relatively strong, electrically conducting samples. Specimen preparation is fairly complex, but not more so than electron microscope specimen preparation. Experiments take a long time, of the order of a day per tip. Finding widely dispersed features is partly a matter of luck, as for example in the case of the $M_{23}C_6$ particle discussed in this paper. On the other hand, the data obtained can be of very high reliability. Also, all atomic species can be detected with apparently equal sensitivity. Second-phase particles can usually be studied without significant interference from the matrix. The eventual practical limitations on the achievable data are the time available for the experiment, and the volume of material available for study at the maximum voltage attainable in the instrument.

(b) STEM
The STEM can be used to study any material which can be formed into electron-transparent samples. Electron energy loss spectroscopy (which has not been discussed in this paper) can provide semi-quantitative information about the light element content of specimens, whilst X-ray analysis can be used for all other heavier elements. Sometimes peak interferences can occur, and can be resolved only in certain instances by looking at other peaks in the spectrum, but this difficulty is of only limited importance. X-ray data must be interpreted with care, as discussed above, and thickness measurements are difficult and time-consuming. If, however, slightly less precise information is acceptable (for example the identification of precipitates by class only), or if relative information (for example, concentration profiles, variations with different treatments, or segregation in different phases) is needed, then the speed of operation of the STEM is invaluable. In one day, up to ten specimens can be scanned, features of interest identified, analyses performed, and data processing completed.

CONCLUSION
Both atom probe analysis and field emission STEM X-ray analysis can produce, with care, accurate analysis of a specimen. Which one to use is a matter of the application. The atom probe should be used for the more detailed, quantitative study of the finer-scale features in a specimen. STEM can well be used for semi-routine analyses, where the important information is contained in specimen-to-specimen variation in details with modest dimensions. It is clear that, at least in the majority of applications, the two techniques do not compete. There are often difficulties or uncertainties associated with one or other of the methods of analysis, and for a complete characterization of many materials there are strong arguments in favour of using both techniques in combination.

ACKNOWLEDGMENTS
The authors are grateful to Mr S. J. Barnard, Miss K. M. Delargy, and Dr P. R. Williams for permission to quote their unpublished results, and to former colleagues Drs P. A. Beaven and M. K. Miller for their contributions. The expert technical assistance of Mr T. J. Godfrey (Oxford) and Mr J. Nakos (MIT) is also acknowledged with thanks. The work in Oxford was supported by the UK Science Research Council, and that at MIT by the National Science Foundation under grant no. DMR-78-24185. This collaborative work was also assisted by a grant from the Office of Naval Research, which enabled G. D. W. Smith to visit MIT for a period of three months in the spring of 1980.

REFERENCES
1 M. K. MILLER, P. A. BEAVEN, and G. D. W. SMITH: *Surface and Interface Analysis,* 1979, **1**, 149
2 J. A. PANITZ: *Rev. Sci. Instrum.*, 1973, **44**, 1034
3 I. R. M. WARDELL, J. MORPHEW, and P. E. BOVEY: in 'Scanning electron microscopy, systems and applications', Inst. Physics Conf. Series No. **18** (1973), p. 182
4 G. CLIFF and G. W. LORIMER: *J. Miscrosc.*, 1975, **103**, 203
5 A. J. GARRATT-REED: these proceedings

6 J. I. GOLDSTEIN: 'Introduction to analytical electron microscopy' (eds. J. J. Hren, J. I. Goldstein, and D. C. Joy), Plenum Press, 1979

7 S. W. K. SHAW: *Met. Progr.*, March 1979, 47

8 S. W. K. SHAW: 4th Int. Symp. Superalloys, Seven Springs, USA, September 1980 (publ. AIME, in press)

9 P. A. BEAVEN, M. K. MILLER, and G. D. W. SMITH: in 'Electron microscopy and analysis', Inst. Physics Conf. Series No. **36**, 1977, 199

10 K. M. DELARGY: DPhil thesis, Oxford University, in preparation

11 B. G. REISDORF: *Trans. AIME*, 1963, **227**, 1334

12 J. GORDINE and I. CODD: *J. Iron Steel Inst.*, 1969, **207**, 461

13 W. S. OWEN: *Trans. ASM*, 1954, **46**, 812

14 M. K. MILLER and G. D. W. SMITH: *J. Vac. Sci. Technol.*, in press (May/June 1981 issue)

15 S. J. BARNARD: DPhil thesis, Oxford University, in preparation

16 N. A. RAZIK, G. W. LORIMER, and N. RIDLEY: *Acta Metall.*, 1974, **22**, 1249 and *Metall. Trans. A*, 1976, **7A**, 209

17 M. K. MILLER and G. D. W. SMITH: *Met. Sci.*, 1977, 249

18 P. R. WILLIAMS: DPhil thesis, Oxford University, 1980

19 S. A. AL-SALMAN, G. W. LORIMER, and N. RIDLEY: *Metall. Trans. A*, 1979, **10A**, 1703

20 J. CHANCE: DPhil thesis, University of Manchester, 1981

21 A. J. GARRATT-REED and J. B. VANDER SANDE: Proc. 38th Annual Meeting EMSA, 1980, 102

22 E. L. HALL, D. IMESON, and J. B. VANDER SANDE: *Philos. Mag.*, 1981, in press

23 N. ZALUZEC: 'Introduction to analytical electron microscopy' (eds. J. J. Hren, J. I. Goldstein, and D. C. Joy), Plenum Press, 1979, 121

24 S. M. ALLEN: *Philos. Mag. A*, 1981, **43A**, 325

DISCUSSION of paper by Smith, Garratt-Reed, and Vander Sande

A. J. PORTER:
There is a simple check which can be made to ensure that analyses of γ and γ' are reasonable. If one adds together the compositions obtained in the right proportions (given by the γ' volume fraction) one should get to the composition of the material as quoted by the manufacturer. This works very well in practice, especially if some allowance is made for the elements tied up in carbides, and so on.

G. D. W. SMITH:
I agree entirely. We have applied this check to our results on IN939, and it works very well. There can be problems with this form of checking if the volume fraction of γ' is around 50%, however, since failure to detect the full extent of partitioning of (e.g.) chromium between γ and γ' phases would not significantly affect the overall apparent composition of the alloy.

Analytical electron microscopy of carbide precipitates in a stabilized austenitic stainless steel

T Thorvaldsson, H Rubinsztein-Dunlop, H-O Andrén, and G L Dunlop

The precipitation of MC and $M_{23}C_6$ in an austenitic stainless steel containing stabilizing additions of Ti and Nb has been investigated as a function of time during aging at 750°C. Detailed measurements of dispersion parameters for the MC precipitates showed that no significant coarsening occurred for aging times up to 5000 h. Analysis by both STEM/EDS and atom probe field-ion microscopy showed that these precipitates had a mixed $(Ti,Nb,Cr)C_{1-x}$ composition. The Cr content of these precipitates decreased somewhat during aging. $M_{23}C_6$ precipitated intergranularly and analysis of this phase by STEM/EDS showed an increasing Cr content and decreasing Fe content during aging.

The authors are in the Department of Physics, Chalmers University of Technology, Göteborg, Sweden.

The effect of different stabilizing additions (V, Nb or Ti) on precipitation processes in 17%Cr-12.5%Ni austenitic stainless steels has been discussed in a recent paper.[1] One of the main results of this previous work was that, although the precipitation of $M_{23}C_6$ was not avoided in the early stages of high temperature aging, additions of Nb or (Nb+Ti) were found to provide considerably better stabilization against the persistence of $M_{23}C_6$ grain boundary precipitation than did additions of Ti or (Ti+V). It was also found that the MC precipitate distributions, which form in alloys containing Nb or (Nb+Ti), had a very strong resistance to coarsening during prolonged high temperature aging. In the work described here the behaviour of the alloy stabilized by additions of (Nb+Ti) has been studied in more detail.

Detailed precipitate size measurements were made using TEM and the precipitates were microanalysed using both STEM with EDS and atom probe field-ion microscopy. The data which were accumulated by atom probe analysis provide a useful check on the accuracy of small precipitate microanalysis by STEM/EDS. The atom probe results also include the light elements which could not otherwise be detected.

Previously reported microanalytical work by atom probe field-ion microscopy[2] and STEM/EDS[1] has shown that MX carbides, nitrides or carbonitrides precipitated in steels of this type contain a considerable amount of dissolved Cr which diminishes during high temperature aging. The carbides are also carbon-deficient but the metal/carbon atom ratio increases with prolonged high temperature aging.[2]

EXPERIMENTAL

Material

The experimental alloy had the composition given in Table 1. The composition was chosen so that the interstitial content (C+N) and also the (Ti+Nb) content was approximately 0.5 at.-%. The alloy was vacuum melted and hot forged. After solution treatment for 1 h at 1250°C followed by hot rolling and a further 15 min at 1250°C, the material was water-quenched and deformed 5% in tension at room temperature. This produced a relatively high dislocation density ($\sim 10^{14}\ m^{-2}$) and thus a large number of nucleation sites for the precipitation of MC. The average linear intercept grain size was 65 μm. Aging treatments were carried out on specimens in evacuated silica capsules for times of up to 5000 h at 750°C.

Analytical electron microscopy

Thin foils were prepared by jet polishing in an electrolyte of 15% perchloric acid in ethanol. The specimens were examined at 100 kV in a Philips EM300 and at both 120 and 200 kV in a JEOL 200CX. Precipitate number densities and size distributions were measured on dark field micrographs taken using a precipitate reflection. In order to limit the amount of inelastic scattering and thus improve the resolution, the size measurements were generally made in specimen areas which were thinner than 120 nm. The smallest clearly resolved precipitates were approximately 2 nm in size. For each heat treatment condition the sizes of over 100 precipitates were measured in each of 3-5 fields of view giving a total of 300-600 measured particles for each condition. For the number density measurements the thicknesses of the thin foils were determined by counting the number of thickness fringes in two beam conditions.

TABLE 1 Composition of (Ti+Nb) stabilized austenitic stainless steel (wt-%)

C	N	Si	Mn	Cr	Ni	Ti	Nb
0.12	0.009	0.48	1.52	17.1	12.5	0.23	0.41

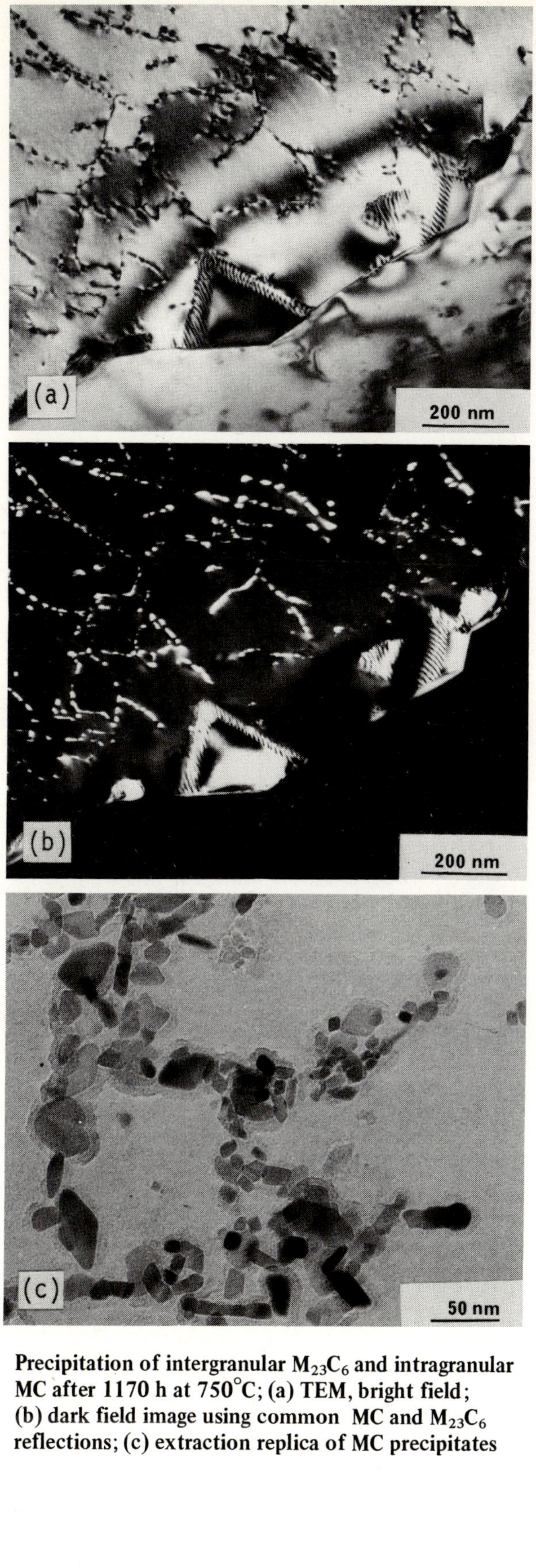

1 Precipitation of intergranular $M_{23}C_6$ and intragranular MC after 1170 h at 750°C; (a) TEM, bright field; (b) dark field image using common MC and $M_{23}C_6$ reflections; (c) extraction replica of MC precipitates

Microanalyses of the substitutional elements contained in precipitates extracted onto carbon replicas were carried out with a Link 860 EDS system attached to the TEM/STEM instrument. Individual $M_{23}C_6$ precipitates were analysed using a stationary electron beam while the MC precipitates were analysed both in this way and also by area scan over a number (in the range 20-200) of particles. A live time of 100 s was used for point analysis and at least 1000 s for area scan analysis. In all cases a beam diameter of ~7 nm was employed. The results were quantified by using the Link Systems RTS-2/FLS computer program[3] which applied the thin foil approximation[4] with corrections for absorption. The estimated error, given by the program for all elements of interest, was less than 1 at.-% for $M_{23}C_6$ and less than 3 at.-% for MC in point analysis. Due to increased background radiation, analysis of MC precipitates by area scan gave an estimated error of about 6 at.-%.

Atom probe field-ion microscopy

The instrument used in this investigation has been described in detail elsewhere.[5] Rods of approximately 0.5 mm square cross-section were cut out with a diamond saw from heat-treated material and electropolished to field-ion specimen tips using a technique involving a layer of floating electrolyte. Precipitates to be analysed were located using neon as an imaging gas. All analyses were carried out in ultra-high vacuum ($<10^{-7}$ Pa). The typical standard deviation of the statistical counting error of a single particle analysis was 3 at.-% for the main elements of interest.

RESULTS

Microstructure

Intergranular precipitation of $M_{23}C_6$ began almost immediately upon aging at 750°C. At the same time a fine distribution of MC carbides precipitated on dislocations within the grains (see Fig. 1). These were identified by electron diffraction as having an fcc crystal structure with a lattice parameter of 4.40 ± 0.03 Å. This value lies between that of pure TiC (a_0 = 4.33 Å) and NbC (a_0 = 4.47 Å) quoted by the ASTM powder diffraction file. Only one MC carbide phase was detected throughout the whole aging sequence of up to 5000 h at 750°C and, within the limits of experimental error, the lattice parameter was found to be constant. Atom probe analysis (Fig. 2) and STEM/EDS analysis showed that the MC precipitates had a mixed composition of the form $(Ti,Nb,Cr)C_{1-x}$. It should be noted that, in contrast to previous work on TiC in similar austenitic steels,[6] no Fe was detected in any of the MC precipitates. This is also in contrast to the case of MC precipitated in a low-alloy ferrite steel where Fe was found at short aging times.[7]

Figure 3 shows the variation of the average MC precipitate size with aging time. As can be seen, there was virtually no increase in particle size during 5000 h aging at 750°C. The lack of coarsening is reflected in the narrow MC particle size distributions (Fig. 4) which did not change significantly during the aging sequence. During the aging period the precipitate number density was virtually constant (Fig. 5), leading to an approximately constant volume fraction of 0.11%.

Detailed compositions of MC precipitates and comparison of analytical methods

(a) Point analysis by STEM/EDS

In order to obtain a reasonable signal-to-noise ratio, precipitates which were in general larger than about 10 nm were analysed by this method. As can be seen from Fig. 4 this means that most of the precipitates which were analysed

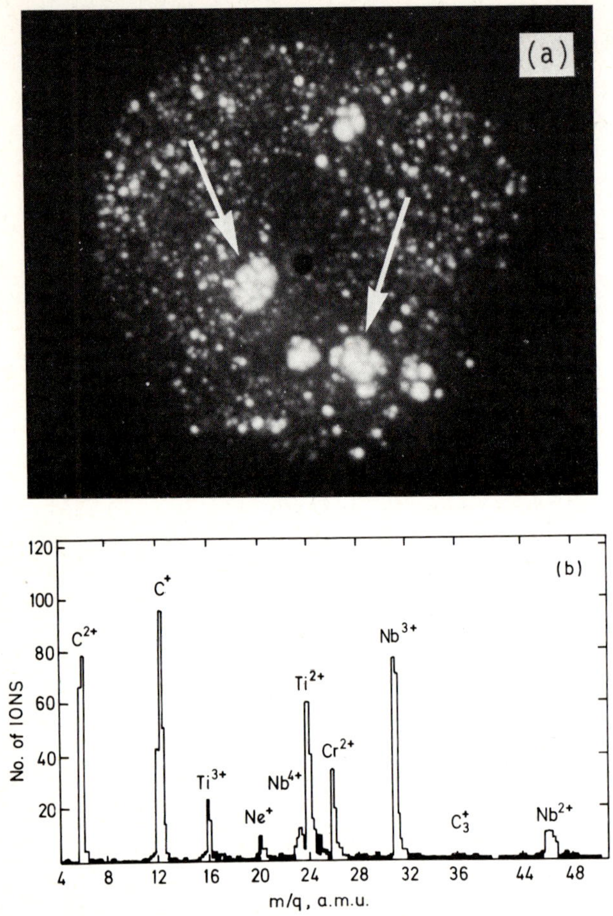

2 (a) Field-ion micrograph of specimen containing MC precipitates (*arrowed*); for analysis specimen was tilted so that precipitate image coincided with probe hole at centre
(b) atom probe spectrum from one MC precipitate after 1170 h at 750°C. Peaks for C, Ti, Cr and Nb were recorded; small Ne peak due to residue of imaging gas

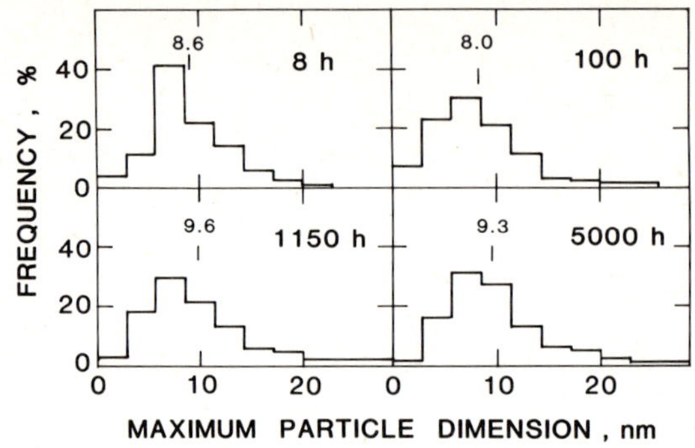

4 Particle size distributions for MC precipitates for various aging times up to 5000 h at 750°C; mean particle sizes for each aging time are indicated

had sizes larger than the average. However, the few analyses which were carried out on smaller precipitates gave results which lay within the scatter of the other measurements. The substitutional element compositions averaged over a number of particles are shown in Fig. 6*a*. The numbers beside the experimental points indicate the number of precipitates analysed at each aging treatment and the bars indicate the standard deviation of the spread between different precipitates. The estimated error for each analysis was less than 3 at.-%. It can be seen from Fig. 6*a* that there was an apparent decrease in the Cr content from around 20 at.-% to about 10 at.-% during aging. The Ti and Nb contents were about 40% and Ti apparently increased with increasing aging time.

(b) Area scan analysis by STEM/EDS

The whole size range of precipitates was analysed by scanning the beam over areas of carbon extraction replicas containing between 20 and 200 precipitates. Thus, this method gave directly an averaged composition. However, the signal-to-noise ratio was generally rather poor so that the estimated error associated with each analysis varied up to 6 at.-%. The results shown in Fig. 6*b* indicate approximately the same levels for Cr, Nb and Ti as given by point analysis. The sharp fall indicated in Fig. 6*b* for the Cr content and coincident increase of the Ti content between 3 and 8 h aging time is also present in Fig. 6*a* although the estimated error in 6*a* makes it less apparent. It can be noted that resistance measurements indicate that the precipitation of MC from supersaturated solid solution ends at some time between 3 h and 8 h.

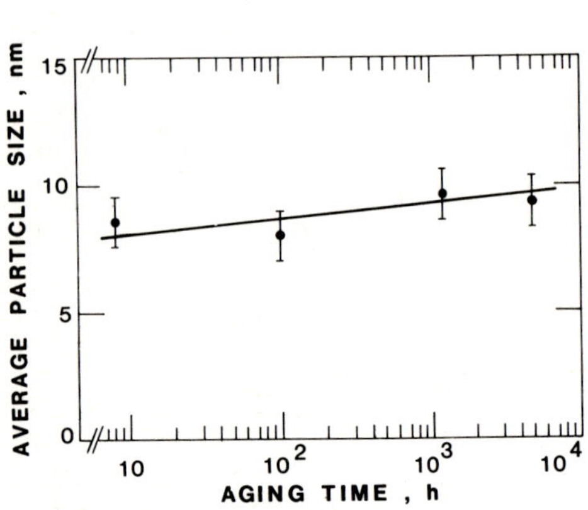

3 Average MC precipitate sizes as function of aging time; bars indicate spread of values obtained in different areas of thin foils

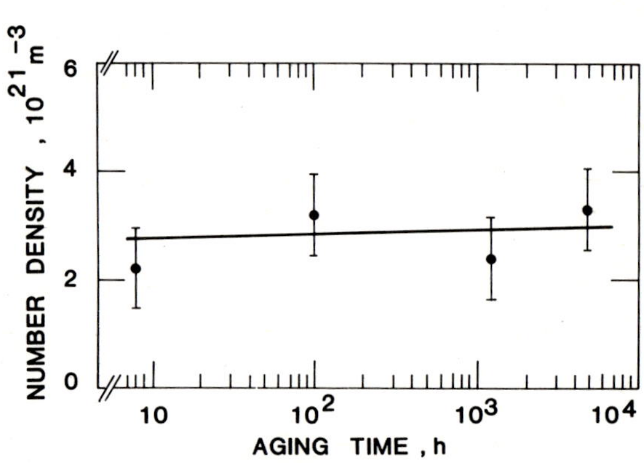

5 Number densities of MC precipitates as function of aging time (bars indicate spread of values obtained in different areas of thin foils)

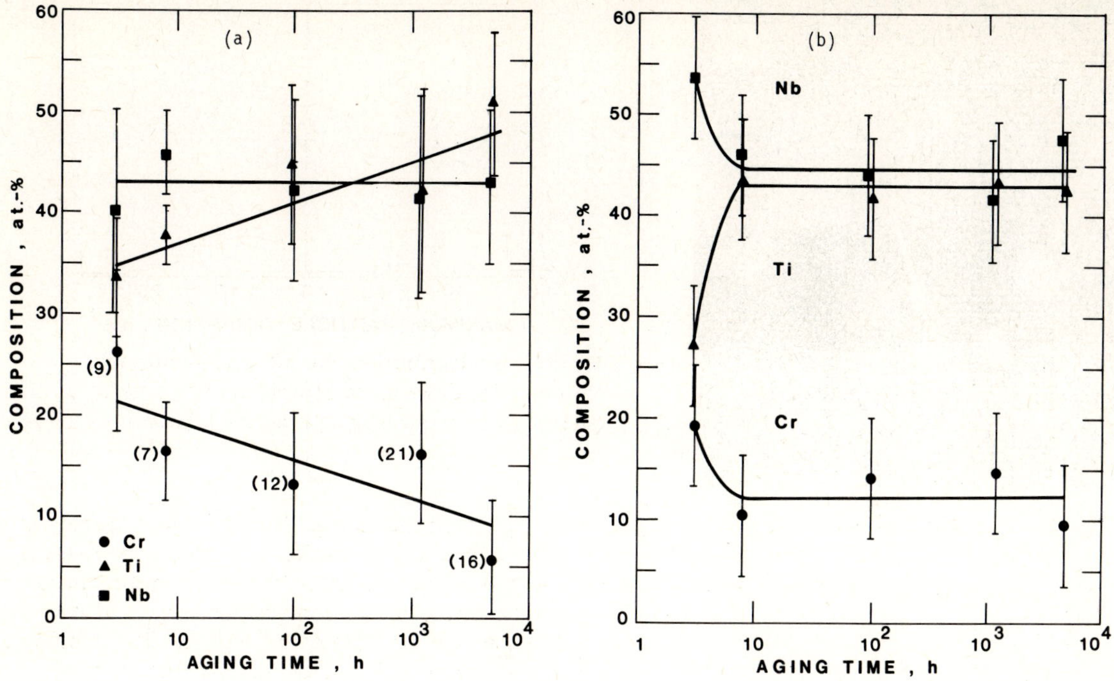

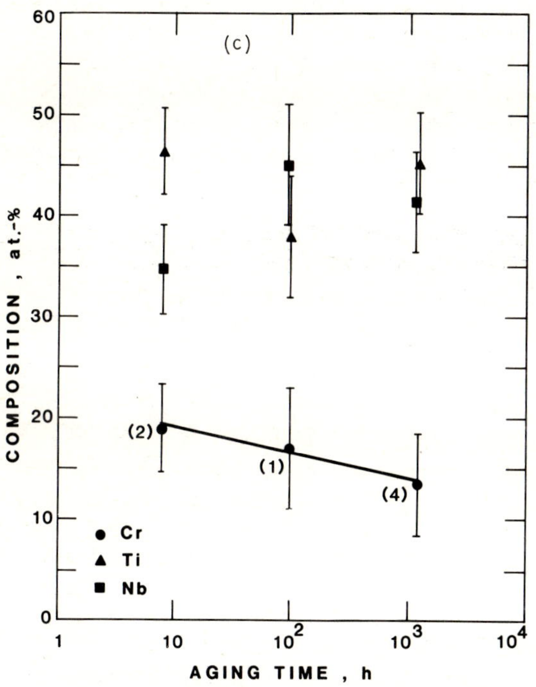

6 Substitutional element composition of MC precipitates as function of aging time (numbers in brackets indicate number of precipitates analysed at each aging time)
(a) Point analysis of STEM/EDS (bars are standard deviation of spread between individual analyses; error of individual analysis less than 3 at.-%)
(b) Area scan analysis by STEM/EDS (bars are estimated error in analysis)
(c) Atom probe analysis (bars are average standard deviation of statistical counting error; spread between individual analyses ~2 at.-%)

(c) Atom probe analysis

Because of the time-consuming nature of atom probe field-ion microscopy only a limited number of precipitates was analysed at each aging time. The number of atoms extracted and analysed from each precipitate varied between 150 and 1100. The precipitates which were analysed were estimated to be generally about 10 nm in size, i.e. about average. Although only a small number of precipitates were analysed by this method, it can be seen from Fig. 6*c* that within the limits of the estimated errors the results of the substitutional element analyses are compatible with those obtained by STEM/EDS.

The C content of the precipitates could readily be determined by atom probe. For the 1170 h aging treatment four precipitates were analysed; three of these were situated on the same dislocation (Fig. 7). As can be seen the C content was always below 50 at.-%.

Composition of $M_{23}C_6$ grain boundary carbides

No $M_{23}C_6$ grain boundary carbides were encountered during atom probe field-ion microscopy but these precipitates were easily analysed by STEM/EDS of extraction replicas. The change in Cr, Fe and Ni content with aging time is shown in Fig. 8. These three elements provide about 96% of the substitutional atom content and the rest consists of Mn, Ti and Nb. The bars indicate the spread between individual precipitates at each aging time and as can be seen the variation is very small. The Ni content was constant at about 3 at.-% for the whole aging sequence while the Cr content increased from about 73 to 84 at.-%. At the same time the Fe content decreased from about 24 to 13 at.-%.

DISCUSSION

The MC carbides which precipitate in this alloy exhibit a strong resistance to coarsening. Microanalysis by both STEM/EDS and atom probe showed that the carbides had a mixed composition of the type $(Ti,Nb,Cr)C_{1-x}$. It is interesting to note that, although some changes in composition occur during aging for times up to 5000 h at 750°C, little change was observed in the dispersion parameters.

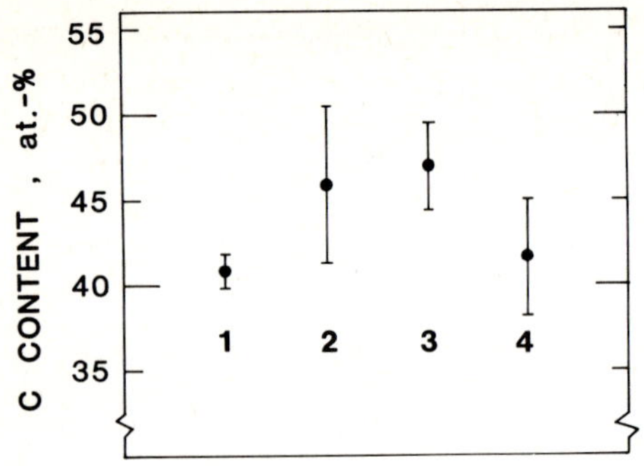

7 **C content of four MC precipitates after 1170 h at 750°C (bars are standard deviation of statistical counting error)**

As can be seen from Fig. 6 the three different ways in which microanalysis of MC precipitates was carried out gave broadly the same results for the Cr, Ti and Nb levels. Because of the relatively large statistical errors involved it is difficult to make a detailed comparison between the results obtained by the different methods. It has been shown previously that the Cr content of (Ti,Cr)C_{1-x} precipitates in a Ti-stabilized austenitic stainless steel decreased with increased aging time.[2] It would seem that the same effect occurs in the present alloy.

It was found by both point STEM/EDS and atom probe that there were considerable fluctuations in the Ti and Nb contents measured in different precipitates in the same heat treatment and even within very localized areas. Since TiC and NbC are fully soluble in each other[8] these fluctuations were probably due to local variations in solute content prior to precipitation. Because of these large variations, it is important that a number of precipitates are analysed if the average composition of precipitate dispersions of this type is required. This has the additional advantage of improving counting statistics, especially for atom probe analysis where the number of counts is generally very low for small precipitates.

The STEM/EDS analyses of $M_{23}C_6$ provided very consistent results which compare well with earlier electron microprobe measurements.[9] As shown in Fig. 8 there is a clear increase in the Cr content of this phase and a consequent decrease in Fe content with increasing aging time at 750°C.

CONCLUSIONS

1 The MC carbides which precipitate in a (Nb+Ti) stabilized austenitic stainless steel have a mixed (Ti,Nb,Cr)C_{1-x} character.

2 These MC precipitate dispersions are highly resistant to coarsening.

3 Within the limit of the errors involved, analysis of MC precipitates by atom probe and STEM/EDS with both stationary and scanning probes gave similar results.

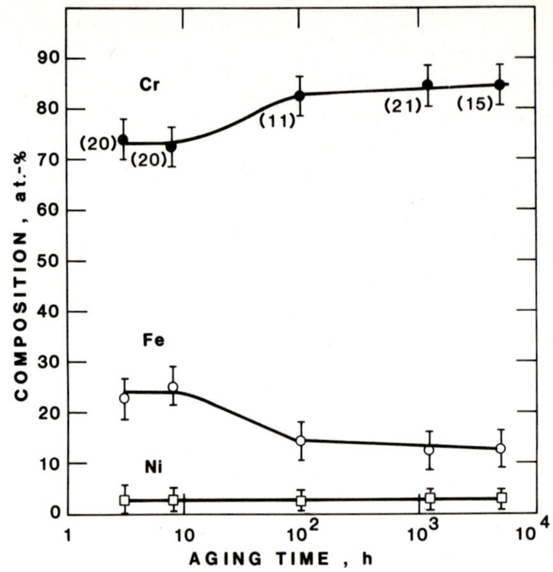

8 **Substitutional element compositions of intergranular $M_{23}C_6$ precipitates as function of aging time as determined by point STEM/EDS analysis (numbers in brackets are number of precipitates analysed; bars are standard deviation of spread between individual analyses)**

4 The Cr content of the MC precipitates decreased during high temperature aging.

5 The Cr content of intergranular $M_{23}C_6$ increased and the Fe content decreased during aging at 750°C.

ACKNOWLEDGMENTS

The experimental material was provided by Uddeholm AB. Financial support was received from the Swedish Board for Technical Development and also from the Swedish Natural Science Research Council. Discussions with H. Nordén and M. H. Loretto are gratefully acknowledged.

REFERENCES

1 T. THORVALDSSON and G. L. DUNLOP: to be published in *Metal Science*
2 H.-O. ANDRÉN, A. HENJERED, and H. NORDÉN: *J. Mater. Sci.*, 1980, **15**, 2365
3 P. STATHAM: Link Systems Ltd
4 G. CLIFF and G. W. LORIMER: *J. Microsc.*, 1975, **103**, 203
5 H.-O. ANDRÉN and H. NORDÉN: *Scand. J. Metallurgy*, 1979, **8**, 147
6 J. BOURGEOT, G. HENRY, B. MICHANT, and B. THOMAS: IRSID Rep. RE 334, 1975
7 G. L. DUNLOP and P. J. TURNER: *Met. Sci.*, 1975, **9**, 370
8 J. T. NORTON and A. L. MOWRY: *J. Met.*, 1949, **1**, 133
9 J. PHILIBERT, G. HENRY, M. ROBERT, and J. PLATEAU: *Mém. Sci. Rev. Mét.*, 1961, **58**:**8**, 557

DISCUSSION of paper by Thorvaldsson, Rubinsztein-Dunlop, Andrén, and Dunlop

R. D. DOHERTY:
What was the exponent in the log-log plot of carbide size against coarsening time? This might give some insight into the mechanism of inhibited coarsening, which could be slow diffusion with a diffusion control or some inhibition of diffusion control.

T. THORVALDSSON:
Fitting of the experimental data by the least-mean-squares method gives an exponent in the relation

$$t = A(\bar{X}^n - \bar{X}_0^n)$$

of $n = 26$. This corresponds to extremely slow coarsening kinetics. To our knowledge such an exponent cannot be explained by current coarsening theories. High values of n have also been obtained for the coarsening of MC carbides in other austenitic stainless steels with different stabilizing additions (Thorvaldsson and Dunlop: *Met. Sci.*, Oct. 1981).

However, when the coarsening of MN nitrides in the absence of C is considered, more rational kinetics with $n = 3$ or 5 (corresponding to lattice diffusion control or dislocation pipe diffusion control respectively) are observed. In the case of MC carbides the coarsening behaviour is complicated by the competition for carbon provided by chromium carbide ($M_{23}C_6$) and, as Dr Thomas suggests, the changing composition of the MC phase.

B. J. E. THOMAS:
Similar work on simple Ti-stabilized stainless steel carried out at IRSID and presented at the Liverpool meeting on microanalysis (1975) showed similar changes in the carbide composition to that presented by Thorvaldsson *et al.* We also observed the slow coarsening kinetics presented in their paper and similar particle size distributions. It may well be that one should include in the coarsening theories the chemical free energy change associated with composition changes during aging.

C. Y. F. COLLIEX:
When you use atom probe techniques for the quantitative analysis of such complex alloys, how are you sure that you have the same probability of evaporation for different impurity atoms?

G. D. W. SMITH:
Different chemical species and different phases certainly do have different evaporation fields. However, the field evaporation end-form is self-adjusting and compensates for this, at least to some extent. (For example, different local radii of curvature will develop over different phases.) The critical thing seems to be to maintain the DC holding field at a low enough value so that slow field evaporation does not occur in the periods between the application of successive high-voltage pulses to the specimen. The analyses obtained by pulse evaporation should then be representative of the true composition of the sample, as is confirmed by experiments.

T. THORVALDSSON:
Almost all atoms at the specimen tip are field-evaporated (i.e. removed from the tip as positive ions), since possible ways of escaping field evaporation (such as surface diffusion, neutral atom formation) are extremely unlikely at the temperatures and electric field strengths employed. It is essential, however, as Dr Smith points out in his reply, that field evaporation occurs only during the high-voltage pulses and that no evaporation occurs at the DC voltage between the pulses. In the present work we have used a pulse amplitude of 15% of the DC voltage, which is sufficient for the analysis of carbide precipitates (an increase to 20% pulse fraction does not change the result of analysis). The two main reasons why no sensitivity corrections need to be applied to atom probe results are that both the probability of signal generation (by field evaporation) and the probability of signal detection (by the ion detector) are mass-independent.

Combined STEM/atom probe investigation of precipitation in powder metallurgy high speed steel

L-E Svensson, P R Howell, H-O Andrén, H Nordén, and G L Dunlop

The microstructure of a complex high speed steel has been investigated by combined analytical electron microscopy and atom probe field-ion microscopy. This combination of techniques has allowed the identification and analysis of microstructural features covering an extremely wide size range. Large primary carbides (1–2 μm dia.) were found to be M_6C and MC while the secondary carbides (2–5 nm) present when tempered to maximum hardness at 560°C are MC and, probably, M_2C. Consistent results were obtained when similar microstructural features were analysed by both STEM/EDS and atom probe.

The authors are all in the Department of Physics, Chalmers University of Technology, Göteborg, Sweden, except P.R.H. who is in the Department of Metallurgy and Materials Science at the University of Cambridge.

The utilization of atom probe field-ion microscopy in association with analytical electron microscopy provides a powerful combination of microanalytical techniques which can be extremely effective in certain types of microstructural investigation. An example of such an investigation is presented here.

The microstructure of high speed steels consists of large primary carbide particles (diameter 1-2 μm) dispersed in a tempered martensitic matrix containing a much finer dispersion of secondary carbides which precipitate during tempering.[1] The steels contain large amounts of alloying elements and it is important for further alloy development to understand the distribution of these elements in the microstructure. The fine secondary carbides cannot be satisfactorily analysed by STEM/EDS of thin foils and the large primary carbides make it very difficult to extract the fine precipitates onto carbon replicas. On the other hand, the very finely dispersed nature of the secondary precipitates makes them a suitable object for investigation by atom probe field-ion microscopy.[2] Thus we have found it convenient to apply both techniques to the overall problem of microstructural analysis of steels of this type.

Depending on alloy composition, the primary carbides which form during the solidification of high speed steels can be either MC, M_6C, M_7C_3, or $M_{23}C_6$.[2,3] Because of the very fine nature of the secondary carbide dispersions at maximum hardness, it is often difficult to determine the crystal structure and composition of these precipitates. Work to date suggests that they can be either MC, M_2C, or M_3C depending upon the exact V, Mo, and W content of the particular steel.[1,4]

The alloy investigated here was ASP 60, manufactured by Uddeholm Tooling, Söderfors, Sweden (for composition see Table 1). This is the most highly alloyed commercially available high speed steel. The powder metallurgical technique used in its production enables high concentrations of alloying elements to be used while keeping the microstructure free from the large-scale heterogeneities (banding, segregation) which are common to cast and wrought steels of this type.[5]

EXPERIMENTAL

The composition of the ASP 60 high speed steel which was investigated is shown in Table 1. The steel was solution-treated at 1180°C, quenched in a molten salt bath held at 550°C, and subsequently cooled in air to room temperature. Most of the material investigated had received the standard tempering treatment (3 times 1 h at 560°C with intermediate air coolings) which eliminates retained austenite and provides maximum hardness. Other heat treatments were carried out at 700°C to produce overaged microstructures.

Thin foil specimens for transmission electron microscopy were prepared by low-speed diamond saw cutting of thin slices, trepanning of 3 mm discs by spark machining, and subsequent jet electropolishing at room temperature in a solution of 10% perchloric acid, 15% glycerol and 75% ethanol with a current of approximately 0.4 A. The thin foils were examined in a JEOL 200CX instrument fitted with a Link 860 EDS system. For microanalysis by EDS a

TABLE 1 Composition of experimental material ASP 60

	C	Si	Mn	Cr	Mo	W	Co	V	Fe
wt-%	2.26	0.41	0.30	4.2	7.0	6.7	10.3	6.4	62.4
at.-%	10.41	0.81	0.30	4.5	4.0	2.0	9.7	6.9	61.6

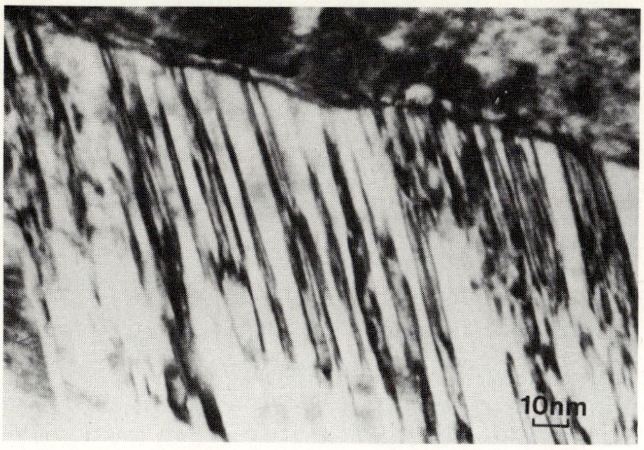

1 Twinned plate martensite after step quenching from 1180°C

typical live-time of 100 s was used. The energy scale of the spectra was calibrated against a zero strobe peak and the Ni K_α peak of a spectrum which was taken up at regular intervals from a Ni standard specimen. The microanalyses were made quantitative by using the Link Systems RTS-2/FLS computer program[6] which compares peak shapes with those from standard spectra of pure elemental standards and applies the thin foil approximation[7] with corrections for absorption. The estimated statistical error in composition for all elements of interest as calculated by the computer program was typically less than 1%. However, when very small particles ($\lesssim$0.2 μm) lying at the edge of the hole in a thin specimen were analysed, the statistical error was typically 2%.

Specimens for atom probe field-ion microscopy were prepared from pins which were cut by a low-speed diamond saw from heat-treated material. The desired needle-shaped specimens were prepared by first producing a necked profile by electropolishing in a 6 mm layer of 15% perchloric acid in ethanol floating on trichloroethylene. Final polishing was carried out in a bath of 2% perchloric acid in 2-butoxyethanol at room temperature. All specimens were examined by transmission electron microscopy before being placed in the atom probe FIM (which is fully described in Ref. 8). The material was field-ion imaged in neon at 92 K and analyses were performed atom-by-atom in ultra-high vacuum (less than 10^{-7} Pa).

RESULTS

Microstructure of quenched material

After quenching into molten salt at 550°C and air cooling, the microstructure consisted of coarse primary carbides

TABLE 2 Average STEM analysis and atom probe analysis of substitutional element composition (at.-%) of martensite after quenching

	STEM	Atom probe
Si	(0.4±0.4)	1.4±0.1
Mn	–	2.1±0.1
V	1.9±0.4	1.0±0.1
Cr	4.6±0.4	4.3±0.2
Fe	73.6±2.1	75.7±0.4
Co	12.5±0.8	11.0±0.3
Mo	3.3±1.7	3.4±0.2
W	3.2±1.0	1.1±0.1
C (at.-% of total)	–	3.4±0.2

Number of atoms analysed: 10 649

TABLE 3 Average analyses of substitutional elements (at.-%) in primary carbides

	MC (STEM)	M_6C (STEM)	M_6C (atom probe)
Si	(0.3±0.3)	4.2±1.5	2.8
V	61.0±7.3	5.8±0.7	3.3
Cr	6.5±1.3	5.8±1.0	4.7
Fe	5.5±2.9	39.3±5.4	41.3
Co	(0.5±0.5)	5.8±1.5	7.1
Mo	16.3±5.5	23.4±7.1	30.0
W	8.1±1.0	15.9±2.3	10.9

Indicated error limits are standard deviations of spread between individual measurements. Results in parentheses are not statistically significant. Atom probe results are average of analyses of two M_6C precipitates (total number of atoms analysed: 2336).

dispersed in heavily twinned plate martensite (Fig. 1). The martensite plates were generally separated by a film of retained austenite. Prior austenite grain boundaries were often decorated by MC carbides. Analyses of the martensite matrix were made by both STEM/EDS and atom probe field-ion microscopy. These results are compared in Table 2, where it can be seen that fairly good agreement is obtained. It can be noted that variations of 1–2 at.-% were found between different areas by STEM/EDS analysis.

Primary carbides

A typical dispersion of primary carbides as seen on an electropolished surface is shown in Fig. 2. The presence of both M_6C and MC types of primary carbide was confirmed by electron diffraction and EDS microanalysis. It was found that these two carbide types had characteristic compositions and, once these were established, identification of carbides in areas such as Fig. 2 was most rapidly carried out by using EDS spectra from the carbides as 'fingerprints'. Average compositions of primary carbides are given in Table 3. It can be noted that the M_6C carbides are rich in Fe, Mo and W while the MC carbides are rich in V and Mo. In both types of precipitate the concentration of Cr was only slightly higher than given by the bulk composition, while the Co content was slightly lower than bulk in M_6C and very much lower in MC. Occasionally, M_6C precipitates were observed which had a significantly increased W

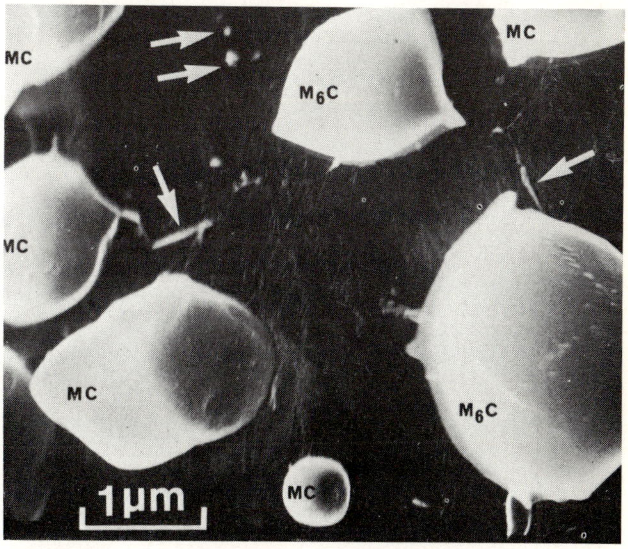

2 Primary carbides seen in electropolished surface of as-quenched material (secondary electron micrograph)

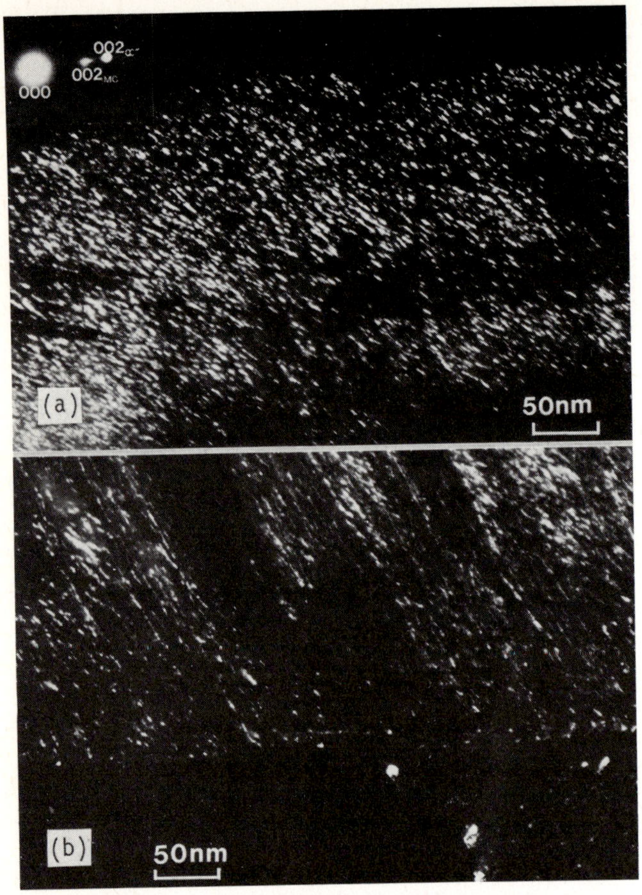

3 **Secondary precipitation during 3 × 1 h tempering treatment at 560°C: (a) general precipitation; (b) precipitation localized to martensite twin boundaries (precipitate dark field micrographs taken using MC reflection: see insert in (a))**

content, but there the Mo content was correspondingly reduced. No significant differences were noted in the compositions of the primary carbides as between the as-quenched and tempered conditions but the detailed compositions of different MC and M_6C precipitates measured by STEM/EDS could vary by a few percent.

Two field-ion microscope specimens were obtained with M_6C primary carbides at their tips. Atom probe analysis of over 2000 atoms from these carbides gave compositions which are also shown in Table 3. It can be seen that these results are in good agreement with those obtained by EDS.

Secondary carbides after tempering at 560°C

The three-stage tempering treatment at 560°C with intermediate air coolings resulted in the transformation of all retained austenite to plate martensite and the precipitation of a very fine dispersion of secondary carbides. Figure 3*a* shows a dark field micrograph of these precipitates together with the associated diffraction pattern. The precipitate reflections corresponded to MC oriented so that it obeyed the Baker–Nutting[9] orientation relationship with the matrix. It is normal for thin plate-shaped precipitates of MC which obey this relationship to lie on {100} ferrite habit planes and therefore the strong streaking of the MC reflection in the $[002]_\alpha$ direction is also consistent with this interpretation. The average size of the precipitates was estimated to be about 2.5 nm.

In many cases the secondary carbides appeared to be precipitated mainly on the martensite twin boundaries (Fig. 3*b*). Here the predominant precipitate habit plane of the three possible cube planes was the {100} plane situated at the lowest angle to the {112} twin boundary plane.

Plate-shaped carbide precipitates were observed by field-ion microscopy. Figure 4*a* shows an example of some relatively large precipitates which, it is considered, correspond to those observed on twin boundaries by electron microscopy. Atom probe analysis of about 100 atoms from one of these precipitates resulted in the composition given in Table 4. The composition suggests that the precipitate is of the MC type with the approximate formula

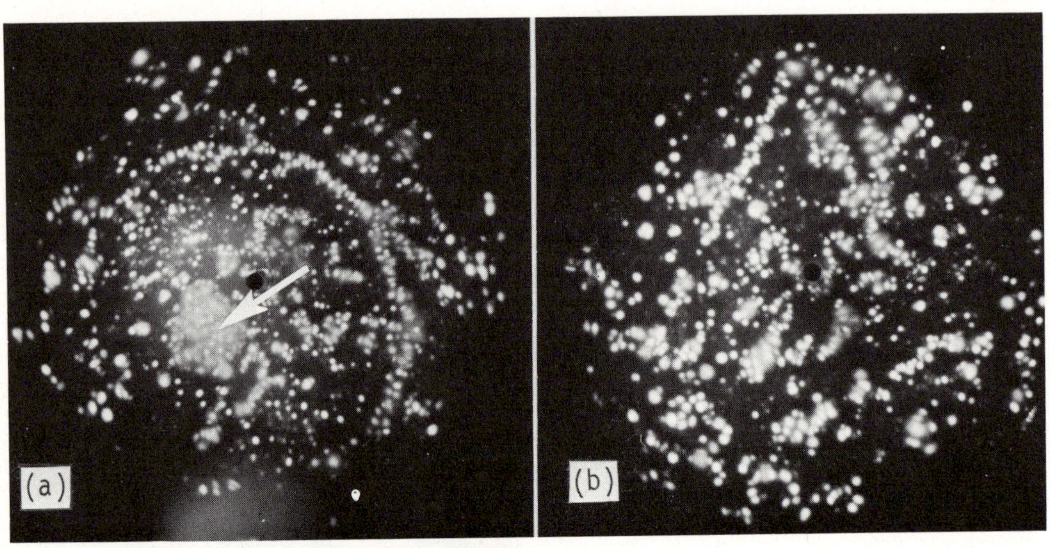

4 **Secondary precipitates observed by field-ion microscopy: (a) larger plate-shaped carbides (*arrowed*) with composition corresponding to MC; (b) smaller carbides with compositions corresponding to either M_3C or M_2C**

TABLE 4 Atom probe analyses (at.-%) of small secondary carbides after tempering 3 × 1 h at 560°C

	MC	M_2C/M_3C
C	46	24
Si	–	0.5
V	17	4
Cr	12	16
Mn	–	0.5
Fe	–	31
Co	1	3
Mo	21	16
W	4	4

Total number of atoms analysed: 104 (MC) and 964 (M_2C/M_3C)

$(V_{0.32}Mo_{0.39}W_{0.07}Cr_{0.2})C_{0.86}$. This composition lies within the range for MC-type precipitates.[10] It is also interesting to compare this composition with that of the long-range ordered vanadium carbide V_6C_5, which has been shown to appear under certain circumstances in low-alloy ferritic steels.[11,12]

Field-ion microscopy also showed the presence of another family of extremely fine precipitates (Fig. 4*b*) which were not identified by transmission electron microscopy. These precipitates appeared to be thin platelets only a few atomic layers thick. Sometimes the precipitates appeared to be oriented along certain directions in the matrix. The average of analyses made on a total of 960 atoms from eight such precipitates is given in Table 4. The composition suggests that the precipitates correspond either to M_3C (i.e. cementite) or to hexagonal M_2C (with a carbon deficiency). The low V content almost certainly eliminates the possibility of their being MC.

Tempering at 700°C

Tempering for 1 h at 700°C resulted in a rapid recovery of the plate martensite to produce a fine grained ferritic structure with a relatively high density of subgrains (Fig. 5). The carbides which precipitated during tempering at this temperature were M_6C and $M_{23}C_6$ and as can be seen in Fig. 5 these were much coarser than the fine carbides which precipitated during tempering at 560°C. The secondary carbides at 700°C often precipitated at the interfaces of primary carbides with the ferrite matrix or else they pinned the subgrain network in the ferrite grains. EDS analysis showed them to have the compositions given in Table 5. Specimens which have been tempered at 700°C for 1 h have not yet been studied by atom probe field-ion microscopy.

DISCUSSION

The extremely fine secondary carbide precipitates, which are responsible for the hardness maximum in this type of steel, have been resolved with the help of both transmission electron microscopy and field-ion microscopy. In the TEM work the secondary precipitates after tempering at 560°C were too small to be analysed by EDS but were identified from diffraction patterns as MC. The atom-probe analyses give good reason to believe that either M_3C or M_2C is also probably present. When M_2C precipitates in tempered martensite the crystallography is such that it generally obeys the Pitsch–Schrader[13,14] orientation relationship with the matrix. Figure 6*a* indicates the expected positions

TABLE 5 Average STEM/EDS analyses of substitutional element compositions (at.-%) of secondary carbides and matrix after tempering 1 h at 700°C

	M_6C	$M_{23}C_6$	Matrix
Si	4.9±2.6	0	1.8±1.2
V	5.7±1.0	6.8±1.0	0.5±0.3
Cr	8.9±2.0	51.1±5.5	2.1±0.2
Fe	33.3±2.2	32.7±2.3	82.0±1.5
Co	6.5±0.6	(1.7±0.6)	12.5±0.6
Mo	28.4±3.0	6.8±3.0	0.8±0.4
W	12.2±1.9	(1.0±0.7)	(0.2±0.1)

Indicated error limits are standard deviations of spread between individual measurements. Results in parentheses are not statistically significant.

of MC and M_2C reflections in a schematic 100 ferrite zone diffraction pattern.[15] A selected area pattern taken from the 560°C tempered martensite and which is of a similar orientation but tilted about an $\{020\}_\alpha$ axis is shown in Fig. 6*b*. It can be seen that the {002} MC reflections in 6*b* are highly streaked in the direction of their habit plane normal and that this streaking overlaps the expected position of the corresponding M_2C reflection. Furthermore, the streaking of the MC reflection is non-symmetric, strongly suggesting the existence of an overlapping reflection at approximately the position where an M_2C reflection should occur.

Because of its large lattice parameter, M_3C should give rise to a large number of reflections which do not overlap with those from other phases. However, no reflections were observed in selected area diffraction patterns which could be identified as coming from M_3C. Therefore, if M_3C is present at the condition of maximum hardness then the volume fraction of this precipitate must be extremely small.

It is of some interest to reflect on the role of different alloying elements in the microstructure of this steel. W, Mo, and V all strongly contribute to the primary M_6C and MC carbides which give the alloy its good wear resistance. The secondary carbides are important for strength at both low and high temperatures. According to the atom probe analysis V, Cr and Mo are important elements in the MC secondary precipitate which is present at maximum hardness (tempered 3 × 1 h at 560°C). The other secondary precipitate which is present in this condition is probably

5 **Fine-grained recovered ferrite microstructure resulting from tempering 1 h at 700°C**

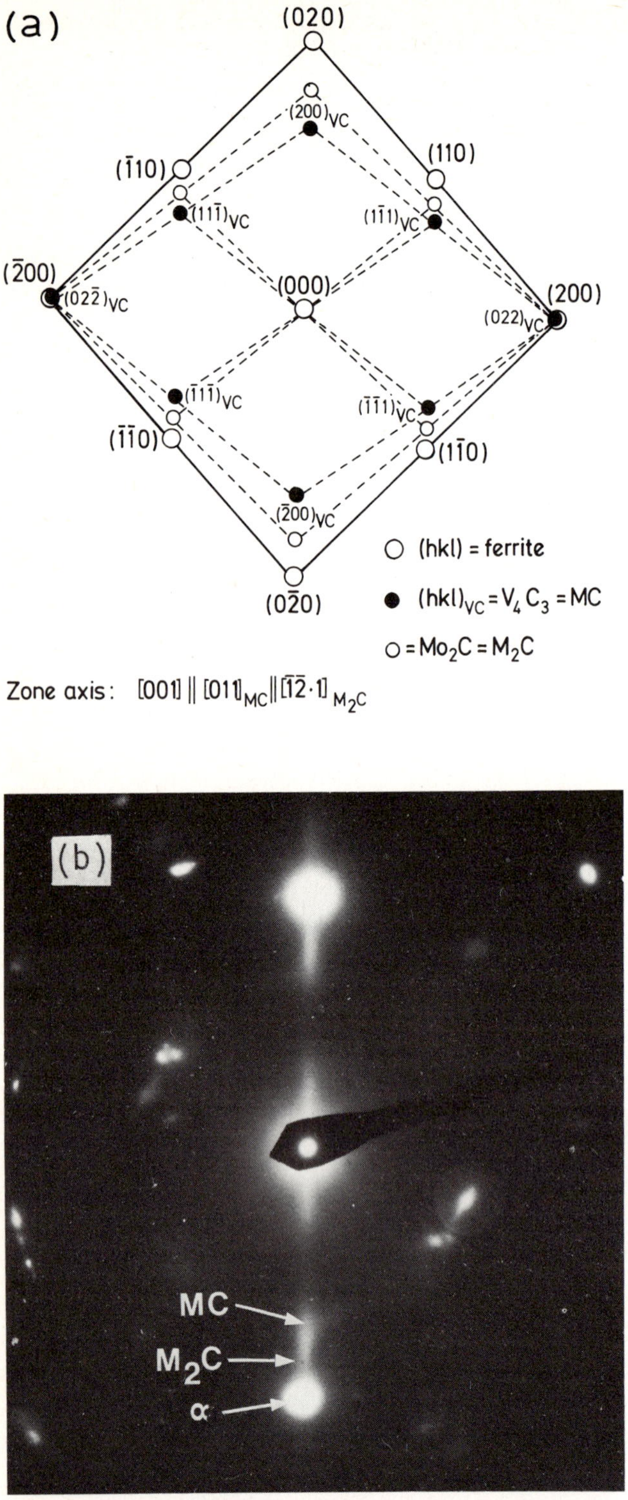

6 **(a) Expected positions in 100 ferrite zone of reflections from MC and M_2C obeying Baker-Nutting and Pitsch-Schrader orientation relationships respectively (b) Selected area diffraction pattern from tempered martensite (3 × 1 h 560°C) containing secondary carbide precipitates; pattern is of similar orientation to (a) but tilted about $\{020\}_\alpha$**

M_2C and here atom probe analysis shows that Fe, Mo, and Cr are the major metallic elements present. Of the major alloying elements listed in Table 1 only Co is not a carbide former and is generally only present in small amounts in the various types of precipitate. Thus, Co is concentrated in the matrix. It has been found in alloy development work that Co reduces the amount of retained austenite which is present after quenching.[16] Thus, Co promotes a microstructurally more evenly tempered final product with a higher secondary hardness which can be retained at higher temperatures. The exact reasons for these effects are still not fully understood.

It is clear that the combination of atom probe field-ion microscopy with TEM/STEM provides considerable advantages in an investigation of this type where the microstructure is heterogeneous and the size of precipitates to be analysed varies by several orders of magnitude. Although it is a time-consuming analytical method, judicious application of the atom probe instrument can provide information which is a valuable supplement to that obtained by analytical electron microscopy. This is particularly so for the analysis of small precipitates and light elements. In cases where direct comparisons can be made between microanalyses made on both instruments on similar microstructural features (see Tables 2 and 3) the agreement is good.

CONCLUSIONS

1 The compositions of large primary M_6C and MC carbides have been determined by STEM/EDS.

2 Electron microscopy combined with atom probe field-ion microscopy suggests that the secondary precipitates present at maximum hardness are MC and M_2C. Compositions of these precipitates have been determined by atom probe analysis.

3 Overtempering at 700°C results in precipitation of M_6C and $M_{23}C_6$ secondary carbides, the compositions of which have been determined by STEM/EDS.

4 Consistent results have been obtained when similar microstructural features have been analysed by both STEM/EDS and atom probe.

ACKNOWLEDGMENTS

This project is financed by the Swedish Board for Technical Development. Further support has been received from the Swedish Natural Science Research Council and from the Science Research Council (England). The experimental material was supplied by Uddeholm AB. Discussions with H. Wisell, P. Hellman, and M. H. Loretto are gratefully acknowledged.

REFERENCES

1 T. MUKHERJEE: ISI Spec. Rep. No. 126, 1970, 'Materials for metal cutting', 80
2 H.-O. ANDRÉN: *Scr. Metall.*, July 1981 (in press)
3 E. HORN: *DEW Techn. Ber.*, 1973, **13**, 171
4 R. LÉVÊQUE and A. CONDYLIS: *Mém. Sci. Rev. Métall.*, 1971, **68**, 87
5 P. HELLMAN: *Werkstoff u. Betr.*, 1976, **109**, 523
6 P. STATHAM: Link Systems Ltd
7 G. CLIFF and G. W. LORIMER: *J. Microsc.*, 1975, **103**, 203
8 H.-O. ANDRÉN and H. NORDÉN: *Scand. J. Metall.*, 1979, **8**, 147
9 R. G. BAKER and J. NUTTING: ISI Spec. Rep. No. 64, 1959, 1
10 W. B. PEARSON: 'Handbook of lattice spacings of metals', 1964, Pergamon Press
11 G. L. DUNLOP and D. A. PORTER: *Scand. J. Metall.*, 1977, **6**, 19

12 G. L. DUNLOP and R. W. K. HONEYCOMBE: *Met. Sci.*, 1978, **12**, 367
13 W. PITSCH and A. SCHRADER: *Arch. Eisenhüttenwes.*, 1958, **29**, 715
14 D. J. DYSON, S. R. KEOWN, D. RAYNOR, and J. A. WHITEMAN: *Acta Metall.*, 1966, **14**, 867
15 I. SARE: PhD thesis, University of Cambridge, 1975
16 S. KARLSSON, B. LEHTINEN, and B. ARONSSON: Swedish Institute for Metals Research Report No. IM-996, 1974

Direct comparison between atom probe and STEM/EDS X-ray analysis of thin specimens

H Nordén, H-O Andrén, M Hellsing, and A Henjered

In a STEM-based microanalytical system lateral resolution is limited to 10–20 nm by beam spreading while depth resolution is determined by the penetration depth of the electrons; this leads to an averaging over ill-defined volumes when performing analyses with high spatial resolution. It was considered useful to compare the results from STEM with those from another technique, atom probe microanalysis, in which signal generation and detection rely upon other physical principles. Both lateral and depth resolution of the atom probe are about 1 nm and relative sensitivity for different atomic species seems to be constant and independent of specimen composition. The use of atom probe specimens (shaped as thin tips) in STEM/EDS analysis has the additional advantage that the local thickness, depth and size of specific features, and the thickness of any surface layer, can be measured directly. Direct comparisons, applying both techniques to the same specimens, are made for homogeneous alloys and large particles with a high content of light elements. In addition, measurements of M *line* k_{XSi} *values for Ta, W and Ir are performed.*

The authors are in the Department of Physics, Chalmers University of Technology, Göteborg, Sweden.

The modern analytical electron microscope, equipped with an energy-dispersive X-ray spectrometer (EDS) system for quantitative analysis of thin foils with a spatial resolution down to about 10 nm, is an extremely powerful instrument which will no doubt be applied to a great variety of microanalytical problems in the future. It is therefore important to be able to estimate the accuracy of EDS analysis by comparing its results with those of other analytical techniques. Unfortunately this is not easily done. Because of the high spatial resolution of the EDS analysis, a meaningful comparison with methods of lower resolution requires extremely homogeneous specimens, and also preparation methods which give clean thin foils of the same composition as the bulk. Differences between results obtained by EDS and other analytical methods are therefore often explained by compositional variations in the material. However, in many cases there are several equally probable explanations, such as: inaccurate absorption corrections (foil thickness and orientation is often less well known, and sometimes the effect of a surface layer is significant); inaccurate detector efficiency factors; or even inaccurate X-ray generation factors (ionization cross-sections and fluorescence yields).

We have therefore found it useful to compare the results of two microanalytical methods of high spatial resolution, EDS and field-ion atom probe analysis, applied to the *same specimen volume*. In addition to allowing direct comparison of the two methods, the use of tip-shaped specimens in the analytical electron microscope has two other advantages: firstly, the geometry of the specimens can be determined very accurately (and so can, for example, the absorption corrections); secondly, the specimen surface condition can easily be checked, both before and after analysis.

EXPERIMENTAL

The materials used in this investigation were either obtained in the form of wire of about 0.5 mm dia. or cut into pins of 0.5 mm square cross-section. In both cases, a length of wire or pin was clamped to a 1 mm dia. aluminium handle and electropolished to tip shape using the floating layer technique. After inspection in the Philips EM300 transmission electron microscope (using a special holder for specimen tips[1]) smooth specimens of sufficient sharpness (tip radius smaller than about 50 nm) were selected for atom probe and EDS analysis. Atom probe analyses were carried out in ultra-high vacuum (less than 10^{-7} Pa) at 92 K, using pulse fractions of 15–20%. A detailed description of the atom probe instrument used is given elsewhere.[2]

For EDS analyses, the specimens were inserted in a specially designed graphite insert to the common holder for the Jeol 200CX Temscan, equipped with a Link EDS system. Quantitative calculations of the microanalyses were made using the Link 8000 computer system. Before analysis, the system was energy-calibrated against a Ni standard and a zero energy strobe peak, and spectra were recorded using a 7 nm probe and a live-time of 100 s. The computer compares the shape of the peaks in the spectra with standard spectra from pure elements, and applies the Cliff–Lorimer thin film approximation[3] with corrections for absorption.[4]

The geometry in the Jeol microscope is such that the EDS analyses obtained from specimen tips of circular cross-section can be corrected for absorption in the specimen, if the standard computer procedure is used with the exception that the 'foil thickness' is replaced by the actual tip thickness at the point of analysis, divided by 3.5. This computed effective thickness value was confirmed experi-

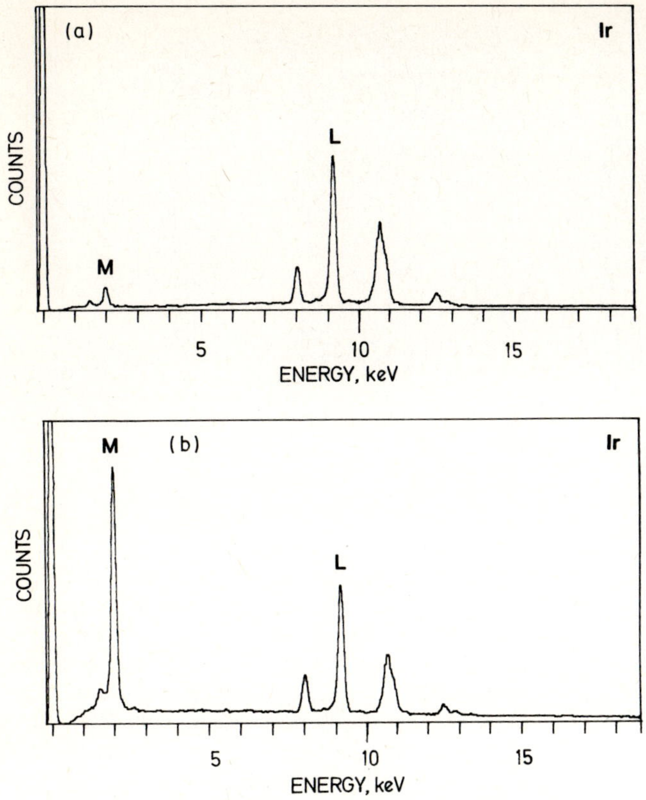

1 **EDS spectra obtained from iridium specimen tip (a) before and (b) after ~100 nm surface layer had been removed by ion etching; note large increase in *M* line intensity**

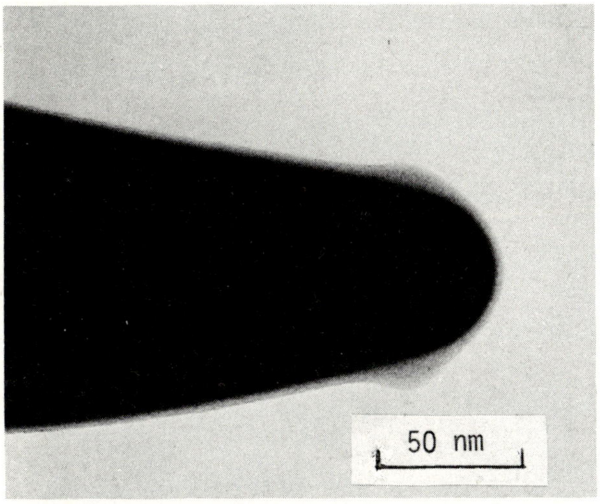

2 **200 kV transmission electron micrograph of tungsten specimen tip after EDS analysis (live-time 100 s) in microscope specially equipped to give low contamination rates; tip rotated 90° around axis to show contamination deposits produced by electron probe**

mentally by measuring the ratio between *K* and *L* or *L* and *M* intensities for different thicknesses in pure materials (see below).

EFFECT OF SURFACE CONDITION

As a simple example of the importance of using clean specimens if accurate EDS analyses are to be obtained, a specimen tip of pure iridium was analysed with the EDS system before and after a ~100 nm layer of surface corrosion was removed by ion etching. This layer was the result of preparation and subsequent storage in dry air for several years. Figure 1 shows the EDS spectra obtained in the two cases. Clearly, if the specimen had been an alloy, the analysis obtained in the uncleaned case could easily have been of a very poor accuracy.

The tip-shaped specimens are also excellent in detecting and measuring the contamination produced during analysis. Simply by tilting 90° around the tip axis, the extent of contamination can be accurately determined. Figure 2 shows one example of this, obtained in a microscope where great efforts to reduce the contamination rate had been made. In an ordinary analytical electron microscope, much worse conditions can generally be expected.

DIRECT COMPARISONS BETWEEN EDS AND ATOM PROBE ANALYSES

A number of specimens have been analysed both in the atom probe instrument and with the STEM/EDS system. Usually at least 2000 atoms were recorded from each specimen in the atom probe in order to obtain good counting statistics, and also in order to measure any small-scale compositional variations present. The EDS analyses were made within 100 nm from the tip apex, in some cases before atom probe analysis (which means that exactly the same volume of material was probed), in other cases after atom probe analysis (in which case the distance between points analysed with the two methods was less than 90 nm). Because the EDS count rate at the tip was usually very low, additional analyses further up the shank were also made.

Molybdenum alloy

A specimen tip of a molybdenum-base alloy containing vanadium and tantalum (alloy and specimen prepared by Dr A. Melmed, NBS, Washington DC) was analysed by EDS and then atom probe. The results obtained are given in Table 1. Since the background in the atom probe spectrum is virtually zero (Fig. 3), the standard deviation given for the atom probe analysis is that of the sampling process only, whereas the standard deviation given for the EDS analyses, as obtained from the Link computer, estimates

TABLE 1 Analyses of molybdenum alloy specimen

	EDS	EDS	Atom probe
Distance from tip apex, nm	100	300, 600, 900 (mean of 3 analyses)	
Thickness at point of anal., nm	100	250, 280, 350	
Composition, wt-%:			
V	1.90±0.52	1.87±0.29	1.71±0.18
Mo	90.75±6.80	89.95±3.46	91.69±0.48
Ta	7.35±2.65	8.18±1.39	6.60±0.67

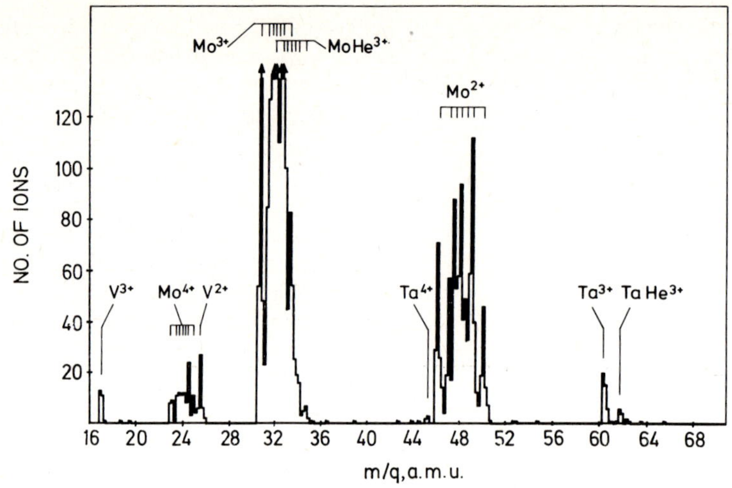

3 **Atom probe spectrum of molybdenum alloy containing vanadium and tantalum. Since atom probe measures mass-to-charge ratio of field-evaporated ions, several isotopes (Mo) and ion charge states are seen in spectrum; note also occurrence of $MoHe^{3+}$ and $TaHe^{3+}$ molecular ions (He used as field-ion imaging gas)**

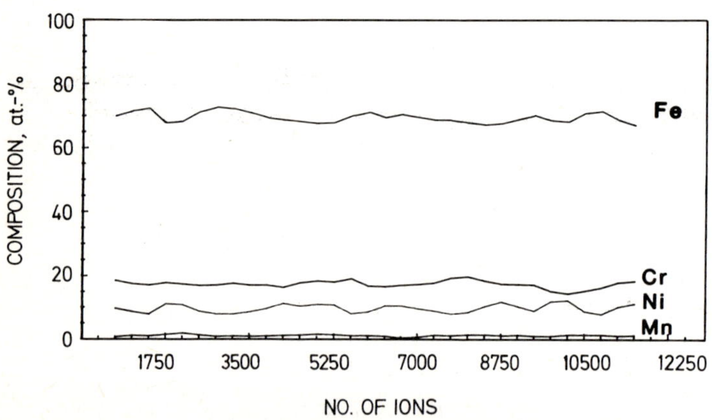

4 **Field evaporation (depth) profile obtained when analysing stainless steel specimen in atom probe instrument; sequence shown corresponds to depth of ~75 nm; all compositional variations seen lie within ±2 standard deviation limits**

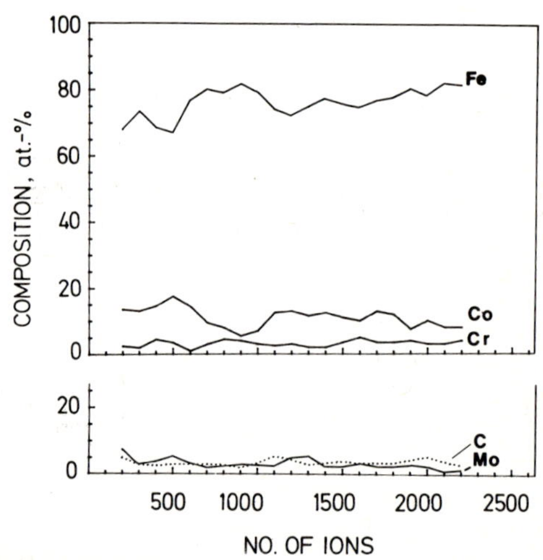

5 **Field evaporation profile from high speed steel specimen, corresponding to depth of analysis of ~20 nm; in this case, compositional variations seen are greater than those expected from counting statistics only**

TABLE 2 Analyses obtained from stainless steel specimen (result of bulk chemical analysis given as comparison)

	EDS	EDS	Atom probe	Chemical analysis
Distance from tip apex, nm	20	80		
Thickness at point of anal., nm	70	150		
Composition (wt-% of subst. elements):				
Si	(0.65)*	0.92±0.45	0.61±0.05	0.11 – 0.44
Cr	18.71±1.81	19.44±0.83	16.68±0.33	17.00
Fe	70.78±2.94	70.32±1.45	70.07±0.44	71.19
Mn	(0.00)	(0.00)	1.18±0.10	1.58
Ni	8.87±1.72	9.20±0.74	10.25±0.29	9.96
Mo	(0.99)	(0.12)	1.21±0.13	0.24
C (wt-% of total)	–	–	0.05±0.01	0.072

* Figures in parentheses: *not significant*

TABLE 3 Analyses of high speed steel specimen

	EDS	EDS	Atom probe
Distance from tip apex, nm	50	100	
Thickness at point of anal., nm	210	70	
Composition (wt-% of subst. elements):			
Si	1.15±0.46	1.13±0.52	0.82±0.13
Cr	4.92±0.69	5.42±0.88	3.31±0.36
Mn	(0.08)	(0.00)	1.47±0.25
Fe	75.73±1.77	75.27±2.15	72.75±0.86
Co	12.45±1.13	12.90±1.41	11.04±0.67
Mo	2.89±1.20	(2.16)	4.98±0.60
W	(1.43)	(0.95)	4.99±0.84
V	1.17±0.52	1.40±0.68	0.62±0.16
C (wt-% of total)	–	–	0.88±0.09

Figures in parentheses: *not significant*

the effects of both the background and the counting statistics.[5] The low count rate at the tip makes the standard deviation rather large, and the atom probe values fall well within the error limits. A comparison with analyses from further down the shank also shows a very good agreement. There is possibly a slight tendency for EDS to overestimate the minor elements Ta and V, but this tendency is by no means statistically significant.

Austenitic stainless steel

A specimen tip of an austenitic stainless steel was analysed in the atom probe, and a field evaporation profile of more than 11 000 ions, corresponding to a depth of analysis of about 75 nm, is shown in Fig. 4. Each point on this diagram corresponds to the composition of 700 atoms in sequence, and the compositional variations seen all fall within the ±2 standard deviation (95% confidence level) limit for 700 atoms. Thus, over the 75 nm distance analysed, the specimen was homogeneous to within the experimental uncertainty.

The specimen was then analysed by EDS both close to (20 nm) and 80 nm away from the tip apex. The results are given in Table 2. The EDS count rate at the tip was not sufficient to give statistically significant values for Si, Mn, and Mo. Within the experimental uncertainty, the results obtained for the major alloy components Fe, Cr, and Ni at the tip agree with the atom probe data. There is a tendency for an increased Cr content in the EDS analysis, and in the analysis 80 nm away from the tip apex, where the specimen thickness was 150 nm, the Cr value is so much higher that it disagrees with the atom probe value. It is possible that this is due to fluorescence induced by the large Fe peak; similar effects have been found in ~200 nm thin foils at 100 kV.[6]

For comparison, the result of a bulk chemical analysis is given in the table.

High speed steel

A powder metallurgical high speed steel was hardened by step quenching from 1180°C, and a specimen containing matrix (untempered, heavily twinned martensite) at the tip was analysed in the two instruments. A field evaporation profile obtained during atom probe analysis is shown in Fig. 5. In this case, the compositional variations exceed the ±2 standard deviation limits, so the composition can be expected to vary over distances as small as perhaps 10 nm (this suggests that clustering or even some tempering occurred during the 550°C salt bath quenching and subsequent air cooling).

The specimen was analysed with EDS twice, at 50 and 100 nm from the tip apex. The results are given in Table 3. Of the statistically significant values, those for Cr are clearly significantly higher than their corresponding atom probe value. Whether this is an effect of compositional variation or Fe-induced fluorescence is not possible to say. Also the discrepancies for Fe and Mo between one of the EDS analyses and the atom probe values could very well represent real compositional differences. Obviously the method of comparison employed in this work requires a specimen homogeneity over a few analysing volumes, say 50 nm, and this requirement was not fulfilled in the hardened high speed steel.

M_6C-type carbide particle

A high speed steel specimen containing a primary M_6C-type carbide at the tip was analysed first in the atom probe (Fig. 6), then with the EDS. The results are given in Table 4. The agreement is in this case very good, except in the case of Si where too few counts were recorded by EDS analyses for a meaningful comparison. This may be due in part to difficulties in accurately separating the overlapping Si *K* and W *M* lines. Apart from this, the similarity in results is rather impressive particularly in view of the many elements present.

MEASUREMENT OF *M* LINE k_{XSi} VALUES

Since the geometry of specimen tips can be accurately determined, the absorption corrections are known to be correct, so that a useful check of the theoretical *M* line (effective) *k* values used in computing compositions (usually referred to Si according to the formula $C_X/C_{Si} = k_{XSi}(I_X/I_{Si})$) can be made. If the *L* and *M* (or *K* and *L*) peaks from the same element in a specimen are treated as originating from separate atomic species, each peak is expected to represent exactly one-half of the true element content in the specimen, provided that the corrections for absorption have been properly done. Data collected on analysing specimen tips of pure W, Ir, and (for comparison)

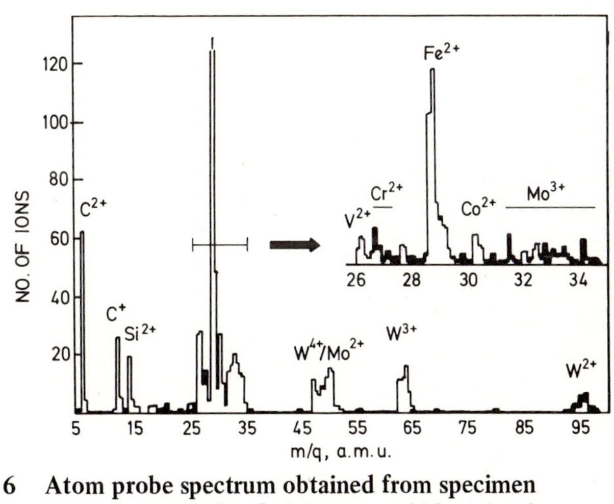

6 Atom probe spectrum obtained from specimen consisting entirely of M_6C-type carbide at its tip

TABLE 4 Analyses obtained from M_6C primary carbide in steel

	EDS	EDS	EDS	Atom probe
Distance from tip apex, nm	90	120	200	
Thickness at point of anal., nm	150	175	200	
Composition (wt-% of subst. elements):				
Si	(1.86)	(0.00)	(1.26)	1.30±0.21
V	2.71±1.29	2.89±0.64	2.76±0.49	2.55±0.41
Cr	5.25±1.42	3.12±0.68	3.42±0.55	4.00±0.51
Fe	25.00±2.32	25.65±1.26	25.47±1.06	24.03±1.12
Co	5.18±1.75	5.45±0.90	5.13±0.75	4.00±0.55
Mo	26.70±8.43	28.08±4.21	26.42±3.52	31.62±2.07
W	33.30±6.21	34.81±3.36	35.54±2.71	32.40±2.67
C (at.-% of total)	–	–	–	14.50±1.29

Figures in parentheses: *not significant*

TABLE 5 Apparent composition (corrected for absorption) obtained by treating *L* and *M* (or *K* and *L*) peaks from same elements as stemming from two separate elements

		200 kV		120 kV	
Peak	Specimen	Apparent composition	Error	Apparent composition	Error
Mo *K*	Pure Mo	49.8	1.7	49.7	2.0
Mo *L*		50.2	0.9	50.3	1.0
W *L*	Pure W	62.3	1.3	63.0	1.5
W *M*		37.7	0.4	37.0	0.5
Ir *L*	Pure Ir	61.6	1.4	61.2	1.2
Ir *M*		38.4	0.6	38.8	0.5
Ta *L*	Large MC carbide	62.7	1.6		
Ta *M*		37.3	0.7		
W *L*	Large MC carbide	62.0	2.8		
W *M*		38.0	1.1		

Mo as well as large cubic carbides containing Ta and W are given in Table 5.

In all cases, the apparent (absorption-corrected) compositions obtained were found to be independent of specimen thickness. However, only in the case of Mo was the apparent content in the two peaks (Mo *K* and Mo *L*) close to 50%. In all cases involving *L* and *M* lines, the *L* line contents were much higher than those of the *M* line. Since the *K* and *L* line k_{XSi} values are now well known,[7] and since the Ta, W and Ir *M* lines all fall around 2 keV, i.e. in a region where the detector efficiency values are expected to be accurately known, the conclusion is that the values of k_{XSi} for Ta, W and Ir *M* lines used at present in the computer evaluation procedures are too small. Assuming the k_{XSi} values for Ta *L*, W *L* and Ir *L* to be 1.88, 1.91, and 2.00 respectively at 200 kV, the following values for *M* line k_{XSi} are obtained at 200 kV: Ta 1.86, W 1.80, and Ir 1.86. The corresponding *M* line values found for 120 kV were: W 1.87 and Ir 1.85, assuming *L* line k_{XSi} values of 2.05 and 2.17 for W and Ir respectively at 120 kV.

CONCLUSIONS

1 Methods have been developed to analyse the same tip-shaped specimen in both field-ion atom probe and STEM/EDS instruments.

2 Contamination and surface layers on tip-shaped specimens can easily be detected in the electron microscope. A 100 nm surface layer on an Ir specimen tip was found to alter the relative Ir peak intensities in EDS spectra significantly.

3 Specimens of a number of materials were analysed both in the atom probe and with EDS. Provided that the specimens were homogeneous over a distance of about 50 nm (Mo alloy, stainless steel, M_6C particle), good or excellent agreement between the two instruments' results were obtained. When the homogeneity criterion was not fulfilled (high speed steel) the results differed slightly.

4 The geometry and therefore the X-ray absorption of tip-shaped specimens can be determined accurately. This makes comparisons between computed and measured k_{XSi} values possible. Measurements of *M* line k_{XSi} values for Ta, W, and Ir at 200 kV and for W and Ir at 120 kV were made.

REFERENCES

1 B. LOBERG and H. NORDÉN: *Arkiv för fysik*, 1969, **39**, 383
2 H.-O. ANDRÉN and H. NORDÉN: *Scand. J. Metall.*, 1979, 8, 147
3 G. CLIFF and G. W. LORIMER: *J. Microsc.*, 1975, **103**, 203
4 J. I. GOLDSTEIN, J. L. COSTLEY, G. W. LORIMER, and S. J. B. REED: SEM/1977, 1977, **1**, ed. O. Johari, IITRI, Chicago, Ill., 315
5 P. J. STATHAM: *Anal. Chem.*, 1977, **49**, 2149
6 G. W. LORIMER, S. A. AL-SALMAN, and G. CLIFF: EMAG 77, 1977, Inst. Phys. Conf. Ser. No. **36**, 369
7 N. J. ZALUZEC: PhD thesis, 1978, Univ. of Illinois
8 Link Systems: RTS 2/FLS program manual, 1978

LIST OF PARTICIPANTS

Adam, P. F.
University of Glasgow

Ahmadzadeh, Dr M.
University of Sussex

Arrowsmith, Dr D. J.
University of Aston in Birmingham

Baker, Dr T. N. (*author*)
University of Strathclyde

Ball, Dr M. D.
Alcan International Ltd, Banbury

Barber, Professor D. J.
University of Essex

Beaven, Dr P. A.
GKSS Forschungszentrum, Geesthacht, Federal Republic of Germany

Bell, J. M.
ICI Ltd, Runcorn

Bourdillon, Dr A. J. (*author*)
University of Cambridge

Bovey, P. E. (*author*)
VG Microscopes Ltd, East Grinstead

Boyes, Dr E. D. (*author*)
University of Oxford

Brooks, Dr J. W.
Inco Europe Ltd, Birmingham

Brown, Dr L. M. (*author*)
University of Cambridge

Buggy, T. W. (*author*)
University of Glasgow

Capers, Dr M. J.
Hitachi Scientific Instruments, Reading

Carruthers, R. B. (*author*)
CEGB, Harrogate

Cawley, J.
Sheffield City Polytechnic

Champness, Dr P. E.
University of Manchester

Chescoe, Miss D. (*author*)
Microstructural Studies Unit, University of Surrey

Cliff, G. (*author*)
Joint University of Manchester/UMIST

Cobbold, A. J.
ICI Ltd, Welwyn Garden City

Cochrane, Dr R. C.
British Steel Corporation, Teesside Laboratories

Collett, S. A. (*author*)
Cavendish Laboratory, Cambridge

Colliex, Dr C. Y. F.
CNRS, Orsay, France

Cooper, J. F.
Admiralty Marine Technology Establishment, Poole

Craenen, J. A.
State University of Ghent, Belgium

Craven, Dr A. J. (*author*)
University of Glasgow

Crawley, P. G.
JEOL (UK) Ltd, London

Crook, Dr A.
Gillette UKRDL, Reading

Cytermann, Dr R. C.
CEA/CEN Saclay, Gif-sur-Yvette, France

Davies, Dr S. T.
University of Warwick

Devenish, R. W.
University of Liverpool

Doherty, Dr R. D.
University of Sussex

Doig, Dr P. (*author, Working Party*)
CEGB Scientific Services Dept, Gravesend

Dons, Miss A. L.
SINTEF, Trondheim, Norway

Doorbar, P. J.
Rolls-Royce Ltd, Derby

Drost, Ir. N.
Metaalinstituut TNO, Apeldoorn, The Netherlands

Drummond, M. J.
University of Glasgow

Duckworth, S. P.
University of Strathclyde

Duncumb, Dr P. (*speaker*)
TI Research Laboratories, Saffron Walden

Eades, Dr J. A.
University of Bristol

Evans, D. W.
Edward Day & Baker Ltd, Birmingham

Faulkner, Dr R. G. (*author*)
Loughborough University of Technology

Ferguson, Dr I. F.
Microstructural Studies Section,
UKAEA Springfields, Preston

Fisher, Dr S. B.
CEGB Berkeley Nuclear Laboratories

Flewitt, Dr P. E. J. (*author*)
CEGB Scientific Services Dept, Gravesend

Freund, Dr E. (*author*)
Institut Français du Pétrole, Rueil Malmaison, France

Garratt-Reed, Dr A. J. (*author*)
Massachusetts Institute of Technology,
Cambridge, Mass., USA

Gee, Dr M. G.
Division of Materials Applications,
National Physical Laboratory, Teddington

Gentsch, Dr P.
NV Philips' Gloeilampenfabrieken, Eindhoven,
The Netherlands

Goldstein, Professor J. I. (*author*)
Lehigh University, Bethlehem, Pa, USA

Goodhew, Dr P. J. (*author*)
University of Surrey

Green, Miss S. M.
UKAEA, Warrington

Greenwood, R. M.
University of Birmingham

Griffiths, J. G.
JEOL (UK) Ltd, London

Guyot, Professor P. A.
Institut National Polytéchnique de Grenoble,
Saint Martin d'Heres, France

Haemers, Ir. G. J. E.
NV Bekaert SA, Zwevegem, Belgium

Hagemann, Dr P.
NV Philips' Gloeilampenfabrieken, Eindhoven,
The Netherlands

Hall, N. M.
Gillette, UKRDL, Reading

Høier, Dr R. H.
SINTEF, Trondheim, Norway

Howell, Dr P. R. (*author*)
University of Cambridge

Hyam, Dr E. D.
Windscale Nuclear Power Development Laboratories,
UKAEA, Seascale

Izraël, Mme A.
CNET, Bagneux, France

Jacobs, Dr M. H. (*author, Working Party*)
TI Research Laboratories, Saffron Walden

Jones, Dr G. P.
Division of Materials Applications,
National Physical Laboratory, Teddington

Jones, Dr I. P. (*author*)
University of Birmingham

Joy, Dr D. C. (*author*)
Bell Telephone Laboratories, Murray Hill, NJ, USA

Keown, Dr S. R.
Consultant, Sheffield

Kerr, R. T. (*author*)
University of Oxford

Kirkwood, Dr D. H.
University of Sheffield

Krahl, Dr D.
Fritz-Haber-Institut der Max-Planck-Gesellschaft, Berlin

Lachowski, Dr E. E.
University of Aberdeen

Law, T. J. (*author*)
TI Research Laboratories, Saffron Walden

Lee, P. J.
University of Birmingham

Long, Dr N. J. (*author*)
VG Microscopes Ltd, East Grinstead

Loretto, Dr M. H. (*author*)
University of Birmingham

Lorimer, Dr G. W. (*author, Working Party*)
Joint University of Manchester/UMIST

Love, Dr G. (*author*)
University of Bath

Lynch, Dr J. P. (*author*)
Institut Francais du Pétrole, Rueil Malmaison, France

Mallon, H. R.
Deloro Stellite, Swindon

Mansfield, J. F.
University of Bristol

Maurice, Miss F.
CEA/CEN Saclay, Gif-sur-Yvette, France

McCormick, N. E.
Tracor Europa Ltd, Inverkeithing, Fife

McFarlane, A. A.
UKAEA, Warrington

McGill, R. J. (*author*)
Department of Geology, University of Dundee

Metcalfe, Dr E. (*author*)
CERL, Leatherhead

Murray, G. E.
JEOL (UK) Ltd, London

Nicholls, A. W. (*author*)
University of Birmingham

Nordén, Dr H. F. (*author*)
Chalmers University of Technology, Göteborg, Sweden

Norman, C. J.
British Aluminium Co. Ltd, Gerrards Cross

Owen, H. T. (*author*)
Electricity Council Research Centre, Capenhurst, Chester

Piller, R. C.
UKAEA, AERE Harwell

Porter, A. J. (*author*)
University of Cambridge

Porter, Dr D. A. (*author*)
University of Luleå, Sweden

Porter, Dr J. R.
Lehigh University, Bethlehem, Pa, USA

Pountney, Dr J. M.
University of Birmingham

Rae, D. A. (*author*)
University of Bath

Raison, Miss J. M.
British Steel Corporation, Teesside Laboratories

Reed, Dr S. J. B.
University of Cambridge

Richards, Dr B. P.
GEC Ltd, Hirst Research Centre, Wembley

Ricks, Dr R. A. (*author*)
University of Cambridge

Ridley, Dr N. (*author*)
Joint University of Manchester (UMIST)

Riley, B. F.
UKAEA, Springfields, Preston

Roebuck, D. J. (*author*)
Pye Unicam Ltd, Cambridge

Rühle, Dr M.
Max-Planck-Institut für Metallforschung, Stuttgart, Federal Republic of Germany

Shannon, Dr M. D.
ICI Ltd, Runcorn, Cheshire

Skarnulis, Dr A. J.
University of Oxford

Slattery, Dr G. F.
UKAEA, Warrington

Smith, Professor D. G. W.
University of Alberta, Edmonton, Canada

Smith, Dr G. D. W. (*author*)
University of Oxford

Statham, Dr P. J. (*author*)
Link Systems Ltd, High Wycombe

Steeds, Dr J. W. (*author*)
University of Bristol

Stephenson, T. A. (*author*)
University of Birmingham

Stohr, Dr G. F.
ONERA, Chatillon/Bagneux, France

Swann, P. R. (*author*)
Gatan Inc., Warrendale, Pa, USA

Tasker, M. P.
Ministry of Defence, RARDE/Fort Halstead, Sevenoaks

Tatlock, Dr G. J.
University of Liverpool

Taylor, C.
Windscale Nuclear Laboratories, UKAEA, Seascale

Thomas, Dr B. J. E.
IRSID, St. Germain-en-Laye, France

Thornton, Dr C. H.
BNF Metals Technology Centre, Wantage

Thorvaldsson, T. (*author*)
Chalmers University of Technology, Göteborg, Sweden

Titchmarsh, Dr J. M. (*author*)
UKAEA, AERE Harwell

Tixier, Dr R (*author*)
IRSID, St. Germain-en-Laye, France

Venables, Dr J. A. (*author*)
University of Sussex

Voice, W. E. (*author*)
Loughborough University of Technology

Von Harrach, Dr H. S.
Cameron Iron Works Ltd, Livingston, Scotland

Walker, J. A. J.
UKAEA, Warrington

Watson, J. M. (*author*)
University of Manchester/UMIST

Watt, I. M.
Johnson Matthey Services Ltd, Sonning Common, Reading

Wenger, Dr D.
CISE, Milano, Italy

Whiteman, Dr J. A.
University of Sheffield

Whitty, S.
UKAEA, Springfields, Preston

Williams, Dr B. G.
University of Cambridge

Williams, Professor D. B. (*author*)
Lehigh University, Bethlehem, Pa, USA

Williams, Dr R. G.
VG Microscopes Ltd, East Grinstead

Wolters, H. B. M.
Kon./Shell Laboratory Amsterdam, The Netherlands

AUTHOR INDEX

Page numbers followed by asterisk refer to discussion contributions

SUBJECT INDEX

Page numbers followed by asterisk refer to discussion contributions